Quantitative Business Modeling

JACK R. MEREDITH
Wake Forest University

SCOTT M. SHAFER
Wake Forest University

EFRAIM TURBAN
City University of Hong Kong

Australia · Canada · Mexico · Singapore · Spain · United Kingdom · United States

Quantitative Business Modeling
by Jack R. Meredith, Scott M. Shafer, and Efraim Turban

Vice President/Editor-in-Chief:
Jack W. Calhoun

Team Leader:
Melissa S. Acuña

Acquisitions Editor:
Charles E. McCormick, Jr.

Senior Developmental Editor:
Alice C. Denny

Marketing Manager:
Joseph A. Sabatino

Production Editor:
Salvatore N. Versetto

Manufacturing Coordinator:
Diane Lohman

Compositor:
Shepherd, Inc.

Printer:
R.R. Donnelley & Sons Co.
Willard Manufacturing Division

Design Project Manager:
Rick A. Moore

Internal Designer:
Michael H. Stratton

Cover Designer:
Grannan Graphic Design,
Ltd./Cincinnati, OH

Cover Images:
© PhotoDisc, Inc.

Photography Manager:
Deanna Ettinger

Photo Researcher:
Feldman & Associates, Inc.

Media Developmental Editor:
Christine A. Wittmer

Media Production Editor:
Robin K. Browning

Printed in the United States of America
1 2 3 4 5 04 03 02 01

For more information contact South-Western,
5191 Natorp Boulevard,
Mason, Ohio 45040.

Or you can visit our Internet site
at: http://www.swcollege.com

Library of Congress Cataloging-in-Publication Data
Meredith, Jack R.
Quantitative business modeling / Jack R. Meredith, Scott M. Shafer, Efraim Turban.
p. cm.
Includes bibliographical references and index.
ISBN 0-324-01600-X
1. Decision making—Mathematical models. 2. Management—Mathematical models. 3. Management science. I. Shafer, Scott M. II. Turban, Efraim. III. Title.
HD30.23.M46 2001
658.4'033—dc21 2001049153

To Job, Mary, Robert Jun., Hugh, and Joshua
Children of Robert Meredith, in the year 1727,
Kent County, Delaware

J.R.M.

To Nikki, a model wife and partner

S.M.S.

To my wife Lina, and my daughters Daphne and Sharon

E.T.

Brief Contents

Preface xvii

About the Authors xxiii

Chapter 1
Decision Making and Quantitative Modeling 1

Chapter 2
Data Collection and Analysis 38

Chapter 3
Statistical Models: Regression and Forecasting 97

Chapter 4
Optimization and Mathematical Programming 148

Chapter 5
Decision Analysis 221

Chapter 6
Queuing Theory 279

Chapter 7
Simulation 317

Chapter 8
Implementation and Project Management 372

Appendix A
Mathematics 435

Appendix B
Tables 441

Index 451

Contents

Preface xvii

About the Authors xxiii

Chapter 1
Decision Making and Quantitative Modeling 1

1.1 Quantitative Business Modeling 7

Definition of a Model 9 Benefits and Drawbacks of Modeling 10 Types of Models 11 Effective Modelers 14

1.2 The Modeling Process 14

A Five-Step Modeling Process 16 Step 1: Opportunity/Problem Recognition 17 Step 2: Model Formulation 17 Step 3: Data Collection 21 Step 4: Analysis of the Model 23 Step 5: Implementation and Project Management 25

1.3 Detailed Modeling Example 28

Step 1: Opportunity/Problem Recognition 28 Step 2: Model Formulation 29 Step 3: Data Collection 30 Step 4: Analysis of the Model 30 Step 5: Implementation and Project Management 30

1.4 Software for Modeling 33

Questions 33 Experiential Exercises 34 Modeling Exercises 35 Case: Henry Ford Hospital 36 Endnotes 37 Bibliography 37

Chapter 2
Data Collection and Analysis 38

2.1 Data Collection 39

2.2 Summarizing Data 42

Descriptive Statistics 42 Statistical Displays 44

2.3 Probability and Random Variables 47

Subjective Probablility 48 Logical Probability 48 Experimental Probability 48 Event Relationships and Probability Laws 48 Probability Distributions 51

2.4 Common Probability Distributions 52

The Binomial Distribution 53 The Poisson Distribution 54 The Exponential Distribution 55 The Normal Distribution 56 The t Distribution 58

2.5 Distributions of Sample Statistics 58

2.6 Chi-Square Goodness of Fit Test 60

2.7 Point and Interval Estimation 64

Interval Estimation of a Mean 65 Determining the Size of the Sample for a Normal Distribution 68 Interval Estimation and Determination of Sample Size for a Proportion 69

2.8 Hypothesis Testing 71

Hypothesis Tests for Means 73 Comparing Multiple Means—Analysis of Variance (ANOVA) 77

2.9 Detailed Modeling Example 81

Step 1: Opportunity/Problem Recognition 81 Step 2: Model Formulation 82 Step 3: Data Collection 82 Step 4: Analysis of the Model 85 Step 5: Implementation 86

Questions 89 Experiential Exercise 89 Modeling Exercises 90 Case: Fiberease Inc. 93 Case: InterAccess Inc. 95 Case: eApp Inc. 95 Endnote 96 Bibliography 96

Chapter 3 Statistical Models: Regression and Forecasting 97

3.1 The Modeling Process for Statistical Studies 99

3.2 The Simple Linear Regression Model 100

Calculating the Regression Model Parameters 103 The Coefficient of Determination and the Correlation Coefficient 105 Regression Analysis Assumptions 109 Using the Regression Model 110

3.3 The Multiple Regression Model 112

3.4 Developing Regression Models 115

Step 1: Identify Candidate Independent Variables to Include in the Model 115 Step 2: Transform the Data 117 Step 3: Select the Variables to Include in the Model 118 Step 4: Analyze the Residuals 118

3.5 Regression Hypothesis Tests 119

3.6 Time Series Analysis 121

Components of a Time Series 121 Time Series Models 123

3.7 Detailed Modeling Example 130

Step 1: Opportunity/Problem Recognition 130 Step 2: Model Formulation 131 Step 3: Data Collection 131 Step 4: Analysis of the Model 131 Step 5: Implementation 135

Questions 140 Experiential Exercise 140 Modeling Exercises 141 Case: Resale Value of Long's Automobile 144 Case: Lewisville Crate Company 144 Bibliography 147

Chapter 4 Optimization and Mathematical Programming 148

4.1 The Modeling Process for Optimization Studies 153

Optimization 153 The Modeling Process 154 Structure of the Chapter 156

4.2 Linear Programming 156

The Output-Mix Problem 157 The Blending Problem 157 Formulating the Linear Programming Model 157 Output-Mix and Blending Problems: Two Examples 158 Example: The Blending (Minimization) Problem 160 The General LP Model 161 Advantages, Assumptions, and Solution Methods 162 Distribution Problems: Transportation, Transshipment, Assignment 164

4.3 Analysis of the Model by the Graphical Method 165

Example 1: A Maximization Problem 165 Example 2: A Minimization Problem 172 Utilization of the Resources—Slack and Surplus Variables 174 Special Situations 175

4.4 Solving Linear Programming Models with Excel 177

Using Excel's Solver 177 Solving Large Problems 181 Back to Startron's Dilemma 185

4.5 Sensitivity ("What-If") Analysis 189

Why a Sensitivity Analysis? 189 Sensitivity Analysis: Objective Function 190 Sensitivity Analysis: Right-Hand Sides 192 Sensitivity Analysis with Excel 192

4.6 Integer Programming 196

Overview of Integer Programming 196 Example: Southern General Hospital 197 The Zero–One Model 200 Example: The Fixed-Charge Situation 201

4.7 Detailed Modeling Example 203

Step 1: Opportunity/Problem Recognition 203 Step 2: Model Formulation 203 Step 3: Data Collection 203 Step 4: Analysis of the Model 205 Step 5: Implementation 208

Questions 210 Experiential Exercise 211 Modeling Exercises 211 Case: The Daphne Jewelry Company 217 Case: Hensley Valve Corp. (A) 219 Case: Hensley Valve Corp. (B) 219 Bibliography 220

Chapter 5
Decision Analysis 221

5.1 The Modeling Process for Decision Analysis Studies 222

The Modeling Process 223 Structure of the Chapter 224

5.2 The Decision Analysis Situation 224

Mary's Dilemma 224 The Structure of Decision Tables 225 Classification of Decision Situations 228

5.3 Decisions Under Certainty 228

Complete Enumeration 229 Example: Assignment of Employees to Machines 229 Computation with Analytical Models 230

5.4 Decisions Under Uncertainty 230

Equal Probabilities (Laplace) Criterion 231 Pessimism (Maximin or Minimax) Criterion 231 Optimism (Maximax or Minimin) Criterion 232 Coefficient of Optimism (Hurwicz) Criterion 233 Regret (Savage) Criterion 237

5.5 Decisions Under Risk 237

Objective and Subjective Probabilities 238 Solution Procedures to Decision Making Under Risk 238 Notes on Implementation 242 Sensitivity Analysis 242

5.6 Decision Trees for Risk Analysis 243

Structure of a Decision Tree 243 Evaluating a Decision Tree 245 The Multiperiod, Sequential Decision Case 246

5.7 The Value of Additional Information 250

Information Quality: Perfect Versus Imperfect Information 250 The Value of Perfect Information 251

5.8 Imperfect Information and Bayes' Theorem 253

Bayes' Theorem 253 Using Revised Probabilities with Imperfect Information 254 Calculating Revised Probabilities 259 Computing the Revised Probabilities 260

5.9 Detailed Modeling Example 262

Step 1: Opportunity/Problem Recognition 262 Step 2: Model Formulation 262 Step 3: Data Collection 263 Step 4: Analysis of the Model 263 Step 5: Implementation 265

Questions 270 Experiential Exercises 270 Modeling Exercises 271 Case: Maintaining the Water Valves 276 Case: The Air Force Contract 277 Endnotes 278 Bibliography 278

Chapter 6
Queuing Theory 279

6.1 The Modeling Process for Queuing Studies 282

Step 1: Opportunity/Problem Recognition 282 Step 2: Model Formulation 282 Step 3: Data Collection 283 Step 4: Analysis of the Model 283 Step 5: Implementation 283

6.2 The Queuing Situation 284

Characteristics of Waiting Line Situations 284 The Structure of a Queuing System 285 The Managerial Problem 286 The Costs Involved in a Queuing Situation 287

6.3 Modeling Queues 288

Queuing Model Notation 288 Deterministic Queuing Systems 289 The Arrival Process 290 The Service Process 292 Measures for the Service 293 The Waiting Line 294

6.4 Analysis of the Basic Queue (*M/M/*1 *FCFS/∞/∞*) 295

Poisson-Exponential Model Characteristics 295 Measure of Performance (Operating Characteristics) 296 Managerial Use of the Measures of Performance 298 Using Excel's Goal Seek Function 298

6.5 More Complex Queuing Situations 298

Multifacility Queuing Systems (M/M/K FCFS/∞/∞) 299 Example: Multichannel Queue 301 Example: Multichannel Queue at Macro-Market 301 Serial (Multiphase) Queues 304 Example: Serial Queue—Three-Station Process 304

6.6 Detailed Modeling Example 306

Step 1: Opportunity/Problem Recognition 306 Step 2: Model Formulation 306 Step 3: Data Collection 306 Step 4: Analysis of the Model 307 Step 5: Implementation 308

Questions 309 Experiential Exercise 310 Modeling Exercises 310 Case: City of Help 315 Case: Newtown Maintenance Division 315 Bibliography 316

Chapter 7
Simulation 317

7.1 General Overview of Simulation 319

Types of Simulation 320 Uses of Simulation 322 Advantages and Disadvantages of Simulation 322

7.2 The Modeling Process for Monte Carlo Simulation 323

Step 1: Opportunity/Problem Recognition 323 Step 2: Model Formulation 323 Step 3: Data Collection 324 Step 4: Analysis of the Model 324 Step 5: Implementation 327

7.3 The Monte Carlo Methodology 327

The Tourist Information Center 327 Simulation Terminology 328 Generating Random Variates in the Monte Carlo Process 330

7.4 Time Independent, Discrete Simulation 332

Example: Marvin's Service Station 333 Solution by Simulation 333

7.5 Time Dependent Simulation 339

Simulation Analysis with Discrete Distributions 240 Simulation with Continuous Probability Distributions 342

7.6 Risk Analysis 342

7.7 Detailed Modeling Example 344

Step 1: Opportunity/Problem Recognition 344 Steps 2 and 3: Model Formulation and Data Collection 344 Step 4: Analysis of the Model 347 Step 5: Implementation 348

Appendix: Crystal Ball 2000 350

Questions 350 Experiential Exercise 359 Modeling Exercises 360 Case: Medford Delivery Service 366 Case: Warren Lynch's Retirement 366 Case: Cartron, Inc. 369 Endnotes 371 Bibliography 371

Chapter 8
Implementation and Project Management 372

8.1 Implementation and Project Modeling 373

The Project Modeling Process 373 Structure of the Chapter 374

8.2 Implementing the Modeling Study 375

Soft Aspects 375 Rational Issues and Reconsideration 377 The Role of Project Management 378 Example: Moose Lake 378

8.3 Planning the Project 381

Step 1: Analysis of the Project 382 Step 2: Sequence the Activities 382 Step 3: Estimate Activity Times and Costs 382

8.4 Scheduling the Project 383

Step 4: Construct the Network 383 Step 5: Event Analysis 385 PERT/CPM Network Characteristics 391 Estimating Activity Times in PERT 393 Finding the Probabilities of Completion in PERT 394 Example: Finding the Probability of Completion within a Desired Time, D 397 Example: Finding the Duration Associated with a Desired Probability 399 Determining the Distribution of Project Completion Times with Simulation 399

8.5 Step 6: Monitoring and Controlling the Project 403

Monitoring the Project 403 Controlling the Project 403 Example: Resource Allocation Schedule 405 Critical Path Method (CPM): Cost–Time Trade-Offs 406 Example: Finding the Least-Cost Plan 409 Example: Least-Cost Plan for 22 Days 441 Analyzing Cost–Time Trade-Offs with Excel's Solver 314

8.6 Detailed Modeling Example 418

Step 1: Opportunity/Problem Recognition 418 Step 2: Model Formulation 418 Step 3: Data Collection 421 Step 4: Analysis of the Model 423 Step 5: Implementation 424

Questions 426 Experiential Exercise 426 Modeling Exercises 426 Case: NutriTech 431 Case: Dart Investments 432 Bibliography 433

Appendix A
Mathematics 435

Appendix B
Tables 441

Index 451

Preface

Quantitative Business Modeling was written to accommodate the tremendous changes that are occurring in the basic quantitative courses in business and management programs. These changes include decreased course time, standardized course content, and the use of spreadsheet software.

Perhaps the most significant change has been the substantial reduction in time allocated for teaching quantitative materials in business and management curricula. In some cases, the statistics and management science courses have been combined. In other situations, management science has been combined with a functional course such as operations management or marketing. And finally, there are some situations where management science has simply been removed from the curriculum.

These new limitations mean that we must make the time spent on quantitative materials more productive. We must realize that most of our students are not going to become statisticians or quantitative analysts. For today's students, the necessary skills include the ability to recognize opportunities, formulate appropriate models, obtain data, and communicate and implement the results. Thus, our new book is about quantitative business *modeling,* not quantitative *models.*

Importance of Quantitative Business Modeling

We firmly believe that the topic of quantitative business modeling is relevant to all business professionals. It is common for decision makers in today's modern business world to be almost overwhelmed by the vast amounts of data computer systems accumulate about customers, products, markets, and internal operations. This textbook will acquaint students with important data analysis and modeling tools that will enhance their ability to use data effectively to better understand customers and markets, and to improve their products and services.

The effective use of data to model a decision-making situation requires the analyst to consider:

1. How the data will be obtained
2. The most relevant forms of analysis
3. What assumptions are being made
4. The meaning of the results
5. The limitations of the analysis
6. How best to implement and monitor the results

We address these critical aspects of modeling with as much attention as we spend in solving the models. Current research describing how expert consultants go about modeling is used to derive the modeling process recommended here. We describe this modeling process, list its several steps, and then illustrate these steps both within each chapter and across the chapters of the book.

Pedagogical Features Foster Quantitative Modeling

To move the student from a passive reader to an active participant in the modeling process, every chapter includes pedagogical features that foster appreciation of the role of modeling. These features and their benefits include:

- **Chapter Opener** Each chapter opens with an opportunity or problem with which the student can easily relate. This problem provides an introduction to the chapter's specific content.
- **Influence Diagrams** For the model formulation step of the modeling process, we use influence diagrams for every situation to show how an early understanding of relationships facilitates model formulation and helps determine the data requirements.
- **Detailed Modeling Example** Each chapter ends with a detailed modeling example that goes through the entire process of identifying some opportunity or problem related to the topic in that chapter, formulating an appropriate model(s), discussing issues commonly involved in data collection, and then solving the model. The example concludes with a memo from the modeler to a decision maker describing the results and meaning of the study, discussing its limitations, and offering appropriate recommendations for implementation.
- **Experiential Exercises** To provide additional insight into the realities of modeling, each chapter contains one or more experiential exercises. These exercises help guide students through the modeling process, including actually collecting their own data.
- **Questions** Discussion questions are also included at the end of each chapter to probe the students' understanding of the issues discussed in each chapter, because understanding rather than solving problems is the purpose of the text.
- **Modeling Exercises** The modeling exercises at the end of each chapter provide another opportunity to increase intuitive understanding through actual model formulation and solution.
- **Cases** Cases involving yet more complex modeling situations conclude each chapter.
- **Glossary** A glossary with definitions of key terms (**boldfaced** in the text) can be found on our Web site at http://meredith.swcollege.com.

Each chapter also contains an up-to-date bibliography. There are two appendices at the end of the text, one a refresher on mathematics and the other a collection of useful tables.

Efficient and Effective Topical Content

Effective quantitative business modeling requires a combination of traditional statistical and management science topics. This, along with the critical need for efficient use of course time, supports our contention that these topics should be integrated, not divorced from one another.

- **Business Statistics** We cover the basic statistics for analyzing, summarizing, testing, and validating data in Chapter 2. Regression analysis is covered as our first quantitative business model in Chapter 3. Thus, our textbook places statistics material where it makes sense with respect to the overall modeling process. For students who have already had statistics, these chapters may be skipped or used as a quick refresher. Moreover, for those courses that combine statistics and management science, as many M.B.A. courses are currently doing, we believe this placement of the statistical material is natural and appropriate.
- **Management Science** The second major change that has occurred in the quantitative area is the near-standardization of the topics covered in management science. We have limited our coverage here to four basic management science topics starting with Optimization (linear and integer programming) in Chapter 4, Decision Analysis in Chapter 5, Queuing Theory in Chapter 6, and Simulation in Chapter 7.
- **Project Management** As the basis for our final modeling topic, Implementation, we address Project Management in Chapter 8. Not only does project management allow us to discuss the managerial issues involved in implementing the results of a quantitative modeling project; it also provides an opportunity to integrate earlier material on optimization, simulation, and statistics.

Microsoft® Excel Provides Spreadsheet Support

The third major change in quantitative courses today is the incorporation of spreadsheets. Being able to shift the emphasis from performing tedious calculations toward interpreting and understanding the analysis has made spreadsheet software virtually mandatory for all quantitative courses. In the past, highly specialized and often complex software packages were used in statistics and management science courses. Unfortunately, these software packages were often different from the ones the students would have access to in both their academic and professional lives.

Our approach here is to confine our attention primarily to Microsoft® Excel. This software provides several important advantages: its use is highly intuitive and many students are already familiar with it; it is widely used; and it can do a wide range of quantitative processing and analysis. Thus, class time can be spent *understanding the modeling process* and its meaning instead of learning how to make the software work.

We do include two other software packages when relevant. We include an appendix on ***Crystal Ball® 2000 Professional Edition*** in Chapter 7's coverage of simulation, and show output from ***Microsoft® Project 2000*** in Chapter 8. As a convenience for instructors who would like to use these two packages in more detail, trial versions are included on CD-ROMs that accompany the text. All inquiries concerning availability of Crystal Ball for members of the academic community should be addressed to Decisioneering's Academic Program Administrator at 800-289-2550, ext. 208.

We have intentionally stayed away from using Excel add-ins throughout the text because these add-ins may not be available to students at the organizations where they will eventually work. Also, we believe that adequately covering the functionality provided by Excel is sufficiently challenging for a one-term course. The Modeling Exercises at the end of each chapter include both exercises to be done by hand, for intuitive understanding, as well as exercises to be done using Excel.

Ancillary Teaching Materials

The ***Instructor's Resource CD*** (ISBN: 0-324-01629-8) provides all instructor ancillaries. Adopters may request a copy online at http://www.swcollege.com. The following are included in this convenient format:

- ***Instructor's Manual*** The *Instructor's Manual,* prepared by the text authors, contains an introduction to the chapter, answers to the end-of-chapter questions, and solutions to the modeling exercises and cases.
- ***PowerPoint™ Presentation Slides*** Prepared by Jeff Heyl of Lincoln University in New Zealand, the presentation slides contain graphics to help instructors create stimulating lectures. The slides may be adapted using PowerPoint software to facilitate classroom use.
- ***Test Bank*** Prepared by Edward Ward of St. Cloud State University, the *Test Bank* includes objective questions and problems for each chapter. The *Test Bank* is provided in Microsoft® Word format on the *Instructor's Resource CD.*

Other teaching and learning materials will be available on our Web site: http://meredith.swcollege.com.

Acknowledgments

We would like to acknowledge the contribution of our reviewers who provided comments and suggestions that helped improve our text. Thanks to our colleagues:

Umit Akinc, Wake Forest University
Tony Arreola-Risa, Texas A&M University
Q. Chung, Villanova University
Abe Feinberg, California State University, Northridge
V. Daniel Guide, Duquesne University
James Hoyt, Troy State University
Jane Humble, Arizona State University East
Gordon Johnson, California State University, Northridge
Xenophon Koufteros, Florida Atlantic University
Henry Mead, University of British Columbia, Canada
Charles T. Mosier, Clarkson University
Mary Ann Murray, St. Mary's University
Richard A. Paulson, St. Cloud University

David Pentico, Duquesne University
Michael Small, University of Illinois, Springfield
Chwen Sheu, Kansas State University
Keah-ChoonTan, University of Nevada, Las Vegas
Edward Ward, St. Cloud State University
Zhiwei Zhu, University of Louisiana at Lafayette

There are many people within the South-Western/Thomson Learning™ publishing group who we would like to thank for their assistance on this project. In particular, we would like to cite Charles McCormick, Jr., senior acquisitions editor, Joe Sabatino, senior marketing manager, Alice Denny, senior development editor, and Sam Versetto, production editor.

Jack Meredith
Scott Shafer
Efraim Turban

About the Authors

Jack R. Meredith

Jack Meredith is Professor of Management and Broyhill Distinguished Scholar and Chair in Operations at the Babcock Graduate School of Management at Wake Forest University. He received undergraduate degrees in engineering and mathematics from Oregon State University, and his M.B.A. and Ph.D. from the University of California, Berkeley.

His current research interests are in the areas of research methodology and the strategic planning, justification, and implementation of advanced manufacturing technologies. He is currently the editor-in-chief of the *Journal of Operations Management,* an area editor for *Production and Operations Management,* and was the founding editor of *Operations Management Review.*

Jack and his wife Carol have three children, and Carol is a freelance writer. Jack is a tennis fanatic and studies genealogy as his spare-time hobby.

Scott M. Shafer

Scott M. Shafer is an Associate Professor of Management at the Babcock Graduate School of Management at Wake Forest University. He received a B.S. in Industrial Management, a B.B.A. in Marketing, and the Ph.D. in Operations Management from the University of Cincinnati.

His current research interests are in the areas of cellular manufacturing, operations strategy, business process design, organizational learning, and technology. His publications have appeared in the *Journal of Operations Management, Decision Sciences, International Journal of Production Research,* and many others. He is certified in Production and Inventory Management by the American Production and Inventory Control Society.

Scott and his wife Nikki have three children. For fun he enjoys working out, martial arts, tennis, golf, snow skiing, concerts, and playing guitar.

Efraim Turban

After a long and distinguished teaching career at California State University, Long Beach, Efraim Turban is now a visiting professor with the City University of Hong Kong. He received his Ph.D. from the University of California, Berkeley.

A prolific researcher and writer, he has authored 14 books and over 100 articles. Efraim Turban is an internationally known authority on expert systems. He serves on the editorial board of several scholarly journals, reviews research proposals for the NSF, and maintains scholarly exchange with information systems scholars and experts all over the the world.

Decision Making and Quantitative Modeling

1

© PHOTODISC, INC.

INTRODUCTION

Congratulations! You just received a job offer from a top consulting firm and the only thing standing between you and the big bucks is your final semester of school. Your new job will require you to travel at least 80 percent of the time, and the jalopy you currently use is well beyond its prime. You have worked hard, and purchasing a new sports car is certainly warranted.

Anxious to begin the quest for a new car, you connect to the Web, and after a brief search at your favorite search engine you identify a Web site called CarPoint (http://carpoint.msn.com). Having been more than a little preoccupied with school, you have not really kept up with all the new models of sports cars currently available. Fortunately, CarPoint has a category titled Sports Cars. Clicking on this category provides a list with links to more than 35 models of sport cars. After following the links, looking at the pictures of the cars, and briefly reading the reviews for the models, you identify the following seven models as candidates worth further investigation: BMW Z3 1.9L, Chevrolet Corvette, Porsche Boxster, Ford Mustang GT Convertible, Pontiac Firebird Convertible, Mercedes-Benz SLK, and the Volvo C70.

Now you begin to think about what criteria you should use to select your new car. Immediately, cost comes to mind. To compare the cars on cost, you add the destination charge plus $500 dealer profit to each car's invoice price.

Warranty is another important factor. To compare the cars in terms of their warranty, you decide to collect data on the length of the basic warranty offered by each manufacturer. The cost to insure the car is also important to you. Further, since the onramps of many of the state roads in your city are quite short, acceleration is a key consideration. You decide that comparing how long it takes the cars to go from zero to 60 miles/hour provides a good measure of acceleration. Finally, you think that it is important to include a couple of practical considerations, including how comfortable the car is and its fuel economy. You determine that leg room is a good surrogate indicator of comfort, while miles/gallon (MPG) captures the issue of fuel economy.

As will be demonstrated throughout this book, developing diagrams often greatly facilitates the quantitative modeling process. One approach that is useful, particularly early in the modeling process, is to develop an influence diagram like that shown in Exhibit 1.1. As its name suggests, an influence diagram is used to help identify the important factors and their relationships in a particular decision-making situation. Seeing the relationships in graphical form can help in formulating a mathematical model that duplicates these relationships in the way it is suspected they operate in the real world. And then, when it is invariably found that the relationships are different from what was initially envisioned, the diagram can be modified to better reflect the newfound reality. It is much easier to experiment with a simple diagram than with a mathematical model. Once the influence diagram properly reflects the reality that is to be modeled, the quantitative model can be restructured to duplicate the new graphic model.

In this book we classify the elements of an influence diagram into one of three categories. The first category corresponds to the objective for the situation being modeled and is identified by a diamond in the influence diagram. Referring to Exhibit 1.1, we see the objective of our car purchase decision is to maximize satisfaction. The second category corresponds to the decision variables or factors that the modeler or manager has control over. The decision variables are identified by rectangles in the influence diagram. In Exhibit 1.1 we see that the only decision is the actual selection of the car. The arrow from the rectangular box labeled "select car" to the diamond labeled "maximize satisfaction" indicates our belief that the selection of a particular car will have an impact on our overall satisfaction.

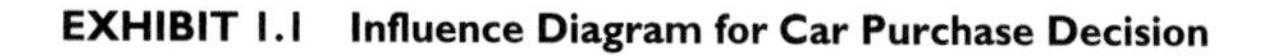

EXHIBIT 1.1 Influence Diagram for Car Purchase Decision

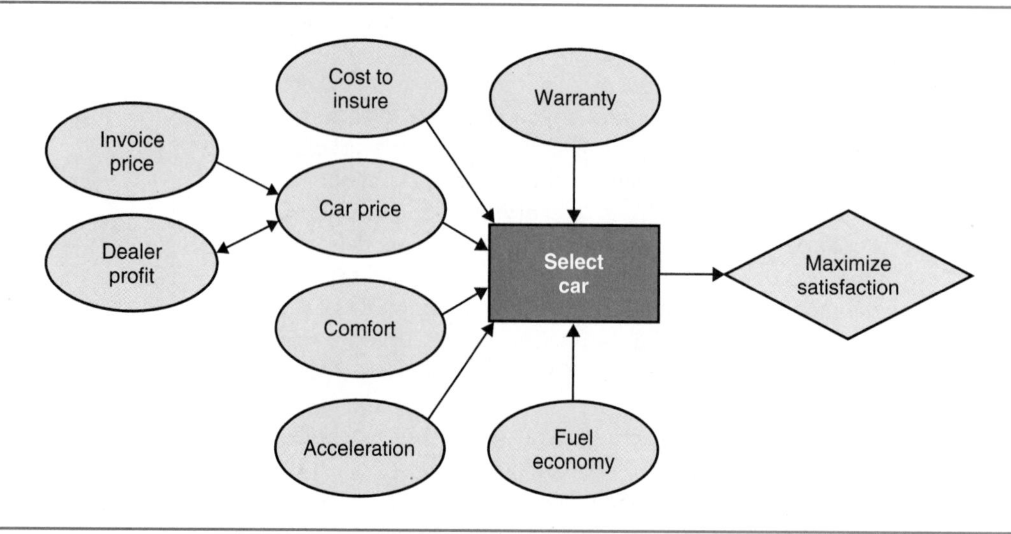

The third type of information shown in an influence diagram corresponds to the parameters that are given—in other words, the factors that the decision maker does not have control over. These are represented by ellipses in the influence diagram. Again, arrows are used to capture relationships among the elements of the diagram. For example, the influence diagram suggests that the price of the car will influence the choice of car selected for purchase. It might be noted that there could also be reciprocal relationships between variables so that, for example, the dealer profit may affect the car price but the car price also may affect the dealer profit. Such instances are shown in an influence diagram with a double-headed arrow. One last available refinement of influence diagrams is labeling the arrows with a + or – sign to indicate whether an increase in one variable is expected to lead to an increase or decrease, respectively, in the related variable. Just as relationships may be initially incorrect, the directions of increase may also have to later be changed if, for example, it is found that increasing one variable does not lead to an increase in another variable but rather a decrease. We will not use this level of refinement in our influence diagrams here, but we note their possible value to those who have more experience in modeling.

Not only do influence diagrams aid us in formulating a model of a situation we are interested in, they also help us to better understand the actual situation. Hence, they will typically evolve over time as the situation itself becomes better understood. Thus, the process of developing an influence diagram is typically iterative. To illustrate, after developing the influence diagram in Exhibit 1.1 and upon closer examination, it may occur to you that there are important relationships between "car price" and "cost to insure," as well as, perhaps, "acceleration" and "fuel economy," "warranty" and "car price," and so on. With more experience concerning the situation, you may decide to incorporate these relationships into your model also. On the other hand, as more such relationships are included, the more complex your model becomes and perhaps also, more difficult to solve. Thus, you may decide not to include some of these relationships and stay with a simpler model that can more easily give you the basic information you want.

Now that you have decided on the sports cars you are interested in and the criteria you will use to evaluate them, the next step is to collect this data for each car. If you had checked the msn® CarPoint Web site a while back, the date would have appeared as shown in Exhibit 1.2. The only problem encountered is that there was no information available for the Mustang in terms of its time to accelerate from zero to 60. One option

EXHIBIT 1.2
Data Collected for Car Purchase Decision[1]

Car	Insurance Rating	MPG	Leg Room (Inches)	Warranty (Months)	Cost	Speed, 0 to 60 (Seconds)
BMW Z3 1.9L	Better than average	20	41.8	48	$27,625	7.1
Corvette	Average	18	42.7	36	34,502	5.0
Porsche Boxster	Average	19	41.6	48	37,160	7.5
Mustang GT convertible	Much worse than average	17	41.9	36	23,604	n.a.
Firebird convertible	Worse than average	19	43.0	36	23,791	8.7
Mercedes SLK	Average	21	42.7	48	35,895	7.5
Volvo C70	Average	20	41.3	48	36,564	6.5

n.a. = not available.

would be to try to find this information from another source. However, this raises the issue of whether the data from a different source is consistent and directly comparable with the data you have already collected. For example, different drivers, tracks, and/or measurement devices could lead to different results being obtained. Another option is to eliminate acceleration as one of your criteria. However, you are not really comfortable with this option. Since the Mustang is the only car that has an insurance rating of "much worse than average," you instead decide to eliminate it from further consideration at this time.

Next, you need to decide how important each criterion is to you, say, out of 100 points total. Cost is clearly your most important consideration, and out of the 100 points you decide that 25 points should be allocated to it. You then decide that warranty is the next most important criterion and assign it 20 points. Looking over the remaining criteria on your list, you decide that insurance rating, speed, and comfort are all equally important to you and assign each 15 points, leaving 10 points to be assigned to fuel economy.

The last thing you need to do is decide how you want to score the cars on each option. Since there are only a few cars being considered, you determine that a three-point scale will best meet your needs.[2] As you look over the data, you note that the range of the cost of the cars is $23,791 to $37,160. Thus, you decide to assign a score of 3 to cars that cost less than $28,000; a 2 to cars that cost between $28,000 and $36,000; and a 1 to cars that cost more than $36,000. Although these ranges are not equal, you are happy with this method of assigning scores since an equal number of cars will fall into each category, thus distinguishing the cars from one another. Alternatively, if all or most of the cars received the same score on cost, your approach would not do a good job of distinguishing among the cars. Continuing in this fashion, you develop a scoring method for the other criteria as shown in Exhibit 1.3.

To rank the cars you then score each car on each criterion and enter the results in a spreadsheet similar to the one shown in Exhibit 1.4. Next you calculate a total score for each car by multiplying the car's score on a particular criterion by the weight you assigned to that criterion and summing these numbers up across all criteria. To illustrate, the total score for the BMW Z3 is calculated as follows:

$$\text{BMW Z3}_{\text{Total Score}} = (3 \times 15) + (2 \times 10) + (1 \times 15) + (3 \times 20) + (3 \times 25) + (2 \times 15) = 245$$

The scores for the other cars are calculated in a similar fashion. As shown in the table, the BMW is the top scorer, followed by the Mercedes.

Although the model you have developed can certainly help you rank your car choices, perhaps the real power of this model is helping you perform sensitivity or "what if" analysis. Indeed, a great deal of information and insight into a decision-making situation can be obtained by performing a sensitivity analysis and this analysis is often of greater value to the decision maker than simply using the model once to make a decision.

EXHIBIT 1.3 Method for Scoring Cars on Each Criterion

Criterion	1 point	2 points	3 points
Cost	> $36,000	$28,000–$36,000	< $28,000
Warranty	36 months		48 months
Insurance rating	Worse than average	Average	Better than average
Zero to 60 MPH	> 8 seconds	7 to 8 seconds	< 7 seconds
Leg room	< 42 inches	42–42.99 inches	≥ 43 inches
MPG	< 20 MPG	20–20.99 MPG	≥ 21 MPG

Entering the Total Score Formula into a Spreadsheet

There are several different ways the total score formula can be entered into a spreadsheet. Perhaps the easiest way is to simply follow the example and enter a formula to calculate the total score for the BMW Z3 in cell H7 as follows.[3]

=(B7*B4) + (C7*C4) + (D7*D4) + (E7*E4) + (F7*F4) + (G7*G4)

Since entering this formula five more times would be a bit tedious, you think to copy this formula from cell H7 to cells H8:H12. If you copy this formula and then move the cursor to cell H8 you will notice that cell H8 contains the following formula:

=(B8*B5) + (C8*C5) + (D8*D5) + (E8*E5) + (F8*F5) + (G8*G5)

$ sign keeps it in place essentially.

In actuality, the formula you want in cell H8 is:

=(B8*B4) + (C8*C4) + (D8*D4) + (E8*E4) + (F8*F4) + (G8*G4)

Why did Microsoft® Excel not copy the formula correctly? The answer has to do with a distinction between absolute and relative cell addresses. By default, when you enter a formula into a spreadsheet, the cell references are assumed to be relative cell addresses. Thus, when you originally entered the formula into cell H7, Excel interpreted the first term of the formula to mean take the value six cells to the left and multiply it by the value of the cell six cells to the left and up three rows. When you copied this cell down to cell H8, the formula still said to take the value six cells to the left and multiply it by the value of the cell six cells to the left and up three rows. Although we do want the first part of this term to always reference the cell that is six cells to the left, we do not want the second part of this term to always reference the cell six cells to the left and three rows up. Rather, we want the second part of this term to always reference the weights that are given in row 4. You can tell Excel to always reference a particular row by entering a dollar sign ($) in front of the row number in the cell address. Likewise, you can tell Excel to always reference a particular column by entering a dollar sign in front of the column letter in the cell address. To illustrate, in order to be able to copy the formula entered in cell H7 to cells H8:H12 we would need to enter the formula as follows:

=(B7*B$4) + (C7*C$4) + (D7*D$4) + (E7*E$4) + (F7*F$4) + (G7*G$4)

It might have occurred to you that there may yet be an easier way to calculate the total score for each car. Indeed, as will be demonstrated throughout this textbook, Excel provides a number of powerful functions and capabilities that greatly facilitate the analyst's job. In this particular case, Excel has a function called SUMPRODUCT.

(continued)

EXHIBIT 1.4 Spreadsheet to Score Sports Car Options

	A	B	C	D	E	F	G	H
1	**Car Purchase Decision**							
2								
3	Criteria:	Ins. Rating	MPG	Leg Room	Warranty	Cost	Speed	
4	Weights:	15	10	15	20	25	15	
5								**Total**
6	**Options**	**Ins. Rating**	**MPG**	**Leg Room**	**Warranty**	**Cost**	**Speed**	**Score**
7	BMW Z3 1.9L	3	2	1	3	3	2	245
8	Corvette	2	1	2	1	2	3	185
9	Firebird Conv.	1	1	3	1	3	1	180
10	Mercedes SLK	2	3	3	3	2	2	230
11	Porsche Boxster	2	1	1	3	1	2	170
12	Volvo C70	2	2	1	3	1	3	195
13								
14	**Key Formula**							
15	Cell H7	=SUMPRODUCT (B$4:G$4,B7:G7) [copyto cells H8:H12]						

Entering the Total Score Formula into a Spreadsheet—*cont'd*

This function multiplies the values in two ranges and then returns the sum of these products. As an alternative to the brute force approach described above, a simpler and more eloquent way of entering the formula in cell H7 using the SUMPRODUCT function is as follows:

=SUMPRODUCT(B7:G7,B$4:G$4)

In English, this formula has two arguments, each defining a range of six cells. The SUMPRODUCT function takes the first cell in the first range and multiplies it by the first cell in the second range. This is repeated until each cell in the first range is multiplied by its corresponding cell in the second range. Finally, the sum of these products is calculated and this is the value returned by the function. (The order of the arguments is immaterial.)

To illustrate, suppose that after test driving a couple of the cars you determine that comfort is as important as cost. The weights of the criteria were adjusted in the spreadsheet shown in Exhibit 1.5. Comparing Exhibits 1.4 and 1.5 it can be observed that the scores for the BMW, Corvette, Boxster, and Volvo all decreased. The score for the Mercedes SLK remained unchanged, and the score for the Firebird actually increased. After making this change, we see that the BMW and the SLK are now tied for first.

Of course, a model is not reality. It is only a tool to help make a decision or gain insight into a situation For example, suppose that, based on the results of your original analysis, you decide to negotiate with a local dealer for a BMW Z3. Unfortunately, because of recent demand for the Z3, the dealer is not willing to sell you the car for $500 over invoice. The best price you can get changes the Z3's score from a 3 to a 2 on cost. As can be seen in Exhibit 1.6, this change reduces the score of the Z3 from 245 to 220, and the SLK now becomes the preferred choice. As you can see, performing a sensitivity analysis can provide a great deal of information and insight into the nature of a particular decision.

This example overviews the focus of this book, namely, the process of modeling decision-making situations. The example began with the recognition and definition of an opportunity (in this particular case, the need for a new sports car). This was then followed by data collection. In our example, the data was collected using the Web.

The next step involved formulating a model to conceptualize the opportunity. While the model we developed may have appeared to be rather intuitive to you—simply multiplying scores on particular criteria by the weights of the criteria and adding

EXHIBIT 1.5 Impact of Changing Criteria Weights

	A	B	C	D	E	F	G	H	I
1	**Car Purchase Decision**								
2									
3	Criteria:	Ins. Rating	MPG	Leg Room	Warranty	Cost	Speed	Changed weights	
4	Weights:	10	10	25	20	25	10		
5								**Total**	
6	**Options**	**Ins. Rating**	**MPG**	**Leg Room**	**Warranty**	**Cost**	**Speed**	**Score**	
7	BMW Z3 1.9L	3	2	1	3	3	2	230	
8	Corvette	2	1	2	1	2	3	180	
9	Firebird Conv.	1	1	3	1	3	1	200	
10	Mercedes SLK	2	3	2	3	2	2	230	
11	Porsche Boxster	2	1	1	3	1	2	160	
12	Volvo C70	2	2	1	3	1	3	180	
13									
14	**Key Formula**								
15	Cell H7	=SUMPRODUCT(B$4:G$4,B7:G7) [copy to cells H8:H12]							

EXHIBIT 1.6 Impact of Changing a Particular Car's Score

	A	B	C	D	E	F	G	H	I
1	**Car Purchase Decision**								
2								Changed cost	
3	Criteria:	Ins. Rating	MPG	Leg Room	Warranty	Cost	Speed		
4	Weights:	15	10	15	20	25	15		
5								**Total**	
6	**Options**	**Ins. Rating**	**MPG**	**Leg Room**	**Warranty**	**Cost**	**Speed**	**Score**	
7	BMW Z3 1.9L	3	2	1	3	2	2	220	
8	Corvette	2	1	2	1	2	3	185	
9	Firebird Conv.	1	1	3	1	3	1	180	
10	Mercedes SLK	2	3	2	3	2	2	230	
11	Porsche Boxster	2	1	1	3	1	2	170	
12	Volvo C70	2	2	1	3	1	3	195	
13									
14	**Key Formula**								
15	Cell H7	=SUMPRODUCT(B$4:G$4,B7:G7) [copy to cells H8:H12]							

these products up—this is actually a formal model called a weighted scoring model. Mathematically, this model can be represented as follows:

$$\text{Total weighted score} = \sum_{i=1}^{n} W_i S_i$$

where

i = index for the criteria
W_i = weight of criterion i
S_i = score of the option being evaluated on criterion i
$\sum_{i=1}^{n}$ = summation from 1 to n
n = number of criteria

Then we solved the model by calculating the scores for the six cars being considered for purchase. In addition, sensitivity analysis was performed. The final steps in the modeling process are evaluating the results and then implementing the chosen solution. Since you would be both the analyst and decision maker in this example, a formal memo, report, or presentation of the results would not likely be required. However, in many business situations those who perform the analysis are distinct from those who actually make the decision, and in these cases the results of the analysis need to be communicated, either formally or informally. In the remainder of this chapter, we discuss the process of modeling for business decision making in more detail.

1.1 Quantitative Business Modeling

So what is **quantitative business modeling?** Well, first we recognize that almost every thought-out action we take is probably based on a model—usually intuitive—of some sort. If we are late to a meeting, we run some model based on our previous experience through our heads to give us an idea of what the consequences might be and how we might react to them. If we want to go from one place to another, we use some intuitive map, or perhaps a real map (which is a type of model), to guide us. If we are trying to decide where to eat tonight, we use some form of model to help us make a decision—type of food, cost, level of service, convenience, time available, and so on. Moreover, we might combine our models, such as a mathematical model of the cost of a fancy restaurant with a conceptual model of the atmosphere and food quality, to give us an overall sense for the resulting value of our choice.

In terms of business modeling, business and its environment are more complex today than ever, and this complexity is increasing for many reasons. First, the number of alternatives is much larger, but the time available to make decisions is shorter. An example of this is the drastic shortening of product life cycles in the midst of greater variety. Second, the cost of making errors has become larger due to the size of operations, intensified worldwide competition, and the chain-reaction propagation of errors via new, computerized interrelationships. Third, the consequences of decisions are more difficult to predict due to increased international, competitive, and governmental uncertainty and environmental ambiguity. Finally, the role of technology and its complexity has had a major impact on businesses. Technology has brought both tremendous benefits to business as well as heightened risks, costs, and complexity. In addition, we now have information overload in many decision situations, and the need is great for a tool that can help filter and summarize all this data. As a result of these trends, it is more important than ever that managers employ formal models to help them make their decisions.

The quantitative business modeling approach assumes that business decision making consists of analyzing phenomena that can be measured, identifying relationships that can be represented quantitatively, and determining cause-and-effect relationships whose internal consistency can be tested experimentally. As opposed to many other texts, we do not restrict our discussion here to certain *types* of quantitative data, such as probabilistic (statistics), algebraic (management science), deterministic (e.g., mathematical programming), or computerized (spreadsheets)—all data that are, or can be, quantified are useful in modeling.

The objective of quantitative business modeling (QBM) is to bring as many management phenomena as possible into the domain of "standardized" decisions that can be treated with standardized tools. For example, for certain types of allocation situations, a tool named *linear programming* was developed. For other managerial situations, additional tools such as *decision analysis, simulation,* and *queuing theory* were developed. Other tools were borrowed from sister disciplines such as probability and statistics. We believe that all managers should be aware of such tools and should know the basics of how to use and interpret them as an aid in making managerial decisions.

Although using quantitative models to help make business decisions won't guarantee success, the use of such models can provide managers with significant competitive benefits in understanding the nature of the situation being analyzed and the tradeoffs involved between the different alternatives. Models can also help managers understand and improve the interrelationships between the various functional areas in their organization such as finance, engineering, R&D, marketing, operations, human resource management, and so on. Moreover, although models can't eliminate the risk and uncertainty that plague business decisions, they can help the manager understand the effect of business decisions, which should lead to better choices. And in this age of information overload, through the marvels of computers and the Internet, a manager who can use models to make effective use of this mass of information and data will obtain a significant advantage in the competitive marketplace.

Not only is QBM an effective approach to addressing important business opportunities and problems, it is a creative, challenging, and even fun process as well. Modelers who are well skilled in the QBM process can address, through their experience and creativity, many more situations than managers generally think is possible. Models can help with the "messes" that managers deal with, as well as the clear-cut opportunities and problems. And models can be used at all levels of the business—the strategic, top management level as well as the tactical, shop floor level.

Modeling is both art and science, but it is more appropriately described as a craft. Such a craft can be learned and finely honed over time and with experience. This text is the first step in that process and should be useful to both those who hope to do modeling in their career as well as those who will be the users and implementers of such models. Let us now look further into what this idea of a model is all about.

Definition of a Model

Formally, a **model** is a simplified representation or abstraction of reality. You might think of the modeling process as having the shape of an hourglass, broad at the top and bottom but narrow in the middle where the sand falls through. The top of the hourglass represents the reality of the complex situation being modeled and the narrow middle is the model that captures only certain, critical elements of that reality. The bottom then represents the various solutions provided by the model—sometimes only a few but sometimes a great many—for the situation being analyzed.

For decision and understanding purposes, the model must capture the key factors and relationships that are important to the decision at hand. An example is the spreadsheet model in the chapter opener that represented the characteristics of the cars you were interested in. The model is usually simplified because reality is too complex to copy exactly and because much of the complexity is actually irrelevant to the specific decision being considered. But these characteristics of *simplification* and *representation* are difficult to simultaneously achieve in practice. For example, a model can be simple but not represent the true situation. In the car example, we overlooked many characteristics, and even those we considered were represented in a simple manner, such as leg room as a surrogate for comfort.

One scholar of modeling (Willemain 1994) has identified the most important qualities of an effective model, based on interviews with a dozen expert modelers. In order of importance, the top three qualities are (1) the *validity* of the model—that is, how well it represents the critical aspects of the situation under consideration; (2) the *usability* of the model in terms of whether it can eventually be used for the purposes intended; (3) the *value* of the model to the client. Of interest in this list are the items that were *not* identified: the cost of the model, its sophistication, the time involved in formulating the model, and various others characteristics that might be thought to be critical to its effectiveness.

But more important than the formal definition of a model is the informal one that applies to all of us—a tool for thinking and understanding before taking action. The truth is, we use models all the time, even though most of them are subjective. We formulate a model when we think about what someone will say if we do something, when we try to decide how to spend our money, when we attempt to predict the consequences of some activity (either ours, someone else's, or even a natural event). As you can see, models are the basis of our life. We wouldn't be able to drive or take any purposeful action if we didn't form a model of the activity first. QBM uses this natural tendency to create models but forces us to think more rigorously and carefully about the models we are using.

Perhaps surprisingly, there is an interesting philosophical argument about modeling that concerns something called *ontology*—how we believe we perceive reality (Meredith 2001). Some modelers, called *positivists,* believe reality is fixed in the sense that we all see the same reality (unless we have vision or other perceptual problems). These modelers thus feel that once they have accurately modeled a situation, they can then manipulate the model to try to find a better managerial alternative for a business. These positivist modelers have made great strides in science, engineering, and some aspects of business.

Other modelers, called *relativists,* believe that there is not one reality but many possible realities, depending on the perceptions of the observers—different observers may perceive different realities. Thus, for relativists, when modeling a managerial situation it is important to view the situation from the perspective of all stakeholders because the resulting model may be invalid in the eyes of some stakeholders, or even in the eyes of the same stakeholder but at different points in time. Hence, model validity is a relative rather than absolute concept, and it is vital to constantly check back with managers and other stakeholders to keep the model valid. Relativist modelers have made equally great strides in psychology, sociology, and other aspects of business. In this text, we employ both views of ontology. We take a positivist view of reality when we analyze our models but a relativist view when we formulate and implement our models. We discuss these viewpoints in more detail later, but the important point to remember is that people may well have different perceptions of the same situation.

Benefits and Drawbacks of Modeling

The following are the major reasons why QBM employs models in general, and mathematical models in particular.

- Models enable the compression of time—years of operation can be simulated in seconds of computer time.
- Manipulating a model is much easier than manipulating the real system; therefore, experimentation is easier. In some cases, such as the use of dummies during automobile crash tests, experimenting with the real system is not even feasible.
- The cost of making mistakes during a trial-and-error experiment is much smaller when done on a model.
- Today's environment involves considerable uncertainty. The use of modeling allows a manager to consider risk in the decision-making process.
- The cost of the modeling analysis is much lower than conducting experiments on the real system.
- Models enhance and reinforce learning.
- Using the QBM process forces the use of rigorous thinking.
- The use of mathematical models enables quick identification and analysis of a very large number of possible solutions.
- Often noted as the most important reason of all, the process of modeling provides a much better understanding of the real situation being modeled.

However, QBM, like any other management tool, is no substitute for good management practice. The manager must still decide what to investigate, what to do about the factors that cannot be quantified, and how to interpret the results of the analysis. It also frequently consumes a great deal of time and may become very costly. Attempting shortcuts by superficially examining the problems, or using inappropriate models or inaccurate data, may produce results that, if applied, will be far more costly than simply using the dictates of subjective judgment. The following are some of the potential drawbacks to using QBM:

- It is time consuming, especially if the model is embellished beyond the level of detail needed.
- Managers might be reluctant to accept the model results if the model has not been explained well.
- Obtaining the necessary data may be difficult, time consuming, expensive, or possibly not even feasible.

- It may be difficult to assess uncertainties.
- It entails the risk of constructing an oversimplified model of reality, possibly leading to erroneous recommendations.
- QBM can be expensive to undertake, relative to the size of the problem.
- Studies may be abandoned for various reasons, or the results ignored, resulting in an unproductive expense.
- The common managerial perception is that if done on a computer, it must be correct.

Moreover, many managerial opportunities and problems are complex to model, involving numerous interrelated variables. The search for and evaluation of alternative solutions can become interminable. Thus, many models are solvable only with the aid of computers and software packages, such as spreadsheets. In this text we use Excel to illustrate these capabilities. In addition to the use of the computer for the execution of the necessary calculations in solving the models, computers are often used in data collection, storage, retrieval, analysis (e.g., identifying frequency distributions), and even in the validation of the models. Computers are also used to assist in implementation, using "what-if" capabilities to perform sensitivity analyses and graphic presentations that make it easier for managers to understand and use the model results.

Types of Models

Models can be classified in terms of the various forms in which they appear or in terms of the uses we have for them. We will consider their various forms first.

Forms The representation of problems/opportunities through models can be done at various degrees of abstraction or realism. Models are generally classified along this dimension into four groups. The least abstract and most realistic form of model is a *physical* replica of the situation, usually based on a different scale than the original. These may appear in three dimensions, such as airplane, car, or bridge models made to scale. Photographs, blueprints, and computer images are other types of physical scale models, but in only two dimensions.

The next most abstract, *analog* models, *do not look like the real situation* but *represent* or *behave* like it. For example, the oil dipstick in a car represents the amount of oil in the crankcase (in quarts or liters, typically). Similarly, two-dimensional charts or diagrams differ in their shape but represent the situation in an alternative way. Some examples of analog models are as follows:

- An hourglass, where the change in the amount of sand on the top or bottom represents the passing of time
- Organization charts that depict structure, authority, and responsibility relationships with boxes and arrows
- Maps where different colors represent water or mountains
- Stock market charts where vertical height indicates value
- Graphs, such as interest rates over the last 100 years, or expected sales as a function of product pricing

The complexity of relationships in some systems cannot be represented with a physical or analog model, or the representation may be cumbersome and take time to construct or manipulate. Therefore, a more abstract model is created with the aid of mathematics. Most QBM analysis is executed with the aid of *mathematical models.* They can describe diverse situations, yet be easily manipulated for purposes of experimentation and prediction. A new form of mathematical model that we will use extensively here is a spreadsheet model. Mathematical models are typically characterized by three sets of variables:

- **Independent variables** are those we can control. They are often our decision or policy variables.
- **Uncontrollable parameters** are governed by nature or outsiders and are beyond our control.
- **Dependent variables** are our measures of interest, such as profit or productivity.

If the parameters and dependent variables, given a set of independent variables, are known with certainty, the mathematical model is called *deterministic.* If the parameters and/or dependent variables can take a range of values according to some probability distribution, the model is called *probabilistic.*

Last, the *subjective* or *intuitive* model, discussed previously, is the most abstract. However, it suffers from many problems such as inadequate formulation, limited data, and insufficient testing. Because it is often not explicitly identified, a critical analysis cannot be made of it, and its accuracy is thus inadequate for most QBM purposes.

Uses We generally consider two primary uses of models. One use is to help us determine the *best* action to take. These kinds of models are termed *prescriptive* (or *normative*). Their purpose is to find what is termed an *optimal* solution to a problem or opportunity, within the assumptions and structure of the modeling scenario. Due to the factors excluded in the simplification of reality to fit the model structure, there may well be other considerations relevant to the problem or opportunity at hand. For example, in the chapter opener, you might also have some preferences for the image each automobile conveyed. These weren't considered in the rather practical model that was constructed, or perhaps you want to include the opinion of a "significant other" in your decision. These softer factors are usually incorporated after the more formal analysis is complete, thus leading to decisions that, based on the model, are "nonoptimal" but actually are more complete and useful. Clearly, in business situations there may be many other factors that will be relevant to a decision other than just those identified in the quantitative model.

Prescriptive techniques consist of either optimal or nonoptimal trial-and-error comparisons of several proposed alternatives. The prescriptive models that yield optimal solutions consist of those that are based on *enumeration* and those that are based on *algorithms,* which evaluate only selected alternatives. When one checks *every possible* alternative, this is called a *complete, exhaustive enumeration.* This technique is useful only when the number of alternatives is relatively small; otherwise, it is a lengthy, tedious, or even impossible approach. When we used an equation to derive a score for our automobile selection opportunity, we used the technique of enumeration to identify the best car for our situation. In contrast, an *algorithm* is a step-by-step process of searching for an optimal solution by gradually improving each solution. Thus, in contrast to complete enumeration, an algorithm checks only a *portion* of all solutions. *Linear programming,* presented in Chapter 4, is an example of an algorithmic, prescriptive model.

Rather than using an optimizing approach to prescribe a course of action, a *descriptive* model can be equally, if not more, valuable to a manager. Descriptive models characterize things *as they are.* Their major use in QBM is to investigate the outcomes or consequences of various alternative courses of action, as reflected by the results of the situation. However, because the descriptive analysis checks the consequences only for given conditions (or given alternatives) rather than for *all* conditions, there is no *guarantee* that an alternative selected with the aid of descriptive analysis is optimal. Descriptive models are usually applied in decision situations where optimizing models are not applicable. They are also used when the objective is to define the problem or to assess its seriousness rather than to select the best alternative. Descriptive models are especially useful in *predicting the behavior* of a system under various conditions.

Simulation is an example of a descriptive technique for conducting experiments with a system by checking the performance of different configurations or scenarios of the system. Simulation usually is based on a computerized mathematical model of a management system operating for an extended period of time. We often use it to help us decide how to alter a system to make it work better, but the simulation itself does not prescribe the best way to make the changes. A major shortcoming of simulation is that we may not identify the *best* alteration to try in the simulation model and thereby miss the optimum solution to our situation. Simulation is described in Chapter 7.

To illustrate the use of descriptive techniques, one might consider a control box, as in Exhibit 1.7. The box has knobs on it representing different independent controllable variables, and dials (gauges) representing the dependent variables that measure the system's effectiveness. The uncontrollable parameters are built into the operation of the box through its internal wiring. When the manager wants to explore the consequence of a given alternative course of action, he or she merely turns the knobs (each combination setting of the knobs represents an alternative course of action) and watches the dials.

The illustration in Exhibit 1.7 represents a control box for an inventory model. The box has only one dial, representing the cost of inventory, the chosen measure of the system's effectiveness. The knobs represent two controllable variables: x_1 is the number of orders of a fixed size placed each year and x_2 is the quantity of safety stock the company keeps. The manager, sitting in front of this box, would manipulate the knobs. For example, the manager may set $x_1 = 2$ orders per year and $x_2 = 400$ units of safety stock. Then the dial that shows the resulting cost, $7,000, can be observed.

To find a solution in the case of *complete enumeration,* the manager would have to experiment with *all* combinations of the knobs. For example, $x_1 = 2$ and $x_2 = 400$ is just the one combination shown in Exhibit 1.7. Each time a combination is tried, the inventory cost is recorded. When *all* possible combinations have been tried, the *optimal* solution can be identified.

EXHIBIT 1.7 A Control Box for an Inventory Situation

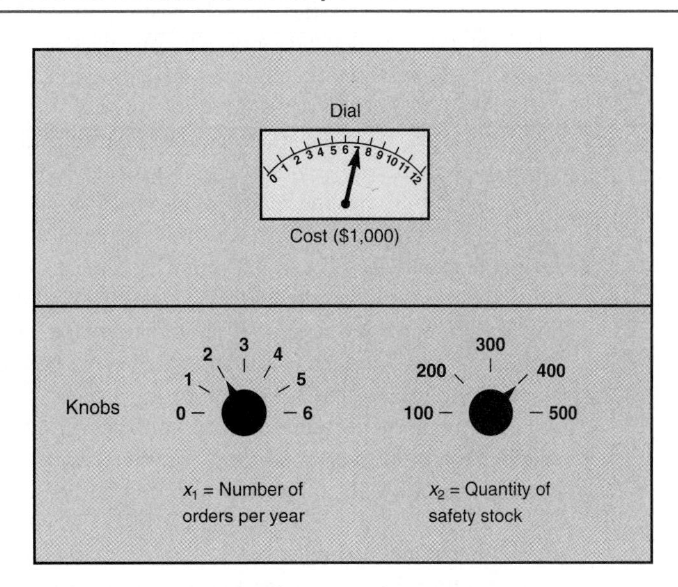

Effective Modelers

Willemain (1994) also queried his dozen expert modelers on their perceptions of the most important qualities of effective modelers. Their responses agreed on four major characteristics. The first and most important was a set of internal skills and characteristics: creativity, sensitivity to the client, and persistence. Clearly, these qualities are partly inherent and partly learned, even creativity. Being sensitive to the client relates back to our earlier philosophical discussion of relativism.

The second set of characteristics was interpersonal: communication and teamwork skills. This included both the customer and contractors/suppliers, as well as colleagues and fellow workers. Good communication and teamwork skills are always valuable but, more than that, they are apparently absolutely necessary to be an effective modeler. The third characteristic was expertise in quantitative business modeling. It is interesting that this characteristic was neither first nor second. Last was expertise in the subject being investigated.

It should be encouraging that most of these qualities of effective modelers can be learned, with experience. Even someone not planning to do quantitative modeling as a career can benefit from these qualities in their occasional modeling efforts, or when working with modelers. Beyond the characteristics of effective modelers, Willemain also investigated the modeling *process* that effective modelers use, which we describe next.

1.2 The Modeling Process

Willemain (1994, 1995), in interviews with a dozen modeling experts, identified some important characteristics of the process these experts used when doing modeling, the primary steps of the process, and the amount of time they spent on each of the steps. The experts agreed that the three most critical characteristics of an effective modeling process, in order of importance, were (1) discovering the *real* opportunity/problem; (2) validating and verifying the model; and (3) identifying the key variables in the model. When facing an actual business situation, it is not at all clear what the real opportunity or problem is. As noted when discussing the relativistic philosophy, the opportunity or problem may differ significantly depending on who one speaks with about it, or whether it even exists at all. For example, marketing may see a drop in sales as an advertising problem whereas operations may believe it is due to low quality. It is crucial not to accept a single viewpoint on the nature of the situation to be investigated but to see the situation from the perspective of all the stakeholders and, if possible, how it changes over time and conditions.

Making sure that the model is internally consistent and correct is obviously important. But it is also crucial to make sure that the model being formulated for validation and verification purposes accurately represents the true situation under study, again as perceived by the various stakeholders. There may be differences of opinion among the stakeholders about what the model should represent, or even what the true opportunity/problem is, and this needs to be straightened out before detailed data are collected and the model is formulated and analyzed.

And again, identifying the key variables in the model totally depends on correctly understanding the real opportunity/problem, as agreed on by the various stakeholders. This applies not only to the most appropriate dependent variables to consider—profit, revenues, quality, productivity—but also to the most appropriate independent, or managerial decision, variables. For example, if a firm allocated money to advertising to counter a drop in sales, when the real problem was the low quality of the product, this would be a major waste of valuable firm resources, as well as time in correcting the problem. Similarly, if a model were constructed to determine managerial policies to maximize revenue, but it was later found that cash flow is even more important, then the wrong dependent variables were identified. Or if the model were formulated to determine what product lines to add or abandon, but it was later decided that governmental or customer requirements would constrain management options in adding or dropping product lines, the opportunity should have been defined more carefully.

The experts also exhibited some common characteristics in their own modeling processes. The most common characteristic they displayed was an orientation in their modeling efforts toward identifying changes in the organization's existing, typically complex, systems. That is, they did not work toward creating new systems or new ways of managing, but rather, toward identifying some changes in existing managerial systems that would improve upon the current values of the study objectives. The next most common characteristic they exhibited was that they developed a unique model for each situation, though every model was highly quantitative. That is, they did not take an existing model, either theirs or the organization's, and make incremental changes in it. Third, their modeling efforts took place over an extended period of time, with frequent client communication about various facets of the modeling process. They did not typically engage in short, quick approaches to simple, straightforward situations. Perhaps in-house consultants handled these types of studies for the businesses.

The next most common characteristic was that the experts tended to develop more than one model for every situation. These various formulations were guided by analogies, drawings such as influence diagrams, and even the seemingly random doodlings of the modelers while working on the study or talking with the client. Last, the experts typically started with a small model and added to it, eventually resulting in a rather complex model.

In terms of the actual steps the experts followed in their modeling process, there was some consistency, but the process for all of them was highly iterative, with a lot of jumping around between the steps as well as following different sequences of the steps. They frequently went back to earlier steps, jumped ahead to upcoming steps, rechecked previous steps, and so on. We saw the same effect in the chapter opener when we decided to add weights to the various criteria after they had been formulated, which then required obtaining the data for the weights. And then again, we decided to do some sensitivity analyses after analyzing our model, in case values changed, or the right car was unavailable, or other factors arose to negate our initial choice.

The general steps in the process, and the percentage of time spent on each step, in the most common order (but this order varied) were opportunity/problem recognition (14 percent), model formulation (59 percent), data collection (9 percent), analysis of the model (16 percent), and implementation (2 percent). As noted, however, sometimes the steps are conducted in another sequence, such as when data mining is used with customer preference data to reveal marketing opportunities, in which case the data collection step comes first. In addition, there were many iterations between the steps, primarily between model formulation and one of the other steps. Thus, as data were collected, the modeler might find it necessary to change the formulation of the model, or even the definition of the opportunity/problem. And as the analysis proceeded, it might be easier to solve by reformulating the model, or as the initial opportunity was better understood, the model might be reformulated.

In the next subsection, we present each of these steps in their most frequently used order and discuss them in detail. In each of the following chapters where we discuss one of the major types of models, we follow the same sequence of steps but emphasize the formulation and analysis step for that model. Our discussion of the other steps will primarily address how that step may differ from what we discuss here for that particular model.

A Five-Step Modeling Process

All the standard business functions of planning, organizing, implementing, and controlling involve decision making and thus are amenable to QBM. A major premise of QBM is that decision making, regardless of the situation involved, is a generic process involving the following five major steps, generally in the order shown:

1. Opportunity/problem recognition
2. Model formulation
3. Data collection
4. Analysis of the model
5. Implementation and project management

A pictorial representation of this process is illustrated in Exhibit 1.8. Generally, a real-life opportunity or problem arises, such as buying a new car in the chapter opener. Note that it may be some time before the opportunity or problem is recognized as such by the manager, if ever. An initial model may then be roughly formulated, based on an influence diagram. Data are then collected for the variables and parameters in the model to validate the model and to use in later analysis. The model is checked, or *validated,* against the real-life situation to assure that it correctly represents the situation we are investigating. We often test it with data that result in outcomes for which we already know the correct answer and then, if necessary, we correct our influence diagram and model. Model analysis includes generating alternatives, determining the outcomes of these alternatives under different conditions, comparing them, and so on. It also includes evaluating how the decision might change if some of the parameters or variables were to change. Finally, the results are communicated to those who can take action for implementing the ultimate decision, and preparations are begun for project implementation. Each of these five steps will now be discussed in detail.

EXHIBIT 1.8 The Generic Quantitative Business Modeling Process

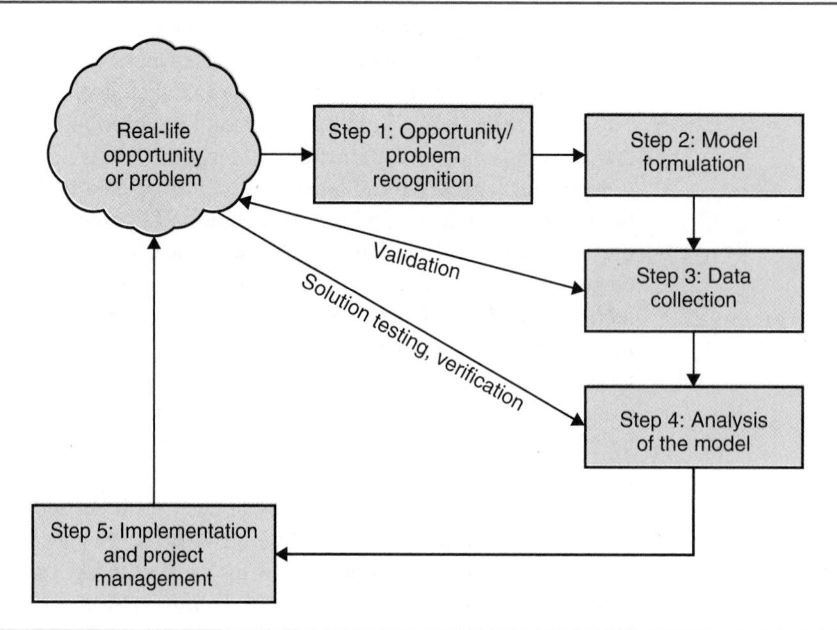

Step 1: Opportunity/Problem Recognition

By **opportunity/problem recognition** we mean recognizing that an opportunity or a problem exists, determining its magnitude, defining it precisely (perhaps with the aid of an influence diagram), and noting what its symptoms are. However, defining the opportunity or problem precisely is easier said than done, for a number of reasons. As many have said before, defining the situation well in the beginning takes you halfway toward solving it. Thus, this is the *most* important step of the entire modeling process; if you solve the wrong problem, the result is an ineffective solution and the entire effort is wasted.

Although QBM is often talked about in terms of "solving" problems, it is important to recognize that problems are usually the flip side of opportunities, and proactively capitalizing on opportunities can be much more valuable to managers than just reactively solving problems. As an example, some years ago the large department stores noted that their appliances were selling extremely well but their clothing was not. They thus focused their attention on their clothing problem and didn't concern themselves with appliances. Sensing an opportunity, large discount chains were formed to offer appliances at lower costs and with more selection, thereby stealing these profitable sales from the department stores.

To begin the recognition step, the modeler should be careful to fully investigate the situation of concern, because what one person thinks is "the problem" may not be the real problem (or *a* problem) at all. Also, others typically see "the problem" differently, and may have a completely different perspective on the situation. Moreover, what is often considered to be a problem (e.g., excessive costs) may only be a *symptom* of the real problem (e.g., improper inventory levels). Because so-called real-world problems are usually complicated by many interrelated factors, it is sometimes difficult to distinguish between symptoms and problems. Further, problems tend to evolve over time. Thus, what may have been the right problem at the beginning of the investigation may turn out to be the wrong problem by the end of the study. Thus, the modeler needs to stay in touch with the managers and the situation, monitoring the opportunity/problem context to make sure she or he is still addressing the correct situation. Careful initial investigation and continuing awareness can keep the modeler from solving the "wrong" problem and thereby save precious time and energy. Staying in touch with the managers will help immeasurably when it comes time to implement the solution.

Step 2: Model Formulation

Model formulation involves abstracting and transforming the complex opportunity or problem—the top of the hourglass—into a mathematical form—the neck of the hourglass—based on a well-thought-out influence diagram. All of the relevant variables are identified and the equations describing their relationships are established. Simplifications are made, whenever necessary, through a set of *assumptions.* For example, a relationship between two variables may be assumed to be linear. It is necessary to find a proper balance between the level of simplification of the model and the degree of representation of reality. The simpler the model, the easier the manipulations and the solutions, but the less representative it will be of the real problem, as noted earlier.

The task of modeling involves a multitude of interrelated activities and methodological issues. The most important of these are determining the components and key variables of the model when the influence diagram is drawn. It may well take a number of iterations to arrive at the final diagram, or as also happens, a set of influence diagrams representing a range of different models may be constructed for investigation. After formulating the model(s), we can then check to see if it (or they) fits one of the standard classifications described in Chapters 2 through 8 that have special solution procedures already developed for that type of model.

The Components of Mathematical Models As noted earlier, all mathematical models are composed of three basic components: *dependent variables, independent (decision) variables,* and *uncontrollable parameters.* These components are connected by mathematical (logical) relationships, as shown in Exhibit 1.9. Examples of these components are given in Exhibit 1.10.

The Dependent Variables. The dependent variables reflect the *level of effectiveness* of the system. That is, they tell how well the system performs or attains its goals. The variables are dependent because for the event described by this variable to occur, another event must occur first—in this case, the independent variables and the uncontrollable parameters.

The Independent Variables. Independent variables describe elements in the problem for which a choice must be made. These variables are manipulable and controllable by the manager. Examples are the quantities of products to produce and the number of tellers to use in a bank (others are shown in Exhibit 1.10). Independent decision variables are also

EXHIBIT 1.9 The General Structure of a Model

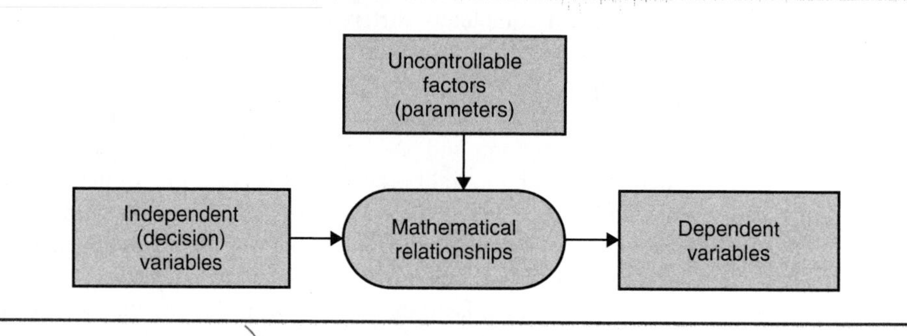

EXHIBIT 1.10 Examples of the Components of Models

Area	Decision Variables	Dependent Variables	Uncontrollable Variables
Financial investment	Investment amounts Period of investment Timing of investment	Total profit Rate of return Earnings per share Liquidity	Inflation rate Prime rate Competition
Marketing	Advertising budget Number of models Zonal sales reps	Market share Customer satisfaction	Disposable income Competitor's actions
Manufacturing	Production amounts Inventory levels Incentive plan	Total cost Quality level Spoilage	Machine capacity Technology Materials prices
Accounting	Audit schedule Use of computers Depreciation schedule	Data processing cost Error rate	Legal requirements Tax rates Computer technology
Transportation	Shipments	Total transport cost	Delivery distance Regulations
Services	Number of servers	Customer satisfaction	Demand for service

called *unknowns* and are commonly denoted by the letters x_1, x_2, and so on, or by x, y, z. Often, the aim of QBM is to find the best (or good enough) values of these independent decision variables.

The Uncontrollable Parameters. In any decision situation, there are parameters (variables, constants) that affect the dependent variables but are not under the control of the manager. Examples are the prime interest rate, building codes, tax regulations, and prices of supplies (others are shown in Exhibit 1.10). Most of these parameters are uncontrollable because they emanate from the environment surrounding the manager.

The components of a mathematical model are tied together by sets of mathematical expressions such as equations or inequalities. Exhibit 1.11 is an example of a model of a manufacturing system. In QBM, the arrows in the picture are replaced by mathematical expressions, as illustrated in Exhibit 1.12. The model can be interpreted as: Find the values of the independent decision variables x_1 and x_2 such that the total revenue R (dependent variable) is maximized, subject to the marketing limitation of 50 total units and market prices of 5 and 2, which are uncontrollable by the manufacturer.

Classification Throughout the history of management, certain types of situations have been encountered repeatedly. The structure of these special situations has been abstracted and standardized and a set of special solution procedures has been developed. Thus, whenever a management situation is recognized as being one of these standard forms, an appropriate tool can be applied for its solution. Although many managerial situations have been standardized for quantitative analysis—sequencing tasks, routing vehicles, maintaining facilities, replacing machines, ordering inventory, scheduling workers, searching for items, bidding on contracts—we restrict our attention here to the four we consider to be the most important.

Allocation situations. These situations arise when (1) there are a number of activities to be performed, (2) there are multiple ways to perform these activities, and (3) resources or

EXHIBIT 1.11 A Manufacturing Systems Model

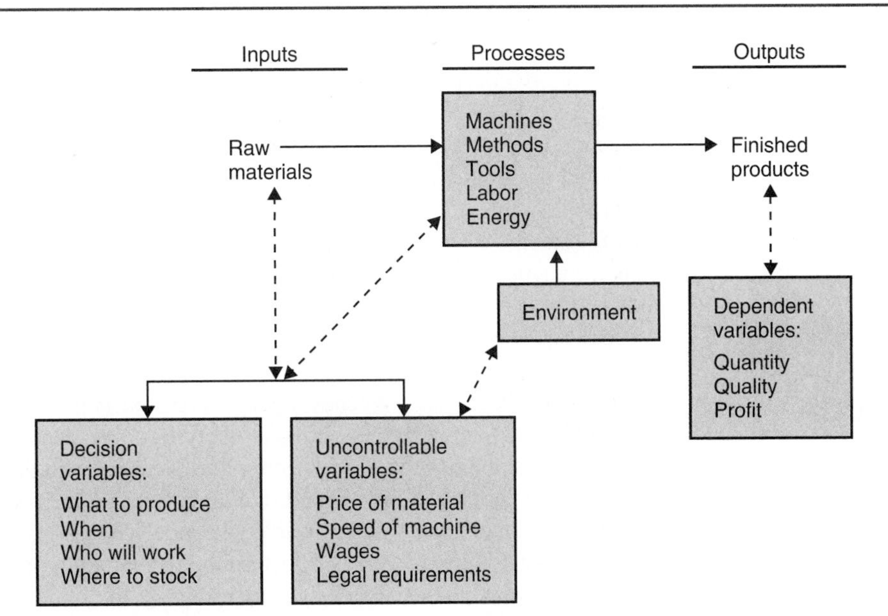

EXHIBIT 1.12 A Simplified Model of a Manufacturing Situation

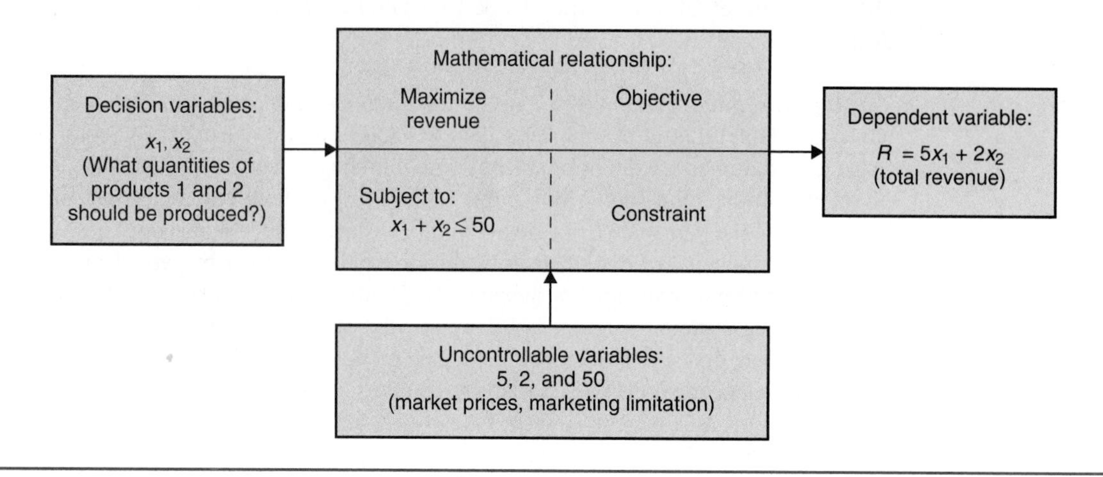

facilities are limited. The task then is to find the best utilization of resources; namely, which activities to pursue and in what magnitude, so that effectiveness will be maximized. The car purchase decision described in the chapter opener was an allocation situation. In business, advertising and investment decisions are other examples. *Optimization* is one of the most popular techniques for addressing allocation situations, in that it attempts to maximize the attainment level of the goal subject to a set of requirements or limitations. Optimization is considered a *prescriptive* modeling technique in the sense that it tells the manager what the best course of action is. Typical managerial situations amenable to optimization models are finding the best combination of resources to maximize profits, or minimize costs. In Chapter 4, "Optimization and Mathematical Programming," we describe the techniques of linear and integer programming for allocation situations. In Chapter 8 we consider allocating scarce resources to project activities to expedite completion of a project.

Decision situations. Many managerial situations involve selecting a course of action among a limited number of alternatives when the outcome is either deterministic or probabilistic. The data in this situation can be presented in a tabular form known as a *decision table* and analyzed with a technique called *decision analysis,* covered in Chapter 5. The car selection opportunity employed a type of decision table. The extension of decision tables for situations involving several decision periods takes the shape of a *decision tree.* This technique is primarily descriptive because there are only a limited number of alternatives that are usually considered, but if these represent all the possible alternatives, the process can be considered prescriptive.

Waiting-line situations. Whenever persons or objects that require service arrive at a service facility, a waiting line or *queue* is likely to occur, especially in rush periods. Typically in such situations there is a waiting line during certain times, but the facility may be idle during slack periods. In general, the larger the service facility, the costlier its operation, but the smaller the waiting time for service. The task is to find the appropriate size of the service facility as well as to determine its operating procedures (e.g., give priority to certain customers) in such a way as to minimize the sum of the relevant costs. For certain types of waiting-line situations, special models have been developed to predict the performance of service systems given their operating characteristics. These models, covered in Chapter 6, "Queuing Theory," are almost always descriptive in nature in the sense that, given certain model inputs they predict what the result will be. That is, they do not tell the manager what the best course of action or inputs are, they simply predict an outcome based on given inputs.

Predicting the behavior of a system. Management is frequently interested in predicting the behavior of a system, such as a project (Chapter 8), under different conditions or scenarios. Although this is not classified as a standard situation by itself, it can often be analyzed with the help of tools such as probability, statistics, and simulation. Probability and statistical analyses (Chapter 2) are powerful descriptive generic tools that can by employed in a wide variety of circumstances, both deterministic and probabilistic. They are commonly used in forecasting systems (Chapter 3) and quality control systems. For the analysis of more complex systems, particularly where other models are inadequate, simulation is another powerful descriptive tool. Simulation modeling, covered in Chapter 7, can include many factors of interest to management and has often been used to analyze situations concerning inventory stocks, scheduling dilemmas, facility utilization, project planning, and other such business issues of managerial importance.

Step 3: Data Collection

Acquiring data for the modeling study can be the most difficult, tedious, and time-consuming step of the entire modeling process when data are not regularly collected and archived. This is frequently disconcerting for novice modelers, who think that the difficult mental tasks of model formulation and analysis should take precedence over the mostly mechanical task of **data collection.** However, without accurate data, the modeler faces the common situation of "garbage in, garbage out." And more often than not, the data will be "fuzzy," incomplete, excessive, ambiguous, inaccessible, purpose-dependent, out of date, mislabeled, scattered throughout the organization (or even in different sites), listed in different formats, or even in conflict with other sources of the same information.

Most of these difficulties, though frustrating, are understandable, but let's take one that is less clear: data that depend on what they are used for. For example, your dollar-per-hour wage rate as an employee can depend on who is asking. To you, the total hours you spend at your employer's disposal is the relevant time, but to the employer, your lunch and rest breaks are nonproductive and may not be included. And what about training, sick, or vacation days? And when the employer is reporting to government agencies such as state workers' compensation, or Social Security, the hours change again. Similarly, you may be interested primarily in your wages, but accounting must also tabulate the costs of workers' compensation insurance, Social Security payments, and other such fringe benefits.

Excessive data is another common problem these days. Computers spew out reams of data without context. A particular danger here is the temptation to use this readily accessible data instead of giving careful thought to the particular data needed to produce a meaningful study. Again, the use of QBM to make sense of this data provides the manager with a tremendous competitive advantage. As an example of another difficulty, outcome variables may occur over an extended period of time, with revenues and expenses being recorded at different points of time. To overcome this difficulty, a *present-value* approach might be selected for all calculations involving past and future funds. But this then raises the difficulty of identifying the appropriate interest rate to use for discounting. So one data problem is solved, but another is raised.

Another difficulty is when objective data do not exist. Then it may be necessary to use subjective data estimates obtained from workers or managers. Although subjective, this does not mean that the data are less accurate, and may well be more so. Objective and subjective data are often equated with *accurate* and *inaccurate.* However, *objective* formally means taken by reference to a standard external to the system, and *subjective* means taken by reference to an internal standard. Using a yardstick as equivalent to a meter measure is objective but inaccurate, whereas using the eye, or pacing the distance off, is subjective but may be more accurate. However, being subjective also means that the data may be biased in terms of the frame of mind of the person providing it. Also, it is assumed that the data used for the assessment and modeling are representative of future conditions. If not, it is necessary to predict the probable nature of future changes and include this in the analysis.

In all these situations, it may become necessary for the modeler to install a data collection system as discussed in Chapter 2 in order to obtain adequate data to use in the modeling study. As might be expected, this is a major, expensive, time-consuming task. As a general rule of thumb, the modeler might be wise to suspect any data or performance measures that are relatively easily obtained, such as utilization rates. To give the student experience with this aspect of QBM, we include some experiential exercises and cases at the end of each chapter of this text that will require them to collect their own data.

Decision situations are also frequently classified on the basis of what the manager knows (or believes) about the situation. It is customary to divide this degree of knowledge into four categories (see Exhibit 1.13), ranging from complete knowledge, on the right, to ignorance, on the left. Specifically, these categories are (1) certainty, or complete knowledge, (2) risk, (3) uncertainty, (4) ignorance, or total uncertainty. We describe each in turn.

In decision making under *certainty,* it is assumed that complete information is available so that the manager knows exactly what the outcome of each course of action will be. For example, the choice to invest in U.S. Savings Bonds is one in which it is reasonable to assume complete availability of information about the future return on the investment. Such situations are also termed *deterministic.* They occur most often with short time horizons. For example, it is more reasonable to assume certainty with decisions whose impact will be felt after three weeks than with decisions whose impact will be felt after two years.

A decision under risk (also known as a *probabilistic* decision situation) is one in which there could be two or more possible results for each alternative action. For example, the Federal Reserve may or may not raise interest rates at its next meeting, depending on the state of the economy. The reason there could be more than one result is that the possible states of nature are uncontrollable by the manager. If we assume that the chance of occurrence of each of the states of nature is known, then we say that the decision situation is *under risk;* otherwise, we call it *uncertainty.*

In risk situations, it is assumed that the long-run probabilities of occurrence of the given states of nature are known or can be estimated. A classic example of such a situa-

EXHIBIT 1.13 Decision-Making Categories

Ignorance	Uncertainty	Risk	Certainty

Increasing knowledge →

tion is roulette. The roulette board is divided into 37 equal parts: 18 are black, 18 are red, and one is marked with zero. The player knows the probabilities of each state of nature represented by parts of the roulette field (e.g., 18/37 for red, 1/37 for zero). In making a decision, the player knows the long-run probability of winning the bet and therefore can assess the degree of risk assumed (termed a *calculated risk*).

In decision making under *uncertainty,* the manager considers situations in which several outcomes are possible for each course of action. However in contrast to the risk situation, the manager *does not know,* or cannot estimate, the probability of occurrence of the possible states of nature. This is not necessarily the case of *ignorance,* however, because he or she at least knows the *possible* states of nature. For example, it may be impossible to assess the probability of the success of a brand new product. Thus, uncertainty situations contain even less information than risky situations. For situations under ignorance, no modeling is possible.

Depending on which of these conditions characterize the manager's decision situation, the modeling approach will differ. As we shall see, some standard modeling approaches have been developed for each of these kinds of situations. Once we have collected the data and determined whether we are facing a situation of certainty, risk, or uncertainty, we can attempt to analyze the model.

Step 4: Analysis of the Model

After a model has been formulated and the data collected, it is important to test how well the model represents reality. Validation requires answering such questions as these: Are the predictions made by the model empirically accurate? Is the model representative of the system's behavior under real-world circumstances?

Validation may be viewed as a two-step process. The first step is to determine whether the model is *internally* correct in a logical and programming sense. For example, the modeler may input simple values of the variables and solve the model manually, comparing the manually obtained results to the computer-calculated results. The second, more difficult, step is to determine whether the model represents the system (or phenomenon) it is supposed to represent.

One way to execute this second step is to try different possible sets of data and see if the solutions resemble the historical behavior of the system. For example, if a model describes sales behavior as a function of interest rates, then it can be tested by "plugging in" several values of interest rates (e.g., 4, 5, or 19 percent), examining what level of sales is predicted in each case, and comparing this to historical data. Obviously, if the model was unable to successfully describe historical occurrences, it should not be considered valid for making future predictions; therefore, changes are necessary in the model. Another way to validate a model is to solve small problems and see if the results make sense to the manager.

After validating the model, the next step is to conduct the **analysis of the model** to find specific sets of values for the independent decision variables that provide a desirable result. As previously stated, analysis procedures for standard problems are well developed, as described in Chapters 2 through 8, and computer codes exist for executing all of them. The manager, therefore, is relieved of the task of developing these procedures, but it is important for the manager to understand some of the concepts and the methodological issues that are involved in this step. This is particularly true for formulas encoded into spreadsheet models, since a change in one place of the spreadsheet is typically transmitted throughout the spreadsheet, whether correct or erroneous.

Using the Model to Select an Alternative In general, the process to identify the best decision involves five steps, as follows.

1. **Generate Alternatives** The decision-making process described in the chapter opener involved the analysis of alternative candidate solutions for the opportunity. In QBM, such alternatives may be either given or else generated by the model. In the first case, the model is used to *evaluate* the given alternatives; in the second case, it is used to generate and then evaluate the alternatives, as is done in optimization (see Chapter 4). At the very least, the search process requires resources such as money, labor, and time. Because these are usually limited, the search must be terminated sooner or later.
2. **Predict the Outcome of Each Alternative** In order to evaluate an alternative, it is necessary to predict its outcome in the future. This is the primary task of the quantitative model as discussed earlier and addressed again in each of the Chapters 2 through 8.
3. **Relate Outcomes to Goals** The value of an outcome is judged in terms of the goal's attainment. Sometimes, an outcome is expressed directly in terms of a goal. For example, *profit* is an outcome, whereas *profit maximization* is the goal, and both are expressed in dollar terms. In other cases, an outcome may be expressed in terms other than those of the goal. For example, the outcome may be in terms of customer waiting time, but the goal can be expressed in terms of customer satisfaction. In such cases, it is necessary to transform the outcome so it is expressed in terms of the goal.
4. **Compare the Alternatives** Once the previous activities have been completed, the manager can compare the alternatives and select one, or perhaps decide to investigate some new alternatives. Some difficulties may be encountered in this stage of the process, such as the following: (a) the models described in this book are frequently based on the analysis of one goal, such as "maximization of profit," but in reality, several goals may exist simultaneously; (b) the dependent variables may be particularly sensitive to changes or errors in some of the independent variables or parameters. The model builder should check this sensitivity to avoid highly sensitive alternatives, as described in the next subsection; and (c) in comparing alternatives, one may find that alternative A will bring $323,200, while alternative B will bring $323,150. Is alternative A superior to B? In theory, the answer is yes. Practically speaking, the two alternatives are basically the same. The accuracy of the data may be such that $50 out of $323,000 is not a significant difference. In general, it is important to determine when an alternative is indeed superior. Because we deal with models, we simplify reality. Alternative A may bring $50 extra, but it might also make some employees unhappy. In this text, a choice is usually made based on the quantitative results, but other factors should also be considered, especially when the quantitative difference is small.
5. **Select an Alternative** The process ends with a choice, namely the selection and recommendation of a solution or alternative course of action. Solutions can be classified as being *optimal* or *nonoptimal,* and *unique* or *multiple.* An *optimal* solution is the best of all solutions. For a solution to be declared optimal there must be proof that *all* solutions were checked and that the proposed solution is better than any other solution. A solution that cannot be classified as optimal is considered nonoptimal, though it may still be acceptable. If there exists only one optimal solution, it is called *unique.* If two or more equivalent optimal solutions can be identified, then there exist *multiple* (or alternate) optimal solutions. Multiple solutions are usually preferred by managers because it gives them greater flexibility to consider other factors when implementing a solution.

Sensitivity ("What-If") Analysis **Sensitivity analysis** is an attempt to help managers when they are uncertain about the accuracy of their information. In sensitivity analysis, the information in question is altered to find what effects, if any, changes or data errors will have on the proposed solution. In other words, the purpose of sensitivity analysis is to determine the effect of *changes* in the independent variables on the values of the dependent variables. For example, in analyzing the car purchase decision in the chapter opener, we considered changes in the factors that might affect the decision situation, such as changing the importance weights and reflecting recent price changes.

These questions are of special interest:

- What change can occur in a certain independent variable or uncontrollable parameter before a change occurs in the recommended solution?
- What is the magnitude of a change in the proposed solution resulting from a change in an independent variable(s) or parameter(s)?
- Which independent variables and parameters are most sensitive? That is, which variables will, when changed only slightly, cause the value of the dependent variables to change significantly? Which independent variables and parameters are insensitive?
- Is a proposed solution highly sensitive? That is, does the solution include variables that, when changed slightly, will alter the solution so that it will no longer be optimal? Conversely, *robust* (insensitive) solutions will hold with a wide range of variations in the independent variables and parameters. Robust solutions are usually easier to implement because their predicted results are more certain to occur and because management can modify the proposed solution with only a small loss in the effectiveness of the system.

Step 5: Implementation and Project Management

Implementation of a decision involves three activities: communicating the results of the study and the modeler's recommendations for action, actually applying the managerially approved actions through project management, and evaluating the results. This topic is discussed in more detail in Chapter 8, as well.

The outcomes generated by a model represent a solution to a simplified scenario of reality based on many assumptions and limitations. Also, in many cases the solution appears in an abstract format (e.g., mathematical symbols) or in technical terminology. Therefore, before the solution is submitted to management for implementation consideration, it needs to be managerially analyzed and then *translated* into managerial language. Managerial analysis involves determining the *meaning* and *appropriateness* of the proposed solution in terms of management's opportunity/problem. Does the solution make sense? Is it realistic? Can the solution be easily integrated into the organization under current conditions and operating procedures, or will changes be required? Are the changes minimal or radical? Is the solution commensurate with the future plans and actions of the organization? These kinds of questions need to be considered in the managerial analysis.

The solution also must be translated into common, everyday language the manager can understand. It should be clear, understandable, and even somewhat intuitive to the manager. The recommendations and actions that need to be taken should be clearly specified. In drafting the final report or presentation, it is helpful to think of the managerial decision maker as the customer of the modeler whose job it is to communicate the results of the study in as straightforward and clear a manner as possible. Obviously, this objective is not met by reports that are highly technical and serve more to highlight the expertise of the modeler than to aid the manager in choosing an appropriate course of action.

It is often useful if the final report includes easy-to-understand graphics, figures, and tables. The implications of these exhibits and the insights the modeler would like the manager to draw from them should be clearly stated. Exhibit 1.14 provides an outline and additional guidelines for effectively communicating the results of a modeling project in the form of a memo.

EXHIBIT 1.14 Suggested Memo Format

MEMO

To: Quantitative Business Modeling Students
From: Scott M. Shafer, Author
Subject: Effectively communicating the results of a modeling project

Introduction

Your introductory paragraph should begin with a clear statement of the managerial problem/opportunity addressed. After the problem/opportunity has been clearly defined, next briefly state your specific recommendation, overview the benefits of your recommendation, and the impact of your recommendation. A memo is not a mystery novel. Therefore, there is no good reason to keep the managerial decision maker(s) in suspense by withholding your recommendations to the very end of the memo. Moreover, numerous benefits accrue when the recommendations are mentioned up front including getting the attention of the decision maker and providing a road map to where the analysis is headed.

Perhaps the most important consideration to keep in mind in drafting a memo is how helpful the decision maker will find it. Therefore, it is important that you use language that a typical manager would understand and not use terminology to impress him or her. Also avoid providing unnecessary details. This includes excessive details about the methodologies employed and details about approaches that were attempted but did not work out. In general, you should include just enough detail to provide the reader with an appreciation of the types of data used and analyses performed so that they can have confidence in the rigor of the study.

Analysis

Begin this section by stating, defining, and describing the methodology you employed and then explain why this methodology is appropriate. Do not include background information, such as when the company was founded, the current sales level, or the organization's products/services. After overviewing the methodology employed, present the results of your analysis. Be sure to include relevant exhibits to support your discussion. All exhibits should be clearly labeled with a sequential number and a descriptive caption. Furthermore, to the extent possible, exhibits should be included in the body of the memo in portrait orientation. Memos that require the reader to turn the page sideways, or leaf through a number of pages to find an exhibit, are not particularly reader-friendly. Finally, highlight any key insights you want the reader to gain after inspecting your exhibits. Do not make the reader guess what an exhibit is showing or what conclusion he or she should reach, based on the exhibit.

To illustrate: "The performance on the second quiz was considerably better than the performance on the first quiz. On average, students scored a full point higher on the second quiz with the average increasing from 7.2 on the first quiz to 8.2 on the second quiz. Figure A shows the distribution of scores for the two quizzes. According to the figure, the improvement in performance came largely from the middle of the distribution. More specifically, on the first quiz, considerably more students scored a 5, 7, or 8 than on the second quiz, while more students scored a 9 or more (M) on the second quiz. Also noteworthy is the fact that the category with the largest number of students on the second quiz was the more-than-9 category. On the first quiz, the largest number of students scored an 8."

FIGURE A Distribution of Scores for First Two Quizzes

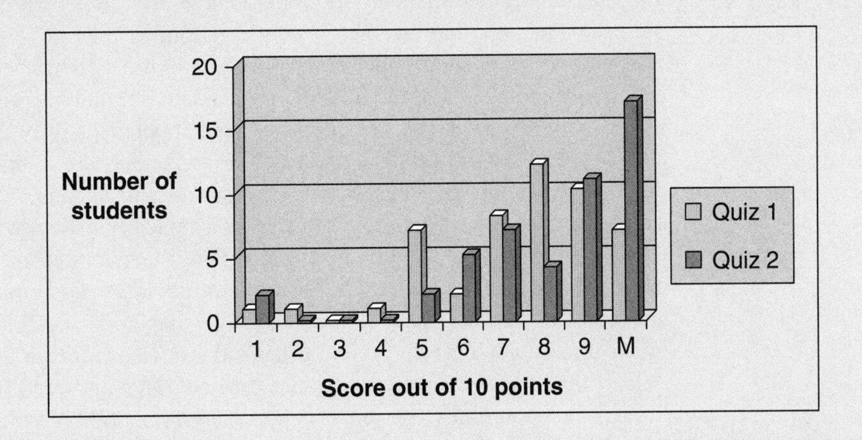

Also include your sensitivity analysis of the factors that could potentially impact your recommendations, the implications for managing these factors, and additional opportunities for the decision maker to consider.

Recommendations

In this section provide the details of your recommendations based on the results of your analysis. Be sure to include a preliminary project plan of action/implementation. At either this time, or in a later, more detailed project proposal, include the expected benefits and risks of the project, a list of potential implementation problems, a suggested schedule with milestones, a budget, and a list of required resources and staff. The proposal should also consider training for those who will be responsible for implementing the proposed solution; without their buy-in, the solution will fail in the implementation stage.

Assumptions/Limitations

The final section of your memo should include a discussion of the key assumptions and limitations associated with your analysis. Examples of the types of issues to address in this section include key model assumptions, the method of data collection, data availability, and the extent that the results can be generalized to other situations.

Next, the action approved by management must be applied, or carried out, typically through some form of project. Applying a solution is a difficult process; it is typically harder to apply a solution through a project than to build and solve the model. It is helpful to bear in mind that managers will prefer a range of alternatives to the situation that was addressed, rather than just a single suggested solution. They always have "what-if" questions concerning changes in assumptions, exploiting other opportunities, alterations in the model, and so forth; studies that anticipate such questions and provide answers to them are more likely to be applied in some form than those that do not anticipate such questions. Moreover, it is important that the managers considering the study results understand what is being recommended because managers will prefer to keep problems they understand than apply solutions they do not understand. Last, the key to successful application is maintaining contact with both management and the staff users right at the beginning and throughout the study duration. These are the people who are the most familiar with the situation, and changes in the situation, as time passes. They are also the ones who will probably supply the relevant data, help the modeler better understand the actual situation, and implement the final recommendation through a project.

Finally, even when a solution has been implemented in a project, the work is not done. First, given the dynamic nature of organizations, the solution will need updating, or possible expanding if it becomes highly successful. Second, solutions for opportunities and problems in one part of an organization typically give rise to opportunities or problems in other parts of the organization. It is not uncommon for a solution in one department of an organization to create a problem in another department. Thus, problems usually don't disappear; they often just change shape and location. The same is true with opportunities; capitalizing on one usually doesn't have the anticipated project payoff, or at least the full payoff, because an unexpected glitch appeared from somewhere else that limited the opportunity. Hence, it is necessary to continually reconsider the situation and adjust the solution(s) to fit the changing circumstances.

1.3 Detailed Modeling Example

Typically we will conclude each chapter with a detailed example to illustrate how the quantitative models presented in a particular chapter are applied in the broader modeling context. Although the discussion in most chapters emphasizes model formulation and analysis, it is important to keep in mind that these activities are only a portion of the modeling process. Hopefully, it is now clear to you that the early steps of opportunity/problem recognition and data collection and the last step of implementation are of at least equal importance to the model formulation and analysis steps. We will conclude each chapter with a detailed example to reinforce this point.

In this first detailed example, we overview a research project with which one of the authors was involved.[4] As you will see, this particular project touches on a number of the topics that will be discussed in later chapters, in addition to providing an overview of the modeling process.

Step 1: Opportunity/Problem Recognition

An automobile manufacturer's production plant that assembles audio equipment for a popular line of their automobiles was having difficulty keeping up with demand. The manufacturer wanted to determine what factors affected worker productivity levels, and also the potential benefit of implementing a new technology that would deliver video-based, task-specific content on demand to computer terminals located at each inspection station. It was speculated that as a training aid such a system might help increase the rate at which workers learned a new task and could help reduce the negative impact of worker forgetting.

Step 2: Model Formulation

To help better understand the situation, the influence diagram shown in Exhibit 1.15 was developed. According to the diagram, worker productivity is identified as one factor that directly impacts assembly-line capacity. The diagram further identifies worker learning rates (see Experiential Exercise I at the end of this chapter), worker forgetting, and task tenures (i.e., the consecutive amount of time a worker performs a particular task) as factors that directly impact worker productivity. Also, observe that the decision of whether to adopt the video-based technology is expected to affect learning rates and worker forgetting.

Two other general types of models were formulated. First, similar to the statistical models that will be described in Chapter 3, curves were fit to each individual worker's performance data to help understand observed changes in the workers' productivity levels. Using an influence diagram like the one shown in Exhibit 1.15 helps determine what data are needed for such curve fitting. Then, based on the curves fit to the worker performance data, simulation models (discussed in Chapter 7) were developed to investigate how the factors identified in Exhibit 1.15 influence worker productivity and ultimately assembly-line capacity. It is also worth noting that the scope of the analysis of a particular situation does not have to include all the relevant factors identified. To illustrate, although process technology and product design were identified as having an impact on the capacity of the assembly line in Exhibit 1.15, they were considered beyond the scope of the current study. Of course, management retains the option of investigating these factors in future studies.

EXHIBIT 1.15 Influence Diagram for Audio Equipment Assembly Line

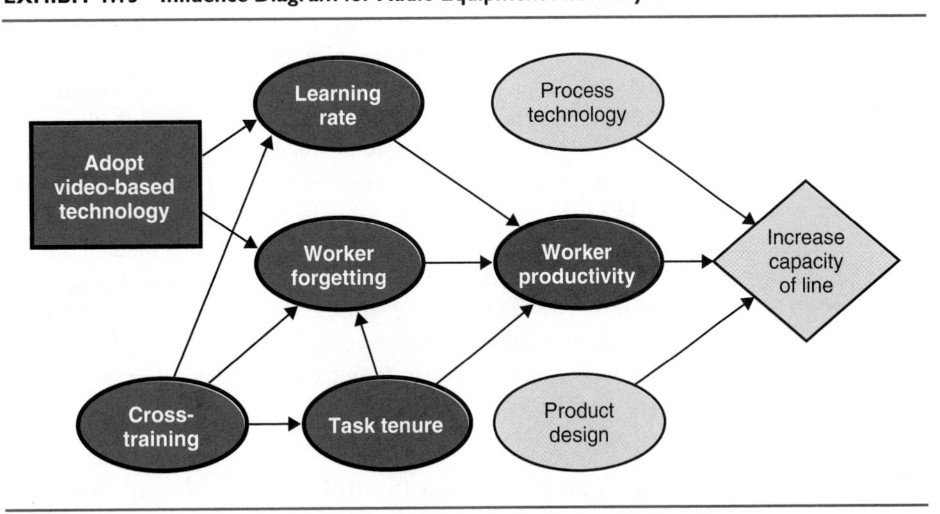

Note: The scope of this study was limited to the factors with bold borders.

Step 3: Data Collection

Analysis of the production line that assembled the audio equipment revealed that the final test and inspection stations at the end of the line were limiting the production rate of the entire line. In its current configuration, there are eight duplicate final test and inspection stations. The final assembly test process itself is a combination of machine-paced and human-paced activities. The machine-paced activities are computer controlled and thus exhibit minimum variability. Each station is also staffed by a worker who performs a number of supplementary functional and cosmetic inspections. In total, the inspection procedure requires 130 operational steps.

A typical inspection cycle begins when the operator picks a radio from the adjacent conveyor and begins a series of cosmetic tests. After completing these manual tasks, the unit is placed in a fixture where a computer tests the internal electronic functions. When this test is completed, the operator places a shipping label on the units that pass or routes the failures back to a rework area.

The data for this study came from header records that are generated when the electronic test equipment begins its cycle. The data collected included the operator number, the test station number, the part number, the serial number, and a timestamp corresponding to the instant the unit begins the electronic test. Unit total inspection times were estimated by taking the difference between successive header records for a given operator at a given test station. This inspection time data was collected for 176,000 units produced between April and August of 1996 by 148 different workers.

Step 4: Analysis of the Model

Upon fitting the curves to the worker performance data, descriptive statistics (discussed in Chapter 2) such as the mean and standard deviation were calculated to gain further insights into the situation. As an example, this analysis indicated that the average worker had the equivalent experience of inspecting 1,150 radios prior to the data being collected for this study. Further analysis indicated that several workers had no prior experience, while one worker had prior experience equivalent to inspecting more than 13,000 radios. As another example, it was determined that on average the workers could inspect 29.7 radios per hour. Interestingly, the fastest worker could inspect 45 radios per hour, while the slowest worker inspected slightly less than 20 radios per hour.

In addition to calculating these descriptive or summary statistics, the correlations (discussed in Chapter 2) between several variables were also investigated. This analysis suggested that workers with more prior experience tended to have higher productivity levels. Also, consistent with the tortoise and hare fable, the results suggested that slower-learning individuals ultimately attained higher levels of productivity. And perhaps most important, it was found that cross-training had a *negative* impact on productivity.

In terms of simulation, a total of 810 unique simulation models were developed. Each of these simulation models was run for one year of simulated time. The results of the simulation study were analyzed using a statistical procedure called analysis of variance (discussed in Chapter 2). Based on this analysis, a number of interesting results were observed. For example, one result of the study highlighted the importance of evaluating a given worker's tendency to forget a task in determining the best set of tasks on which to cross-train the worker.

Step 5: Implementation and Project Management

A memo providing additional details and implementation recommendations to management is shown in Exhibit 1.16.

EXHIBIT 1.16 Sample Memo to Management of Automobile Manufacturer

MEMO

To: George Samuelson, Plant Manager
From: Brianna Regan, Senior Analyst
Subject: Audio equipment line capacity shortage

Introduction

The production rate of our audio equipment line must be increased immediately in order to meet current demand levels for the minivan line. Analysis of the production line indicates that the final test and inspection stations are the bottleneck operation. In an effort to increase the production rate in the short term, it is recommended that the workers that are assigned to the inspection station be dedicated to this operation and not cross-trained to perform other production activities. The results of my analysis indicate that this change alone will increase the capacity of the line by 12 percent. One explanation for this improvement is that by dedicating employees to the inspection stations, the experience associated with performing this task is concentrated in a fewer number of workers leading to higher levels of proficiency among this group of workers. While this recommendation may seem to contradict the benefits of having a more flexible worker pool, note that this change is being recommended only for the workers assigned to the final inspection stations.

Analysis

Inspection time data were collected for 176,000 units produced between April and August of 1996 by 148 different workers. Curves were then fit to each individual worker's performance data to help understand observed changes in the workers' productivity levels. Finally, simulation models were developed to investigate how various factors influence worker productivity such as the rate of worker learning, the consecutive amount of time a worker is assigned to the test station, the amount of cross-training, and the degree to which workers forget information.

Analysis of the curves fit to the worker performance data suggests that the workers studied had the equivalent experience of inspecting 1,150 radios on average prior to my study period. The worker with the most experience had the equivalent experience of inspecting more than 13,000 radios, while twelve of the workers appeared to have no equivalent past experience. In terms of the workers' rate of learning, on average the workers needed to inspect 921 radios in order to reach half their potential productivity rate. The slowest learner in the group required the experience of inspecting almost 15,000 radios to achieve this level of proficiency, while the fastest learner in the group could achieve this proficiency after inspecting 2 radios.

Further analysis of the results indicated significant relationships between the variables. For example, higher levels of previous experience were strongly correlated with higher potential productivity levels. Also, workers with more previous experience tended to learn at a slower rate but ultimately achieved higher productivity levels.

(continued)

EXHIBIT 1.16 Continued

In terms of worker cross-training, the results of this study suggest that the use of cross-training effectively increases the pool of workers that can perform a particular task. While this certainly increases the flexibility of the work force, it also increases the time between successive assignments to the inspection station. As the time between successive assignments to the inspection station increases, worker productivity tends to decrease due to increased forgetting.

Recommendations

To increase the capacity of the production line in the very near term, it is recommended that the following actions and projects be initiated.

- Action: Assign workers with the most previous inspection experience to the final inspection stations.
- Project: Investigate ways to manage task assignments to the final inspection station more effectively. Consider offering appropriate incentives, instituting policies on seniority and job bumping, and negotiating with the labor union to decrease turnover at the final inspection station.

Over the longer-term, the following additional projects are recommended:

- Project: Test on a pilot basis the video-based computer system that provides workers with task-specific on-demand content at the final inspection stations to determine the extent to which it can help mitigate the impact of worker forgetting and increase the rate of worker learning.
- Project: Utilize the methodology developed in this study to create profiles for all workers related to the extent to which they have a tendency to forget various tasks. Use these profiles to determine the best set of tasks for each worker to be cross-trained on.
- Project: If worker cross-training continues to be pursued, utilize the methodology developed in this study to assess the forgetting rate of new hires on a variety of tasks during a probationary period and use this information to help in making permanent hiring decisions.

Limitations

In the course of conducting this study a number of assumptions were made that may limit the generalizability of my findings. First, it was assumed that the employees assigned to the inspection stations worked independently. Therefore, I did not address social factors related to individual learning, such as the impact of familiarity with other group members. Second, the data used in this study were limited to the completion of a single task. Hence, consideration was not given to possible carryover benefits that may accrue when workers perform multiple tasks. Finally, all workers cross-trained to perform the inspection operation were assumed to be equally likely to be selected to fill vacancies in the final inspection area. If this final assumption is not valid, the simulation models could be easily enhanced to incorporate a more realistic worker assignment procedure.

1.4 Software for Modeling

We do not include in this text the many special types of software packages that can be used for QBM because we believe that learning statistics, management science, and how to apply these techniques using a familiar software package such as Microsoft Excel is enough for one semester. However, we do want students to be aware that many more specialized packages are available for these various techniques. Needless to say, packages come and go, get upgraded and discontinued, and are always in a state of change, so that any description of a package (including Excel) is obsolete before the book describing it is published. For a timely update on many such packages including forecasting, statistical analysis, simulation, decision analysis, optimization, and spreadsheet add-ins, consult the newsletter *OR/MS Today*'s online software survey at http://lionhrtpub.com/softwaresurveys.shtml. *OR/MS Today* is published by INFORMS, the international society for operations research and management science (http://www.informs.org).

For example, PrecisionTree (http://www.palisade.com/html/ptree.html) is appropriate for Chapter 5, "Decision Analysis," and SAS (Statistical Analysis System, http://www.sas.com) is an excellent package for both simple statistical studies as well as advanced research investigations. Crystal Ball (http://www.decisioneering.com) and @Risk (http://www.palisade.com/html/risk.html) are very helpful simulation packages that can be used for the situations in Chapter 7, "Simulation," and Microsoft Project2000 is the most popular software for project management (Chapter 8). Current information and links for modeling software will be maintained at this book's companion Web site.

QUESTIONS

1. Why is it impossible to avoid the issue of modeling in business? That is, why *must* the use of models be faced?
2. Why is it important for managers to be familiar with QBM?
3. List some managerial opportunities or problems that would seem amenable to QBM techniques. List some that would not.
4. Do you consider yourself a positivist or a relativist? Identify situations where you might be primarily one or the other.
5. Analyze a managerial situation of your choice that would be amenable to modeling and identify the following:
   a. Whether the situation is one of certainty, risk, or uncertainty.
   b. Whether the situation is one of allocation, waiting-line analysis, decision analysis, or prediction.
   c. The formal models that could be applied, and if mathematical, whether each is deterministic or probabilistic.
   d. The uses of each model in terms of being prescriptive or descriptive. If the former, would it be optimal or nonoptimal, and if optimal, based on enumeration or an algorithm?
   e. The independent and dependent variables, and the uncontrollable parameters.
6. Why do you think expertise in QBM was not the first or second most important quality of effective QBM experts?
7. How would you reconcile the fact that effective modelers create a *unique* model for every situation versus the QBM aim of identifying managerial situations that are amenable to *standardized* solution techniques?
8. Why do you think expert modelers tended to develop multiple models for each situation they studied?
9. Use an example you are familiar with to describe the major steps associated with modeling.
10. Why is it important to think of situations in terms of opportunities instead of problems?

11. What advantages do influence diagrams offer in the model formulation process?
12. Describe a situation that will demonstrate the meaning of sensitivity analysis.
13. Why would a manager rather live with a continuing problem than employ a solution he or she doesn't understand?
14. What is the key to successful implementation, in your view?

EXPERIENTIAL EXERCISES

I. Learning Curve Exercise

It was mentioned in the detailed example (Section 1.3) that curves were fit to the worker performance data to help understand observed changes in the workers' productivity levels. More specifically, a learning curve model with three parameters was selected from the literature to model the patterns of learning exhibited in the performance data for each worker.

A wide variety of models have been proposed in the literature to model patterns of worker learning. One of the more popular models was based on observations of the airframe manufacturing industry, where it was observed that each time the output doubled, the labor hours per plane decreased to a fixed percentage of their previous value. In the case of the airframe industry, it was 80 percent. Thus, if the first plane in a series required 100,000 labor-hours to produce, the second plane would require 80,000 (100,000 × 0.80) labor hours, the fourth 64,000 (80,000 × 0.80) labor hours, and so on. Mathematically, this relationship can be described by the negative exponential function as

$$T_n = T_I N^r$$

where

T_n = the labor hours for the Nth unit
T_1 = the labor hours for the first unit
N = unit number of interest
r = log(learning ratio)/log(2) = log(learning ratio)/0.693

As an example, the time required to build the third plane can be calculated as:

$$T_3 = 100{,}000(3)^{\log(0.8)/.693}$$
$$= 70{,}210.4 \text{ labor hours}$$

Assignment Obtain a child's puzzle designed for 3- to 6-year-olds. Working in teams of at least two, select one person to be the puzzle solver and another team member to serve as the timer. Have the timer dump the puzzle on a table and mix up the pieces, facing upward. Then have the timer announce to the solver to begin solving the puzzle. The timer should then monitor the progress of the puzzle solver, and note and record the time required to solve the puzzle. Repeat this process three additional times.

1. Use the times required to complete the puzzle the first and second times to estimate the learning percentage. Estimate from this the time it should have taken on the third and fourth trials. In a similar fashion, use the completion times for the second and fourth trials to estimate the learning percentage. Would it make sense to use the average of these two learning rates to estimate the learning percentage? Why or why not?
2. Use your estimates based on solving the puzzle on trials 1 and 2, on trials 2 and 4, and the average of these two learning rates to estimate the time it will take the puzzle solver to complete the puzzle a fifth, sixth, seventh, and eighth time.
3. Have the puzzle solver work the puzzle four more times and record the times. How do the actual times of trials 5–8 compare to the estimated times?
4. Using a spreadsheet, plot the actual times and the estimated times for all eight trials (leave the estimated times blank for the first and second trials). Which learning rate worked best to predict trials 5–8? How good a fit does the learning curve model provide to the observed pattern of learning?
5. Based on this exercise, do you see any implications for organizations that must bid on contracts? What about for organizations that mass produce consumer electronic products?

II. Scoring Model Exercise

Develop a weighted scoring model in Excel to help you select a desktop computer.
Requirements:

- Identify specific desktop computer models either by visiting one or more retail stores, scanning computer magazines, and/or searching the Web.
- Define what criteria you will use to evaluate the computers. A list with a brief definition of

potential technical criteria is available on the Web site that accompanies this book. Draw an influence diagram to help you construct your weighted scoring model.

- Decide how much weight each criterion should be given.
- Collect data on all criteria for each computer model you are evaluating. You may find the Web, computer magazines, store personnel, and manufacturer sales reps through 800 numbers as good sources for this information.
- Decide how you will score the computers on each criterion.
- Develop a spreadsheet that computes the total scores for each computer.
- For the computer that ranks number three on your list, perform a sensitivity analysis on its scores and weights to determine what changes would be needed to give it the highest total score. What are the implications of this to the manufacturer of this computer?

MODELING EXERCISES

The exercises in this section describe typical management situations in order to introduce you to some of the actual complexities of QBM. Our purpose is to show you why quantitative modeling can be of help in addressing such situations.

1. Amuse Corporation is planning to build a theme park in the city of MidAmerica. They have also discussed a plan for the parking lot, which is to be an independent business unit. They are discussing the size of the parking lot and the pricing policy. It is estimated that 20 to 25 percent of the users of the parking lot will *not* go to the amusement park. Among the options discussed are:

   - Very high parking fees with reimbursement to the amusement park guests
   - Free parking for everybody
   - Low fees and no reimbursement

   All these options are being used in shopping centers, theaters, and other theme parks in town. Amuse Corporation is a private corporation whose objective is to maximize net revenues from the park as a whole.

   a. What is the opportunity/problem here? What might happen under each option?

   b. Construct an influence diagram and determine how to formulate the model to address this opportunity/problem. What assumptions would be required? What types of models would be most useful for this situation?

   c. What data would you need? Where would you get the data? What might be a difficulty in obtaining it? How would the availability of data influence both the model formulation as well as its analysis?

   d. How would you go about making a recommendation for implementation to management? What would be the key points of such a recommendation?

2. Before the days of telemarketing, e-commerce, traffic jams, and complex shopping centers, salespersons used to travel a lot. Assume that you work for the MBI Corporation as a salesperson. You visit 10 cities each time you go on the road. Obviously, you want to return home at the end of your trip. Also, corporate policy allows you to visit each city once and only once. (*Note:* This is known as the "traveling salesperson problem.")

   a. Describe how you might formulate the problem through an influence diagram. Are there any alternative formulations that are equally reasonable? Is the situation one of certainty, risk, or uncertainty?

   b. Identify some important data concerning the problem such as distances and the relative locations of the cities, including your home city.

   c. Build an analog model of this problem—that is, show it on a map for 10 cities.

   d. Show one possible solution. How many feasible alternative solutions would you guess exist for such a problem? What changes in the data might alter the optimum alternative solution? How?

   e. Describe the type of model this is (deterministic vs. probabilistic, etc).

   f. List some factors that may be encountered in real life that would make this problem more complicated than what you have just considered. Explain the impacts of each of these factors.

3. Draw an influence diagram to help you model how to select a job from a variety of offers. Consider not just the major decision factors but also identify the elements that impact upon these factors, and the elements making up these impacts. Describe what type of model could help you with this opportunity. Would simulation be of value? Optimization?
4. Develop an influence diagram that would help a home builder determine in which neighborhoods to purchase land for the construction of new homes. In addition to considering the decision of what neighborhoods, also model the builder's objective, the key factors that affect this decision, and the secondary factors that impact the key factors.

CASE

Henry Ford Hospital

At the Henry Ford Hospital, a key aspect of planning and scheduling is to match available capital, workers, and supplies to a highly variable pattern of demand. The hospital has 903 beds arranged into 30 nursing units, with each nursing unit containing 8 to 44 beds. For purposes of planning, each of the nursing units is treated as an independent production facility. However, a number of factors complicate the planning process at the hospital. First, as noted, demand exhibits a high degree of variability. For example, while the average number of occupied beds in 1991 was 770, in one eight-week period it was 861 and in another it was 660. A second complicating factor is the large penalty incurred by the hospital for HMO patients who require care but cannot be admitted because a bed is not available so the patient must be sent to another hospital. In these cases, not only does the hospital lose the revenue from the HMO, but it must also pay another hospital for the patient's stay. A third complication is the tight labor market for registered nurses, making it difficult and expensive to change the rate of production. On average, it takes the hospital 12 to 16 weeks to recruit and train each nurse, at a cost of approximately $7,600 per nurse. A final complication is the high costs associated with idle facilities. The hospital estimates that the fixed cost of one eight-bed patient module exceeds $35,000 per month.

Planning is particularly important at the Henry Ford Hospital because if a sufficiently long-term view of the organization isn't taken, short-run decisions can be made that adversely affect the hospital in the long run. For example, during one period at the hospital, a decision was made to reduce the staff. However, shortly after the staff was reduced, it was determined that more staff was needed and thus new staff members had to be recruited. The net result was that the hospital incurred both the costs associated with reducing its staff and the costs associated with recruiting and training new staff a short time later.

Questions

1. What is the opportunity/problem facing the Henry Ford Hospital? Is this decision making under certainty, risk, or uncertainty? Why?
2. Draw an influence diagram to help formulate this situation. What are the dependent variables, decision variables, and uncontrollable parameters?
3. What data would you need to collect as an analyst to model this situation? From where would you obtain this data? What would be some of the major limitations associated with the data?
4. Develop a spreadsheet model that reflects the mathematical relationship among the variables and factors. How would you go about validating your spreadsheet model?
5. What criteria would you use to evaluate alternative solutions?
6. How could the model you developed be used to perform sensitivity analysis? Of what value would sensitivity analysis be to the planner at Henry Ford Hospital?

Source: W.R. Schramm and L. E. Freund, "Application of Economic Control Charts by a Nursing Modeling Team," *Industrial Engineering* (April 1993): 27–31.

ENDNOTES

1. msn® CarPoint, http://carpoint.msn.com, May 8, 1999.
2. It is important to scale all criteria with the same number of points or the weighting scheme you develop will be thrown off. Thus, if you score the cars on cost using a three-point scale, you must also use a three-point scale for warranty, insurance rating, speed, comfort, and MPG.
3. Formulas entered into Excel must begin with an equal sign (=).
4. Based on S. M. Shafer, D. A. Nembhard, and M. V. Uzumeri, "The Effects of Cross-Training and Task-Tenure on Assembly Line Productivity with Empirical Learning and Forgetting, Working Paper, Babcock Graduate School of Management, Wake Forest University, October 1999.

BIBLIOGRAPHY

Ackoff, R. L. Beyond Problem Solving. *Decision Sciences,* 5:2, 1974, 51–54.

Checkland, P. B. Model Validation in Soft Systems Practice. *Systems Research,* 12:1, 1995, 47–54.

Clemen, R. T. *Making Hard Decisions: An Introduction to Decision Analysis,* 2e, Pacific Grove, CA: Duxbury Press, 1996.

Etzioni, A. Humble Decision Making. *Harvard Business Review,* July–August, 1989.

Evans, J. *Creative Thinking in the Decision and Management Sciences.* Cincinnati, OH: South-Western Publishing, 1991.

Gass, S. I. Managing the Modeling Process: A Personal Reflection. *European Jr. of Operational Research,* 31, 1987, 1–8.

Geoffrion, A. M. An Introduction to Structured Modeling. *Management Science,* 33, 1987, 547–588.

Harpell, J. L., M. S. Lane, and A. H. Mansour. Operations Research in Practice: A Longitudinal Study. *Interfaces,* 19:3, 1989, 65–78.

Keys, P. Approaches to Understanding the Process of OR: Review, Critique and Extension. *Omega,* 25, 1997, 1–13.

Keys, P. Creativity, Design and Style in MS/OR. *Omega,* 28, 2000, 303–312.

Little, J. D. C. On Model Building. In W.A. Wallace (ed.), *Ethics in Modeling,* Oxford, England: Pergamon, 1994.

Mehra, S. Applying MS/OR Techniques to Small Businesses. *Interfaces,* 20:2, 1990, 38–41.

Meredith, J. R. Reconsidering the Philosophical Basis of OR/MS. *Operations Research,* 46:1, 2001, 1–9.

Miser, H. Avoiding the Corrupting Lie of a Poorly Stated Problem. *Interfaces,* 23:6, 1993, 114–119.

Morris, W. T. On the Art of Modeling. *Management Science,* 13:12, 1967, B707–717.

Pidd, M. *Tools for Thinking: Modelling in Management Science.* Chichester, England: Wiley, 1996.

Pidd, M. Just Modeling Through: A Rough Guide to Modeling. *Interfaces,* 29:2, 1999, 118–132.

Pidd, M. and R. N. Woolley. Four Views on Problem Structuring. *Interfaces,* 10:1, 1980, 51–54.

Rivett, B. H. P. *The Craft of Decision Modelling.* Chichester, England: Wiley, 1994.

Saaty, T. L. and J. H. Alexander. *Thinking with Models.* Elmsford, NY: Pergamon, 1981.

Schellenberger, R. Criteria for Assessing Model Validity for Managerial Purposes. *Decision Sciences,* 5, 1974, 644–653.

Shafer, S. M. and J. R. Meredith. *Operations Management: A Process Approach with Spreadsheets.* New York: Wiley, 1998.

Smith, G. F. Defining Managerial Problems: A Framework for Prescriptive Theorizing. *Management Science,* 35:8, 1989, 963–981.

Tsoukas, H., and D. Papoulias. Creativity in MS/OR: From Technique to Epistemology. *Interfaces,* 26:2, 1996, 73–79.

Volkema, R. Managing the Process of Formulating the Problem. *Interfaces,* 25:3, 1995, 81–87.

Willemain, T. R. Insights on Modeling from a Dozen Experts. *Operations Research,* 42:2, 1994, 213–222.

Willemain, T. R. Model Formulation: What Experts Think About and When. *Operations Research,* 43:6, 1995, 916–932.

2 Data Collection and Analysis

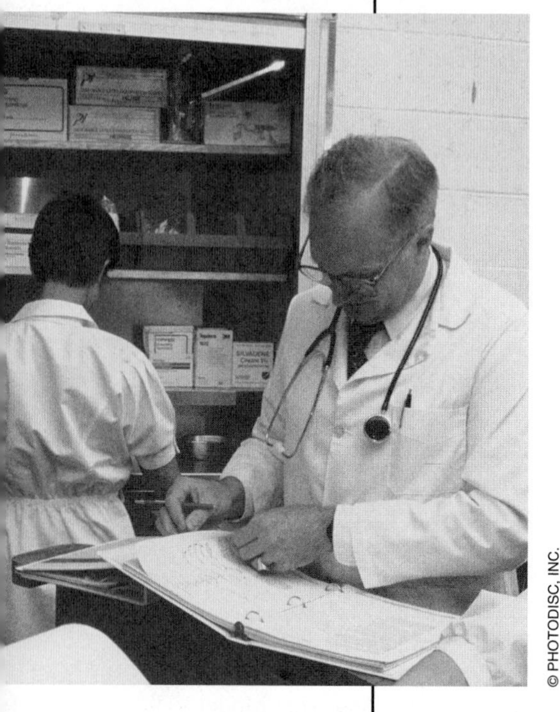

© PHOTODISC, INC.

INTRODUCTION

Hospital.com is a (fictitious) online retailer of health supplies and equipment, primarily for the hospital market. Public hospitals, under pressure to hold down costs, are trying to minimize their stocks of medical goods by contracting with suppliers who can deliver items just-in-time as they are needed. Recently, Hospital.com has been frustrated by its inability to manage its inventory effectively. More specifically, in a typical week it had excessive amounts of some products and was short on a number of other products. The problem is that both having too much and too little inventory costs the company money. For example, having too much invested in inventory ties up capital that could be used for other purposes such as advertising. Also, excessive inventory requires renting more warehouse space than would otherwise be needed. On the other hand, because quick and reliable customer service are hallmarks of online retailing, any time the company stocks out of an item it incurs extra costs to expedite the order (e.g., overnight shipping costs). Also, the company thinks that there are many significant intangible customer-related costs associated with stockouts, such as reduced future purchases and negative word of mouth.

In its annual report to shareholders, the president of Hospital.com stated that developing a better system for managing its inventory was a top priority for the coming fiscal year. A team was thus immediately formed to resolve this problem.

To begin the process, weekly sales data for the last year were collected by the team for a representative product, stethoscopes (see Exhibit 2.1). Hospital.com receives weekly shipments from its suppliers. The company's current policy is to maintain an inventory level 25 percent greater than the average weekly demand for a product. Then, if the level of on-hand inventory is below this amount when orders are being placed, Hospital.com places an order to replenish its inventory up to this target level. This approach effectively maintains an extra 25 percent in inventory above what would be needed during an average week.

To illustrate this using the data in Exhibit 2.1, Hospital.com's average weekly demand for stethoscopes is about 24 units. Increasing this by 25 percent yields a target inventory level of 30 stethoscopes. If on the day that an order is to be placed it is determined that the current on-hand inventory level is 20 stethoscopes, then Hospital.com would place an order for 10 additional stethoscopes to bring its inventory up to the target level.

What is your initial reaction to Hospital.com's approach to managing its inventory? Do you think attempting to carry 25 percent more inventory than the average weekly demand is sufficient? Or do you think this might be a bit excessive? The focus of this chapter is on the collection and analysis of data to specifically address these types of issues. We will return to the dilemma facing Hospital.com in the detailed modeling example discussed at the end of this chapter.

2.1 Data Collection

This chapter primarily details the first and third steps of the QBM process described in Chapter 1; that is, problem/opportunity recognition and data collection. In large part, this chapter deals with basic statistical issues about data, including probability estimates, but the analysis of this data is often the basis for recognizing an opportunity or problem. Although the following chapter on regression continues our treatment of data and statistics, it is more related to model formulation and analysis, the second and fourth steps in the QBM process. Following this, the next four chapters treat other modeling and analysis techniques. The final chapter is then concerned with the last step in the QBM process, implementation and project management.

As noted in Chapter 1, the collection of appropriate data can be a difficult and time-consuming task because the data may be unavailable, excessive, in conflict, or any of the other concerns described earlier. Collecting data is also expensive, and any modeling study must specify early on how much data are worth collecting. Clearly, the more data available for analysis, the greater the confidence in the results, but often there isn't enough money, or time, to collect all the data that one would like. The less data, the more likely that there is error in the results, of course—typically in inverse proportion to the amount of data collected. But it is often possible to determine how great that error might be and management can thus decide what level of risk they are willing to take for the time and money expended in conducting the study. Thus, it is common in QBM to take only a *sample* of data concerning the situation under investigation rather than trying to collect data on the entire *population* of interest. As a result, the statistics involving a sample of data are somewhat different from those when the entire population has been collected, and we will note those differences as we proceed through our analyses.

Perhaps the most important rule in the entire QBM process, especially throughout the data collection step, is to be absolutely certain that you as the modeler understand what each data element or observation means. That is, where does it come from, how was it

EXHIBIT 2.1 Weekly Sales Data for Stethoscopes

	A	B	C	D	E
1	**Week**	**Sales**			
2	1	22			
3	2	25			
4	3	25			
5	4	24			
6	5	18			
7	6	31			
8	7	22			
9	8	30			
10	9	19			
11	10	19			
12	11	24			
13	12	21			
14	13	27			
15	14	21			
16	15	12			
17	16	16			
18	17	24			
19	18	29			
20	19	31			
21	20	33			
22	21	23			
23	22	18			
24	23	34			
25	24	26			
26	25	20			
27	26	22			
28	27	26			
29	28	32			
30	29	25			
31	30	23			
32	31	20			
33	32	29			
34	33	25			
35	34	30			
36	35	18			
37	36	28			
38	37	21			
39	38	21			
40	39	35			
41	40	23			
42	41	21			
43	42	28			
44	43	26			
45	44	23			
46	45	23			
47	46	25			
48	47	18			
49	48	23			
50	49	20			
51	50	26			
52	51	29			
53	52	23			
54	**Average**	**24.2**			
55					

=AVERAGE(B2:B53)

generated, what did people mean when they gave you this number, what were their understandings of your question, and so on. More errors are made because of the failure to follow this guide than for any other reason in the entire QBM process. Collecting data is fraught with misunderstandings and miscommunication. For example, just imagine how difficult it would be to poll a group to determine how many rooms were in their house, or the square feet of living space. Would the rooms include storage areas? Bathrooms? Would the square feet include a finished basement? Closets? And this would be a relatively objective question compared to some typical business questions such as, "How satisfied were you with our service?"

The following classic example illustrates the danger of biased sampling. In the 1948 presidential race, one newspaper tried to scoop the other papers by not waiting until the final votes were in but instead made a prediction of the winner based on a massive telephone sample of voters. They announced in large headlines that Dewey won the presidency over Truman, but the actual results turned out to be just the opposite, leaving the paper's publishers to look like fools. As it happened, in 1948 only the more well-off people had telephones, so the poll only sampled this biased group, who were not at all typical of the average voters.

The general data collection procedure consists of three segments. First, there is the *issue of interest,* such as the seniority of a group. To determine this, you need a *measure,* also called a *variable.* In this case, one logical measure (sometimes multiple measures are used) would be the age of the people in the group. Then you need to collect the data, or *observations,* that you will use for analysis. If it is a big group, you may take a sample, as described earlier, instead of collecting data from every member of the group.

There are two ways you might decide to collect this data. You might just record the age of every person, which seems most direct, or for reasons of efficiency, you might instead decide to group the data, say by five-year segments: 0–5, 6–10, and so on. We often use grouped data in business because we are interested in general categories rather than individual items. Conducting the statistics for grouped items usually requires less calculation than for individual items.

Another set of data collection approaches based on time is also worth noting. The method just described would be considered *cross-sectional* because all the data were collected for the same time period of interest. Often, however, we are interested in how something is changing over time, and in that case we collect the data over the time period of interest, such as sales over the last five years, or profitability over the next four quarters. In this chapter, we primarily treat cross-sectional data. In the following chapter, we direct our attention to *time series* data.

Finally, we consider the data observations themselves. There are many different kinds of data observations. Although not all of them are amenable to statistical analysis, they may still be relevant to business modeling and decisions. For completeness, we will describe them all here and then confine our attention to those that are easily handled by statistical methods. First, it is useful to distinguish between *categorical* and *numerical* data. We discuss categorical data first.

As the name implies, categorical data would be observations that constitute a grouping of some form. Although it may have a numeric basis, the main concept is that of the group. For example, males and females would be two categories that had no numeric basis, while teenagers would be categorical but with a numeric basis that included anyone between the ages of 13 and 19. Both of these are examples of what is known as *nominal* categorical data because the groups are associated with some name. An example from business might be managerial information in the eastern, southeastern, midwestern, southwestern, and western regions.

However, in contrast to nominal categorical data, there is another type of categorical data known as *ordinal* that ranks the groups in order, relative to each other. For example, a survey about the food in a restaurant may include the categories *poor, acceptable, good,* and *excellent.* Also, common grades in a class will usually be A, B, C, D, and F. Clearly, numbers could be associated with either of these latter two examples. Thus, a student's grade point average, such as a 3.6 out of 4.0, labels that person as a very good student having an A- average. Yet, the student would never get a 3.6 in a particular class; the grade would be a 92 percent, or A-, or something similar. The transformation of ordered categories to numbers is just a convenience when dealing with a larger set of data to determine where the average falls. Note that any numbers associated with nominal data are arbitrary and have no meaning, such as numbering the eastern region as 1, the southeastern as 2, and so on. Finding the average of these numbers is meaningless.

Most of our discussion of data collection and analysis will concern numerical rather than categorical data, and here, too, there are two major categories: discrete and continuous. For example, the number of chairs in a room is a *discrete* number, such as 2 or 5. And the number of refrigerators sold this week might be 37. In these cases, only numeric digits are applicable, since there cannot be a fraction of a chair or refrigerator. In other situations, we may be dealing with *continuous* data such as $237.45 of total sales on Monday, or the length of a piece of wood being 48.7 centimeters. However, the vagaries of humans when reporting data often tend to blur even these relatively clear distinctions such as when a restaurant patron makes a reservation for three and a half people (the half being a child, of course). Thus, we tend to run across data in a variety of forms and again, we need to be absolutely clear that we understand what each observation means.

2.2 Summarizing Data

We now turn to issues concerning the treatment of the raw observations: summarizing them for meaning and understanding, illustrating them, grouping them, identifying their characteristics, and so on. In this section, we look first at summary statistics for representing groups of data and illustrating them in meaningful ways.

Descriptive Statistics

The two most common measures to summarize a group of data are the **mean** (also called the *average*), which is the "center" of the group, and the **variance,** or the "spread" of the data. The mean gives a sense of the "center of gravity" of the set of data, or that point about which the observations tend to fall, with outlying points having a heavier impact due to their longer leverage. The variance measures the spread of the observations about that center, a high value meaning the points are quite spread out and a small value meaning they are rather closely grouped. Often, the **standard deviation** of the group—the square root of the variance—is used instead because it is expressed in the same units as the mean. In addition, it can often be used intuitively as that distance about the mean that will contain a certain percentage of the group's observations.

The mean of a *population* of observations is calculated as

$$\mu = \frac{\Sigma x_i}{N}$$

where μ = the mean (pronounced "mu," in English)
Σ = the summation of what follows over all values of the subscript
x_i = the value of the ith data item in the population
N = the number of data items in the population

The mean of a *sample* of items from a population is given by

$$\bar{x} = \frac{\Sigma x_i}{n}$$

where $\bar{x}$ = the sample mean (pronounced "x bar")
x_i = the value of the ith data item in the sample
n = the number of data items selected in the sample

The variance of a population of observations is calculated as

$$\sigma^2 = \frac{\Sigma(x_i - \mu)^2}{N}$$

where σ^2 = the population variance (pronounced sigma squared).
The *variance of a sample* of items is given by

$$s^2 = \frac{\Sigma(x_i - \bar{x})^2}{n - 1}$$

where s^2 = the sample variance.
The *standard deviation* is simply the square root of the variance. That is,

$$\sigma = \sqrt{\sigma^2} = \sqrt{\frac{\Sigma(x_i - \mu)^2}{N}}$$

and

$$s = \sqrt{s^2} = \sqrt{\frac{\Sigma(x_i - \bar{x})^2}{n - 1}}$$

where σ and s are the population and sample standard deviations, respectively.

If the data for a study have been entered into a spreadsheet, the mean, variance, standard deviation, and other characteristics of the data can be computed by utilizing standard functions in the program. Exhibit 2.2 summarizes a number of the statistical functions available in Excel.

Referring back to our chapter opening example, the data listed in Exhibit 2.1 are sample data because they correspond to the weekly sales data for only the most recent year and do not contain the sales data for *all* previous weeks. Therefore, the variance and

EXHIBIT 2.2 Useful Statistical Functions Available in Excel

Quantity Calculated	Excel Function
Mean*	=AVERAGE(range of values)
Median	=MEDIAN(range of values)
Mode	=MODE(range of values)
Population variance	=VARP(ranges of values)
Population standard deviation	=STDEVP(range of values)
Sample variance	=VAR(range of values)
Sample standard deviation	=STDEV(range of values)

*Note that the same function is used to calculate the mean of a population and the mean of a sample. The AVERAGE function returns the mean of the population if the data range corresponds to the entire population and likewise returns the sample mean when the data corresponds to data sampled from overall population.

standard deviation could be calculated by entering the following formulas into the spreadsheet, respectively:

=VAR(B2:B53)
=STDEV(B2:B53)

Other measures of the center of a group of observations are also sometimes used because they give a different perspective about the group, convey different information, or are easier to determine. The mode and the median are the most common. The *mode* is the most frequent value in a set of observations, and the *median* is the middle value of an *ordered* group of the observations. Thus, in the data set 3, 2, 9, 6, 1, 5, 7, 3, 4, the mode is 3 since it occurs more frequently than any other value. A data set can have more than one mode if there are two or more values that have the same frequency. (If there is a large data set that has two or even three values that exhibit much higher frequencies than the other values, we often say the data is bi- or tri-modal, even if the frequencies are not precisely equal.) The mode conveys a sense of aggregation of the data, even if the data are distributed about this value in one direction more than in the other. For that matter, the mode may even be at the extreme upper or lower end of the data.

To determine the median, we must order the data first: 1, 2, 3, 3, 4, 5, 6, 7, 9. Now we see that 4 is the median since half the observations (i.e., 1, 2, 3, 3) are below this value and half (5, 6, 7, 9) are above this value. If there are an even number of observations in the data set, then it is common practice to specify the median as the average of the two middle values. Thus, if there had been an 8 in the above data set, the median would be $(4 + 5)/2 = 4.5$. The median is determined more easily than the mean (which is 4.44 for the original set of nine values), yet is usually close to it. It also conveys a sense of where most of the data observations tend to lie, without being pulled away by extreme values the way the mean would be. For example, if the last number in this data set were 99 instead of 9, the mean would shift from 4.44 to 14.44, but the median would stay at 4.

Another measure for the dispersion of values in a group of data is simply the *range,* the largest value minus the smallest value. The big advantage of the range is that it is much easier to calculate than the variance or standard deviation, yet it gives a sense for the dispersion of the data. For example, the range of the above data set is $9 - 1 = 8$. Note that the range does not give any sense of the dispersion about the center of the group of data, like the variance does, and it is highly sensitive to outliers (extreme points). Furthermore, the range tends to increase as the size of the sample increases, whereas the standard deviation or variance does not.

Statistical Displays

For most managerial purposes, and for analysis reasons as well, it is useful to display the data in terms of groupings instead of inspecting the raw data. The groupings can be either natural or artificial, examples of the latter being an aggregation of the data into quartiles (one-fourth of the entire data set) or deciles (one-tenth of the data set), or the conversion into percentages of the total. For categorical data, *bar charts* of various forms are most useful. To use our earlier example of an ordinal measure, suppose a week's sample of customer service surveys indicated that 14 people scored service as excellent, 37 as good, 25 as acceptable, and 8 as poor. We might then enter this data into a spreadsheet as a *frequency table* (Exhibit 2.3) and construct a bar chart using Excel's Chart Wizard, as follows:

1. Highlight the range of the chart A2:B5.
2. Select the Chart Wizard by clicking the Chart Wizard button one time.
3. In step 1, Excel's Chart Wizard requires that the type of chart be specified. In this case the first column chart was specified.

EXHIBIT 2.3 Frequency Table and Corresponding Bar Chart for Customer Service Survey

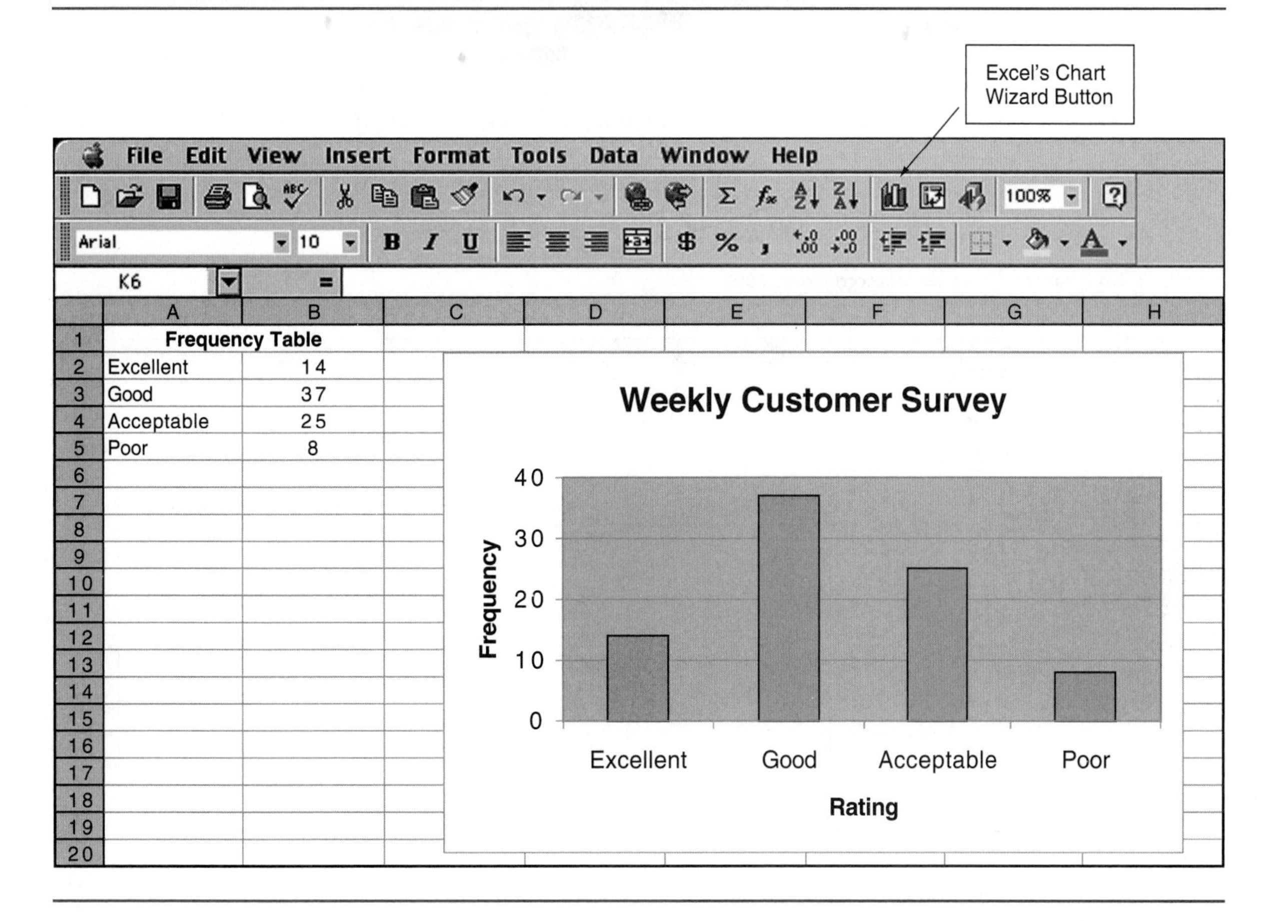

4. No changes are needed in the second step because the range was highlighted prior to beginning the Chart Wizard. Therefore, in step 2 we simply click on the Next button and move on to step 3.
5. In step 3 a chart title and labels for the axes can be entered if desired. In the chart shown in Exhibit 2.3, Weekly Customer Survey was specified for the chart title. Likewise, Rating and Frequency were specified for the x and y axes, respectively.
6. In step 4, the location for the chart is specified. The options are to place the chart in a new sheet or place it in one of the current sheets. In Exhibit 2.3 we placed the bar chart in the same sheet as the data used to create the chart. After specifying this, the Finish button was selected and the chart was created. The chart can easily be resized by simply dragging any of the corners of the chart.

Note that the categories are in a fixed perceptual order because these are ordinal categorical data—good is better than acceptable, which is better than poor. We would do the same thing with the distribution of grades in a class.

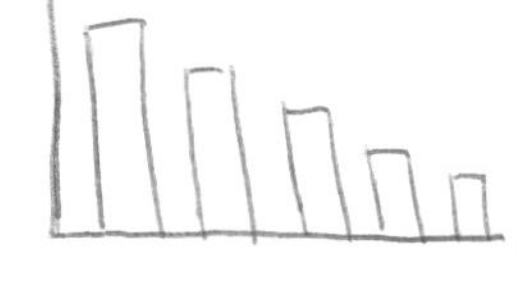

However, with nominal categorical data there is no natural order, and the bars can thus be placed in any convenient order. It is common in this case to display the group as a **Pareto chart** with the highest frequency on the left, then the next highest frequency group, and so on. Using our earlier example of different regions of the nation, suppose the total sales of the Dotcom Corporation for the year 2000 was $83,000,000, composed of sales in each region (in millions of dollars, rounded) as follows: 17—east; 11—southeast; 25—midwest; 8—southwest; and 22—west. This example is illustrated in Exhibit 2.4, where we also calculate and display the mean regional sales and the variance about that mean. Note that Excel can sort the data from highest to lowest for the purpose of creating a Pareto chart by selecting the options Data/Sort . . . Also, note that we could convert the data into proportional or percentage frequency values by dividing each region's sales by the total of 83, as shown in Exhibit 2.5. This *normalized* data could then be plotted as well, as shown on the figure. Clearly, we could also have normalized the earlier examples of service categories and grades in a class. In comparing Exhibits 2.4 and 2.5, note that normalizing the data does not change the shape of the bar graph.

EXHIBIT 2.4 Pareto Chart for Dotcom Corp. 2000

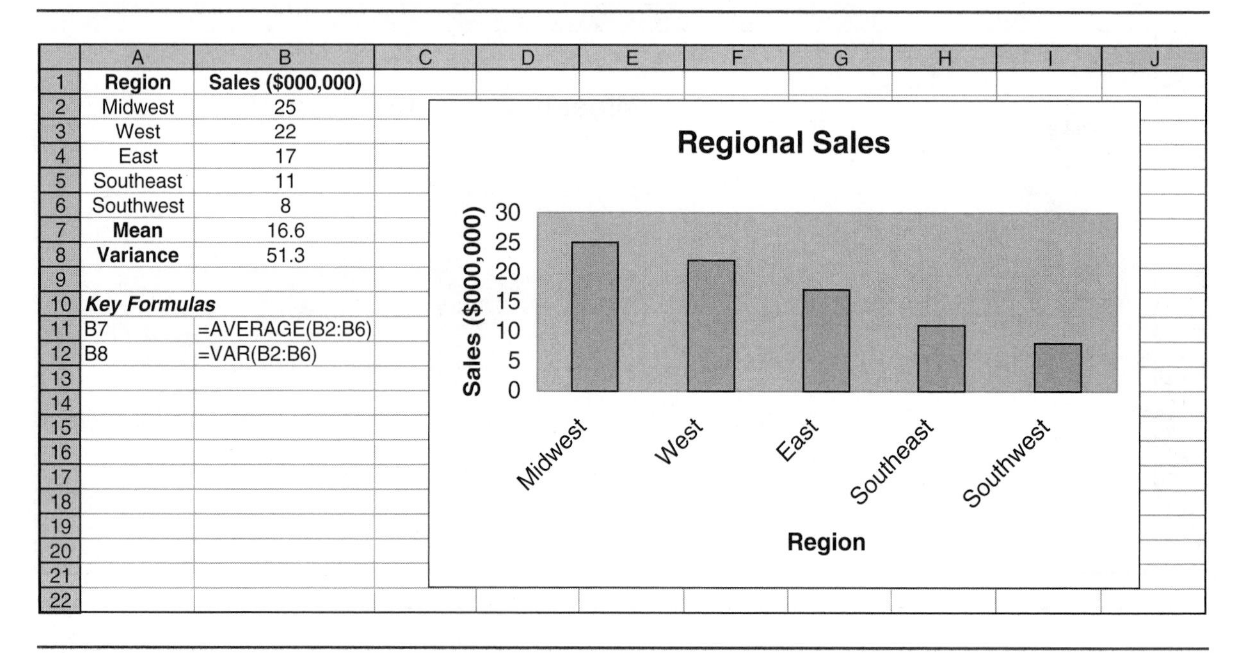

	A	B
1	**Region**	**Sales ($000,000)**
2	Midwest	25
3	West	22
4	East	17
5	Southeast	11
6	Southwest	8
7	**Mean**	16.6
8	**Variance**	51.3
9		
10	***Key Formulas***	
11	B7	=AVERAGE(B2:B6)
12	B8	=VAR(B2:B6)

EXHIBIT 2.5 Percentage of Sales: Dotcom Corp. 2000

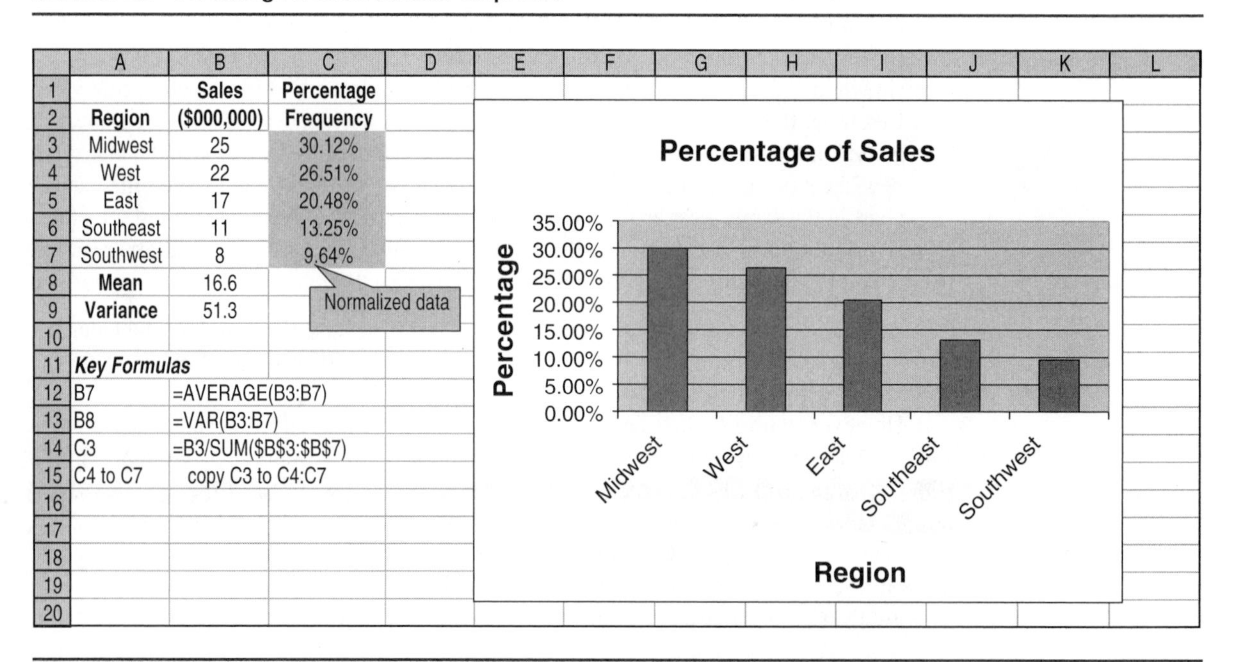

	A	B	C
1		**Sales**	**Percentage**
2	**Region**	**($000,000)**	**Frequency**
3	Midwest	25	30.12%
4	West	22	26.51%
5	East	17	20.48%
6	Southeast	11	13.25%
7	Southwest	8	9.64%
8	**Mean**	16.6	
9	**Variance**	51.3	
10			
11	***Key Formulas***		
12	B7	=AVERAGE(B3:B7)	
13	B8	=VAR(B3:B7)	
14	C3	=B3/SUM($B$3:$B$7)	
15	C4 to C7	copy C3 to C4:C7	

For numerical data, we can use frequency tables, bar charts, and Pareto charts in the same way. However, with numerical data we often have a great range of data, either discrete or continuous. Thus, it may be more convenient to group the data into artificial, or natural, subsets for easier visualization. For example, if we were interested in the weekly sales of refrigerators to determine the number of weeks of both high and low sales during the year, we might divide the data into four groups (quartiles). (Of course, we would probably first want to determine the mean and variance of the data to give us a sense of typical weekly sales and their variability.)

Suppose the largest number of refrigerators sold in any week during the year 2002 was 15 and the least was 2. Then we might divide the weekly sales into the four groups: 1–4, 5–8, 9–12, and 13–16. With 52 weeks in the year, this would give us about 13 data points, on average, in each group, so the distribution, with some having more than 13 and others having less, should be clear to observe. (We could, of course, use more groups with fewer numbers in each, but the shape of the distribution would then be less clear due to random variability among the groups.) If we then constructed a bar chart of the number, or percentage, of weeks during the year that had sales between 1–4 and plotted that on the left, with 5–8 next to it, and so on, this chart would be called a **histogram.** The histogram differs from a categorical bar chart in two ways. Most importantly, the groups have a fixed order on the horizontal scale because the scale is a continuous one rather than a set of categories (even if the categories are ordinal). Thus, the bars might not be in descending order of frequency. Second, it is common practice to leave no space between the bars on a histogram to indicate the relatively continuous nature of the data (even the discrete data are more continuous than categorical data).

2.3 Probability and Random Variables

Uncertainty in organizational decision making is a fact of life. Demand for an organization's output is uncertain. The number of employees who will be absent from work on any given day is uncertain. Whether it will rain tomorrow is uncertain. Each of these *events* is more or less uncertain. If the result of this uncertain event is numerical, we refer to it as a *random variable.* We do not know exactly whether the event will occur, nor do we know the value that the particular random variable (e.g., demand for the output, number of absent employees) will assume. Note that the outcome "rain" is a categorical, not a numerical, result of a random event, and thus would not normally fit the definition of a random variable. However, we can assign the value 1 to a sunny day and 2 to a rainy day (and perhaps 3 to a day that is cloudy, etc.) and thereby convert this categorical variable to a random one.

In common terminology, we reflect our uncertainty in life with such phrases as "not very likely," "not a chance," "for sure." But, while these descriptive terms communicate one's feeling regarding the chances of a particular event's occurrence, they are not precise enough to allow the calculation of statistical measures and probabilities.

Simply put, **probability** is a number on a scale used to measure uncertainty. The range of the probability scale is from 0 to 1, with a 0 probability indicating that an event has no chance of occurring and a probability of 1 indicating that an event is absolutely sure to occur. The more likely an event is to occur, the closer its probability is to 1. This probability definition, which is general, needs to be further augmented to illustrate the various types of probability that modelers can assess. The modeler should be aware of three types of probability:

- Subjective probability
- Logical probability
- Experimental probability

Subjective Probability

Subjective probability is based on individual information and belief. Different individuals will assess the chances of a particular event in different ways, and the same individual may assess different probabilities for the same event at different points in time. For example, one need only watch the blackjack players in Las Vegas to see that different people assess probabilities in different ways. Also, daily trading in the stock market is the result of different probability assessments by those trading. The sellers sell because it is their belief that the probability of appreciation is low, and the buyers buy because they believe that the probability of appreciation is high. Clearly, these different probability assessments are about the same events.

Logical Probability

Logical probability is based on physical phenomena and on symmetry of events. For example, the probability of drawing a three of hearts from a standard 52-card playing deck is 1/52. Each card has an equal likelihood of being drawn. In flipping a fair coin, the chance of "heads" is 0.50. That is, since there are only two possible outcomes from one flip of a coin, each event has one-half the total probability, or 0.50. A third example is the roll of a single die. Since each of the six sides are identical, the chance of any one event occurring (i.e., a 6, a 3, etc.) is 1/6.

Experimental Probability

Experimental probability is based on the frequency of occurrence of events in trial situations. For example, in determining the appropriate inventory level to maintain in a raw material inventory, we might measure and record the demand each day from that inventory. If, in 100 days, demand was 20 units on 16 of the days, the probability of demand equaling 20 units is said to be 0.16 (i.e., 16/100). In general, the experimental probability of an event is given by

Probability of event = number of times event occurred/total number of trials

Both logical and experimental probability are referred to as *objective* probability, in contrast to the individually assessed subjective probability. Each of these is based on, and directly computed from, facts. However, be aware that probabilities developed objectively are not automatically or intrinsically more accurate than subjective probabilities.

Event Relationships and Probability Laws

Events are classified in a number of ways that allow us to further state rules for probability computations. Some of these classifications and definitions follow.

1. **Independent Events** Events are independent if the occurrence of one does not affect the probability of occurrence of the others. For example, whether a coin comes up heads or tails is independent of prior tosses, assuming it is a fair coin.
2. **Dependent Events** Events are dependent if the occurrence of one affects the probability of occurrence of others. For example, the probability of drawing a second card that is a diamond from a deck of cards depends on the suit of the first card drawn (assuming that the first card is not placed back in the deck). If the first card drawn from the deck is a diamond, then the probability that the second card drawn is also a diamond is 23.5 percent (12/51). Alternatively, if the first card drawn was a spade, then the probability that the second card will be a diamond is 25.5 percent (13/51).

3. **Mutually Exclusive Events** Two events are mutually exclusive if the occurrence of one precludes the occurrence of the other. For example, in the birth of a child, the events, "It's a boy!" and "It's a girl!" are mutually exclusive.
4. **Collectively Exhaustive Events** A set of events is collectively exhaustive if on any one trial at least one of them must occur. For example, in rolling a die, one of the events 1, 2, 3, 4, 5, or 6 must occur; therefore, these six events are collectively exhaustive.
5. **Complements Law** If the probability of an event *not* occurring is known, say p, then the probability of the event occurring is $1.0 - p$, and vice versa. For example, suppose the die mentioned above was "loaded" so that the probability of a 6 was 0.3 instead of 0.167 (1/6) for a fair die. Then, the probability that a 6 would *not* turn up on a roll of the die (or equivalently, that a 1, 2, 3, 4, or 5 would appear) would be 0.7.

We can also define the union and intersection of two events. Consider two events A and B. The *union* of A and B includes all outcomes in A or B or in both A and B. This is illustrated in the Venn diagram of Exhibit 2.6 where the union includes all the white space in A, all the shaded space in B, and all the dark (joint) space shown as part of *both* A and B. If A and B completely overlapped, one precisely on top of the other, then the union would consist of only the single space. If they were completely separate, with no dark joint space at all, the union would then include all of A and all of B separately. For example, suppose that in a card game you will win if you draw a diamond or a jack. The union of these two events includes all diamonds (including the jack of diamonds) and the remaining three jacks (hearts, clubs, spades). The "or" in the union is the *inclusive* "or." That is, in our example you will win with a jack or a diamond or a jack of diamonds (i.e., both events).

The *intersection* of two events includes all outcomes that are members of *both* events; that is, only the dark joint space in Exhibit 2.6. Thus, in our previous example of jacks and diamonds, the jack of diamonds is the *only* outcome contained in both events and is therefore the only member of the intersection of the two events.

Let us now consider the relevant probability laws based on our understanding of the definitions and concepts. For ease of exposition let us define the following notation:

$P(A)$ = probability that event A will occur
$P(B)$ = probability that event B will occur

If two events are mutually exclusive, then their joint occurrence is impossible. Hence, $P(A \text{ and } B) = 0$ for mutually exclusive events. If the events are not mutually exclusive, $P(A \text{ and } B)$ can be computed (as we will see in the next subsection); this probability is termed the *joint* probability of A and B. Also, if A and B are not mutually exclusive, then we can also define the *conditional* probability of A *given that* B has already occurred, or the conditional probability of B given that A has already occurred. These probabilities are written as $P(A \mid B)$ and $P(B \mid A)$, respectively.

EXHIBIT 2.6 Venn Diagram

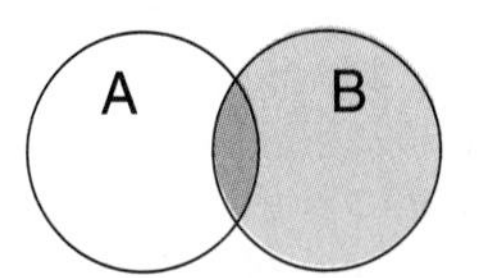

The Multiplication Rule The joint probability of two events that are not mutually exclusive is found by using the **multiplication rule.** If the events are dependent events, the joint probability is given by

$$P(A \text{ and } B) = P(A) \times P(B \mid A), \text{ or } P(B) \times P(A \mid B)$$

For example, the probability of drawing a 3 and a 5 from a deck of cards without replacement would be

$$\begin{aligned} P(3 \text{ and } 5) &= P(3) \times P(5 \mid 3) = P(5) \times P(3 \mid 5) \\ &= 4/52 \times 4/51 = 4/52 \times 4/51 \\ &= 0.0060 \end{aligned}$$

If the events are independent, then $P(B \mid A)$ and $P(A \mid B)$ are equal to $P(B)$ and $P(A)$, respectively, and therefore the joint probability is given by

$$P(A \text{ and } B) = P(A) \times P(B)$$

For example, if the first card were replaced in the deck after drawing it, then the probability of drawing a 3 and a 5 would be:

$$P(3 \text{ and } 5) = 4/52 \times 4/52 = 0.0059$$

From these two relationships, we can find the conditional probability for two dependent events from

$$P(A \mid B) = P(A \text{ and } B)/P(B)$$

and

$$P(B \mid A) = P(A \text{ and } B)/P(A)$$

For example, the probability of a drawn card being a 3, knowing that a heart (H) was drawn, would be:

$$\begin{aligned} P(3 \mid \text{H}) &= P(\text{H and } 3)/P(\text{H}) \\ &= (1/52)/(13/52) \\ &= 1/13 \end{aligned}$$

which is, as it should be, the chance of drawing any single card from a set of all 13 hearts (ace through king). Similarly, the probability of a drawn card being a heart, given that a 3 was drawn, would be:

$$\begin{aligned} P(\text{H} \mid 3) &= P(3 \text{ and H})/P(3) \\ &= (1/52)/(4/52) \\ &= 1/4 \end{aligned}$$

which is, as it should be, the chance of drawing a heart from the set of four 3s (heart, spade, club, and diamond).

Also, $P(A)$ and $P(B)$ can be computed if the events are independent, as

$$P(A) = P(A \text{ and } B)/P(B)$$

and

$$P(B) = P(A \text{ and } B)/P(A)$$

For example, the probability of drawing a 3, say, equals the probability of drawing the 3 of hearts (1/52) divided by the probability of drawing a heart (13/52), which equals 1/13. This, of course, is the same as would be obtained through the logic of knowing there are four 3s in the deck of 52 cards, giving 4/52 =1/13.

The Addition Rule The **addition rule** is used to compute the probability of the union of two events. The probability of A or B is given by

$$P(A \text{ or } B) = P(A) + P(B) - P(A \text{ and } B)$$

We can see the reasonableness of this expression by again looking at the Venn diagram in Exhibit 2.6. If we add all the area of A and all the area of B, we have included the dark joint area twice. Therefore, to get the total area of A or B, we must subtract one of the joint areas of the intersection that we have added. As an example, the probability of drawing either a 3 or a heart on a single draw, two events that are *not* mutually exclusive, is the probability of drawing a 3 (again, 4/52) plus the probability of drawing a heart (13/52) minus the probability of drawing the 3 of hearts (1/52), for a total of 16/52, or 4/13.

If two events are mutually exclusive, then $P(A \text{ and } B) = 0$ as we indicated previously, and there would be no overlap of the circles corresponding to events A and B in the Venn diagram shown in Exhibit 2.6. Therefore, the probability of either A or B or both is simply the probability of A or B. This is given by

$$P(A \text{ or } B) = P(A) + P(B)$$

For example, the probability of drawing either a 3 or a 5 on a single draw, which are mutually exclusive events, is equal to the probability of drawing a 3, that is, 4/52, plus the probability of drawing a 5, also 4/52, for a total of 8/52, or 2/13.

If two events are collectively exhaustive, then the probability of A or B is equal to 1. That is, for two collectively exhaustive events, one or the other or both must occur, and therefore, the probability of A or B must be 1. For example, the probability of drawing either a red (heart or diamond) card (26/52) or a black (spade or club) card (26/52) is 52/52, or 1.

Probability Distributions

For a discrete random variable x, the distribution of the probabilities over all the values of x is determined from a **probability function,** denoted by $f(x)$. This function is simply the probability that a particular value of x will occur and might be determined from experience, logic, or from an assumed common probability distribution, such as the normal or binomial (discussed in Section 2.4). Clearly, $f(x)$ must be greater than or equal to zero, since it is nothing more than a probability, and all the values of $f(x)$ over the range of values that x can take must sum to 1.0. As a simple example, the probability function for the six values that may turn up on a roll of a die would be $f(x) = 1/6$. In this case, the function is independent of any particular value of x, since every value of x has the same likelihood of appearing. More generally, however, the probabilities will depend on the value of x. Suppose the die were loaded so that the probabilities of each value of x appearing were as follows:

x	f(x)
1	1/21
2	2/21
3	3/21
4	4/21
5	5/21
6	6/21

Note that the sum of the probabilities is 21/21 = 1.0, as it must be. The probability function for this distribution would then be simply $f(x) = x/21$.

EXHIBIT 2.7 Probability Function and Cumulative Distribution Function of a Loaded Die

	A	B	C	D	E	F	G
1	*x*	*f(x)*	*xf(x)*	*(x - mu)* ^{2} *f(x)*	*P(X ≤ x)*		
2	1	4.76%	0.05	0.53	4.76%		
3	2	9.52%	0.19	0.52	14.29%	Cumulative distribution	
4	3	14.29%	0.43	0.25	28.57%		
5	4	19.05%	0.76	0.02	47.62%		
6	5	23.81%	1.19	0.11	71.43%		
7	6	28.57%	1.71	0.79	100.00%		
8	*E(x)*		**4.33**				
9	*Var(x)*			**2.22**			
10							
11	***Key Formulas***						
12	B2	=A2/21 {copy to cells B3:B7}					
13	C2	=A2*B2 {copy to cells C3:C7}					
14	C8	=SUM(C2:C7)					
15	D2	=((A2-$C$8)^2)*B2 {copy to cells D3:D7}					
16	D9	=SUM(D2:D7)					
17	E2	=SUM(B$2:B2) {copy to cells E3:E7}					

When we wish to determine the average value, or mean, of a random variable, we call it the expected value $E(x)$ to indicate that we are dealing now with probabilities. Although it is precisely what we described previously, we can employ the probability function to calculate this mean as follows:

$$E(x) = \Sigma x f(x)$$

For example, the expected value of one roll of a fair die would be $1(1/6) + 2(1/6) + 3(1/6) + 4(1/6) + 5(1/6) + 6(1/6) = 3.5$.

Similarly, the variance of the random variable is given by:

$$\text{Var}(x) = \Sigma(x - \mu)^2 f(x)$$

Exhibit 2.7 illustrates the probability function for the loaded die described earlier with the distribution function $x/21$, its expected value, and its variance. (Note that mu = $E(x)$.) In addition, the *cumulative distribution function* (cdf) is tabulated, which shows the probability of getting a value X less than or equal to a given value x: $P(X \leq x)$. (*Note:* ≤ means less than or equal to, ≥ means greater than or equal to, < means strictly less than, and > means strictly greater than.) For example, the probability of getting a value less than or equal to 2 would be $1/21 + 2/21 = 0.1429$. This, or its opposite, the probability of getting a value greater than X (which would equal 1—the cdf) is often of interest for managers and modelers.

2.4 Common Probability Distributions

In this section we describe the most common statistical distributions used in quantitative business modeling, both discrete and continuous. Our descriptions will be more in terms of what situations the distributions represent than the statistics of the distributions, though we will give their probability functions. To determine values or probabilities using these functions, we again defer to the common spreadsheet that has all these standard functions available for your use.

As you recall, if a random variable may take only certain specific numerical values such as integer numbers, then the probability distribution that characterizes the process that generated that random variable is called a *discrete distribution.* However, if the random variable may take any value (within a specified interval), then the probability distribution is labeled *continuous.* In this section, we will discuss the following distributions: the binomial (discrete), Poisson (discrete), the exponential (continuous), the normal (continuous), and the Student t (continuous).

The Binomial Distribution

The **binomial distribution** is a discrete distribution representing situations where there is a sequence of n independent but identical events resulting in only one of two possible outcomes, such as success or failure. The probability of a success is labeled as p (and thus a failure is $1 - p$) and is the same for every one of the events. Hence, this distribution could represent the number of successes in, say, ten attempts to win a race, the number of times a machine produces defects when sequentially molding 35 items, or the number of females born among five births. The probability function for the binomial distribution is:

$$f(x) = [n!/(x!\,(n - x)!)]p^x(1 - p)^{(n - x)}$$

where $n!$ means $n(n - 1)(n - 2) \ldots 1$ (e.g., $4! = 4(3)(2)(1) = 24$), and the term $f(x)$ is the probability there will be x successes in the n trials.

An example of getting seven heads in 10 tosses of a fair coin is given in Exhibit 2.8. As is demonstrated in the figure, the Excel function BINOMDIST can be used to calculate probabilities based on the binomial distribution. The syntax for this function is

=BINOMDIST(# of successes, number of independent trials,
the probability of success, cumulative)

EXHIBIT 2.8 Probability of Getting Seven Heads in Ten Tosses of a Coin

	A	B	C
1	Independent Trials	10	
2	Probability of Heads	50%	
3			
4			Cumulative
5	Number of Heads	Probability	Probability
6	0	0.10%	0.10%
7	1	0.98%	1.07%
8	2	4.39%	5.47%
9	3	11.72%	17.19%
10	4	20.51%	37.70%
11	5	24.61%	62.30%
12	6	20.51%	82.81%
13	7	11.72%	94.53%
14	8	4.39%	98.93%
15	9	0.98%	99.90%
16	10	0.10%	100.00%
17			
18			
19	*Key Formulas*		
20	Cell B6	=BINOMDIST(A6,B$1,B$2,FALSE) {copy to cells B7:B16}	
21	Cell C6	=BINOMDIST(A6,B$1,B$2,TRUE) {copy to cells C7:C16}	

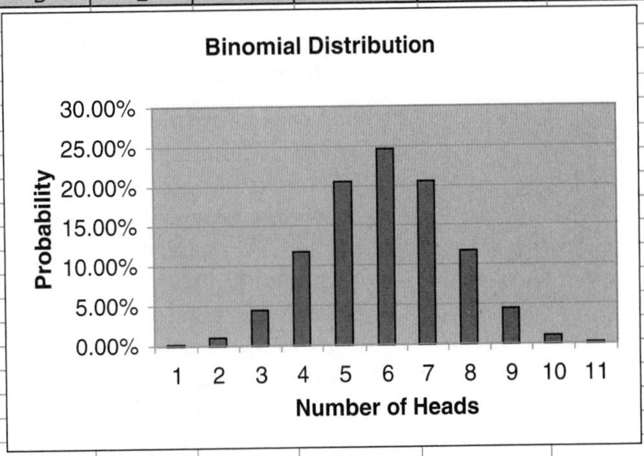

The first parameter corresponds to the number of successes. In Exhibit 2.8, a success was defined as getting a heads on a given coin toss. The second parameter corresponds to the number of independent trials, 10 in this case, as shown in cell B1. In the example presented in Exhibit 2.8, we are interested in determining the probability that we get a given number of heads in 10 tosses of a coin. Therefore, the appropriate cell in column A and cell B1 were specified for the first two parameters, respectively. The third parameter is the probability of succeeding, or in our case the probability of getting a heads. Based on the assumption that we are using a fair coin, 50 percent was entered in Cell B2 and specified for the third parameter. Finally, the last parameter determines whether the probability of getting *exactly* the number of successes specified in the first parameter is calculated (cumulative = false) or no *more* than the number specified (e.g., ≤ 7) for the first parameter (cumulative = true). Referring to Exhibit 2.8, we observe that there is an 11.72 percent chance of getting exactly seven heads in 10 tosses of a fair coin and a 94.53 percent chance of getting seven or fewer heads in 10 tosses.

The Poisson Distribution

The **Poisson distribution,** another discrete distribution, describes situations where there is a succession of events happening over some interval, such as the number of items sold per day, the number of defects per meter of fiber optic cable, or the number of crashes of a computer system in the next month. The events occur individually, or one at a time, and their timing or spacing is unpredictable. In addition, the occurrence of an event in one interval is independent of the occurrence of events in another interval of equal length. Last, there is no upper limit on the possible number of occurrences in an interval. Clearly, it is important to the manager to have adequate capacity to service the events occurring in every interval.

For this distribution, the average number of occurrences per interval is labeled λ (lamda). The probability function for the Poisson distribution is:

$$f(x) = e^{-\lambda}\lambda^{x}x!$$

where $e = 2.71828$ (the base of the natural logarithms) and the term $f(x)$ is the probability there will be x occurrences in an interval.

An example of the use of this distribution is given in Exhibit 2.9. In a similar fashion to the binomial function, Excel provides a function for computing probabilities based on the Poisson distribution. The syntax for the Poisson function is as follows:

=POISSON(number of occurrences, average number of occurrences, cumulative)

The first parameter corresponds to x, the number of times an event occurs, while the second parameter corresponds to λ, the average or expected number of occurrences per interval. The third parameter is the same as with the binomial distribution, *false* meaning exactly the specified number and *true* meaning less than or equal to the specified number. In Exhibit 2.9 the probability that various numbers of customers arrive at a particular ATM in a one-hour interval is calculated given that it is known that 10 customers arrive at this ATM each hour, on average.

For example, according to the calculations shown in Exhibit 2.9 there is a 3.47 percent chance that exactly 15 customers will arrive at the ATM in a given one-hour interval. Likewise, there is a 95.13 percent chance that 15 or fewer customers will arrive at the ATM during a particular hour interval. This further implies that there is a 4.87 percent (1.0 – 0.9513) chance that 16 or more customers will arrive at the ATM during a given hour interval. (Note that, as opposed to the binomial situation in Exhibit 2.8, not *all possible* arrivals (e.g., 0, 3, 21, and such others) are listed in Exhibit 2.9.)

EXHIBIT 2.9 Arrivals at a Bank ATM

	A	B	C
1	Mean Arrival Rate (lambda)	10	
2			
3			Cumulative
4	Number of Arrivals	Probability	Probability
5	4	1.89%	2.93%
6	5	3.78%	6.71%
7	6	6.13%	13.01%
8	7	9.01%	22.02%
9	8	11.26%	33.28%
10	9	12.51%	45.79%
11	10	12.51%	58.30%
12	11	11.37%	69.68%
13	12	9.48%	79.16%
14	13	7.29%	86.45%
15	14	5.21%	91.65%
16	15	3.47%	95.13%
17	16	2.14%	97.30%
18	17	1.28%	98.57%
19	18	0.71%	99.28%
20			
21	*Key Formulas*		
22	Cell B6	=POISSON(A5,$B$1,FALSE) {copy to cells B5:B19}	
23	Cell C6	=POISSON(A5,$B$1,TRUE) {copy to cells C5:C19}	

Poisson Distribution

Probability: 0.00%, 2.00%, 4.00%, 6.00%, 8.00%, 10.00%, 12.00%, 14.00%

4 5 6 7 8 9 10 11 12 13 14 15 16 17 18

Number of Arrivals per Hour

The Exponential Distribution

The **exponential distribution** is a *continuous* distribution for events that occur randomly over some continuous interval. It is the inverse of the Poisson in the sense that if a random variable x has a Poisson distribution, then $1/x$ has an exponential distribution. Thus, it describes the *length of an interval* between events in situations where there is a succession of independent, random events occurring. To illustrate, if the number of arrivals at an ATM follows a Poisson distribution with an average of 10 arrivals per hour, then the time between arrivals would follow an exponential distribution with a mean of 0.1 hours, or 6 minutes. Once again, the events occur individually, or one at a time, their timing or spacing is unpredictable, the occurrence of an event in one interval is independent of the occurrence of events in another interval of equal length, and there is no upper limit on the possible number of occurrences in an interval. Common situations in business that fit this distribution would be the length of time between customer arrivals at a store or the length of a production run until a failure. For this distribution, the average interval between occurrences is $1/\lambda$, and the *probability density function* is

$$f(x) = e^{-\lambda x}\lambda$$

where x is the interval of interest and $e = 2.71828$. Note here that since we are dealing with a continuous distribution, $f(x)$ no longer gives the probability of an interval of exact length x (which would be zero) but rather, the relative height, or *density,* of the function at the value x. Probabilities with a continuous distribution such as we are dealing with here are determined by the *area* between two values x_1 and x_2.

The Excel function EXPONDIST (x, λ, cumulative) can be used to calculate the probability of a specified time interval between events. The first parameter specifies the time interval, the second again corresponds to the average number of occurrences per interval, and the last parameter is used to specify whether the probability density function value is returned (FALSE) or the cumulative distribution function is returned (TRUE). In problems of the type discussed here you will most often want to use the cumulative distribution function (to get a probability) and thus specify TRUE for the third parameter.

An example of the use of this distribution is given in Exhibit 2.10. In the figure, the probability that the time between successive customer arrivals for a range of interarrival times is calculated given that the average customer arrival rate is 10 customers per hour (cell B1). Note that column A was created because it is more intuitive to specify the interarrival times in minutes rather than as fractions of an hour. Then in column B, the times specified in column A were converted to hours by dividing the times in column A by 60. In column C, the cumulative probability calculated using the EXPONDIST function indicates the likelihood of a customer arrival sometime within the corresponding interarrival interval given in column A (or B).

As shown in the exhibit, if the average rate of arrivals to the ATM machine is 10 per hour (average interarrival time of 6 minutes), then there is a 90.3 percent chance that a customer will arrive within a 14 minute (or 0.23 hour) interval. This also implies that there is a 9.7 percent chance (1 – 0.903) that the time between two successive customers will be greater than 14 minutes.

The Normal Distribution

The **normal distribution** is a continuous distribution discovered more than 250 years ago. It was then considered to be the law governing distributions of all natural phenomena, but this belief has been modified as other natural distributions were discovered. Yet, it well describes the amount of rainfall in a season, the dimensions of standard items (such as the heights of people), the times to do standard tasks (such as processing customers), and so on. Another important phenomenon the normal distribution characterizes

EXHIBIT 2.10 Interarrival Times at a Bank ATM

	A	B	C	D
1	Arrivals per Hour	10		
2				
3	**Interarrival Time**	**Interarrival Time**	**Cumulative**	
4	**(minutes)**	**(hours)**	**Probability**	
5	2	0.03	28.35%	
6	4	0.07	48.66%	
7	6	0.10	63.21%	
8	8	0.13	73.64%	
9	10	0.17	81.11%	
10	12	0.20	86.47%	
11	14	0.23	90.30%	
12	16	0.27	93.05%	
13	18	0.30	95.02%	
14	20	0.33	96.43%	
15	22	0.37	97.44%	
16	24	0.40	98.17%	
17	26	0.43	98.69%	
18	28	0.47	99.06%	
19	30	0.50	99.33%	
20				
21	***Key Formulas***			
22	B5	=A5/60 {copy to B6:B19}		
23	C5	=EXPONDIST(B5,B$1,TRUE) {copy to C6:C19}		

is the distribution of the *averages* of samples of random variables selected from *any* distribution, a phenomenon we will discuss in more detail in the next section.

The normal distribution is still the most commonly used one in statistics, and we will thus spend more time describing it and its uses than the other distributions. As opposed to the distributions we have discussed so far, the normal distribution requires two parameters to fully describe it: the mean μ and the standard deviation σ. The distribution is shown in Exhibit 2.11 and is a bell-shaped curve symmetrical about its highest point, the mean (which is also its median and mode). As the mean changes, the location of the distribution changes on the horizontal axis. As the standard deviation increases, the curve spreads out further around the mean, and vice versa. As shown on the bell curve, approximately 68.3 percent of the area of the curve is contained within plus or minus one standard deviation of the mean. In addition, the area from point $-\infty$ (left side) to point x_2 is 84 percent (34 percent plus the mean, which is exactly at 50 percent). It is also known that 95.4 percent of the curve lies between $\pm 2\sigma$ around the mean, and 99.7 percent lies between $\pm 3\sigma$.

The probability density function of the normal distribution is

$$f(x) = \frac{e^{\frac{-(x-\mu)^2}{2\sigma^2}}}{\sqrt{2\pi\sigma^2}}$$

where

$\pi = 3.14159$ and $e = 2.71828$, again

A normal distribution whose mean is zero and whose standard deviation is one is called a *standard normal distribution* and is particularly useful for determining the probability of events occurring between two values of x. To use the standard normal

EXHIBIT 2.11 The Normal Distribution

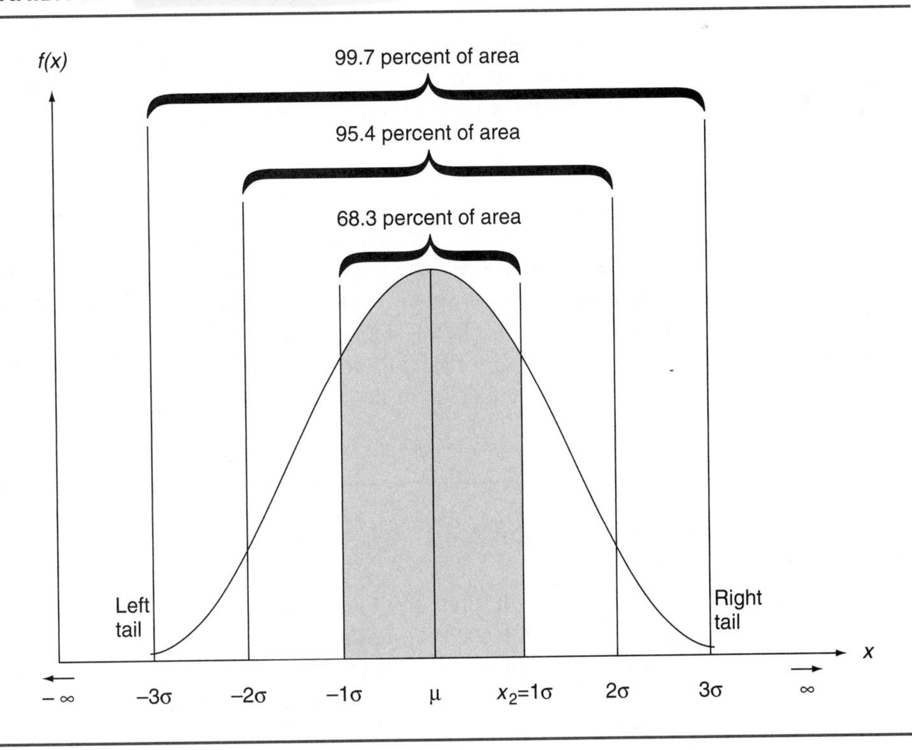

EXHIBIT 2.12 Excel's Normal Distribution Functions

Function Syntax	Description
=NORMDIST(x, μ, σ, cumulative)	Returns the cumulative probability for the specified x value given the specified mean and standard deviation when "TRUE" is specified for the fourth parameter. When "FALSE" is specified for the fourth parameter, this function returns the probability density function given the values of x, μ, σ. For the vast majority of problems you will encounter, TRUE should be specified for the cumulative parameter.
=NORMSDIST(z)	Returns the cumulative probability of z.
=NORMINV(probability, μ, σ)	Returns the x value for the given probability, mean, and standard deviation specified.
=NORMSINV(probability)	Returns the z value from the standard normal distribution for the probability specified.

distribution, we simply convert the value of any normally distributed random variable x to that of the standard normal random variable z with the equation:

$$z = (x - \mu)/\sigma$$

Instead of entering the preceding formula to calculate z values, you can use the Excel STANDARDIZE function. Using the above notation, this function has the following syntax:

=STANDARDIZE(x,μ,σ)

Prior to the widespread use of spreadsheets, it was common for textbooks to include a number of tables that could be used to calculate probabilities from a normal distribution. However, as is shown in Exhibit 2.12, Excel has four functions that make obsolete the need for these tables. Exhibit 2.13 illustrates the use of these functions.

The *t* Distribution

The t distribution is a very special distribution that is primarily used for determining confidence intervals when the population standard deviation is unknown and the number of data observations is small, usually less than 30. It looks almost identical to the normal distribution but is slightly more spread out at the extremes, depending on its parameter: degrees of freedom. We will demonstrate the use of this distribution in Sections 2.7 and 2.8, where we address confidence intervals and hypothesis testing.

2.5 Distributions of Sample Statistics

The modeler will often be in the situation of trying to determine the mean (μ) for some population of interest such as the average number of ounces of liquid in a can, or the average number of errors on a production line. In these cases, a *sample* of size, say n, is taken as representative of the population and the mean determined from this sample. However, the modeler also wishes to know how close this sample mean might be to the

EXHIBIT 2.13
Illustrative Examples for Excel's Normal Distribution Functions

Scenario	Excel Formula	Returned Value
It is estimated that the time it takes to complete a project follows a normal distribution with a mean of 21 weeks and a standard deviation of 2.45 weeks. What is the probability of completing the project in 23 weeks?	=NORMDIST(23,21,2.45,TRUE)	79.3%
It is estimated that the time it takes to complete a project follows a normal distribution with a mean of 21 weeks and a standard deviation of 2.45 weeks. What is the probability of completing the project in 19 weeks?	=NORMDIST(19,21,2.45,TRUE)	20.7%
How much area is there under the standard normal distribution from $-\infty$ to a z value of 1.5?	=NORMSDIST(1.5)	93.3%
How much area is there under the standard normal distribution from zero to a z value of 1.5?	=NORMSDIST(1.5)–0.5	43.3%
It is estimated that the time it takes to complete a project follows a normal distribution with a mean of 21 weeks and a standard deviation of 2.45 weeks. If the project manager would like to be 90% certain of completing the project on time, how much time should be allotted to complete the project?	=NORMINV(0.90,21,2.45)	24.1 weeks
From $-\infty$ to what value of z provides a 93.3% probability?	=NORMSINV(0.933)	1.499

true population mean; that is, what the error might be in this estimate. As it turns out, there is a way to determine this. It is based on the hypothetical distribution of sample means taken were the modeler to take many, many samples of size n, determine the mean ($\bar{x}$) of each sample as a random variable, and then plot these values on a chart.

This distribution of sample means for samples of size n is called, fittingly, the **sampling distribution of the mean.** As might be expected, the mean of this distribution is precisely equal to the mean of the population the samples have been drawn from, μ. Also, the standard deviation of the distribution, called the **standard error of the mean,** equals the population standard deviation divided by the square root of the sample size, that is, $\sigma/\sqrt{n}$. This dependence of the spread of the distribution (i.e., its standard deviation) on the sample size reflects the fact that the *larger* the sample size, the smaller the standard error of the mean, and thus the more sure we are that the sample mean represents the true mean of the population. Indeed, if we sampled the *entire population,* we would be absolutely certain to have the true mean!

But to obtain an estimate of our error, we still need to know the shape of the sampling distribution of the mean. As it happens, if the original population was normally distributed, the sampling distribution of the mean will be as well. More importantly, however, the **central limit theorem** tells us that *regardless of the shape of the original population distribution,* the sampling distribution of the mean will still be approximately normally distributed if the sample size n is large enough. It has been determined through many studies that, for most practical purposes, an n of 30 is "large enough" (but the larger, the better).

In review, we can determine the accuracy of using the mean of our sample as an estimate of the population distribution mean μ by appealing to a hypothetical distribution that is characterized simply by the size of our sample (n) and its derived mean. That is, we don't need to actually construct this sampling distribution of the mean; we simply use the information that we know will characterize it (its mean, its standard deviation, and its shape) had we drawn enough samples to accurately construct it!

Exhibit 2.14 illustrates this process for two different sample sizes. The top graph shows data from an exponential distribution with a mean of 10. The middle graph shows the distribution of the sample means of multiple samples of two observations each, taken at random from the exponential distribution. Notice how, with such a small sample size, the distribution of means reflects the underlying exponential distribution. The bottom graph shows the distribution of sample means of multiple samples but using a sample size of 30 observations instead. This distribution is much closer to that of a normal distribution. Comparing the middle and bottom graphs in the exhibit illustrates how the sample distribution of the mean approaches the normal distribution as the sample size is increased.

2.6 Chi-Square Goodness of Fit Test

Many situations that modelers face involve unknown distributions, and modelers will find that an analysis of the situation is much easier to conduct if they can identify and use the actual distribution that underlies the data. For this case, the modeler will test the data against different possible distributions that seem to be logical candidates for the resulting distribution of data. Before doing any analysis on raw data, however, it is wise to plot it first and examine it visually for outliers, anomalies, and its general dispersion. Usually, a plot of the data will give the modeler some insight into whether the underlying distribution is normal, exponential, Poisson, or some other distribution. In addition, the characteristics of each distribution will often give the modeler a clue as to the correct distribution. For example, the length of time until a light bulb burns out is independent of how long it has been burning so far and instead is a function of the tungsten element, thus indicating a distribution such as the exponential. And the likelihood of a person arriving at a service counter in a particular interval of time is independent of whether someone else just arrived, thereby indicating a Poisson distribution.

It is possible to test how well a set of data conforms to a suspected known distribution with a simple test called the **chi-square goodness of fit test.** Before describing the test for distributions, we will demonstrate how the chi-square test works. Suppose we suspect that a die is loaded and want to test it. We hypothesize that the die is fair, and then test a sample of tosses against the frequencies f_e we would expect if the die were indeed fair; that is, one-sixth of the tosses should show each value on the die.

Let us construct Exhibit 2.15 for recording the number of tosses that results in each face of the die, f_o (note the subscript o is short for observed). We divide the table into six cells in which to place our results. One requirement of the chi-square test is that the expected values in each cell, that is, the f_e, must be at least 5; hence we will have to make at least $5 \times 6 = 30$ tosses. When we complete the experiment, suppose we have the results shown in Exhibit 2.15: six tosses resulted in a 1, three resulted in a 2, and so on.

EXHIBIT 2.14 Impact on the Distribution of the Sample Mean as the Sample Size Is Increased

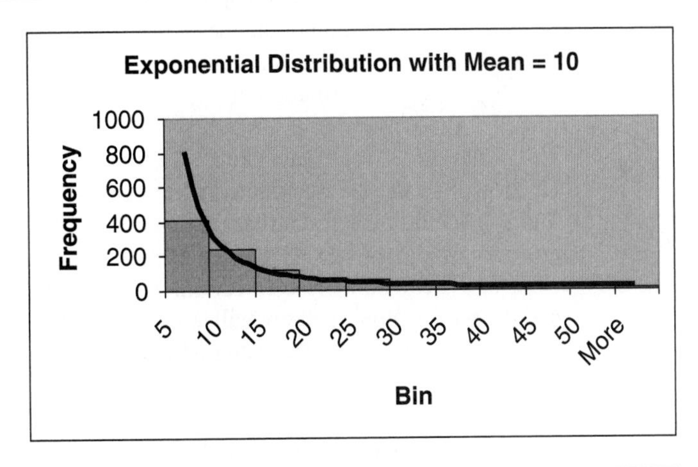

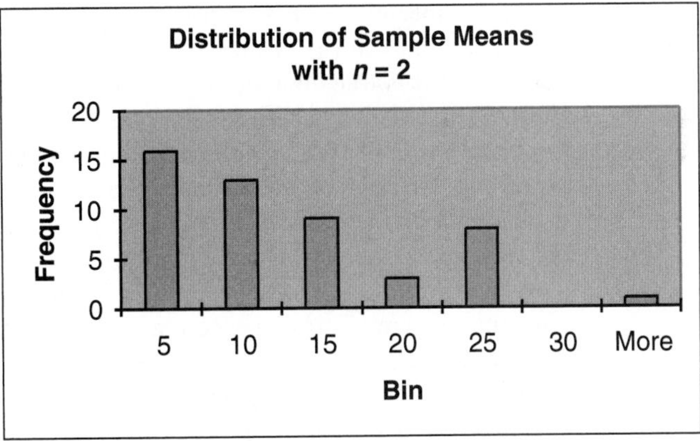

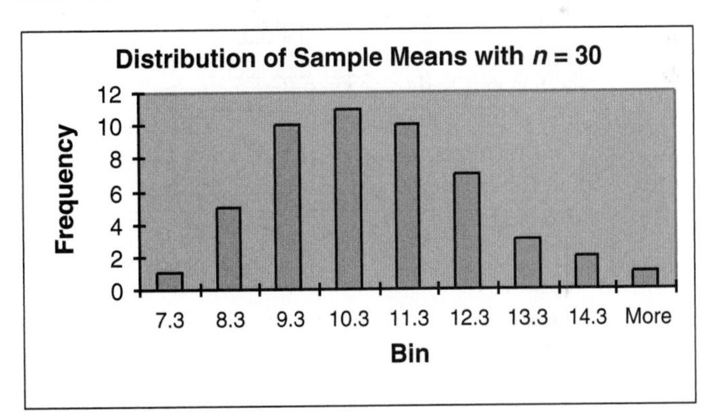

EXHIBIT 2.15 Chi-Square Table for 30 Die Tosses

Value on die	1	2	3	4	5	6
Observed frequency, f_o	6	3	3	6	9	3
Expected frequency, f_e	5	5	5	5	5	5

From this data we construct the following chi-square measure:

$$\chi^2 = \Sigma\ (f_o - f_e)^2/f_e$$

which gives

$$(6-5)^2/5 + (3-5)^2/5 + (3-5)^2/5 + (6-5)^2/5 + (9-5)^2/5 + (3-5)^2/5 = 6.0$$

Clearly, if all the frequencies we observed turned out to be precisely what we expected with a fair die, then the calculated chi-square value would have been zero and we would conclude that the die was indeed fair. The further this calculated value is from zero, the more evidence we have that the die is, in fact, loaded. We thus compare this value, 6, to see if it exceeds a certain critical value obtained from a chi-square table or spreadsheet function. If the calculated chi-square value is larger than the critical value, it indicates that the calculated value is much higher than would be expected by chance if the hypothesis were true.

In determining what critical value to use, we look for a value based on two parameters: the *significance level* we desire for this test, denoted by the symbol α (alpha), and the number of *degrees of freedom* in the chi-square measure. The significance level corresponds to the chance that we will reject the hypothesis that the die is fair when in fact it is true. We would, of course, like to use an alpha level such that this unfortunate event never happens, but this phrasing of the test implies that there is another error we might make, accepting the hypothesis that the die is fair when it is, in fact, wrong. Clearly, we can't win on both these two errors; the more sure we are of not making the first error, the more likely we will be to make the second error. The only way to reduce the chance of not making either error is to increase the sample size, which is generally costly. Thus, it is commonly assumed that $\alpha = 0.05$, which means that we are willing to accept a 5 percent chance that we are incorrectly rejecting the hypothesis, or in the case of our example, that we are incorrectly concluding that the die is loaded when in fact it is fair.

The second parameter, degrees of freedom, is based on the number of cells we used in the test. Although we used six cells, only five of the six cells were independently determined by our sample test since, from the way we segmented the test results, the value in the last cell had to be such that the sum of all the cells added to 30; hence we lost one degree of freedom when we selected the sample size of 30. Thus, the final number of degrees of freedom, as this parameter is called, is 6 – 1 = 5. Checking a chi-square distribution for the alpha value corresponding to 0.05 with five degrees of freedom, we would obtain 11.07. This can be found by using Excel's CHIINV function as follows:

=CHIINV(0.05,5)

where the first parameter corresponds to the level of significance and the second parameter to the degrees of freedom. Since our calculated chi-square value of 6 is less than the critical value of 11.07 as illustrated in Exhibit 2.16, we cannot *reject* the hypothesis that the die is fair. Thus, we conclude that it is indeed fair. However, we may be making the other error we mentioned earlier, accepting a hypothesis when it is in fact false.

Suppose we had sampled twice as many tosses of the die, 60, but got the same proportions as in Exhibit 2.15. That is, 12 rolls showing a 1, 6 rolls showing a 2, and so on. The expected frequencies in this case would again be one-sixth of the total, or 10 for each cell. Calculating the chi-square value for this larger sample, we would get 12. And since the chi-square test value is still at an alpha of 0.05 with 5 degrees of freedom, it would still be 11.07. However, now we would *reject* the fair hypothesis (see Exhibit 2.16 again)! Why is this? Simply because the divergence of the sample from the expected frequencies with twice the number of events is much less likely with a fair die than with a

EXHIBIT 2.16 The Chi-Square Distribution

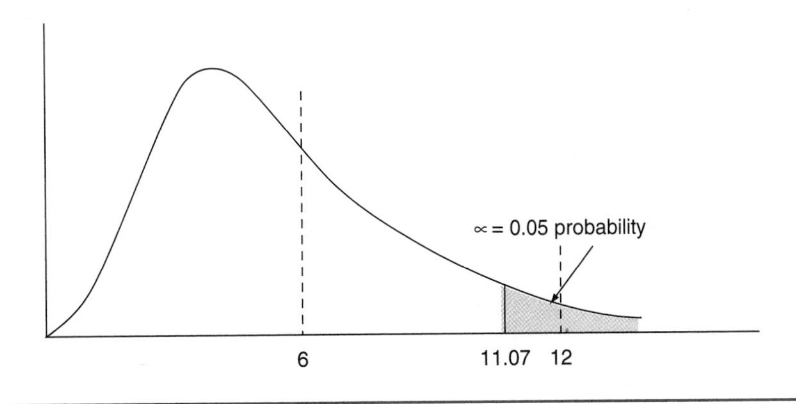

loaded die. (Recall our earlier discussion of the value of larger sample sizes and how the standard error of the mean shrinks with larger values of n.) Indeed, with this large a test, we would have expected the results to be much closer to the expected values if the die were fair, but they were not; hence, we reject the fair hypothesis. Again, at the extreme where all the 60 values were exactly what was expected, the calculated chi-square value would have been zero and we would have accepted the fair hypothesis.

Applying this concept to the testing of data to standard, known distributions, we have to make only one adjustment, and that is to the degrees of freedom corresponding to the test value of the chi-square statistic. In addition to subtracting one from the number of cells being tested, we also use the parameters of the distribution being tested to generate the expected frequencies, and thus must subtract the number of these parameters from the degrees of freedom also. For the case of the Poisson, binomial, and exponential distributions, we need only one parameter: p for the binomial, and λ for the Poisson and exponential distributions. However, for many other distributions, such as the normal, we may need two parameters such as μ and σ. Thus, a test of a sample of data against the normal distribution will have the number of total cells, or *partitions,* of the data, minus 3, as the final number of degrees of freedom.

For example, suppose a modeler divides the data into eight segments (either equal or unequal) along the variable of interest, such as lengths of pipe. Using these eight segments, the modeler then determines how many data points would have been expected to fall in each segment if the underlying distribution were normally distributed with a mean and standard deviation equal to the mean and standard deviation of the sample, the modeler's best estimates of the population parameters. The chi-square test statistic would then be found for an alpha of 0.05 and 8 – 3 = 5 degrees of freedom.

To illustrate, assume that a sample of 100 "six-foot" long pipes were taken and their lengths were measured. Further assume that the specifications for these pipes are that they are to have an average length of 72 inches and a standard deviation of no more than 15 inches. The calculations to test the hypothesis that the pipe lengths are normally distributed with a mean of 72 inches and standard deviation of 15 inches are shown in Exhibit 2.17. Columns A and B correspond to the frequency table for the sample data. The probability of falling in each of these six categories is calculated in column C using Excel's NORMDIST function. In column D, the expected number of pipes that would be observed in each category if a random sample of 100 pipes were drawn and the pipe lengths were normally distributed with a mean of 72 inches and standard deviation of

EXHIBIT 2.17 Chi-Square Test for Pipe Lengths

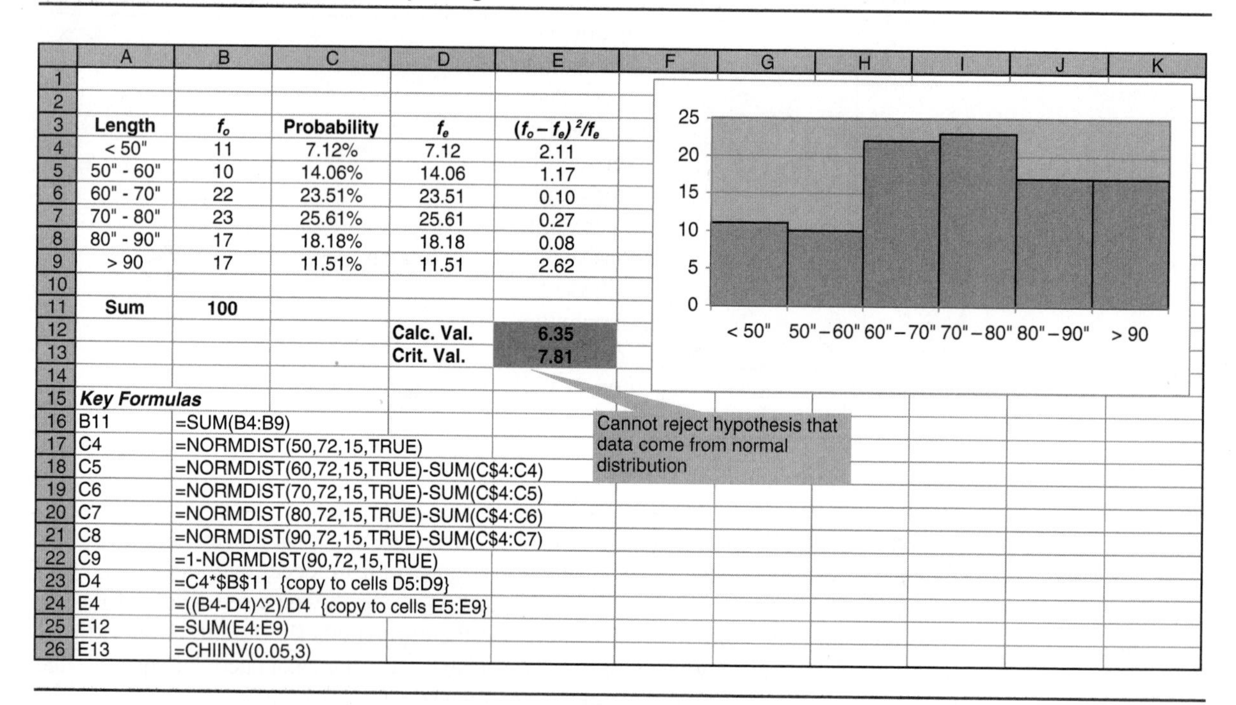

	A	B	C	D	E
1					
2					
3	Length	f_o	Probability	f_e	$(f_o - f_e)^2/f_e$
4	< 50"	11	7.12%	7.12	2.11
5	50" - 60"	10	14.06%	14.06	1.17
6	60" - 70"	22	23.51%	23.51	0.10
7	70" - 80"	23	25.61%	25.61	0.27
8	80" - 90"	17	18.18%	18.18	0.08
9	> 90	17	11.51%	11.51	2.62
10					
11	Sum	100			
12				Calc. Val.	6.35
13				Crit. Val.	7.81
14					

	A	B
15	*Key Formulas*	
16	B11	=SUM(B4:B9)
17	C4	=NORMDIST(50,72,15,TRUE)
18	C5	=NORMDIST(60,72,15,TRUE)-SUM(C$4:C4)
19	C6	=NORMDIST(70,72,15,TRUE)-SUM(C$4:C5)
20	C7	=NORMDIST(80,72,15,TRUE)-SUM(C$4:C6)
21	C8	=NORMDIST(90,72,15,TRUE)-SUM(C$4:C7)
22	C9	=1-NORMDIST(90,72,15,TRUE)
23	D4	=C4*$B$11 {copy to cells D5:D9}
24	E4	=((B4-D4)^2)/D4 {copy to cells E5:E9}
25	E12	=SUM(E4:E9)
26	E13	=CHIINV(0.05,3)

15 inches is calculated. Note that each f_e in column D exceeds 5.0. Finally in column E, the calculated value of the chi-square test statistic is computed.

Referring to Exhibit 2.17, although the histogram of the data does not appear to strongly resemble a normal distribution, we cannot reject the hypothesis that the distribution of pipe lengths comes from a normal distribution with a mean of 72 and standard deviation of 15. This is due to the fact that our calculated chi-square test statistic of 6.35 is less than the critical value of 7.81.

2.7 Point and Interval Estimation

Analyzing the results of a modeling study often requires estimating parameters such as the mean, standard deviation, or slope of a linear regression equation. For example, after developing a queuing or simulation model to study the operations of the branch of a retail bank, estimates of the average customer waiting time, average time in the system, and average number of customers in the system might be used to evaluate alternative operating procedures and policies. As another example, in Chapter 3 we use regression analysis to determine how the size of a newly constructed house affects its price. In both of these examples, estimates of key parameters were used to make inferences about the situation being studied.

In what follows, we will detail various characteristics of these types of **point estimators**—that is, single measures—for model parameters and performance measures and discuss what characteristics are the most important. As might generally be expected, the larger the sample, the better the estimate of a parameter but the more costly and lengthy the data collection process.

In discussing this issue, we will also give some guidance about how much data are needed to make our estimates. Finally, because we are basing our analysis on only a sample of the data, we realize that our estimate of the parameter(s) will probably not be precisely correct due to *sampling error.* Thus, we would like to also obtain, with some level of specified confidence, an **interval estimate** within which we believe our parameter lies.

The most important characteristic of an estimator is that it be *unbiased;* that is, that the *expected* value of the estimator should equal the true population value of the parameter. For example, using the 25th percentile of a sample would obviously be a biased estimator of the population mean, consistently underestimating the correct value. The median or mode of an *exponential* distribution would also be a biased estimator of its mean, though not for that of a normal distribution.

Another important characteristic of an estimator is its *efficiency,* in the sense of having a tight, or narrow, distribution of the estimator about the population parameter. For example, consider sampling from a normal population to estimate its mean. The standard error of the mean (i.e., the standard deviation of the distribution of the sample means taken from the population) is equal to $\sigma/\sqrt{n}$, as we discussed earlier. However, the standard deviation of the distribution of *medians* is equal to $\sigma\sqrt{\pi/2n}$ (where π is the constant 3.14159), which is about 25 percent larger. Thus, the sample mean is a more efficient estimator of the population mean than the sample median for a normal population.

Another characteristic is the *consistency* of an estimator, or its tendency to pinpoint the parameter of interest as the sample size grows without bound. Using our earlier example of the 25th percentile of a distribution as an estimator for the mean would result in an inconsistent estimator because it is a biased estimator. For a consistent estimator, both the bias and the variance must approach zero as the sample size increases to infinity.

The last characteristic we will mention here is the *robustness* of the estimator, or its lack of dependence on the population of interest being normally distributed. For example, in many nonnormal populations, the sample median is a more efficient estimator of the population mean than the sample mean, and hence, depending on the population, could be more robust.

Interval Estimation of a Mean

An interval estimate for a parameter is a set of two numbers, obtained from the analysis of a data sample, within which the parameter can be stated to lie with a specified level of confidence. Thus, this is often called a **confidence interval.** If the interval is very small for a given level of confidence, then we have estimated the parameter with high accuracy. For example, we might state that the 90 percent confidence interval for a mean is 5.4 to 5.7, meaning that there is only a 10 percent chance that this interval does *not* contain the true mean. Confidence intervals of 90, 95, and 99 percent are commonly used. For a given set of data, if a higher level of confidence is desired the result will be a wider interval for the parameter.

To obtain a confidence interval for the population mean of some distribution, we can use the central limit theorem as follows. In Section 2.5 we noted that the standard error of the sample mean was $\sigma/\sqrt{n}$ and if we are dealing with "large" values of n (i.e., greater than 30), the distribution of sample means will be approximately normally distributed. Since two standard deviations of a normal distribution enclose about 95 percent of the distribution, we can thus say that we can be 95 percent confident that the interval of our sample mean plus or minus $2\sigma/\sqrt{n}$, or $\bar{x} \pm 2\sigma/\sqrt{n}$, will contain the true population mean. As a result, there is about a 2.5 percent chance the interval will be below the mean and the same chance it will be above the mean. For a 90 percent confidence interval, the corresponding figure is 1.65 standard deviations (instead of 2) and for a 99 percent interval it is 2.58. (For 95 percent it is actually 1.96 rather than 2.)

To illustrate this, 1,000 random numbers were generated from a normal distribution with a mean of 100 and standard deviation of 20 using Excel's Random Number Generation tool and placed in column A (only the first 23 numbers are shown) of the spreadsheet in Exhibit 2.18. A random sample of 30 observations was taken from column A, and the 90 percent confidence interval calculated as shown in column D. Likewise, independent random samples of 50 and 100 were also taken, and the 90 percent confidence intervals for these samples were calculated as shown in columns E and F. The actual mean of the 1,000 randomly generated numbers is 100.088. As you can see from the exhibit, while each of the three sample means based on the different sample sizes was relatively close to the true mean, the 90 percent confidence intervals get narrower as the sample size is increased from 30 to 100. Intuitively, the reason for this is because we expect to have better information about the underlying population mean when we collect more data and therefore have more confidence that our estimate based on sample data is close to the true unknown population parameter. This is reflected in the way the standard error is calculated and you can easily verify that the standard error will decrease as the sample size n increases.

Referring to Exhibit 2.18 we see that based on our random sample of 30 observations, we would be 90 percent confident that the true population mean is between 94.0 and 103.7. Alternatively, based on the random sample of 100 observations, we would be 90 percent confident that the true population mean is between 96.3 and 103.2. Notice that the width of the confidence interval decreases from 9.7 (103.7 – 94.0) to 6.9 (103.2 – 96.3) as the sample size is increased from 30 to 100.

However, there can be problems in this procedure, especially if we do not know in advance what the population standard deviation σ is. If we are working with large samples (i.e., over 30, as discussed in Section 2.5), we can simply substitute the standard de-

EXHIBIT 2.18 Calculating 95 Percent Confidence Intervals for Large Samples

	A	B	C	D	E	F	G
1	91.3			**Sample 1**	**Sample 2**	**Sample 3**	
2	144.9			***n* = 30**	***n* = 50**	***n* = 100**	
3	86.4		**Sample Mean**	98.8	101.1	99.8	
4	111.9		**Sample Std. Dev.**	16.1	17.7	21.0	
5	131.4		**Standard Error**	2.9	2.5	2.1	
6	109.4		**Lower 90% Limit**	94.0	97.0	96.3	
7	120.5		**Upper 90% Limit**	103.7	105.2	103.2	
8	88.2		**Range of Confidence Interval**	9.7	8.2	6.9	
9	94.9						
10	99.5		**Key Formulas**				
11	112.5		Cell D3	=AVERAGE(A1:A30)			
12	116.2		Cell E3	=AVERAGE(A1:A50)			
13	81.7		Cell F3	=AVERAGE(A1:A100)			
14	85.6		Cell D4	=STDEV(A1:A30)			
15	130.5		Cell E4	=STDEV(A1:A50)			
16	117.6		Cell F4	=STDEV(A1:A100)			
17	62.3		Cell D5	=D4/SQRT(30)			
18	96.9		Cell E5	=D4/SQRT(50)			
19	121.1		Cell F5	=D4/SQRT(100)			
20	153.5		Cell D6	=D3-(1.65*D5) {copy to cells E6:F6}			
21	91.8		Cell D7	=D3+(1.65*D5) {copy to cells E7:F7}			
22	114.3		Cell D8	=D7-D6 {copy to cells E8:F8}			
23	83.2						

viation of our sample of data values, s, for σ. We then proceed to make our interval estimate because we know the sampling distribution of the mean will be approximately normal. If we have only a small sample, then how we proceed depends on what the underlying population distribution is. If the population is *not* normally distributed, then we have no choice but to collect additional data until we have a large sample. However, if we are working with a normal population but only have a small sample of data, the t distribution is the appropriate distribution to calculate our interval. This information is summarized in Exhibit 2.19.

As noted earlier, the t distribution is virtually identical to the normal for sample sizes greater than 30 but is more spread out for smaller sample sizes and diverges from the normal even further as the sample size decreases. One additional piece of information needed to use the t distribution is the degrees of freedom, equal to one less than the sample size: $n - 1$. The reduction of degrees of freedom by one is due to the use of the mean of the sample data to help estimate s. When dealing with confidence intervals, it is common to specify them in terms of the area *outside* the confidence interval, called the *rejection region* and denoted by α. In this case, the level of confidence is simply $1 - \alpha$. Using this notation, the $(1 - \alpha)$ percent confidence interval using the t distribution is calculated as follows:

$$(1 - \alpha) \text{ percent confidence interval } = \bar{x} \pm t_{\alpha/2,n-1} \frac{s}{\sqrt{n}}$$

where $t_{\alpha/2,n-1}$ = the t value with an area in each of the regions above and below the confidence interval of $\alpha/2$ and $n - 1$ degrees of freedom. Note that the total area outside the confidence interval sums to α.

There are a number of ways of determining the t value to use based on α and the degrees of freedom. In the past, the most common approach was to use tables. However, spreadsheets also have the capability to calculate t values. Earlier we described how to use Excel to calculate z values, or values from the standard normal distribution using the NORMSINV function. Excel has a similar function called TINV that can be used to calculate t values. The syntax of this function is

=TINV(1–α, degrees of freedom)

Note that the TINV function returns values based on the assumption that the rejection region consists of both upper and lower regions. Therefore, the TINV function returns the

EXHIBIT 2.19
Distribution and Standard Deviation Guidelines for Interval Estimation and Hypothesis Testing

Population Distribution	Population Standard Deviation	
	Known	**Unknown**
Normal	Use z, σ	Use t, s
Unknown	Increase n	Increase n

(*a*) Small sample

Population Distribution	Population Standard Deviation	
	Known	**Unknown**
Normal	Use z, σ	Use z, s
Unknown	Use z, σ	Use z, s

(*b*) Large sample

t value corresponding to $\alpha/2$ in the tails of the distribution. To illustrate, to calculate a 95 percent confidence interval based on a sample of 20 observations we would enter =TINV(0.05,19) in Excel. Recall that the degrees of freedom are calculated as $n - 1$.

An example of using the t distribution to obtain confidence intervals for the mean is given in Exhibit 2.20, using the same data that were used in Exhibit 2.18. The major difference is that in Exhibit 2.20, smaller sample sizes were used. In this case we know that the data came from a normal distribution, so the use of the t distribution is justified. Had we not known this, we could have used the chi-square goodness of fit test as described earlier to verify this assumption. In comparing the confidence intervals developed in Exhibit 2.20, we observe that as in Exhibit 2.18, the confidence intervals created on the basis of smaller sample sizes are considerably wider than when larger sample sizes are employed. We also point out that in all cases the confidence intervals contained the true population mean.

Determining the Size of the Sample for a Normal Distribution

As noted earlier, once the data have been collected, the determination of the confidence interval is relatively fixed. A certain confidence range will correspond to a fixed interval; if greater confidence is desired, the interval must increase. Thus, it would be useful to have a way of setting the sample size beforehand in order to obtain a maximum *half-interval error, E.* Again appealing to the central limit theorem, suppose we want the half-interval on either side of the sample mean to be E with 95 percent confidence (i.e., about two standard deviations). Then using the previous confidence interval formula $\bar{x} \pm 2\sigma/\sqrt{n}$, we see that we want $E = 2\sigma/\sqrt{n}$, or $n = 4\sigma^2/E^2$, or even greater for more confidence or a smaller interval.

EXHIBIT 2.20 Using the t Distribution to Calculate Confidence Intervals for Small Samples

	C	D	E	F
1		**Sample 1**	**Sample 2**	
2		***n* = 5**	***n* = 15**	
3	**Sample Mean**	91.8	100.9	
4	**Sample Std. Dev.**	20.1	20.3	
5	**Standard Error**	9.0	5.2	
6	**Lower 95% Limt**	81.5	94.8	
7	**Upper 95% Limit**	102.0	107.1	
8	**Range of Confidence Interval**	20.4	12.3	
9				
10	**Key Formulas**			
11	Cell D3	=AVERAGE(A1:A5)		
12	Cell E3	=AVERAGE(A1:A15)		
13	Cell D4	=STDEV(A1:A5)		
14	Cell E4	=STDEV(A1:A15)		
15	Cell D5	=D4/SQRT(5)		
16	Cell E5	=E4/SQRT(15)		
17	Cell D6	=D3-(TINV(.05,4)*D5)		
18	Cell E6	=E3-(TINV(.05,14)*E5)		
19	Cell D7	=D3+(TINV(.05,4)*D5)		
20	Cell E7	=E3+(TINV(.05,14)*E5)		
21	Cell D8	=D7-D6 {copy to cell E8}		

The problem at this point is that we probably again do not know σ, and since we have not yet collected the data either, we don't even know *s* to substitute for σ. The common ways of estimating *s* in this situation are to use previous knowledge/samples/experience, run a pilot study and collect a little data, or estimate the range of the data to be collected and divide this range by 4.

As an example, assume that it is known that a particular population is normally distributed and past experience suggests that the standard deviation for this population is 20. The spreadsheet shown in Exhibit 2.21 calculates the required sample size for alternative half-interval error values. According to the results shown in the exhibit, in order to be 95 percent confident that the true population mean is within ± 1 unit of the sample mean, a sample size of 1,600 would be required. Likewise, if the decision maker wanted to be 95 percent confident that the population mean was within ± 4 units of the sample mean, a sample of only 100 would be required. Note that this procedure can be modified to develop half-interval errors for other levels of confidence and *t* values can be used when a large sample is taken and σ is unknown or when a small sample is taken and the underlying population distribution is known to be normal. Finally, notice that the formula entered to calculate the required sample size utilized Excel's ROUNDUP function to ensure *at least* a 95 percent confidence level. Had the numbers been rounded down or truncated to the nearest integer, the confidence level could have sometimes been below the target 95 percent level. The first argument in the ROUNDUP function is simply the value you want to round, and the second argument is the number of places you desire to the right of the decimal point.

Interval Estimation and Determination of Sample Size for a Proportion

In many situations, we are interested in the proportion *p* of observations that take one of two possible outcomes, such as the proportion of heads in 100 coin tosses, the proportion

EXHIBIT 2.21 Calculating Required Sample Size

	A	B	C
1	**Standard Deviation**	20	
2			
3	**Half-Interval**		
4	**Error, *E***	*n*	**Formula in Column B**
5	1.0	1600	=ROUNDUP((4*(B$1^2))/(A5^2),0) {copy to B6:B23}
6	1.5	712	
7	2.0	400	
8	2.5	256	
9	3.0	178	
10	3.5	131	
11	4.0	100	
12	4.5	80	
13	5.0	64	
14	5.5	53	
15	6.0	45	
16	6.5	38	
17	7.0	33	
18	7.5	29	
19	8.0	25	
20	8.5	23	
21	9.0	20	
22	9.5	18	
23	10.0	16	

of females in 400 births, or the proportion of time a queuing system is empty. We can also include here situations involving multiple outcomes and/or multiple events, such as the proportion of time 2 or 3 customers are waiting for service. An unbiased estimator of p is the sample proportion $\bar{p} = x/n$ where x is number of outcomes involving the event of interest (heads, or females) and n is the sample size. Since the sampling distribution of p follows the binomial distribution (see Section 2.4), the standard error of the proportion is $\sqrt{\bar{p}(1-\bar{p})/n}$. If np and $n\ (1 - p)$ are both at least 5, the sampling distribution of the proportion can be approximated by the normal distribution in the usual way. Thus, once again, if we want an approximately 95 percent confidence interval, we use $\bar{p}$ as an estimate for p and construct the interval $\bar{p} \pm 2\sqrt{\bar{p}(1-\bar{p})/n}$. For other confidence levels, we again use the corresponding values such as 1.65 for 90 percent or 2.58 for 99 percent.

To determine the sample size for the proportion to attain a specified confidence interval and a half-interval error E, we also proceed in the same fashion as before. However, since we know the formula for the standard error, we use it instead of trying to estimate it from the data, resulting in: $n = 4\,\bar{p}(1 - \bar{p})/E^2$.

An example of determining the sample size and calculating a confidence interval follows. Assume a political organization recently took a random sample of 1,200 registered voters in the United States and 335 of the respondents indicated that they considered themselves Independents, 303 as Republicans, 318 as Democrats, and the remaining 244 as "Other." As shown in Exhibit 2.22, we can be 95 percent confident that the actual proportion of Independent voters is between 25.33 to 30.51 percent, resulting in a half-interval error of (30.51 – 25.33)/2 = 2.59 percent. Also shown in Exhibit 2.22 is the number of voters that would need to be sampled to obtain a specified half-interval error, E. For example, to be 95 percent confident that the true percentage of Independent voters was within ± 2 percent of the sample proportion, a sample size of 2,013 voters would have been required.

EXHIBIT 2.22 Confidence Intervals and Required Sample Sizes for a Proportion

	A	B	C	D	E	F	G	H
1	Number of independents, *x*	335						
2	Sample size, *n*	1200						
3								
4	Sample Proportion, *p* bar	27.92%	=B1/B2					
5	Standard error	0.01295	=SQRT((B4*(1-B4))/B2)					
6	Lower 95% limit	25.33%	=B4-(2*B5)					
7	Upper 95% limit	30.51%	=B4+(2*B5)					
8								
9	**Half-Interval**							
10	**Error, *E***	*n*						
11	1.00%	8050	=ROUNDUP((4*B$4*(1-B$4))/(A11^2),0) {copy to B12:B20}					
12	2.00%	2013						
13	3.00%	895						
14	4.00%	504						
15	5.00%	322						
16	6.00%	224						
17	7.00%	165						
18	8.00%	126						
19	9.00%	100						
20	10.00%	81						

2.8 Hypothesis Testing

In some ways, **hypothesis testing** can be considered to be the opposite of estimating a confidence interval. With a confidence interval, we state that we have a certain level of confidence that the population parameter is contained within our derived interval. With hypothesis testing, we check to see if a variable's value as derived from our modeling study falls within a confidence interval representing our hypothesis. If it does not, we state that we are confident, again at a certain level, that the hypothesis is false. However, hypothesis testing is somewhat more powerful than interval estimation. For example, it is common in hypothesis testing to consider one-sided tests, such as testing whether something is greater than or equal to some specified value, instead of just equal to some value. Also, we can test whether two, three, or more items are all equal to each other.

The procedure for hypothesis testing is generally as follows. We first develop the hypothesis we wish to test, called the **null hypothesis,** commonly designated H_0, and the alternative that we suspect is actually true, called the **alternative hypothesis,** designated H_a. Then we select a level of confidence, such as 95 percent, that we wish to have if we reject the null hypothesis. However, in the hypothesis testing process it is common to describe the *complement* of the level of confidence instead, 1.0 – confidence level, which is called the *significance level* of the test. Using the null hypothesis and significance level, we determine a *rejection region* about the value of the null hypothesis. If the rejection region encloses the sample test statistic derived from our modeling study, it constitutes evidence that allows us to reject the null hypothesis as *false* and confirm our suspicions that the alternative hypothesis is correct. Next, we elaborate on this procedure and apply it in a number of example situations.

A null hypothesis can be framed in a variety of ways: that a variable (e.g., a mean) has a certain value, is equal to another variable or sets of variables, and so on. The testing process assumes that the null hypothesis is *true*—for example, a mean equals a particular value—and therefore, a rejection region can be constructed about the variable of interest. (If the null hypothesis had been stated in the form that the mean was *not* equal to a particular value, we would not know how to construct a rejection region because we would not know where to locate it.) If the test statistic falls within the rejection region, we can conclude with the chosen level of confidence that the null hypothesis is false. However, if the statistic does not fall in the rejection region, we *cannot conclude* that the null hypothesis is true and the alternative hypothesis is incorrect; all we can say is that we were not able to reject the null hypothesis based on the results of the study or the sample data.

Thus, the construction of the null hypothesis usually employs a reverse mentality, of sorts. If we wish to show that something is not equal to a certain value, we set up the null hypothesis to say that it is in fact equal to the value. We then hope that we can confidently *reject* that statement, thereby proving with a certain confidence level that it is indeed larger or smaller than that value. If we want to have high confidence in our statement, such as 99 percent, then we would use a 1 percent significance level. This says, in effect, that there is only one chance in a hundred that the null hypothesis would be rejected when it was in fact true. If we wish to be even more certain than that, we could use a 0.1 percent level of significance and state that there is only one chance in a thousand that we might be in error. However, with such small significance levels, we may find that we are unable to reject the null hypothesis based on the results of our analysis and thus have to back off from such confident-sounding results. As noted earlier, 95 percent confidence levels, corresponding to 5 percent significance levels, are most common. In some situations, such as early or exploratory tests, we may even be willing to accept a 10 percent significance level, or one chance in ten of incorrectly rejecting the null hypothesis.

For example, when doing X-ray examinations for cancer *using the null hypothesis that the patient is healthy,* we would use a relatively large significance level, just in case there might be something wrong. This is because in this case it is better to incorrectly reject the null hypothesis that the patient is healthy when in fact he/she *is* healthy than it is to incorrectly not reject the null hypothesis that the patient is healthy when in fact he/she is *not.* Getting a false positive on a cancer test is no doubt disturbing, but not detecting cancer when it is present can be fatal. That is, we would not want to state that we were 99 percent confident (1 percent significance level) that patient A was *not* healthy (i.e., had cancer) but be *unable* to state that patient B was not healthy (had cancer) when at 95 percent we would have also concluded that patient B had cancer.

Notice that all hypothesis tests involve two potential errors. The first is the issue we have been discussing earlier, the declaration that the null hypothesis is false when it is in fact true. This is known as a **type I error,** and its probability is generally denoted by the symbol α, equivalent to the significance level of the test. This is the error that we wish to control at a specified level of confidence using the corresponding significance level in our hypothesis test, such as one chance in a hundred that we are accidentally wrong in rejecting the null hypothesis. However, considering the X-ray situation, we see that there is also another mistake we need to be careful about. That is, concluding that the hypothesis may be correct (more formally, that we cannot reject the hypothesis) when in fact it *is* false but our data weren't powerful enough to make that conclusion at the level of significance we specified. This is called a **type II error,** and is generally denoted by the symbol β. (The value $1 - \beta$ is called the *power of the test* and represents the probability of correctly rejecting the null hypothesis when it is indeed false.) Although more difficult to control, type II error tends to decrease as the significance level, α, and thus the probability of a type I error, increases. The only way to reduce the likelihood of both errors is to take a larger data sample, which costs more.

The rejection region for a hypothesis test tends to fall in the tails of the distribution specified by the null hypothesis and the significance level. If the null hypothesis is of the form that some variable equals a particular value, as in Exhibit 2.23*a*, then we are looking at a two-sided (or two-tailed) rejection region. If we believe that a variable is greater than some particular value, then we would be using a one-sided (or one-tailed) rejection region (same thing if less than). An example is shown in Exhibit 2.23*b* for the case where we set the null hypothesis as "the mean is less than or equal to" some stated value because we believe it is actually greater than the stated value. The opposite case, i.e., the mean is greater than or equal to, would have the rejection region, also of size α, in the left tail. Generally speaking, in one-tailed tests the entire rejection region of size α is concentrated in one tail, whereas in two-tailed tests the rejection region is $\alpha/2$ in each tail.

With the advent of computers and software for automatically conducting hypothesis tests, we can set up a test without arbitrarily specifying the significance level and let the software return the significance level, called the *p value,* at which the null hypothesis would be rejected. (*Note:* This p should not be confused with the binomial proportion p.) For instance, a p value of 0.02 indicates that there is only a 2 percent chance of getting a test statistic as large as the one calculated if the null hypothesis is actually true. Based on this p value we have to decide whether this constitutes sufficient evidence to reject the null hypothesis. Generally, a p value of 0.02 is relatively strong evidence for rejecting the null hypothesis. Of course, it would not satisfy a confidence level requirement of 99 percent, but it would 95 percent. Alternatively, a p value of 0.30 would not generally be considered strong enough evidence to reject a null hypothesis because this implies there is a 30 percent likelihood of getting a test statistic this large or larger just by chance when the null hypothesis is true. We illustrate this process in some of the examples that follow.

EXHIBIT 2.23 One- and Two-Sided Rejection Regions

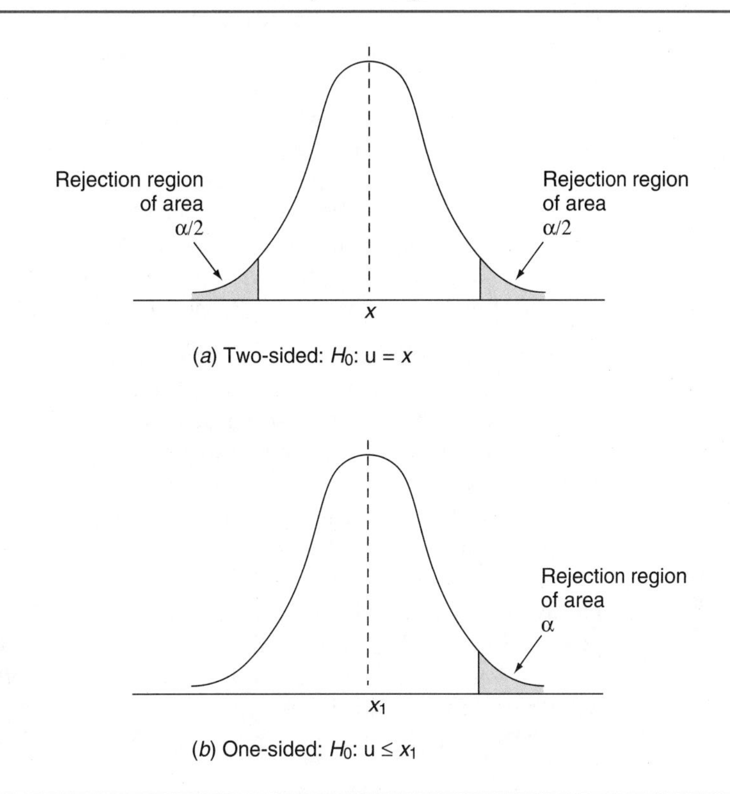

Hypothesis Tests for Means

We now illustrate the process of hypothesis testing. To facilitate our discussion, it is helpful to formally divide the process into five steps:

Step 1 Define the null (H_0) and alternative H_a hypotheses. Earlier we noted that the null hypothesis is the hypothesis we wish to test. The alternative hypothesis states what we suspect may be the true situation. For example, a null hypothesis might state that the proportion of Independent votes is equal to 25 percent. Three possible alternative hypotheses associated with the null hypothesis are (a) that the number of Independent voters is not equal to 25 percent, (b) that the number of Independent voters is greater than 25 percent, and (c) that the number of Independent voters is less than 25 percent. In the first case, we have a two-tailed test and evidence of either more or less than 25 percent Independent voters would be used to reject the null hypothesis. In the second case, we have a one-tailed test where extreme values in the lower tail would be used to reject the null hypothesis. Similarly, in the last case we also have a one-tailed test but we would be looking for extreme values in the upper tail to reject the null hypothesis.

Step 2 Determine the level of significance, α. The level of significance is used to determine the rejection region for the hypothesis test. As α increases, the rejection region increases, and the likelihood of rejecting the null hypothesis increases. However, as the probability of rejecting the null hypothesis increases, so, too, does the probability of rejecting the null hypothesis when in fact it is true (i.e., making a type I error).

Step 3 Define the rejection region. If our calculated test statistic falls in the rejection region, we use this as evidence to reject the null hypothesis. In two-tailed hypothesis tests, the area in each tail is $\alpha/2$. In one-tailed tests, the area in the tail is α.

Step 4 Collect data and calculate the test statistic. The appropriate test statistic to use depends on the hypothesis being tested. Common test statistics require calculations that will include the sample mean and the sample proportion and their standard errors. Earlier in the chapter, the test statistic used in the chi-square goodness of fit test was based on the differences between observed and expected frequencies to test the hypothesis that the collected data came from the specified probability distribution.

Step 5 Make a decision. If the test statistic falls in the rejection region, we reject the null hypothesis. Otherwise, we conclude that the null hypothesis cannot be rejected.

Exhibit 2.24 illustrates the five-step process for a two-tailed hypothesis test using a large (30 or greater) data sample that a population mean equals a specified value. The data used for this example are the same data discussed earlier in conjunction with Exhibit 2.18. In the first step, we define the null hypothesis as "the true population mean is equal to 100," with the alternative hypothesis being that "the population mean is *not* equal to 100." In step 2 we decide to use a 5 percent level of significance.

In the third and fourth steps, the rejection region is specified and the test statistic is calculated. Since our sample size is larger than 30, we can use the normal distribution for our hypothesis test. Also, note that we have a two-tailed test because the alternative hypothesis was stated as "not equal to." Therefore, based on a significance level of 5 percent, we want to find the z values such that 2.5 percent of the area is in the lower tail and 2.5 percent of the area is in the upper tail. This can be easily accomplished using Excel's NORMSINV, function as shown in Exhibit 2.24. According to the exhibit, the null hypoth-

EXHIBIT 2.24 Two-Sided Hypothesis Test for Large ($n \geq 30$) Sample

	A	B
1		**Sample 1**
2		**$n = 30$**
3	**1. H_0: population mean = 100**	100.0
4	**2. Level of Significance, alpha**	5.0%
5	**3. Rejection Region:**	
6	**Lower Tail**	-1.96
7	**Upper Tail**	1.96
8	**4. Collect Data and Calculate Test Statistic**	
9	**Sample Mean**	98.8
10	**Sample Standard Deviation**	16.1
11	**Standard Error**	2.9
12	**Test Statistic**	-0.4
13	**5. Decision**	**Cannot Reject**
14		
15		
16	**Key Formulas**	
17	Cell B6: =NORMSINV(B4/2)	
18	Cell B7: =NORMSINV(1-(B4/2))	
19	Cell B11: =B10/SQRT(30)	
20	Cell B12: =(B9-B3)/B11	
21	Cell B13: =IF(OR(B12<B6,B12>B7),"Reject","Cannot Reject")	

esis should be rejected whenever our calculated test statistic is less than –1.96 or greater than 1.96. The test statistic that we will use in this case is z, which is calculated as

$$z = \frac{\bar{x} - \mu_0}{s/\sqrt{n}}$$

where $\bar{x}$ = the sample mean
μ_0 = the hypothesized value in the null hypothesis
s = the sample standard deviation
n = the sample size

In the fifth and final step, Exhibit 2.24 shows that the calculated test statistic value of z is –0.4. Since the test statistic does not fall in either the lower or upper tail, we cannot reject the null hypothesis, as is indicated in Cell B13. You might find it of particular interest to examine the way Excel's OR function and IF function were combined to automatically determine whether the null hypothesis should be rejected or not rejected in Cell B13. Also, note that different hypotheses can easily be tested using the spreadsheet by simply changing the value entered for μ_0 in cell B3 and/or the level of significance entered in cell B4.

Exhibit 2.25 illustrates a one-sided hypothesis test when we suspect that the population mean is less than 100 under the same circumstances. (Thus, our null hypothesis is that the mean is 100 or more.) In particular, notice how the formulas were changed in cells B6 and B13.

If we know the population standard deviation σ, we use it directly; otherwise, we estimate it with the sample standard deviation s. As discussed in the previous section concerning the estimation of confidence intervals and summarized in Exhibit 2.19, if the sample size is large, we can employ the normal distribution to test our hypotheses, even

EXHIBIT 2.25 One-Sided Hypothesis Test for Large ($n \geq 30$) Sample

	A	B
1		**Sample 1**
2		***n* = 30**
3	**1. H_0: population mean ≥ 100**	100.0
4	**2. Level of Significance, alpha**	5.0%
5	**3. Rejection Region:**	
6	**Lower Tail**	-1.64
7		
8	**4. Collect Data and Calculate Test Statistic**	
9	**Sample Mean**	98.8
10	**Sample Standard Deviation**	16.1
11	**Standard Error**	2.9
12	**Test Statistic**	-0.4
13	**5. Decision**	**Cannot Reject**
14		
15		
16	**Key Formulas**	
17	Cell B6: =NORMSINV(B4)	
18	Cell B11: =B10/SQRT(30)	
19	Cell B12: =(B9-B3)/B11	
20	Cell B13: =IF(B12<B6,"Reject","Cannot Reject")	

if we don't know the population standard deviation and have to estimate it from the sample data.

Moreover, if the sample size is small but we believe the population is normally distributed and we know the population standard deviation, we can again use the normal distribution for our hypothesis tests, as shown in Exhibit 2.19. But if we don't know the population standard deviation under these circumstances and have to use s as an estimate of σ, we must use the t distribution instead. This situation is illustrated in Exhibit 2.26. In this case, our test statistic is calculated as

$$t = \frac{\bar{x} - \mu_0}{s/\sqrt{n}}$$

Perhaps the most significant change with using t as our test statistic is that the rejection region is now calculated on the basis of the t distribution with $n - 1$ degrees of freedom, as opposed to using the normal distribution. As was discussed earlier, Excel's TINV function can be used in a similar fashion as the NORMSINV function. However, keep in mind that the TINV function inherently assumes that a two-sided test is being used and divides α in half; therefore, it is not necessary to divide the significance level in half in Excel as is done with the NORMSINV function. This is illustrated in Exhibit 2.26.

Finally, as illustrated in Exhibit 2.19, if the population distribution cannot be assumed to be normal and we are working with a small sample size, we simply have to increase the size of our sample.

For hypothesis tests involving the population proportion p, we proceed as above but do not have to worry about knowing the population standard deviation, since it can be computed from the hypothesized mean, as explained earlier. However, in order to use the normal distribution as an acceptable approximation to the sampling distribution of $\bar{p}$, both

EXHIBIT 2.26 Two-Sided Hypothesis Test for Small (n < 30) Sample

	A	B
1		**Sample 1**
2		***n* = 5**
3	**1. H_0: population mean = 100**	100.0
4	**2. Level of Significance, alpha**	5.0%
5	**3. Rejection Region:**	
6	**Lower Tail**	-2.78
7	**Upper Tail**	2.78
8	**4. Collect Data and Calculate Test Statistic**	
9	**Sample Mean**	91.8
10	**Sample Standard Deviation**	20.1
11	**Standard Error**	9.0
12	**Test Statistic**	-0.9
13	**5. Decision**	**Cannot Reject**
14		
15		
16	**Key Formulas**	
17	Cell B6: =-TINV(B4,4)	
18	Cell B7: =TINV(B4,4)	
19	Cell B11: =B10/SQRT(5)	
20	Cell B12: =(B9-B3)/B11	
21	Cell B13: =IF(OR(B12<B6,B12>B7),"Reject","Cannot Reject")	

np and $n(1 - p)$ must exceed 5. In cases where the normal distribution approximation can be used, the test statistic z is calculated as

$$z = \frac{\bar{p} - p_0}{\sqrt{\frac{p_0(1 - p_0)}{n}}}$$

where $\bar{p}$ is the sample proportion and p_0 is the hypothesized value of the population proportion.

An example of a hypothesis test regarding whether the true proportion of Independent voters is 30 percent is given in Exhibit 2.27 based on the normal approximation. Based on the sample of 1,200 voters, the hypothesis that 30 percent of registered voters are independent cannot be rejected.

Comparing Multiple Means—Analysis of Variance (ANOVA)

We may sometimes find ourselves in the position where we want to test the possibility that the underlying population means of two or more sets of data are equal. If this is formulated as the null hypothesis and our analysis of the data results in rejecting this hypothesis, the interpretation is simply that not *all* the means are equal, or at least one of the means is significantly different from the other(s), although we don't know which one without doing further tests. This situation can be addressed if the following assumptions are appropriate:

1. The k populations represented by the data are normally distributed. If the sample sizes from each of the k populations are equal, this assumption is not as critical.
2. The variance within each of the k populations is the same.
3. The data observations are independent.

EXHIBIT 2.27 Two-Sided Hypothesis Test of Population Proportion Using Normal Approximation

	A	B
1	**1. H_0: population proportion, p_0**	30.0%
2	**2. Level of Significance, alpha**	5.0%
3	**3. Rejection Region:**	
4	**Lower Tail**	-1.96
5	**Upper Tail**	1.96
6	**4. Collect Data and Calculate and Calculate Test Statistic**	
7	**Number of Independents, *x***	335
8	**Sample Size, *n***	1200
9	**Sample Proportion, *p* bar**	27.92%
10	**Standard Error**	0.0132
11	**Test Statistic**	-1.57
12	**5. Decision**	**Cannot Reject**
13		
14		
15	**Key Formulas**	
16	Cell B4: =NORMSINV(B2/2)	
17	Cell B5: =NORMSINV(1-(B2/2))	
18	Cell B9: =B7/B8	
19	Cell B10: =SQRT((B1*(1-B1))/B8)	
20	Cell B11: =(B9-B1)/B10	
21	Cell B12: =IF(OR(B11<B4,B11>B5),"Reject","Cannot Reject")	

Given these assumptions, we can conduct a procedure termed an **analysis of variance (ANOVA)** by calculating the ratio of the average variance *between* the k sets of data to the average variance *within* each of the k sets of data. The logic here is that if the groups are all from the same population, there will not be much difference between their variances no matter how they are calculated. But if they truly are different populations, then the variance between these different groups will be quite large compared to the variance within the groups. This ratio then becomes the test statistic, which is distributed according to the F distribution, and is represented this way:

$$F = \frac{\text{average variance between the sample means}}{\text{average variance within the samples}}$$

The calculated F ratio has to be quite large to justify rejecting the null hypothesis, again with a specified level of confidence.

To use the F distribution requires two parameters, the degrees of freedom in the numerator and the degrees of freedom in the denominator. The degrees of freedom in the numerator are $k - 1$, where k is the number of data sets, or means being tested. The degrees of freedom in the denominator are $N - k$, where N is the total number of observations across all the data sets.

As an example of this procedure, consider the mid-term exam grades for three sections of a core MBA class shown in Exhibit 2.28. As it turns out, while these courses use the same syllabus and have the same exams, they are taught by three different professors, with each professor grading his or her own exams. The instructors are interested in determining if there are significant differences across the three sections, perhaps due to the way the exams were graded, so that an appropriate adjustment can be made if needed.

ANOVA can be used to test the null hypothesis that there is no difference across the three sections versus the alternative hypothesis that at least one of the sections is different from the others. The following process can be used to have Excel perform an ANOVA on this data:

1. From the menu at the top of the screen, select the Tools option.
2. Next, select Data Analysis and then the Anova: Single Factor option. In the scenario presented here there is only one factor that differentiates the data, namely, the class section. Although beyond the scope of this book, we do note that it is possible to use ANOVA when there are multiple factors. For example, if we added the student's gender, we could test to see whether the student's gender, the section, or the combination of these two factors had an impact on student mid-term grades. In other cases, we may need to control the levels of some factors even though these factors may not be of particular interest in the present study. The interested reader is referred to the Bibliography for more advanced books on experimental design and the use of randomized block designs.
3. After selecting Anova: Single Factor, the Anova: Single Factor dialog box is displayed. In the Input Range field, enter the range in the spreadsheet that contains the data. The data shown in Exhibit 2.28 is entered in the range B1:D31, which includes the labels entered as column headings. Next, the radio button Columns is selected to tell Excel that the data are grouped in columns. After this, we select the check box Labels in First Row to let Excel know that the first row in the data set contains the labels used to identify the data. Using this feature is useful because these labels will be used in the output report Excel generates, as will be seen shortly. Finally, we specify a value for alpha (the level of significance we desire), tell Excel where to put the output report, and click on the OK button. Exhibit 2.29 shows the completed Anova: Single Factor dialog box for this example and Exhibit 2.30 displays the ANOVA output report Excel generates.

EXHIBIT 2.28 Mid-term Exam Grades Across Three Sections of a Core MBA Class

	A	B	C	D
1		Section A	Section B	Section C
2		86	87	92
3		82	84	88
4		89	98	94
5		94	82	91
6		94	92	96
7		97	86	95
8		77	86	93
9		87	80	92
10		93	84	90
11		83	85	88
12		85	89	92
13		80	83	91
14		79	87	93
15		83	83	87
16		84	84	88
17		77	85	95
18		85	89	87
19		86	95	94
20		89	93	91
21		86	95	98
22		86	77	90
23		86	80	90
24		95	86	92
25		88	80	90
26		87	88	85
27		85	81	91
28		98	89	81
29		92	100	94
30		100	80	83
31		85	79	91
32	**Average**	**87.3**	**86.2**	**90.7**
33	**Maximum**	**100**	**100**	**98**
34	**Minimum**	**77**	**77**	**81**

As shown in Exhibit 2.30, Excel's ANOVA report is divided into two sections. The top section provides summary data information such as the number of observations in each category, the sum for each category, and the average and variance for each category. Note that the averages reported for the three sections shown in cells I5:I7 in Exhibit 2.30 are the same as the averages calculated in cells B32:D32 in Exhibit 2.28. Also notice that the actual labels entered in Exhibit 2.28 were used in Exhibit 2.30 (cells F5:F7) because we included these labels in the Input Range.

The bottom section contains the actual ANOVA results. In column H the degrees of freedom are calculated. In this case we have three groups of data (corresponding to the three sections) so $k = 3$ and there are a total of 90 data points (30 students in each section $\times$ 3 sections). Therefore the degrees of freedom are calculated as $k - 1 = 2$ and $N - k = 87$, as shown in cells H12 and H13, respectively.

In cell J12, the test statistic based on the F distribution is calculated from the variances. F ratios are found by taking the sum of the mean squares (MS) between the groups divided by the sum of the mean squares within: 166.67778/27.48390805 = 6.06456. This test statistic is then compared to the critical F value shown in cell L12, which is based on the degrees of freedom discussed earlier and the level of significance specified. In Exhibit 2.30 we observe that the test statistic of 6.06 is larger than the *F critical* value of 3.10, providing evidence that the null hypothesis should be rejected. We thus conclude that the grades in one or more sections differ significantly from some others. In fact, according to the *p* value shown in cell K12, there is less than a 1 percent chance of getting a value of *F* as large as 6.06456 when the null hypothesis that there is no difference among the sections is true.

In this scenario, this analysis suggests that the professors ought to further investigate this situation and determine what caused the differences. First, the professors might conduct three *t* tests (again, at the 0.05 level of significance) to see which sections differed

EXHIBIT 2.29 Excel's Anova: Single Factor Dialog Box

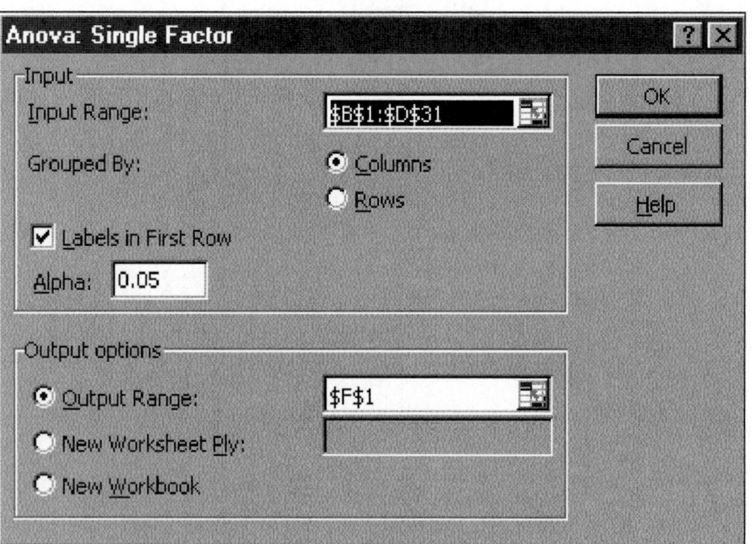

EXHIBIT 2.30 Excel's ANOVA Output Report

	F	G	H	I	J	K	L
1	Anova: Single Factor						
2							
3	SUMMARY						
4	*Groups*	*Count*	*Sum*	*Average*	*Variance*		
5	Section A	30	2618	87.26667	35.16782		
6	Section B	30	2587	86.23333	33.28851		
7	Section C	30	2722	90.73333	13.99540		
8							
9							
10	ANOVA						
11	*Source of Variation*	*SS*	*df*	*MS*	*F*	*P value*	*F crit*
12	Between Groups	333.356	2	166.67778	6.06456	0.00342	3.10129
13	Within Groups	2391.100	87	27.48390805			
14							
15	Total	2724.456	89				

from each other.[1] Furthermore, the instructors might each grade a small number of the other professors' already-graded exams to see if there were systematic differences in the way the exams were graded. Also, the instructors could summarize the students' grades, question by question, in their section to see how different topics were emphasized across the sections. Of course, it is possible that after conducting this follow-up analysis, the instructors are still unable to identify any particular reason for the differences across the sections, in which case no adjustment would be required.

As you can see, ANOVA provides a systematic way for comparing two or more sets of data. It is frequently used to analyze the results of simulation studies (Chapter 7). In this context, each level of the controlled factor included in the study corresponds to a particular data set. Furthermore, ANOVA can easily handle situations where more than one controlled factor is included in the study, and in these cases can highlight the existence of interaction effects between multiple factors. When interaction effects are present, the interpretation of each variable depends on the level of one or more other variables. Because the presence of interaction effects can complicate the interpretation of a modeling study, experienced modelers tend to limit the number of factors included in any given study.

2.9 Detailed Modeling Example

In this section we return to the Hospital.com example in the chapter opener to illustrate data collection and analysis in greater detail and in the broader modeling context.

Step 1: Opportunity/Problem Recognition

As you may recall from the beginning of the chapter, the primary problem facing Hospital.com is its apparent inability to manage its inventory levels effectively. This is exemplified by the fact that the company often has excessive inventory levels of some products while at the same time experiencing shortages for other products. An influence diagram for Hospital.com is shown in Exhibit 2.31.

EXHIBIT 2.31 Influence Diagram for Hospital.com

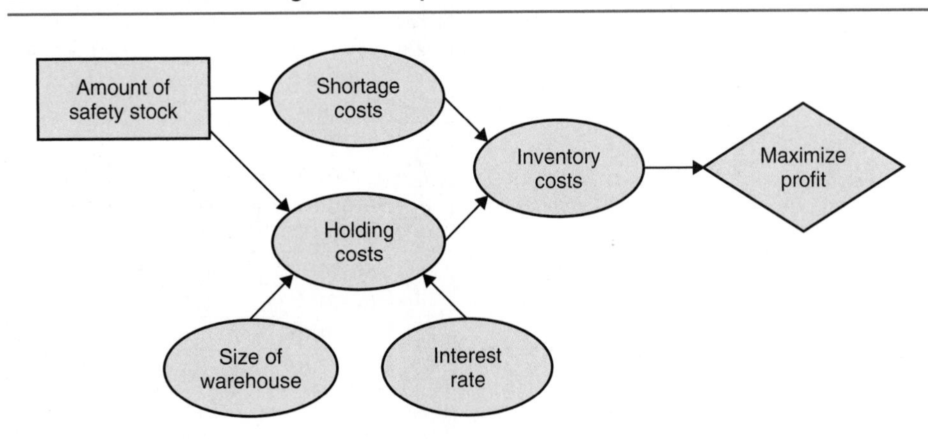

Step 2: Model Formulation

The main tool for evaluating the appropriateness of a 25 percent *safety stock,* as it is called, is the distribution of weekly sales. If the weekly demand rarely exceeds this level of safety stock, calculated earlier as 30 stethoscopes, then we should rarely run out of inventory. In fact, we may be carrying *too much* inventory and should reduce it to save the holding cost of this safety stock. On the other hand, if we frequently exceed 30 stethoscopes, the 25 percent may be far too little, and some other action may be warranted. We might simply raise the safety stock to cover, say, 95 percent of all weekly demands. Or we might try something different, such as ordering more often, or offering a discount if the customer has to wait for new stock to arrive. The first step, however, is to examine the statistics of our weekly demand and plot the frequency distribution for further analysis.

Step 3: Data Collection

Hospital.com has collected weekly sales data on a representative product (see Exhibit 2.1 at the beginning of the chapter) to help analyze its current inventory management policy of setting a target weekly inventory level equal to 25 percent above the average weekly demand. Having collected this data, Excel's Descriptive Statistics and Histogram tools are particularly useful for performing a preliminary analysis of the data. Both of these tools are part of Excel's Data Analysis Toolpak add-in and can be accessed by selecting Tools/Data Analysis . . . If the Data Analysis menu item is not listed after selecting Tools, you can add it by selecting Tools/Add Ins . . . and then selecting the Analysis ToolPak checkbox. If the Analysis ToolPak is not displayed, search Excel's help using the keyword "Analysis ToolPak" to get detailed instructions on how to install and add the Analysis ToolPak.

As shown in Exhibit 2.32, the Descriptive Statistics tool provides a large number of summary statistics for a data set, including all the measures of central tendency and variance that were discussed earlier in this chapter. Other information provided by the Descriptive Statistics tool includes the minimum and maximum values in the set of data, the sum of all the observations, the total number of observations, as well as a number of sta-

EXHIBIT 2.32 Summary Statistics Provided by Excel's Descriptive Statistics Tool

	D	S
1	*Column 1*	
2		
3	Mean	24.17308
4	Standard error	0.66858
5	Median	23.5
6	Mode	23
7	Standard deviation	4.8212
8	Sample variance	23.24397
9	Kurtosis	-0.046846
10	Skewness	0.169088
11	Range	23
12	Minimum	12
13	Maximum	35
14	Sum	1257
15	Count	52

tistics that we have not discussed yet. Although we could have calculated all of these summary statistics using standard Excel functions, it is much more efficient to use this tool. To use Excel's Descriptive Statistics tool to summarize the data shown in Exhibit 2.1, select Tools/Data Analysis/Descriptive Statistics and complete the Descriptive Statistics dialog box as shown in Exhibit 2.33. Be sure to click the Summary statistics check box in the dialog box.

Earlier in the chapter we discussed how Excel's Chart Wizard could be used to develop bar charts and histograms. In situations where we have a large number of data values that have not been placed in categories, we can use Excel's Histogram tool to automatically generate the frequency distribution and create the histogram chart. To use the Histogram tool, we must first define the intervals for each category or group. In general, histograms should have between 5 to 20 groups. Small data sets like the one we are dealing with in Exhibit 2.1 would have fewer groups than larger data sets that can have tens of thousands and even hundreds of thousands of observations. Referring to Exhibit 2.32 we see that the range for the stethoscope weekly sales data is 23. Therefore, it is easy to divide the observations into six categories each, with a width of 4 as follows: 12 to 15, 16 to 19, 20 to 23, 24 to 27, 28 to 31, and 32 to 35.

The upper boundary for each interval or group is entered in a single column in Excel and is called the Bin Range. The highest number in each interval is entered in this column, and the values must be entered in ascending order. In Exhibit 2.34, the Bin Range was entered in cells D2:D7. According to this Bin Range, the number of observations that are less-than-or-equal to 15 is determined as the frequency for the first group or bin. Likewise, the second bin is defined as the number of observations that are greater than 15 but less-than-or-equal to 19 and so on. A frequency table is created when the Histogram tool is used and is displayed in cells F1:G8 in Exhibit 2.34. Also notice that Excel always

EXHIBIT 2.33 Excel's Descriptive Statistics Dialog Box for Data Listed in Exhibit 2.1

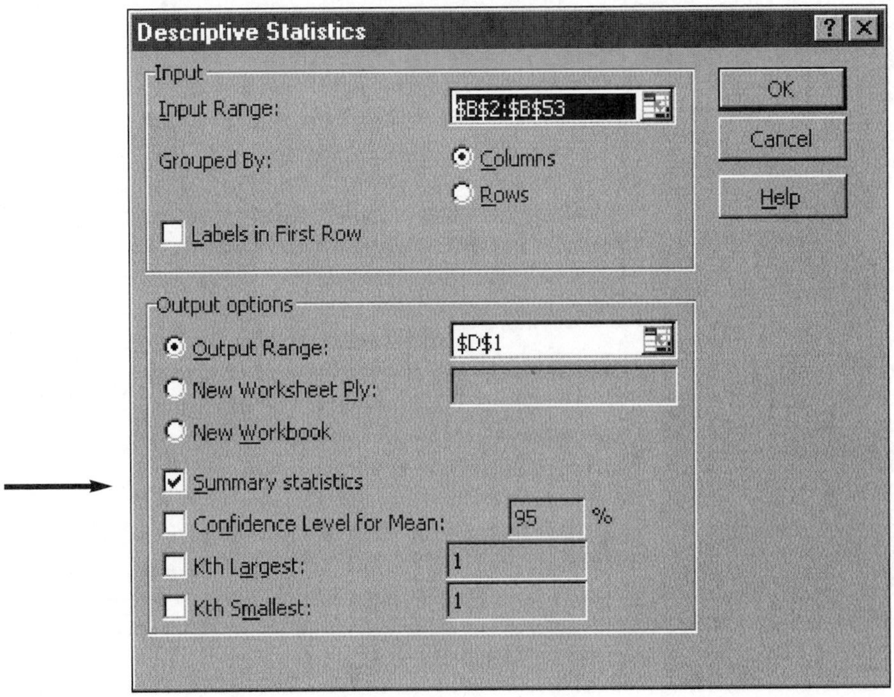

Descriptive Statistics
Input
Input Range: $B$2:$B$53
Grouped By: Columns
Rows
Labels in First Row
OK
Cancel
Help
Output options
Output Range: $D$1
New Worksheet Ply:
New Workbook
Summary statistics
Confidence Level for Mean: 95 %
Kth Largest: 1
Kth Smallest: 1

EXHIBIT 2.34 Output Generated by Excel's Histogram Tool

Row	A: Week	B: Sales
1	Week	Sales
2	1	22
3	2	25
4	3	25
5	4	24
6	5	18
7	6	31
8	7	22
9	8	30
10	9	19
11	10	19
12	11	24
13	12	21
14	13	27
15	14	21
16	15	12
17	16	16
18	17	24
19	18	29
20	19	31
21	20	33
22	21	23
23	22	18
24	23	34
25	24	26
26	25	20
27	26	22
28	27	26
29	28	32
30	29	25
31	30	23
32	31	20
33	32	29
34	33	25
35	34	30
36	35	18
37	36	28
38	37	21
39	38	21
40	39	35
41	40	23
42	41	21
43	42	28
44	43	26
45	44	23
46	45	23
47	46	25
48	47	18
49	48	23
50	49	20
51	50	26
52	51	29
53	52	23
54	Average	24.2

Row	D: Bin Range	F: Bin	G: Frequency
1	Bin Range	Bin	Frequency
2	15	15	1
3	19	19	7
4	23	23	18
5	27	27	13
6	31	31	9
7	35	35	4
8		More	0

Frequency table

Histogram

Frequency: 0, 5, 10, 15, 20

Bin: 15, 19, 23, 27, 31, 35, More

calculates the number of observations that are larger than the last bin boundary specified. Since the value of the last bin in the Bin Range was exactly equal to the maximum value in the data set, no values in the data set are outside of the Bin Range specified, and the frequency in this "more" range is zero and can be ignored. To use the Histogram tool select Tools/Data Analysis . . . /Histogram and complete the Histogram dialog box as shown in Exhibit 2.35. If you want to actually generate the histogram, you must select the Chart Output check box in the dialog box.

Step 4: Analysis of the Model

Based on the preceding, we observe that stethoscopes had average sales of approximately 24 units per week with a standard deviation of approximately 4.8 (see Exhibit 2.32). Furthermore, based on the histogram shown in Exhibit 2.34, it may be reasonable to model the sales process with a normal distribution after taking into account the limited sample size.

As a first attempt to model the situation, we will use the chi-square goodness of fit test to test the hypothesis that weekly sales of stethoscopes follow a normal distribution with a mean of 24 and standard deviation of 4.8. To investigate this hypothesis, the spreadsheet shown in Exhibit 2.36 was developed. According to the calculations shown in Exhibit 2.36, the calculated chi-square value of 1.95 does not exceed the critical value of 3.84, which indicates that this distribution is an acceptable fit to the data. Thus we conclude that the assumed normal distribution with a mean of 24 and standard deviation of 4.8 provides a reasonable fit to the data and will be used to model weekly stethoscope sales.

Having determined that a normal distribution provides a good fit to the weekly sales data, we can use this information to model and further investigate Hospital.com's approach to managing its inventory. Based on the assumption that demand follows a normal

EXHIBIT 2.35 Excel's Histogram Dialog Box

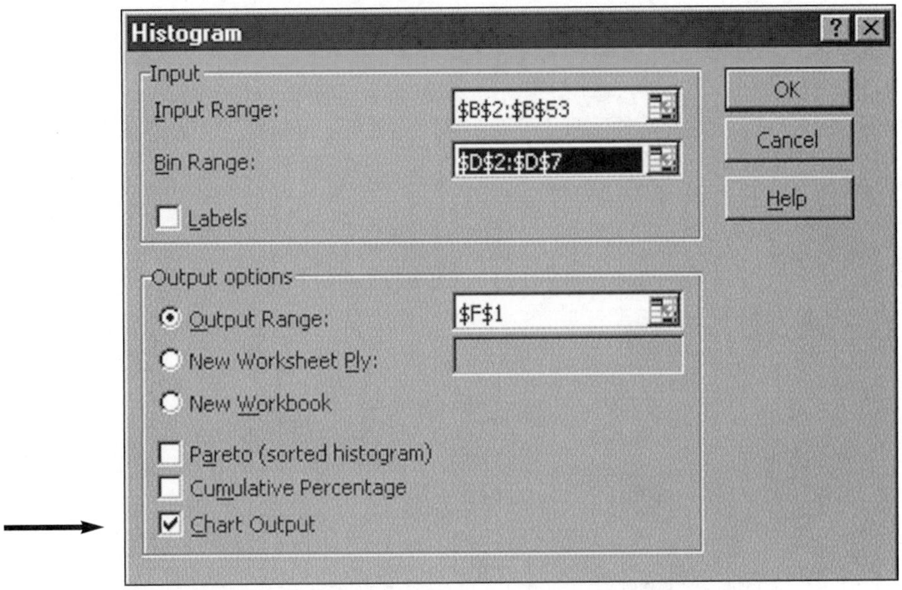

Histogram

Input
Input Range: $B$2:$B$53
Bin Range: $D$2:$D$7
☐ Labels

Output options
◉ Output Range: $F$1
○ New Worksheet Ply:
○ New Workbook

☐ Pareto (sorted histogram)
☐ Cumulative Percentage
☑ Chart Output

OK
Cancel
Help

distribution with a mean of 24 and standard deviation of 4.8, we can easily determine the probability that demand during any week is less than or equal to the target inventory level of 30 stethoscopes in Excel as follows:

=NORMDIST(30,24,4.8,TRUE)

Entering this formula into Excel reveals that there is an 89.4 percent chance that Hospital.com will have sufficient inventory during the week and a 10.6 percent (1 – 0.894) chance of not having enough stethoscopes. Rather than specifying a target inventory level in units, management might find it better to specify a target service level (i.e., the probability that demand will be met during the week). For example, if management would like to carry enough inventory to ensure that there was sufficient inventory 95 percent of the time this could be calculated as follows:

= NORMINV(0.95,24,4.8)

The amount of inventory required to meet demand 95 percent of the time turns out to be 32 units. Similarly, if Hospital.com wanted to have greater than a 99 percent chance of meeting all weekly demand it should increase its weekly target inventory level to 36 units.

Step 5: Implementation

Hospital.com needs to reevaluate its inventory management system. A key problem with its current approach is that it arbitrarily sets a target level for inventory based solely on the mean rate of sales and does not consider demand variability. To illustrate, we determined that for stethoscopes, setting the target inventory level at 25 percent above average weekly demand results in slightly more than a 10 percent chance that the company will stock out of stethoscopes in any given week. However, suppose that the variability in demand increases for stethoscopes to 6.8 from 4.8 while the weekly average demand re-

EXHIBIT 2.36 Chi-Square Test of Weekly Stethoscope Sales Data

	A	B	C	D	E	F	G
1	**Group***	**f_o**	**Probability**	**f_e**	**$(f_o - f_e)^2/f_e$**		
2	19	8	14.88%	7.74	0.01		
3	23	18	26.87%	13.97	1.16		
4	27	13	31.65%	16.46	0.73		
5	31	13	26.60%	13.83	0.05		
6	**Sum**	**52**					
7				**Calc. Val.**	**1.95**		
8				**Crit. Val.**	**3.84**		
9							
10	***Key Formulas***						
11	B6	=SUM(B2:B5)					
12	C2	=NORMDIST(A2,24,4.8,TRUE)					
13	C3	=NORMDIST(A3,24,4.8,TRUE)-SUM(C$2:C2) {copy to cell C4}					
14	C5	=1-SUM(C2:C4)					
15	D2	=C2*$B$6 {copy to cells D3:D5}					
16	E2	=((B2-D2)^2)/D2 {copy to cells E3:E5}					
17	E7	=SUM(E2:E5)					
18	E8	=CHIINV(0.05,1)					

Accept hypothesis that data comes from normal distribution

*Only the upper limit of each group interval was used so that these values could be used in the formulas entered in column C. In reality the first interval is ≤ 19, the second interval is > 19 and ≤ 23, and so on.

mains unchanged. This could occur for a number or reasons such as frequent price promotions, favorable government tests, and so on. In any event, if this increase in variability occurred and Hospital.com continued to maintain its target inventory level of 30 units, the chance of a stockout during any given week would increase to 18.9 percent.

Thus, we see that the problem facing Hospital.com stems largely from the fact that the variability in demand was not considered in setting the target inventory levels. Exhibit 2.37 illustrates in the form of a memo how the results of this analysis might be communicated to the president of Hospital.com. The companion Web site for this book contains a sample PowerPoint presentation of the results of this analysis.

EXHIBIT 2.37 Sample Memo to President of Hospital.com

MEMO

To: Meredith Sloan, President, Hospital.com
From: John Douglas, Senior Analyst
Date: 3/19/02
Subject: Inventory Management Problems

Introduction

I have concluded my preliminary investigation into the problems we are experiencing related to managing our inventory levels effectively. My conclusion is that the problem stems from the fact that our target inventory levels are set purely on the basis of average sales levels with no consideration given to sales variability.

Analysis

To illustrate the problems created by not considering sales variability, consider the stethoscopes that I studied in detail. Based on the weekly sales data that I compiled for an entire year, I was able to conclude that weekly sales of stethoscopes are normally distributed with a mean of 24 units and standard deviation of 4.8. Based on our current approach for setting target inventory levels, the target level for stethoscopes is 30 units. On the basis of its distribution of demand and current target inventory level, I was able to determine that the probability of incurring a stockout during any given week is slightly more than 10 percent. However, I was also able to verify that the risk of a stockout increases as the variability of demand increases. For example, if the standard deviation of demand for stethoscopes increases from 4.8 to 6.8, the chance of a stockout in any given week would almost double to 19 percent.

In a similar fashion, as the standard deviation of demand decreases the risk of a stockout also decreases. This suggests that the variability inherent in the demand should be a primary factor in determining the amount of inventory carried for each product. To further illustrate this, assume that there are seven products that each have an average demand of 24 units per week but have different standard deviations

(continued)

EXHIBIT 2.37 Continued

Average Weekly Demand	Standard Deviation of Weekly Demand	Target Weekly Inventory Level	Probability of a Stockout
24	1.8	30	0.04%
24	2.8	30	1.61
24	3.8	30	5.72
24	4.8	30	10.56
24	5.8	30	15.05
24	6.8	30	18.88
24	7.8	30	22.09

of demand, as shown in the table. Based on our current approach, the target inventory level would be set to 30 units for each product. However, because of differences in the variability of demand as measured by the standard deviation, the risk of stocking out of these seven products varies widely from almost zero percent up to 22 percent. This clearly demonstrates why setting an inventory target level based on some arbitrary percent increase in the average demand puts us in the position of having too much inventory of some products and too little of others.

Recommendations

Therefore, I recommend that target inventory levels be based on both the average sales rate and sales variability. Implementing this approach entails a shift from setting target inventory levels based on an arbitrarily chosen percentage increase of expected sales to setting target inventory levels based on a desired customer service level. Additional analysis of inventory related and shortage costs needs to be undertaken to determine optimal service levels for our products.

Limitations

While considering both the average sales rate and sales variability are important factors in determining optimal safety stock levels, it is important to emphasize that the results of this approach are only as good as the data this analysis is based on. Therefore, it is important to periodically update the estimates of the average sales rates and sales variability as changes in market conditions occur.

QUESTIONS

1. Describe the different ways that errors in collecting data can occur and explain their impact on the results of a QBM analysis.
2. Distinguish between categorical data and numerical data; between ordinal and nominal data; and between discrete and continuous data.
3. Why do we use the variance of a data sample, which is difficult to compute, instead of just the range?
4. What is the difference between a bar chart and a histogram? Could we not plot the grades received in a class as a histogram?
5. Distinguish between mutually exclusive and collectively exhaustive. Distinguish between the union and the intersection.
6. Describe the relation between probability and the cumulative distribution function.
7. What is the most important difference between the Poisson and normal distributions?
8. Describe the sampling distribution of the mean, its parameters, and its importance.
9. Rephrase in your own language the central limit theorem.
10. As a measure of variability, why is the range more sensitive to increases in the sample size than either the standard deviation or the variance?
11. In what situations might subjective probabilities be more accurate than objective probabilities?
12. When considering statistical estimators, describe in your own words the meaning of unbiased, efficient, consistent, and robust. Why are each of these important?
13. How do the standard confidence levels help a manager make a decision between having a low level of confidence due to a small data sample and spending the money to obtain more data to gain higher confidence?
14. Why do you think the sample median would be a more robust estimator of the population mean than the sample mean?
15. With a given set of data, why does having a higher confidence level require a wider parameter interval?
16. What is the logic in estimating the range of a variable to be collected and dividing by 4 to get an estimate of the population standard deviation?
17. The similarity between estimating a confidence interval and conducting a hypothesis test was noted in the chapter. How might a hypothesis test be conducted with an existing confidence interval?
18. For what type of situations are we more concerned about making a type II error than a type I error?
19. What might be a better formulation of a hypothesis when testing for cancer than that stated in the chapter?
20. If an ANOVA test tells us that there is a significant difference between a group of means, what additional tests should we run to determine which of the means are different?

EXPERIENTIAL EXERCISE

Select an ATM machine that you are familiar with and a particular time of the day when you imagine it would be heavily utilized (e.g., lunch hour, when classes end, etc.). Next, define a time interval such as three minutes or five minutes and record the number of arrivals at the ATM machine during each time interval. Collect at least two hours of data either in one consecutive two-hour interval or visiting the ATM on different days at the selected time (e.g., observing the ATM on Monday and Wednesday from 11:30 A.M. to 12:30 P.M.). Of course the more data you collect the better. After collecting the data complete the following tasks:

- Summarize the data you collected using relevant descriptive statistics.
- Create a histogram of the data.
- Perform a chi-square goodness of fit test to investigate whether the Poisson distribution provides a reasonable fit to the data.
- Calculate the probability that various numbers of customers arrive at the ATM during the specified time interval.
- Write a memo to the management of the organization that operates the ATM. In the memo discuss how the data were collected, how the data were analyzed, what key insights were obtained, any recommendations or suggestions that occur to you, and the key limitations associated with this study.

MODELING EXERCISES

Use influence diagrams wherever helpful.

1. The table below shows the length-of-stay distribution for patients in a Toronto hospital, in days.

Days	2	3	4	5	6	7	8	9	10
Probability (%)	3	5	5	10	15	25	20	10	7

The hospital makes $120 net profit per day during the first four days of patients' stay and $40 per day for later stays. How much profit will the hospital make in one year of 365 days if it admits 20 patients per day on average?

2. Al's Auto keeps detailed data on car sales for the previous year. The following table summarizes these sales in terms of the number of days various amounts of cars were sold.

	Number of Days
No cars were sold	15
One car was sold	30
Two cars were sold	87
Three cars sold	141
Four cars sold	27
Total days observed	300

   a. Find the frequency distribution of the sales.
   b. Find the average number of cars sold daily.
   c. Find the annual profit of Al's Auto if the average profit per car sold is $237.

3. You have been offered a chance to play a dice game in which you will receive $10 each time the point total of a toss of two dice is 4. If it costs you $1 per toss to play, should you accept or not?

4. Given the following monthly distribution of automobile accidents on the Santa Monica freeway in Los Angeles, find the mean, median, mode, standard deviation, variance, and range of monthly accidents. Then plot a monthly bar chart, a percentage frequency chart again using months, and a quartile histogram based on the monthly frequencies.

Month	Jan	Feb	Mar	Apr	May	Jun
Accidents	4	7	0	1	4	6
Month	Jul	Aug	Sep	Oct	Nov	Dec
Accidents	9	3	5	2	4	3

5. Last year's sales in Consolidated Megacorp's seven divisions, in $ millions, were: Electronics, 35; Appliances, 22; Reusable Launch Vehicles, 51; Picnic Supplies, 11; Hair Salons, 7; Food and Drugs, 17; and Plumbing, 8. Construct a Pareto diagram for Megacorp's sales and find the mean divisional sales, the standard deviation, and the range. Recalculate the divisional sales in percent of the total and construct a bar chart with the divisions presented in alphabetical order.

6. Find the following probabilities in draws from a deck of cards without replacement:
   a. A red card and a spade in two draws
   b. A red card or a spade in two draws
   c. Two diamonds in two draws
   d. At least one spade in two draws

7. A certain foreign coin has a probability of 0.55 of coming up heads. Find the probability of 7 heads in 10 tosses of this coin and compare it with that found for a fair coin in Section 2.4. How much has the probability increased, percentagewise?

8. Given a worn machine that creates defects at the average rate of 5 units per hour, what's the probability of no defects occurring in any specific hour? What's the probability of 8 or more defects in any hour?

9. Given a pre-teen talking on a cell phone to a friend (where the chance of hanging up is independent of how long the conversation has already lasted), how likely is it that the conversation will be finished within three hours if the average conversation lasts 45 minutes?

10. If the chance for a package arriving within X days is normally distributed with mean 6 and standard deviation 1.3, find the probability the package will arrive before 4 days; after 7 days; between 5 and 6 days; between 4 and 7 days.

11. A sample of 100 items is selected from a population known to be normally distributed. If the mean of the sample is 24 and its standard deviation is 5, what would you expect the mean and standard deviation of the original population to be?

12. Use the chi-square distribution with a significance level of 0.05, determine if a coin that came up heads in 12 out of 20 tosses is fair.

13. Using the chi-square distribution with a significance level of 0.05, determine if the following distribution of power outages is Poisson distributed:

No. of months	2	4	3	1	2	0
Power outages	0	1	2	3	4	5

14. A particular history professor assigns a letter grade of A to students whose course average is 93 percent or higher. In the most recent semester, the overall course average across all students was 80 percent with a standard deviation of 0.078. What is the probability that a particular student in the course will be assigned a letter grade of A assuming the distribution of course averages follows a normal distribution? Assuming a class size of 150 students, how many students will receive an A?
15. Assume that 10 monkeys are seated and ready to compete in the ABC hit TV game show *Who Wants to be a Millionaire?* What is the probability that any one monkey will win the million dollars? Assume that all three lifelines are available, the entire studio audience is made up of other monkeys, and only other monkeys may be called using the telephone lifeline. For those of you who are not familiar with this show, one of the monkeys will win the fastest finger question. This monkey must then answer 15 multiple-choice questions, each with four choices. The monkey also has three lifelines that may each be used one time: he/she may phone a friend, poll the studio audience, or eliminate two of the four choices.
16. A random survey of likely voters was taken regarding the issue of privatizing Social Security. The results of the survey are summarized in the table below.

	Democrats	Republicans	Independents
Support privatization	100	150	200
Don't support privatization	200	25	50

    a. According to the results, what is the probability that a likely voter would be a member of the Republican Party? Of the Democratic Party?
    b. What is the probability that a likely voter selected at random would support the privatization of Social Security?
    c. What is the probability that, given a voter was a Democrat, he/she would support privatization of Social Security?
    d. What is the probability that a voter selected at random is both a Democrat and supports privatization of Social Security?
    e. What is the probability that, given the voter does not support privatization, the person is an Independent?
17. Referring to the data given in Exercise 16:
    a. What is the probability of a voter being a member of one of the major parties (i.e., not an Independent)?
    b. What is the probability of a voter being a Republican or supporting the privatization of Social Security?
    c. What is the probability of a voter being a member of one of the major parties and supporting the privatization of Social Security?
18. A manufacturer of printed circuit boards periodically samples 10 boards at random for inspection. Currently, about 90 percent of the boards pass this inspection. Assuming that the binomial distribution applies, answer the following questions.
    a. What is the probability that exactly 7 of the 10 boards pass the inspection?
    b. What is the probability that at least 7 of the 10 boards pass the inspection?
    c. What is the probability that 6, 7, or 8 boards pass the inspection?
    d. What is the probability that two or fewer boards don't pass the inspection?
    e. Create a frequency chart and cumulative frequency chart showing the probabilities of various numbers of boards passing the inspection.
19. It has been determined that the number of arrivals to a particular Web site follows a Poisson distribution. Analysis of historical data suggests that at peak times, an average of eight new visitors come to the site every minute.
    a. What is the probability that exactly 13 new people visit the site in a particular one-minute interval?
    b. What is the probability that at least 5 new visitors come to the site in a particular one-minute interval?
    c. What is the probability that between 5 to 10 new visitors come to the site in a given one-minute interval?
    d. What is the probability that 8 or fewer new visitors come to the site in a given one-minute interval?
    e. Create a frequency chart and cumulative frequency chart showing the probabilities of various numbers of new visitors coming to the site in a given one-minute interval.
20. Referring to the situation described in Exercise 19:
    a. What is the probability that the interarrival time between two successive visitors will be less than 10 seconds?

b. What is the probability that the interarrival time between two successive visitors will be between 6 and 10 seconds?
c. What is the probability that more than 16 seconds will elapse between two successive arrivals to the Web site?

21. A national pizza chain purchases its pie crusts from one supplier. For its thin-crust large pizzas, the pizza chain has specified that the crusts should be 75 ounces, but considers the crusts acceptable as long as they are between 70 to 80 ounces. A recent audit of the supplier's production process suggests that the average weight of the pie crusts is 73.5 ounces with a standard deviation of 4 ounces.
    a. What is the probability that the supplier will deliver a pizza crust that is too small?
    b. What is the probability that the supplier will deliver a pizza crust that is too large?
    c. What is the probability that a delivered pie crust will meet the pizza chain's specifications?
    d. Should the pizza chain continue to purchase its pie crusts from this supplier? Why or why not?
    e. What recommendations would you offer the pie crust supplier?

22. The following data correspond to the interarrival times (in minutes) between successive customer arrivals to an ATM. Perform a chi-square goodness of fit test to determine whether an exponential distribution provides a good fit to the data.

0.8	4.2	0.6	5.3
0.6	5.2	1.6	5.3
2.4	13.8	0.8	4.7
1.6	5.2	0.1	5.1
2.9	13.6	0.1	7.8
2.5	9.0	1.2	7.8
3.3	8.3	0.3	7.0
3.7	10.0	1.3	6.7
3.5	11.6	0.0	7.0
3.6	11.6	3.9	8.3
2.2	12.0	15.9	14.7
16.2	19.4	21.9	22.6

23. The following data correspond to the amount of time (in seconds) customers spend on hold waiting for the next available technical support person.

216	138	167	158	179
241	276	233	151	167
146	155	218	265	222
154	174	258	111	262
223	178	164	211	113
137	204	221	210	75
144	144	189	154	151
163	150	173	225	243
197	230	262	246	94
149	156	123	216	187

    a. Use Excel or another software package to find the average, standard deviation, maximum, minimum, and range. Explain these values in terms a general manager would understand.
    b. Develop a frequency chart for the data. What distribution does the frequency chart most resemble?
    c. Perform a chi-square goodness of fit test to investigate whether the distribution you specified in b. provides a reasonable fit to the data.
    d. Assuming that a normal distribution provides a good fit to the data, what is the probability that a customer will be on hold less than 100 seconds? The probability a customer will be on hold between 1.5 and 2.5 minutes? The probability a customer will be on hold for more than 4 minutes?

24. In taking a sample from a population, we determine that the mean is 44.6 and the standard deviation of the sample is 5.8. Find the 99 percent confidence interval for the population mean if:
    a. the population distribution of the variable is normal but the sample is small.
    b. the sample is large.
    c. the standard deviation of the population is 8 and the sample is large.
    d. the population distribution of the variable is normal with a standard deviation of 8.
    e. the population distribution of the variable is normal and the sample is large.

25. Determine the sample size needed to be 99 percent confident that the true population mean is within 5 units of the sample mean for a distribution that is believed to range from 31 to 87. What problem occurs if you only needed to be 70 percent confident?

26. If a sample of 1,000 computer chips from an etching machine identified 2 percent as defective, what would be the 99 percent confidence interval for the true population fraction of defectives? What situation arises if the sample size had just been 100 instead? What sample size would be needed if we wanted the half-interval error to be no more than 0.1 percent (i.e., 0.001)?

27. In a test of the hypothesis that a population mean was greater than or equal to 50, a sample of 100

had a mean of 47 and a standard deviation of 12. If the level of significance is 5 percent, what would the decision be concerning the hypothesis? What if the level of significance was 1 percent? What if the hypothesis had been that the population mean was equal to 50 and alpha was 1 percent?

28. In a test of the hypothesis that a normally distributed population had a mean of 30, a sample of 10 had a mean of 33 and a standard deviation of 15. At the 5 percent level, what would the decision be concerning the hypothesis? How would you proceed if the hypothesis had been that the population mean was less than or equal to 30? Greater than or equal to 30?
29. Test the hypothesis at an alpha of 1 percent that a population proportion is 0.2, given a sample of 30 with a proportion of 0.3. Repeat the test with the hypothesis that the proportion is 0.2 or less.
30. Given the following independent data from four normally distributed populations with equal variances, determine if the means are all equal, and if not, which ones are unequal.

Run 1	64	75	81	66	85	72	59	77
Run 2	88	81	73	60	76	63	84	58
Run 3	72	64	61	87	77	74	71	60
Run 4	78	80	72	66	70	66	50	85

31. Referring to Exercise 11, construct a 95 percent confidence interval for the population mean. Explain the interpretation of the confidence interval in terms a typical manger would understand.
32. Referring to Exercise 16, test the two-tailed hypothesis that the true proportion of voters that support the privatization of Social Security is 25 percent, assuming that the normal distribution can be used to approximate the binomial distribution.
33. Referring to Exercise 21, assume that the audit of the pie crust supplier entailed sampling 150 pie crusts.
    a. Construct a 95 percent confidence interval for the true population mean pie crust weight. What are the implications of this confidence interval for the national pizza chain?
    b. Test the hypothesis that the average weight of the pie crusts is 75 ounces.
    c. Explain any similarities and differences in your analysis with respect to parts a and b above.

CASES

Use influence diagrams where appropriate.

Fiberease Inc.

Fiberease Inc. produces fiber optic cable for the telecommunication and computer industry. Due to its success in the United States, it is considering offering its products to overseas customers. However, with its domestic production facility currently operating at capacity, the president of Fiberease has decided to investigate the option of adding approximately 10,000 square feet of production space in Europe at a cost of $5 million.

The project to construct the European facility involves four major phases: concept development, plan definition, design and construction, and start-up and turnover. A team of analysts identified the major tasks associated with each of these phases and then developed probability distributions for each task. The probability distributions were then used to simulate the completion of the project 100 times. The results of the simulation study are shown in Exhibit A and are provided in the supplementary materials accompanying this book.

Questions

See Exhibit A.

1. Summarize the data from the simulation study using relevant descriptive statistics.
2. Create a histogram of the data.
3. Perform a chi-square goodness of fit test to investigate whether the normal distribution provides a reasonable fit to the data.
4. What is the probability that the project to construct the European plant can be completed within 30 months? What is the probability that the project will take longer than 40 months? What is the probability that the project will take between 30 and 40 months?

EXHIBIT A Simulated Completion Time Data to Construct European Plant (Months)

	A	B	C	D
1–3	Simulation Run Number	Time to Complete Project	Simulation Run Number	Time to Complete Project
4	1	38.4	51	24.9
5	2	28.9	52	36.9
6	3	34.6	53	31.5
7	4	32.6	54	39.6
8	5	36.2	55	36.0
9	6	39.8	56	36.4
10	7	29.2	57	37.7
11	8	27.5	58	41.0
12	9	38.0	59	36.6
13	10	34.9	60	30.0
14	11	35.0	61	30.9
15	12	27.7	62	39.0
16	13	28.6	63	36.1
17	14	33.0	64	34.3
18	15	27.2	65	37.6
19	16	31.2	66	35.8
20	17	40.1	67	40.1
21	18	34.9	68	29.5
22	19	37.2	69	42.5
23	20	36.4	70	39.5
24	21	33.6	71	42.5
25	22	33.9	72	40.7
26	23	39.5	73	38.4
27	24	40.1	74	31.7
28	25	37.9	75	32.6
29	26	39.3	76	43.7
30	27	46.0	77	28.3
31	28	35.2	78	30.8
32	29	44.0	79	38.5
33	30	31.1	80	38.5
34	31	38.0	81	44.2
35	32	29.9	82	40.2
36	33	40.6	83	38.9
37	34	37.7	84	32.6
38	35	39.5	85	40.3
39	36	35.3	86	36.3
40	37	33.8	87	35.1
41	38	34.3	88	35.8
42	39	34.4	89	30.9
43	40	35.5	90	39.8
44	41	29.5	91	30.4
45	42	29.9	92	38.6
46	43	32.9	93	37.7
47	44	33.7	94	26.9
48	45	36.9	95	42.2
49	46	31.6	96	31.4
50	47	33.5	97	28.0
51	48	46.2	98	28.8
52	49	29.9	99	31.1
53	50	32.8	100	33.5

InterAccess Inc.

InterAccess, Inc. is an Internet Service Provider (ISP). Beginning in May of this year and continuing through December, the company began getting an unusually large number of complaints from customers regarding the amount of time required to connect to the Internet. This is particularly troubling to management as the company has tried to differentiate itself on the basis of fast connection times. In fact, since its beginning, the company president has monitored this aspect of its service by randomly logging in 10 times each month and recording the time required to connect to the service. This data (in seconds) for the most recent year is provided in Exhibit B.

Upon examining the data, the president of InterAccess was still puzzled. After calculating the average connection time for each month, no noticeable increase in connection times was apparent. In fact, beginning in July average connection times appeared to be quite acceptable.

Questions

1. Do you see any patterns or trends in the data that could help explain the increase in the number of complaints beginning in May?
2. Based on your analysis, write a short memo to the president of InterAccess summarizing your findings and recommendations.

EXHIBIT B
Connection Times

	Jan	Feb	Mar	Apr	May	Jun	Jul	Aug	Sep	Oct	Nov	Dec
	62	61	62	57	98.6	55	70	43.5	80.7	30	50	42
	57	65	59	66	80.5	39	50	56.6	94.7	63	59	87
	63	67	57	53	59.1	58	44	58.6	33.1	87	42	43
	62	47	68	70	34.6	55	50	43.8	26.4	39	70	72
	63	54	63	69	88.8	60	71	50.1	54.9	71	33	35
	60	50	56	55	33.5	65	47	46	61.4	50	18	35
	53	70	68	69	80.4	48	45	49.1	56.2	55	79	59
	60	65	56	53	86.6	75	45	76.4	15.8	41	78	72
	58	62	60	57	89	97	43	61.5	45	62	49	86
	60	61	56	59	86.3	49	53	25.3	62.7	47	42	68
Average	**60**	**60**	**61**	**61**	**74**	**60**	**52**	**51**	**53**	**55**	**52**	**60**

eApp Inc.

eApp Inc. processes and evaluates online mortgage applications for a number of financial institutions. Because of its technical expertise in developing and administering Web-based forms and databases, eApp has been able to significantly reduce the amount of time required to process its clients' mortgage applications. Currently it promises its financial institution clients that it can process loan applications in three weeks (15 working days) or less.

The president of eApp is keenly aware that his clients depend on its ability to quickly process online mortgage applications to gain a competitive advantage. Therefore, the company continuously monitors via random samples the length of time required to process online applications. In the most recent quarter, 100 online applications were randomly selected and the actual processing time recorded in the spreadsheet shown in Exhibit C.

Questions

1. Perform a chi-square goodness of fit test to test the hypothesis that the data are normally distributed.
2. Develop a 95 percent confidence interval for the time to process an online application. Does the confidence interval support eApp's claim that it can process applications in 15 days or less?
3. Perform a hypothesis test to investigate whether eApp's claim that it can process online applications in less 15 days is statistically valid.

EXHIBIT C Random Sample of Time to Process 100 Applications (Days)

	A	B	C	D	E
1	10.8	14.3	6.6	15.3	15
2	7.2	11.2	9.9	10.5	7.9
3	15.1	13.4	13.4	13.2	12.7
4	9.3	7.2	11.2	12.9	9.1
5	5.4	8.7	6.8	16.1	8.4
6	15.2	11.7	17.2	8.6	11.3
7	8.7	15.7	10.4	10.7	11
8	7.9	12.4	17.3	12.8	7.5
9	7.5	12	12.2	9.5	13.4
10	8.5	11.3	12.6	12.6	10.5
11	13.7	11.8	15.5	10.2	7.4
12	9.9	14.6	11.7	13.5	9.2
13	7.4	10.2	6.6	16.3	9.5
14	15.2	6.4	11.3	9.4	13.8
15	17.2	8.4	12.2	17	9.7
16	13.5	9.1	10.6	8.1	11.4
17	8	10.2	12.5	12.8	9.1
18	11	9.8	7.2	10.8	15.1
19	11.3	10.7	14.3	10	6.9
20	13.9	11.5	8.4	10.6	7.8

ENDNOTE

1. Normally, when conducting multiple tests for significant differences at, say, the 0.05 significance level, one must reduce the significance level alpha since out of 100 tests, 5 will be "significant" just by chance and thus are not really significant. This can be done with the *Bonferroni correction* (alpha/number of tests). However, that is not the situation here because we know already that at least one of the groups is significantly from the others and we are just trying to identify which one(s) it is. Another way to check differences among groups is with *multiple comparisons* using Tukey's test, but this is beyond our scope (see Hildebrand and Ott 1998 in the Bibliography, for example).

BIBLIOGRAPHY

Albright, S. C., W. L. Winston, and C. Zappe. *Data Analysis and Decision Making with Microsoft® Excel.* Pacific Grove, CA: Brooks/Cole, 1999.

Anderson, D. R., D. J. Sweeney, and T. A. Williams. *Statistics for Business and Economics,* 8th ed. Cincinnati, OH: South-Western, 2002.

Berenson, M. L. and D. M. Levine. *Basic Business Statistics: Concepts and Applications,* 7th ed. Upper Saddle River, NJ: Prentice-Hall, 1999.

Hildebrand, D. H. and A. L. Ott. *Statistical Thinking for Managers,* 4th ed. Pacific Grove, CA: Brooks/Cole, 1999.

Keller, G. and B. Warrack. *Statistics for Management and Economics,* 5th ed. Pacific Grove, CA: Brooks/Cole, 2000.

Mason, R. D., D. A. Lind, and W. G. Marchal. *Statistical Techniques in Business and Economics,* 10th ed. Boston: Irwin/McGraw-Hill, 1999.

Siegel, A. F. *Practical Business Statistics.* Boston: Irwin/McGraw-Hill, 2000.

Statistical Models: Regression and Forecasting

3

© CORBIS

INTRODUCTION

Since relocating to North Carolina two years ago, Nick Carol and his wife have spent their weekends visiting numerous communities where new houses are being constructed. They have recently decided on a community in which to build their dream house and have reserved a lot. Having obtained the plans for their house, the next step is to negotiate with one or more builders regarding the cost of building the home. To prepare for the negotiation process, Nick collected data on the size and price of all the houses that have been completed in the development over the last two years, assuming that a major influence on the price of a house is its size. The data are summarized in Exhibit 3.1.

Given that the house Nick and his wife would like to build has 3,262 ft^2, how much should they expect to pay?

To begin to answer this question, we might first want to investigate the degree to which the size of a house as measured by its square feet is related to the price of the home. If we determine that there is a strong relationship between these variables, then we may be able to estimate the price of the home based solely on its size. If we determine that there is not a particularly strong relationship between these two variables, we may decide to collect additional data such as the number of bedrooms, the number of stories, the price of the lot, the company that built the home, and so on.

EXHIBIT 3.1 Data for All Completed Houses over the Last Two Years

Lot Number	House Size (ft^2)	House Price
145	2,620	$266,500
144	2,635	266,900
119	3,019	364,500
136	3,049	384,900
7	3,141	389,900
114	3,141	399,900
97	3,264	439,000
90	3,319	405,000
108	3,403	414,500
200	3,578	442,000

EXHIBIT 3.2 House Size Versus House Cost

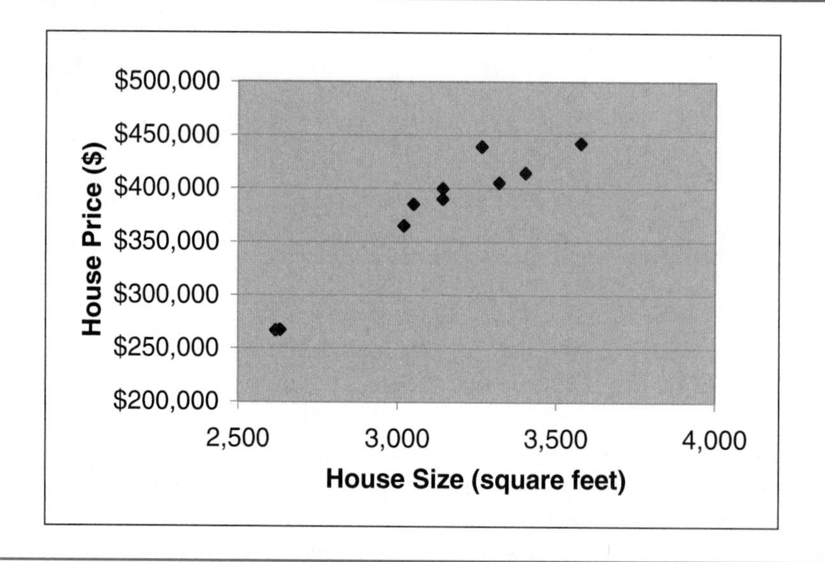

To investigate the extent to which house size and price are related to one another, Nick decided to plot these two variables as shown in Exhibit 3.2. Visual inspection of this graph, called a **scatter diagram,** indicates an apparently strong relationship between the two variables. In fact, as is illustrated in Exhibit 3.3, house prices appear to increase in a relatively linear fashion as house size is increased.

We could, of course, simply draw what appears to be the most appropriate straight line and use that to predict the house price. However, a more accurate statistical modeling approach, *linear regression,* is particularly appropriate for situations such as this one where the decision maker would like to predict the value of one variable based on the values of one or more other variables. Applying **linear regression analysis** to this situation (as will be described shortly) yields the following statistical model that relates the price of a newly constructed home to its size:

$$\text{Price of new house} = -226{,}688.0 + 193.7816 \text{ (house size)}$$

EXHIBIT 3.3 Apparent Linear Relationship Between House Size and Price

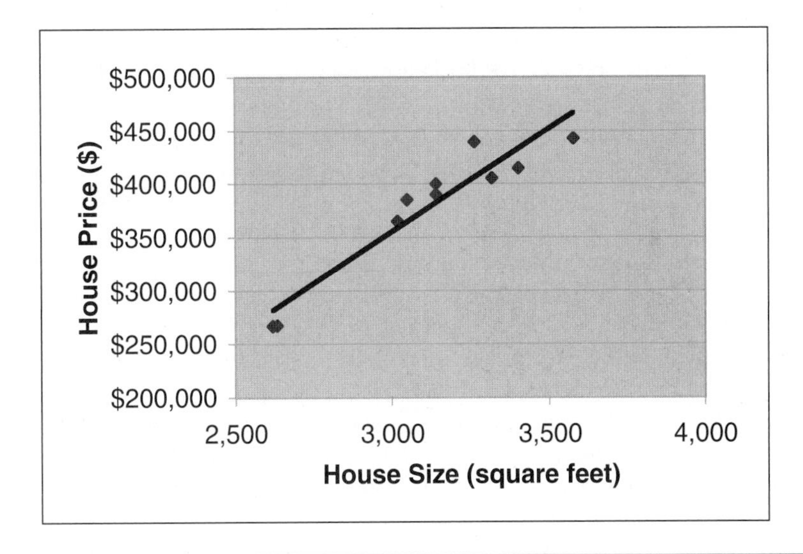

Using this model, Nick can estimate the price of his new house as follows:

$$\text{Price} = -226{,}688.0 + 193.7816(3{,}262) = \underline{\$405{,}427.72}$$

The remainder of this chapter discusses in greater detail the modeling of decision-making situations with statistical models. Our discussion begins with important background material on regression modeling such as the distinction between independent and dependent variables. We will also overview the primary assumptions related to regression analysis and discuss the difference between prediction and explanation. Building on this foundation, later sections in the chapter address the process of developing simple linear and multiple regression models. Our discussion of statistical models concludes with an overview of several approaches for modeling time series data.

3.1 The Modeling Process for Statistical Studies

As the previous example illustrated, statistical models are often used to investigate and identify relationships among two or more variables. As such, these models tend to be used more *descriptively* than *prescriptively*. For example, Nick could have used the regression model discussed earlier to prescribe what he should pay for his new house. However, a more appropriate use of the model is to describe, in a general sense, how the cost of the house is related to its size and then use this information as a general guide in pricing the house. There are a variety of reasons why it would be inappropriate for Nick to use the model to specify the "fair" price. For example, Nick had only a limited number of observations from which to infer a relationship. Moreover, he omitted a number of other important variables, such as quality of materials used, type of basement, number of bathrooms, type of flooring, number of bays in the garage, shape of the house, exterior materials used, type of windows, and so on.

The variables used in statistical models are typically classified, as described in earlier chapters, into two general categories: independent and dependent variables. To reiterate, a variable that is used to predict or explain the value of another variable is referred to as an *independent variable.* Alternatively, a variable whose value we seek to predict or explain is called a *dependent variable.* In the earlier example, Nick attempted to predict the price of building a new house on the basis of the size of the house. In this case, the variable "size of the house" is the independent variable and the variable "price" is the dependent variable. In the statistical models discussed in this chapter there is always exactly one dependent variable, while there may be one or more independent variables.

Generally speaking, the statistical models discussed in this chapter can be used in one of two ways. One use of the models is to simply *predict* the value of the dependent variable based on the values of one or more independent variables. This was Nick's purpose in attempting to predict the price of a house on the basis of its size. Alternatively, models can be used to attempt to *explain* how values of the independent variable(s) actually influence the value of the dependent variable.

To further illustrate this distinction, consider the regression model that Nick developed that related the price of the house to its size. If Nick were truly interested in *explaining* how the independent variable size of the house influenced the dependent variable price, it would be vital that the coefficient for the independent variable in the model (i.e., 193.7816) actually reflect the amount the price of the house changed per unit change in the size of the house. In other words, Nick would want the model to show how a change in the size of the house would affect the price of the house. Alternatively, if Nick's purpose is to simply *predict* what a house of a given size should cost, he is less interested in the actual parameters in the regression model and more interested in obtaining an accurate estimate or prediction of the house price. As you can probably imagine, developing statistical models that actually explain how one or more independent variables affect the dependent variable is a much more daunting task and needs to be based on some underlying *theory* regarding how the variables are related to one another. We discuss both purposes in this chapter.

Modeling with regression includes many of the same opportunity/problem recognition and data collection issues that were described in Chapter 2. However, rather than just describing or summarizing the data, we are now using that data to formulate statistical models of a managerial situation, conduct an analysis by using the model, and recommend and implement some action that addresses the managerial situation.

Thus, this is the first chapter that actually describes some models to use for quantitative business modeling. Following chapters will describe alternative models for other kinds of managerial situations.

3.2 The Simple Linear Regression Model

Simple linear regression analysis involves using the values of a single independent variable to predict or explain the values of the dependent variable. If we wish to include more than one independent variable in our model, we have a *multiple regression model,* which will be discussed later in this chapter. Prior to using simple linear regression analysis, it is appropriate to plot the values of the independent and dependent variables in a scatter diagram, as was done in Exhibit 3.2. The diagram helps us visually verify a key assumption that the variables appear to be linearly related. Exhibit 3.4 illustrates three common ways two variables can be related to one another. If it is discovered that the variables are not linearly related to one another, it may be possible to "transform" one or both of the variables so that they are approximately linearly related. Frequently used transformations include taking the square root, inverse, or logarithm of the data. We will return to the topic of transforming the data later in the chapter.

EXHIBIT 3.4 Example Relationships Between Variables

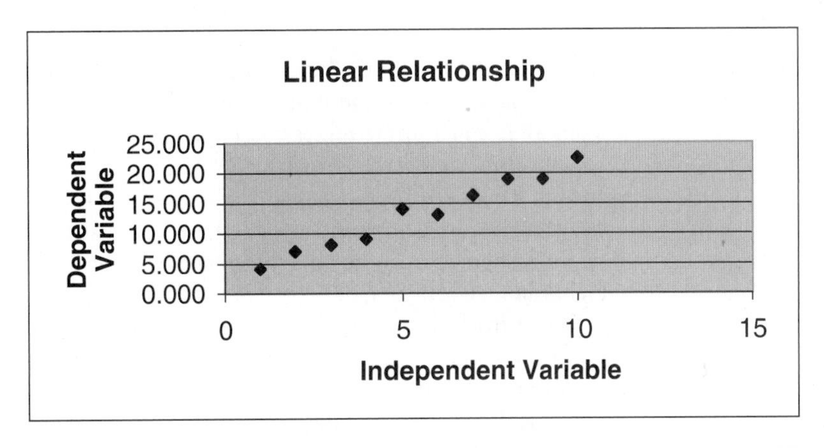

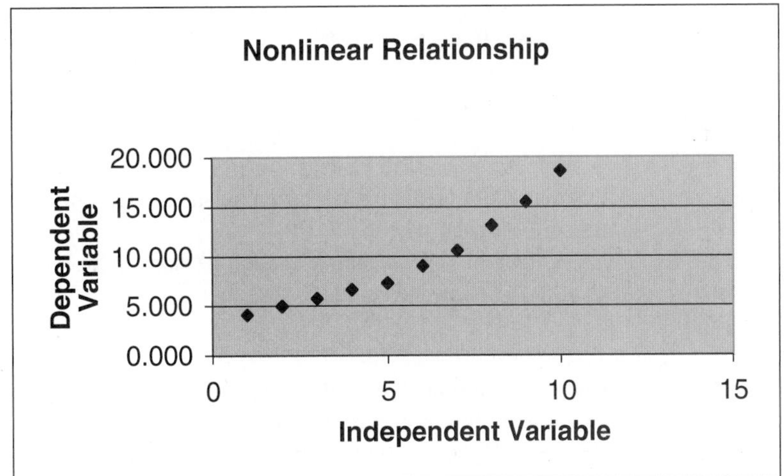

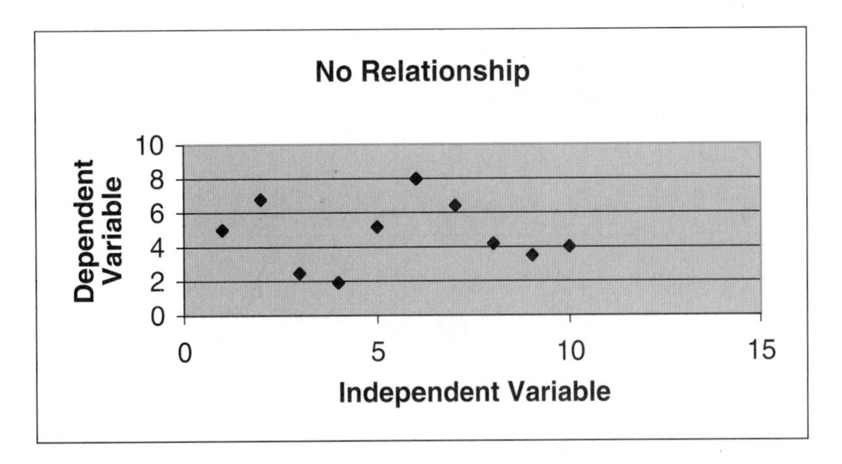

The mathematical form of the simple linear regression model is as follows:

$$Y = \alpha + \beta X + \varepsilon$$

where X corresponds to the independent variable, Y to the dependent variable, and α and β are the parameters of the model. According to this model, the value of the dependent variable Y is equal to the regression model constant α plus the model parameter β times the value of the independent variable X. Also notice that a *residual,* or *error term,* ε, is included in the model to account for the fact that it is typically not possible to determine the exact value of the dependent variable based on just the two model parameters α and β. Therefore, there is likely to be a difference between the predicted value of the dependent variable and the actual value.

Recall from our discussion in Chapter 2 that while in reality the random variable of interest does indeed have a population mean μ and standard deviation σ, determining these parameters requires collecting data on the entire population. Because this is often not practical, we collect sample data from a subset of the population and use the sample statistics $\bar{X}$ and s as estimates of the true but unknown population parameters μ and σ, respectively. In a similar fashion, the true parameters of the regression model are also unknown to the modeler, so sample data are used to estimate these unknown parameters. When our regression model is based on sample data, the model is written as

$$Y = a + bX$$

where a and b are estimates based on sample data of the unknown population parameters α and β, respectively.

Earlier in your academic career, perhaps in an algebra class, you may have seen the equation of a line expressed as

$$Y = mX + b$$

where Y represents the value on the vertical axis, X corresponds to the value on the horizontal axis, m represents the slope of the line (i.e., the amount the line rises for a unit change in X), and b corresponds to the Y intercept, or the point where the line intersects the Y axis (which is also the point on the line where $X = 0$). The regression model presented earlier is completely analogous to this, the only differences being that a is used to represent the Y intercept (b in the standard equation of a line) and b is used in place of m to represent the slope of the line, as is illustrated in Exhibit 3.5.

EXHIBIT 3.5 Parameters of a Regression Line

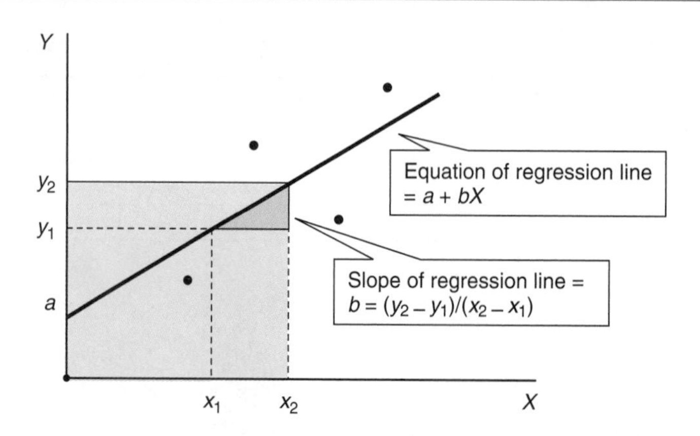

Calculating the Regression Model Parameters

There are many ways that a line can be fit to a set of data. One way is to simply use a ruler and visually determine which line provides the best fit to a set of points plotted on a graph by adjusting the angle of the ruler. The best line could then be drawn and its equation determined. Although this approach often yields good results, statisticians and decision makers often favor the use of more formal, less subjective approaches. The approach most often used is based on minimizing the sum of the squared vertical distances between the data points and the regression line fit to the data points (that is, the *errors* from using the regression line to make a prediction). Because of this, it is often referred to as the **least squares regression** model.

To illustrate how the least squares approach works, consider the small data set consisting of four observations as shown in Exhibit 3.6. In the figure, the vertical distance from each point to the line fit to the data is shown. These vertical distances can be thought of as errors, e_i, since they represent the difference between what the line predicts the value of Y will be for a given value of X and what Y actually is for the given value of X. The least squares approach fits the line to the data such that the sum of the squared errors, $\sum e_i^2$, is minimized. In the example shown in Exhibit 3.6, this means fitting a line to the data so that the sum $e_1^2 + e_2^2 + e_3^2 + e_4^2$ is minimized.

Equations can be easily derived using calculus for computing the parameters a and b in a regression model based on the least squares approach. Although deriving these equations is relatively straightforward, here we simply present the equations for calculating a and b as follows:

$$b = \frac{\sum XY - n\overline{X}\overline{Y}}{\sum X^2 - n\overline{X}^2}$$

$$a = \overline{Y} - b\overline{X}$$

where $\sum XY$ = X times Y for each data point and then summed over all data points
$\sum X^2$ = X squared for each observation and then summed over all data points
$\overline{X}, \overline{Y}$ = the average of the X and Y values, respectively
n = the number of data points

Fortunately, spreadsheets and other software programs have built-in functions that greatly facilitate the calculation of the regression model parameters and thereby eliminate

EXHIBIT 3.6 Least Squares Approach to Fitting Line to Set of Data

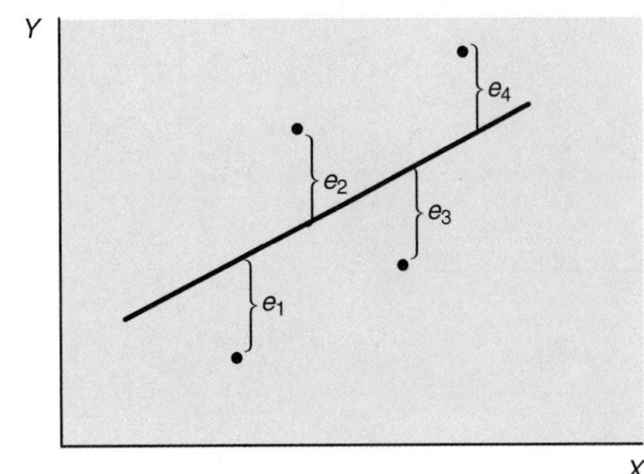

the need to use these formulas. For example, a built-in Excel feature was used to fit the least squares trend line shown in Exhibit 3.3 to the data plotted in Exhibit 3.2. The procedure for fitting a trend line to a set of data points using Excel is as follows:

1. Plot the *X* and *Y* values using Excel's Chart Wizard, as discussed in the previous chapter. In this case, select XY (Scatter) as the type of chart.
2. After creating an XY (Scatter) chart as shown in Exhibit 3.2, select the graph by clicking on it.
3. Next click on any of the data points plotted in the chart.
4. Select Chart from the menu at the top of the screen and then Add Trendline from the next menu.
5. Finally, select Linear and then the OK button. After you click the OK button, Excel fits and plots the least squares linear trend line to the original data as shown in Exhibit 3.3. If you followed this example in Excel you likely noticed that in addition to fitting a linear trend line to the data, Excel can also fit more complex models to a set of data including logarithmic, polynomial, power, and exponential curves. Although these types of models are beyond the scope of this book, it is worth noting that this functionality and power is available to the spreadsheet user.

Of course fitting the trend line to the data set is only half the battle. For example, we would also like to know what the parameters of the trend line or regression model are. One way to calculate these parameters is to use Excel's LINEST function. More specifically, the LINEST function uses the least squares method to calculate the parameters *a* and *b* of a trend line. The syntax of this function is

LINEST(range of *Y* values, range of *X* values)

Note that the LINEST function is a special type of function called an **array function** because it is used to return multiple values (i.e., the parameters *a* and *b*) rather than a single value such as the average or standard deviation of a range of data. To illustrate the use of this function, the labels *b* and *a* were entered in cells B14 and C14, respectively, as shown in Exhibit 3.7. Next, the cells B15 and C15 were highlighted and the following formula entered:

=LINEST(C3:C12,B3:B12)

EXHIBIT 3.7 Using Excel's LINEST and TREND Functions

	A	B	C	D
1	Lot	House	House	
2	Number	Size (ft^2)	Price	TREND
3	145	2,620	$266,500	$281,020
4	144	2,635	$266,900	$283,927
5	119	3,019	$364,500	$358,339
6	136	3,049	$384,900	$364,152
7	7	3,141	$389,900	$381,980
8	114	3,141	$399,900	$381,980
9	97	3,264	$439,000	$405,815
10	90	3,319	$405,000	$416,473
11	108	3,403	$414,500	$432,751
12	200	3,578	$442,000	$466,663
13				
14		*b*	*a*	
15		193.7816	-226,688	

=TREND(C3:C12,B3:B12)

=LINEST(C3:C12,B3:B12)

Because we are using the LINEST function as an array function, when this formula is entered we must press and hold down the Ctrl key and the Shift key as we press the Enter key. After pressing Ctrl + Shift + Enter, the values of *b* and *a* are calculated and displayed in cells B15 and C15, respectively. You may recall that these are the same values of *a* and *b* that Nick Carol used when he developed his model to predict the price of his new home. Note that the LINEST function returns the value for *b* first and then the value for *a,* just as we presented the formula for calculating *b* first and then used the calculated value of *b* to calculate *a.* If you were to look at the contents of either cell B15 or C15, you would see the following formula displayed:

{=LINEST(C3:C12,B3:B12)}

The braces associated with array-based functions are added automatically by Excel and are not entered as part of the equation.

Another useful Excel function is the TREND function. This function fits a straight line to a column of *X* and *Y* values and then returns the values that would appear on the trend line for each value of *X*. The syntax for the TREND function is as follows:

=TREND(range of *Y* values, range of *X* values)

The TREND values were calculated in column D by first highlighting the cells D3:D12 and then entering the formula:

=TREND(C3:C12,B3:B12)

In a similar fashion to the LINEST function, since the TREND function returns multiple values, the Ctrl and Shift keys must be held down while pressing the Enter key. The resulting trend values are the values you would get if the actual values for *X* were substituted into the trend equation. For example, referring to lot 145 in Exhibit 3.7, the trend line fit to the data estimates that the price of a 2,620 ft^2 house will be $281,020 (cell D3). We can verify this result by simply plugging the *X* value into the equation of the trend line as follows:

$$\text{Price}_{2620} = -226{,}688 + 193.7816(2620) = \$281{,}019.79$$

In developing regression models, the analyst must often make judgments about how to handle **outliers,** or extreme data points. In some cases, outliers may be the result of data entry errors and therefore should be corrected. At other times, outliers may be the result of unusual circumstances (e.g., a labor strike, a natural disaster, and so on), and in these cases they can perhaps be justifiably omitted or adjusted. In the remaining cases where no error or unusual circumstance can be discovered, it is difficult to justify eliminating or adjusting an outlier.

The issue of outliers is important because of the impact these data points can have on the regression model. As is illustrated in Exhibit 3.8, an outlier is a data point that has an extreme value on the independent variable dimension, on the dependent variable dimension, or on both dimensions. As shown in the top graph in Exhibit 3.8, an outlier on the independent variable dimension can have a profound impact on the regression line fit to the data, altering both the slope of the line and its *Y* intercept. In contrast, an outlier on the dependent variable dimension primarily shifts the *Y* intercept of the regression line in the direction of the outlier, but generally has little impact on the slope of the line. Thus, the predicted change in the dependent variable for a unit change in the independent variable remains much the same in the case where the outlier is the result of an extreme value of the dependent variable. This is not the case when the outlier is the result of an extreme value of the independent variable, since the slope of the regression line also changes.

The Coefficient of Determination and the Correlation Coefficient

Having gone through the process of fitting a linear trend line to a set of data, it is next logical to consider how well the model fits the data. One way to assess the quality of the

EXHIBIT 3.8 Impact of Outliers on Regression Line Fit to Set of Data

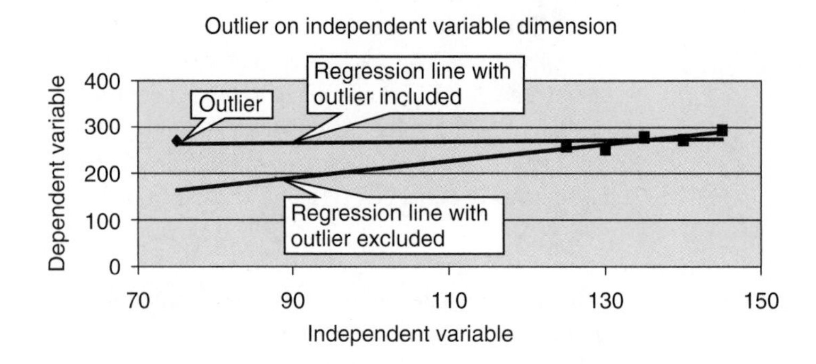

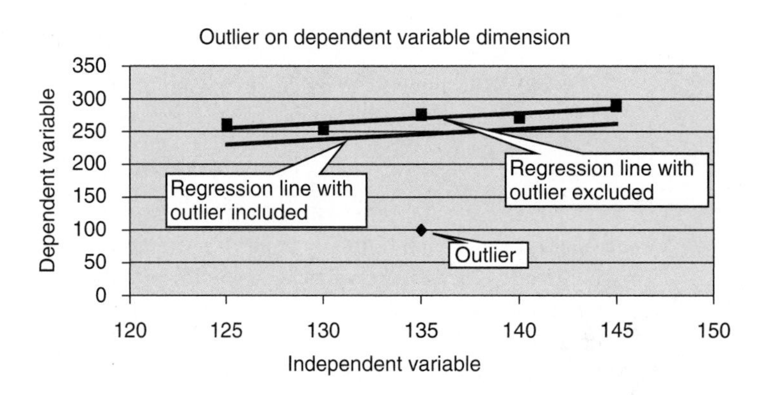

model is to simply plot the trend line and data on the same graph and visually evaluate the quality of the fit. Another, more objective, approach is to determine the proportion of variation in the dependent variable that can be explained by the independent variable. This measure is called the **coefficient of determination,** and is typically represented symbolically as R^2. Since R^2 corresponds to the proportion of variation in the dependent variable that is explained by the independent variable, it should not surprise you to learn that the R^2 will be between zero and one. An R^2 of one indicates that the independent variable completely accounts for the variation in the dependent variable, even though it may not *cause* it. (There could be a third factor causing both, or the direction of causation may be the reverse. For example, it has been observed that overweight people drink more diet soft drinks than others. However, drinking diet soft drinks does not *cause* one to be overweight.) In this case, all data points will fall precisely on the trend line. Alternatively, an R^2 of zero indicates that there is no relationship between the independent and dependent variables.

One way to calculate the R^2 value when using Excel is to follow the procedure discussed earlier for adding a trend line to plotted data. After selecting the data (B3:B12 and C3:C12) and then the menu items Chart/Add Trendline, select the Options tab in the Add Trendline dialog box. As shown in Exhibit 3.9 in the Trendline dialog box, select the "Display R-squared value on the chart" check box and then press the OK button. The value of R^2 will then be calculated and displayed in the chart as shown in Exhibit 3.10.

EXHIBIT 3.9 Excel's Trendline Dialog Box

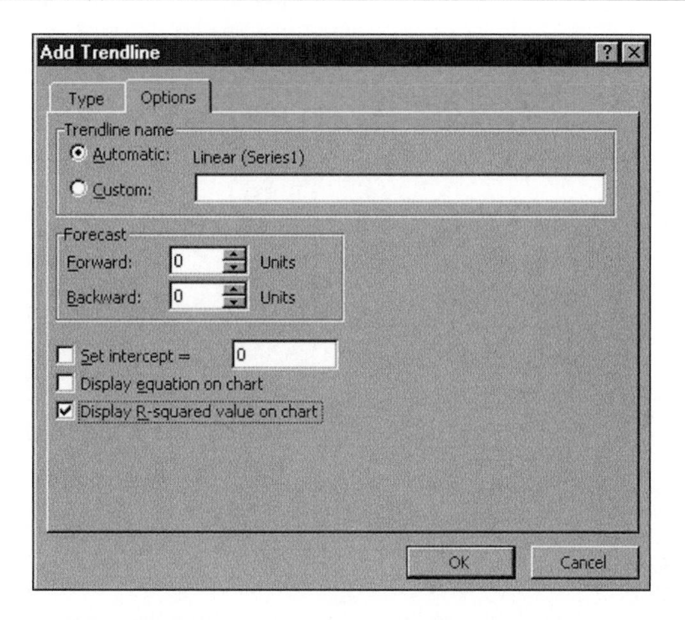

EXHIBIT 3.10 Using Excel to Calculate R^2

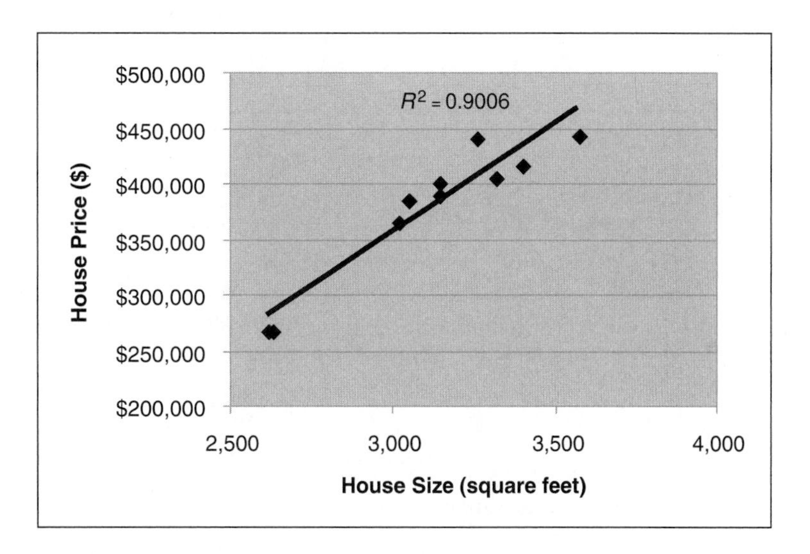

The R^2 of 0.9006 displayed in Exhibit 3.10 suggests that more than 90 percent of the variation in the price of a new house can be explained by its size. The remaining 10 percent of variation in house prices is due to other variables. Clearly, this R^2 is quite high, and in many real-life situations a value half this large would be considered more than acceptable.

Another way to explain R^2 is to divide the variation of the data into its elemental parts as shown in Exhibit 3.11. As was illustrated in Chapter 2, the variance of a set of data is based on the squared distance of the data points from the average. In an analogous

EXHIBIT 3.11 Components of Variation

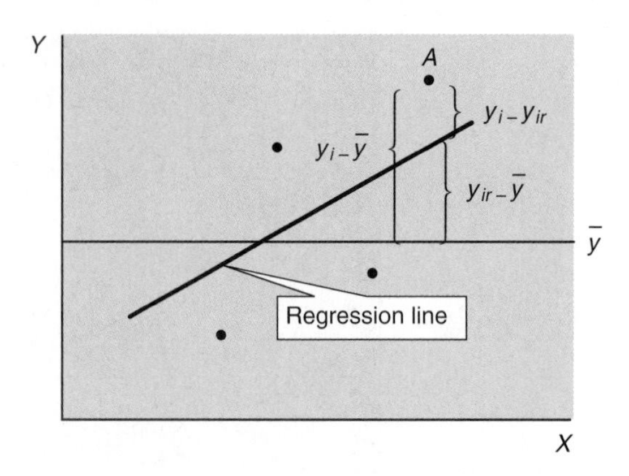

fashion, the variation of the data in a regression study is often assessed using a measure called the **total sum of squares,** or SST. Just as the variance of a set of data is based on the squared distance of each observation from the mean, SST is based on the squared vertical distance of each data point from the average value of the dependent variable. More specifically, if we let y_i represent the value of dependent variable associated with data point i and we let $\bar{y}$ represent the average value of the dependent variable across all n data points, then SST is calculated as

$$\text{SST} = \sum_{n} (y_i - \bar{y})^2$$

As is illustrated in Exhibit 3.11, the total variation of the data from the mean as measured by SST can be divided into two components. For example, referring to point A in Exhibit 3.11, we can observe that the regression line fit to the data helps explain some of its variation from the mean. However, because point A does not fall directly on the regression line, some of its variation still remains unexplained. Thus, the closer the data points are to the regression line, the better the regression line explains the variation in the data. More formally, we can assess the variation in the data explained by the regression line by calculating the **sum of squares of regression,** or **SSR.** Letting y_{ir} be the regression line estimate for data point i, SSR can be calculated as

$$\text{SSR} = \sum_{n} (y_{ir} - \bar{y})^2$$

Finally, the variation around the mean not accounted for by the regression line is called the **sum of squares of error** or SSE and is calculated as

$$\text{SSE} = \sum_{n} (y_i - y_{ir})^2$$

Based on Exhibit 3.11, we can see that SST = SSR + SSE. Earlier we noted that R^2 represents the percent of variation in the dependent variable that is explained by the independent variable. Based on this, one way to calculate R^2 is to assess the percent of the total variation explained by the regression line as follows:

$$R^2 = \frac{\text{SSR}}{\text{SST}} = \frac{\text{SST} - \text{SSE}}{\text{SST}}$$

The **correlation coefficient,** R, is another measure for assessing the extent to which two variables are related to one another. More specifically, the correlation coefficient measures the degree to which there is a linear relationship between two variables and is calculated by taking the square root of R^2 and appending a plus or minus sign, according to whether the slope is positive or negative. The correlation coefficient can thus range between –1 and +1. It is positive if Y tends to increase as X increases and negative if Y tends to increase when X decreases. Like the coefficient of determination, a correlation of zero suggests there is no linear relationship between X and Y, but a large value does not necessarily imply causation. Referring to the spreadsheet shown in Exhibit 3.7, Excel's CORREL function can be used to calculate the correlation coefficient between house price and house size as follows:

=CORREL(C3:C12,B3:B12)

The calculated correlation coefficient using the CORREL function is 0.949. Squaring the correlation coefficient yields the R^2 of 0.9006 (i.e., $0.949^2 = 0.9006$) shown in Exhibit 3.10. Note that the use of the CORREL function does not require defining one range of variables as the dependent variable and the other as the independent variable. Therefore, it does not matter whether the range of the independent or dependent variable is entered as the first parameter in the CORREL function.

Finally, we note that because R^2 provides precise information regarding the percent of Y's variation that can be explained by X, its interpretation is more meaningful than the correlation coefficient, and it is therefore the preferred measure of the two.

Regression Analysis Assumptions

In addition to the assumption that there is a linear relationship between the dependent and independent variables, regression analysis also assumes the following.

- **The Residuals Are Normally Distributed** That is, for a particular value of the independent variable, a plot of all the errors around the regression line at this point would be normally distributed.
- **The Expected Value of the Residuals Is Zero, $E(e_i) = 0$** The plot of the errors would be centered about zero in terms of their mean value. This also implies that the expected value of the dependent variable falls directly on the regression line for each possible value of the independent variable.
- **The Residuals Are Independent of One Another** The value of one error does not have any effect on the value of another error, either positive or negative.
- **The Variance of the Residuals Is Constant** The spread of the errors about the regression line does not vary with the independent variable.

Perhaps the most common way to verify that these assumptions are met is to perform an analysis of the residuals (or errors). For a particular value of X, the residual is calculated by subtracting the trend line estimate of Y for that value of X from the actual Y value corresponding to that value of X. For example, referring to Exhibit 3.12, the residual for the 2,620 ft^2 house is the actual price of the house minus what the trend line estimates a 2,620 ft^2 house would cost, or \$266,500 – \$281,020 = –\$14,520.

Once the residuals are calculated (column E of Exhibit 3.12), it is common to plot them against the independent variable, as shown in the graph at the right of the exhibit. Exhibit 3.13 illustrates several possible scenarios and the implications of these scenarios in terms of the regression model assumptions. In the top graph, the residuals seem to fall randomly around a mean of zero, and no violation of the assumptions is indicated. In the middle graph, the residuals are increasing as the value of X increases leading to what statisticians call **heteroscedasticity,** or nonconstant variance. In the bottom graph, the residuals display a systematic or nonrandom pattern. This suggests that the error terms or residuals are not independent of one another. Patterns such as those exhibited in the

EXHIBIT 3.12 Analysis of Residuals

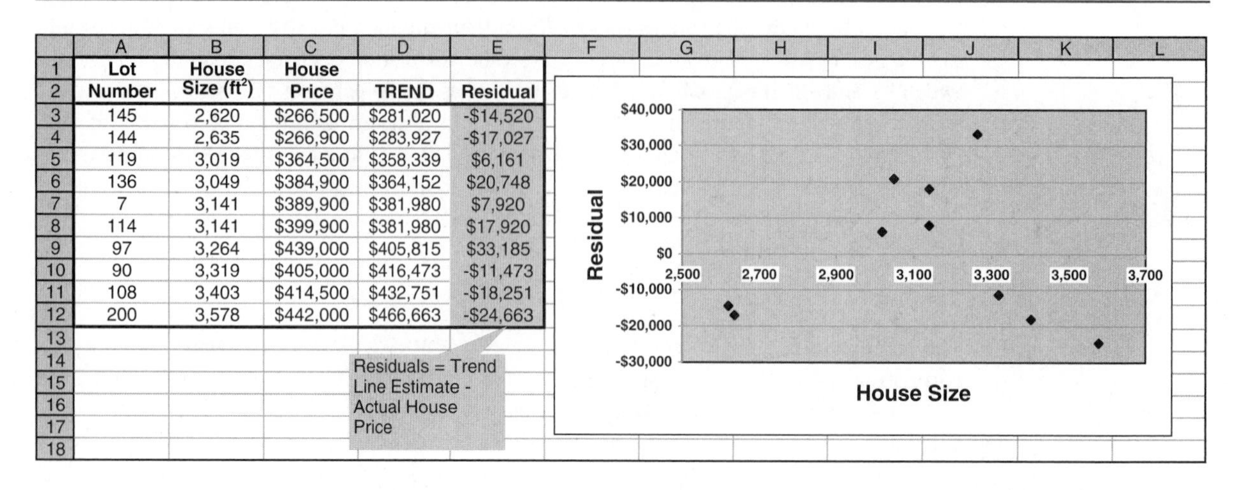

Lot Number	House Size (ft²)	House Price	TREND	Residual
145	2,620	$266,500	$281,020	-$14,520
144	2,635	$266,900	$283,927	-$17,027
119	3,019	$364,500	$358,339	$6,161
136	3,049	$384,900	$364,152	$20,748
7	3,141	$389,900	$381,980	$7,920
114	3,141	$399,900	$381,980	$17,920
97	3,264	$439,000	$405,815	$33,185
90	3,319	$405,000	$416,473	-$11,473
108	3,403	$414,500	$432,751	-$18,251
200	3,578	$442,000	$466,663	-$24,663

middle and bottom graphs are a clear indication that one or more of the regression model assumptions are violated. Such violations of the model assumptions may be the result of inappropriately trying to use a linear model when another type of model is more appropriate, not properly transforming the variables, and/or failing to include other important relevant variables. At any rate, violations of these four assumptions should send the modeler back to reevaluate the model.

It should also be noted that the patterns depicted in Exhibit 3.13 were exaggerated for the purpose of illustration. In reality, the actual patterns exhibited in a plot of the residuals are often much more subtle, as in Exhibit 3.12 where it appears our assumptions are valid. Determining whether a violation of the assumptions exists is largely based on the judgment and experience of the modeler.

Using the Regression Model

Just as it is important to ensure that the regression model assumptions are not violated, it is valuable to understand how to properly use the results of a regression analysis. Generally speaking, new users of regression analysis should be aware of three common pitfalls. The first pitfall is to use a regression model to make predictions outside the range of data that was used to develop the model. As an example, it would be improper to attempt to use the regression model fit to the data in Exhibit 3.1 to predict the price of a 4,500 ft^2 house. This would be improper because none of the observations in the data set is representative of a house of this size. Attempting to use a model to predict values that are not represented in the data set is called **extrapolation.**

A second pitfall is attempting to overly *generalize* the results of a regression model. For example, the data shown in Exhibit 3.1 were collected for new homes built in a particular subdivision in North Carolina. It is not at all clear whether the regression model fit to this data could be used to predict the price of a new house in other subdivisions in the same city. And it is even less likely that the regression model could be used to predict the price of a new house in other parts of the state or regions of the country.

The problem of generalization may, at first glance, appear to be similar to the problem of extrapolation. There is, however, an important difference. Extrapolation happens when we attempt to make a prediction beyond the values in our data set. Alternatively, **generalization** occurs when we attempt to use the model fit to data collected from one population to predict values in another population. Again, referring to the house price

EXHIBIT 3.13 Sample Residual Plots

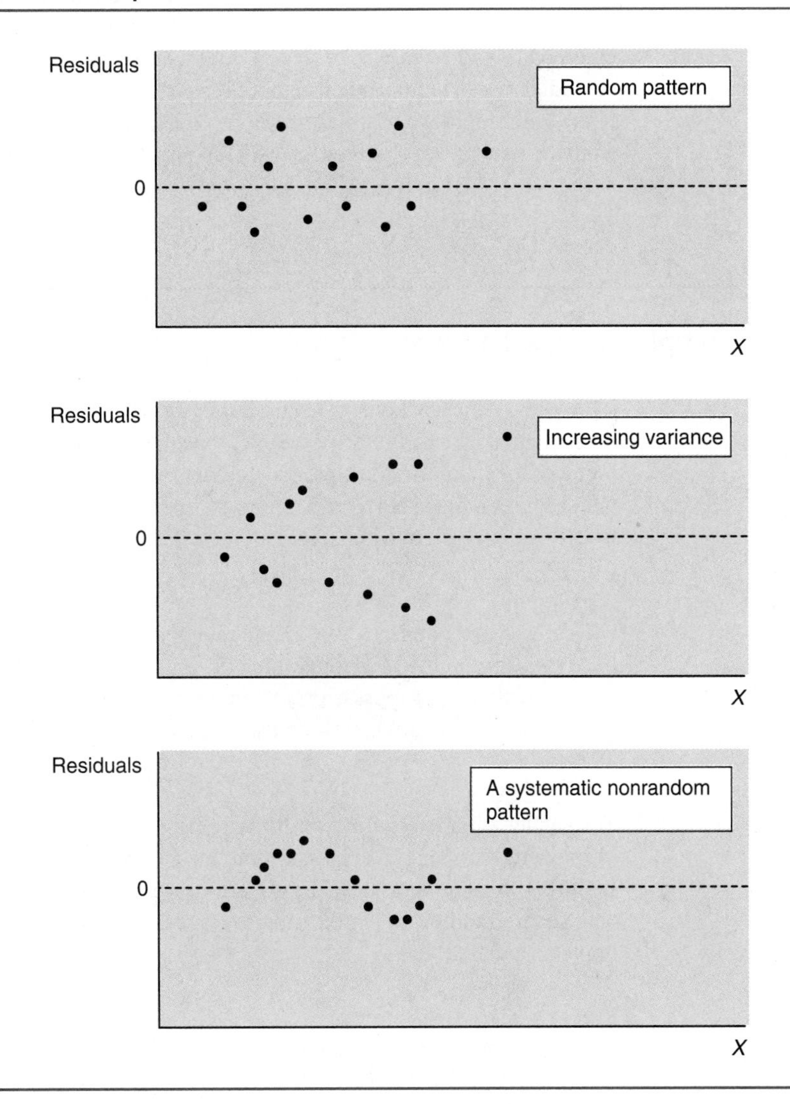

model, we can think of each subdivision or region as a population. Clearly, house construction costs could vary from subdivision to subdivision based on a variety of factors, including the cost of the land, quality of schools, distance to important destinations, available amenities, and so on. Thus, the problem of extrapolation occurs when we attempt to use a model to predict values for the population of interest that are not represented in our data set, while generalization occurs when we are attempting to use the model to make predictions for an entirely different population. Of course, it is possible to make both mistakes at the same time.

The final pitfall is to improperly assume that the development of a regression model proves that there is a cause-and-effect relationship between the independent and dependent variables. Generally speaking, a regression model can be used to help validate that such a cause-and-effect relationship exists, but the actual existence of such a relationship must have its basis in some underlying theory. As a rather extreme example, suppose you collected monthly data for a number of years on ice cream sales and the number of drownings in the United States. If you were to develop a regression model with the number of drownings as the dependent variable and ice cream sales as the independent variable, you would likely get a pretty high R^2. Of course, concluding that the use of ice cream causes drownings (or worse, vice versa) on the basis of this regression model is a bit ludicrous. What is actually happening in this situation is that both variables are correlated with another variable—namely, weather. Thus, we remark that while regression analysis may be well suited to establishing the extent that two variables are correlated with one another, inferring causation between the variables is far more tenuous.

3.3 The Multiple Regression Model

Up to this point we have focused on the use of one independent variable to predict values of the dependent variable. As we demonstrate in this section, it is possible to extend this methodology and use multiple independent variables to predict the value of the dependent variable. Including more than one independent variable in the regression model is called **multiple regression.** Mathematically, the form of the multiple regression model is

$$Y = \alpha + \beta_1 X_1 + \beta_2 X_2 + \cdots + \beta_n X_n + \varepsilon$$

where X_i corresponds to the ith independent variable for $i = 1, 2, \ldots, n$ and $\alpha, \beta_1, \beta_2, \ldots, \beta_n$ are the model parameters.

As with simple regression, when a multiple regression model is developed based on sample data the model is written as:

$$Y = a + b_1 X_1 + b_2 X_2 + \cdots + b_n X_n$$

The model parameters for a multiple regression model are calculated in a similar fashion as they are for a simple regression model. In both cases, the model parameters are chosen such that the summation of the squared errors (or residuals) over all observations in the data set are minimized. In the simple regression model, the error for a particular observation is calculated as

$$e = Y_0 - (a + bX)$$

where e is the error or residual for a given observation, Y_0 is the actual observed Y value, and $(a + bX)$ is the predicted Y value for the observation based on the regression model. In English, the error for a given observation of the dependent variable is its observed or actual value minus the predicted value based on the regression model. Extending this, the error in a multiple regression model for a given observation can be calculated as

$$e = Y_0 - (a + b_1 X_1 + b_2 X_2 + \cdots + b_n X_n)$$

The least squares approach then selects the parameters of the regression model such that the sum of the squared errors or $\sum e_i^2$ is minimized.

While beyond the scope of this book, we do note that equations can be easily derived using calculus to calculate the parameters a and b_i. Instead, we rely on the Excel functions previously discussed to calculate the model parameters for a multiple regression model in a similar fashion to the way they were used to calculate the model parameters for a simple regression model. In fact, the only difference in using these functions to calculate the parameters for a multiple regression model is that the parameter corresponding to the range of X values will include more than one column of data.

To illustrate calculating the parameters for a multiple regression model, assume Nick Carol expanded his idea of the factors influencing a house price to that shown in the influence diagram of Exhibit 3.14. He thus obtained additional data for the price of the lots,

EXHIBIT 3.14 Factors that Influence the Price of a New Home

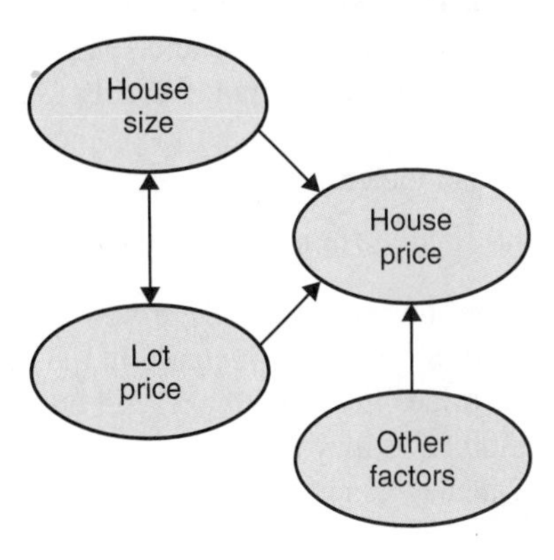

as shown in Exhibit 3.15, ignoring any other factors for the moment. After inserting column C and entering the newly obtained lot price data, the labels b_2, b_1, and a were entered in cells D14:F14, respectively. Next, Excel's LINEST function was used to calculate these model parameters by highlighting the range D15:F15 and entering the formula

=LINEST(D3:D12,B3:C12)

which identifies the first range of data as the dependent variable Y and the second range of data as the independent variables in the order X_1, X_2, and so on. After entering this formula, the Ctrl + Shift + Enter keys were pressed simultaneously, with the resulting model parameters shown in the exhibit. Note that in general, Excel's LINEST function returns the parameters of the independent variables in the following

EXHIBIT 3.15 Multiple Regression Model to Predict New Home Price

	A	B	C	D	E	F
1–2	Lot Number	X_1 - House Size (ft^2)	X_2 - Lot Price	Y - House Price	Trend	Error
3	145	2,620	$51,500	$266,500	$270,804	-$4,304
4	144	2,635	$52,000	$266,900	$274,022	-$7,122
5	119	3,019	$72,500	$364,500	$369,437	-$4,937
6	136	3,049	$82,000	$384,900	$390,252	-$5,352
7	7	3,141	$67,000	$389,900	$379,430	$10,470
8	114	3,141	$72,500	$399,900	$388,734	$11,166
9	97	3,264	$78,000	$439,000	$417,493	$21,507
10	90	3,319	$68,500	$405,000	$410,122	-$5,122
11	108	3,403	$58,000	$414,500	$405,647	$8,853
12	200	3,578	$78,000	$442,000	$467,158	-$25,158
13						
14				b_2	b_1	a
15				1.691573827	158.1708	-230719.6
16						
17						
18						
19						
20						

Regression model parameters with second independent variable (lot price) added.

order: b_n, b_{n-1}, b_{n-2}, . . . , b_1, a, which is the reverse of the order in which they were entered. The TREND function was used in a similar fashion by first highlighting the range E3:E12, entering the formula =TREND(D3:D12,B3:C12) and then pressing the Ctrl + Shift + Enter keys simultaneously. According to the results shown in Exhibit 3.15, the multiple regression model to predict the price of a new house based on the size of the house and the cost of the lot is

$$\text{House price} = -230{,}720 + 158.17\ (\text{house size}) + 1.69\ (\text{lot \$})$$

The interpretation of the new coefficient for lot price is that, other factors staying constant, a one-dollar increase in the lot price will be reflected in a $1.69 increase in the house price. In Nick's case, the cost of the lot that he is interested in building on is $89,500. You may also recall that he wants to build a 3,262-ft^2 house. Based on this, we can calculate a new estimated price using the multiple regression model as follows:

$$\text{House price} = -230{,}720 + 158.17(3{,}262) + 1.69(89{,}500) = \$436{,}485.54$$

Comparing the estimated cost based on the multiple regression model to the estimated cost based on the simple regression model developed earlier, we observe that there is a fairly significant difference of more than $31,000 ($436,486 – $405,428) between the two estimates. Which do you think is more accurate?

There are a number of ways to assess this. One way would be to compare the average absolute error of the two models. We need to compare the absolute value of the error because the sum of the error terms when the least squares method is used is always zero. Since the sum of the errors is zero, dividing this sum by the number of observations to get the average error would also result in zero. Using Excel's AVEDEV function =AVEDEV(F3:F12), it can be easily determined that the average absolute error was reduced by almost 40 percent from $17,187 in the simple regression model to $10,399 in the multiple regression model. In the case of the multiple regression model, this indicates that, on the average, the difference between the trend estimate for a given house's price and its actual price differed by $10,399.

Another way to compare two or more regression models is to compare their R^2 (called the *multiple coefficient of determination* in the case of multiple regression) values. We previously determined that the R^2 for the simple regression model was approximately 90.1 percent. When the cost of the lot is added to the model, the R^2 increases to approximately 95.6 percent, indicating that the combination of house size and lot cost explains over 95 percent of the variation in house price. Calculating R^2 for multiple regression models will be discussed in the detailed modeling example at the end of this chapter.

It should be noted that it would be incorrect to conclude that the dependent variable added (i.e., lot cost) only explains 5 percent of the variation in house price. This is because there is some amount of correlation (shown by the double-headed arrow in Exhibit 3.14) between house size and lot price (known as *multicollinearity*). Therefore, some of the variation that lot price explains in the price of the house has already been accounted for by house size. Using Excel's CORREL function—CORREL(C3:C12,B3:B12), it is easy to determine that the correlation between house size and lot price is approximately 59 percent, and squaring this yields an R^2 of approximately 35 percent. Thus, we see that approximately 35 percent of the variation in lot price is explained by the house size (and vice versa).

Adding additional variables to a regression model therefore only helps explain variation in the dependent variable to the extent that the independent variables are not correlated with one another. If two independent variables were perfectly correlated with one another, fitting a multiple regression model to the data with both variables would provide no better fit than either of the variables alone.

This discussion raises another important issue. Generally speaking, we can interpret b_i as the impact that any changes in the ith independent variable will have on the dependent variable while holding the other regression model parameters constant. However, the individual impact of each independent variable can get blurred when some of the independent variables are highly correlated with others. To check for this problem, the correlation coefficients between all pairs of independent variables can be calculated. As a rule of thumb, only include two independent variables in the model when the correlation coefficient between them is less than 0.80.

3.4 Developing Regression Models

Trying to remember all the issues related to proper regression model development can be overwhelming at times. This challenge can be greatly diminished if the modeler breaks down the regression model development process into four logical and sequential steps. In the remainder of this section, we overview this four-step process for regression model development.

Step 1: Identify Candidate Independent Variables to Include in the Model

Upon defining the dependent variable to be investigated, the first step in the development of a regression model is to identify candidate independent variables to include in the model. Constructing an influence diagram may be particularly helpful at this stage of the regression model development process. Of course developing an influence diagram and selecting candidate independent variables requires knowledge of the dependent variable and the factors that affect it. In Nick Carol's case, he determined that the size of a house affects its price. Of course, depending on the modeler's expertise with the dependent variable being studied, he or she may need to consult with managers and other people to identify the variables that may influence the dependent variable. For example, suppose you were asked to develop a model for predicting the engine emissions of light-duty, diesel-powered engines. Most likely, you would not know what variables affect engine emissions, and therefore, you would need to consult with one or more specialists (e.g., engineers, mechanics, and scientists).

Once a candidate pool of potential independent variables has been identified, it is important to check the correlation among the independent variables. As was discussed in the previous section, when two independent variables are highly correlated, the individual impact of each independent variable can get blurred if both variables are included in the multiple regression model. The easiest way to avoid this problem of multicollinearity is to calculate the correlation coefficients between all pairs of independent variables and not include in the model both independent variables if their correlation coefficient exceeds .80.

As we have seen so far, independent variables can be used to include quantitative continuous data such as house size and lot price in a regression model. Categorical data can also be included in regression models through the use of a special type of independent variable called a **dummy variable.** For example, if Nick Carol wanted to investigate the effect that having a daylight basement (one of the "other factors" in Exhibit 3.14) had on house price he could add a third independent variable defined as follows:

$$X_3 = \begin{cases} 0 \text{ if house does not have a daylight basement} \\ 1 \text{ if house does have a daylight basement} \end{cases}$$

As shown in Exhibit 3.16, this variable was added and each of the houses was coded with either a zero or one, depending on whether it had a daylight basement (DLB—a basement that is partially above ground, with windows for natural light). Using the LINEST function—=LINEST(E3:E12, B3:D12)—we get the new regression model with the dummy variable added:

$$\text{House price} = -215{,}717 + 152\ (\text{house size}) + 1.66\ (\text{lot \$}) + 16{,}296\ (\text{DLB})$$

The interpretation of the dummy variable is relatively straightforward. If a particular house has a daylight basement, then DLB = 1 and the model would increase the price of the house by a constant $16,296. On the other hand, if the house does not have a daylight basement, then DLB = 0, and the $16,296 is not added. Therefore, our regression model suggests that adding a daylight basement to a house increases its price by $16,296.

A final consideration worth mentioning regarding the choice of independent variables is whether *lagged* values of the variable should be used. For example, a company might be interested in determining the impact that its investments in information technology are having on organizational performance. In such cases there is often a time lag between when an investment is made, such as the purchase of a new enterprise resource planning system, and the realization of any benefits. In fact, it is quite common for such investments to have a significant negative impact on performance in the very short run, but a more positive impact over the longer term. In cases such as this it often makes sense to lag the independent variable. In the situation discussed here, management may decide that it makes sense to use dollars invested in information two years ago as the independent variable to predict the change in profitability in the current year. Of course, the com-

EXHIBIT 3.16 Regression Model with Dummy Variable

	A	B	C	D	E
1	Lot	X_1 - House	X_2 - Lot	X_3 - Daylight	Y - House
2	Number	Size (ft²)	Price	Basement	Price
3	145	2,620	$51,500	0	$266,500
4	144	2,635	$52,000	0	$266,900
5	119	3,019	$72,500	1	$364,500
6	136	3,049	$82,000	0	$384,900
7	7	3,141	$67,000	0	$389,900
8	114	3,141	$72,500	1	$399,900
9	97	3,264	$78,000	1	$439,000
10	90	3,319	$68,500	0	$405,000
11	108	3,403	$58,000	1	$414,500
12	200	3,578	$78,000	0	$442,000
13					
14	b_3	b_2	b_1	a	
15	16,295.87	1.66	152.01	-215,717.12	
16					
17					
18					
19					
20					
21					

Regression model parameters with addition of third independent variable (daylight basement)

pany is not limited to including only one year's investment in information technology in its model; it could include the investments made in several of the past periods.

Step 2: Transform the Data

As was discussed earlier, prior to developing a regression model, it is prudent to plot the independent and dependent variables to verify that they are indeed linearly related. In the case of multiple regression, a scatter diagram plotting each independent variable with the dependent variable should be developed to ensure that each independent variable is linearly related to the dependent variable. If one or more of these plots indicates that the variables are not linearly related, it may still be possible to transform the variables in order to obtain a linear relationship.

To illustrate this, Exhibit 3.17 depicts six possible relationships between the independent and dependent variable. In Panel *a,* there is a clear linear relationship between X and Y, so no transformation is required. Panel *b,* however, indicates that there is a quadratic relationship between X and Y, such as $Y = X^2$. If this type of pattern is observed, a linear relationship between X and Y can be obtained by taking the square root of the X values. In other words, the regression model would be fit to the square root of the original X values.

EXHIBIT 3.17 Example Relationships Between the Independent and Dependent Variables

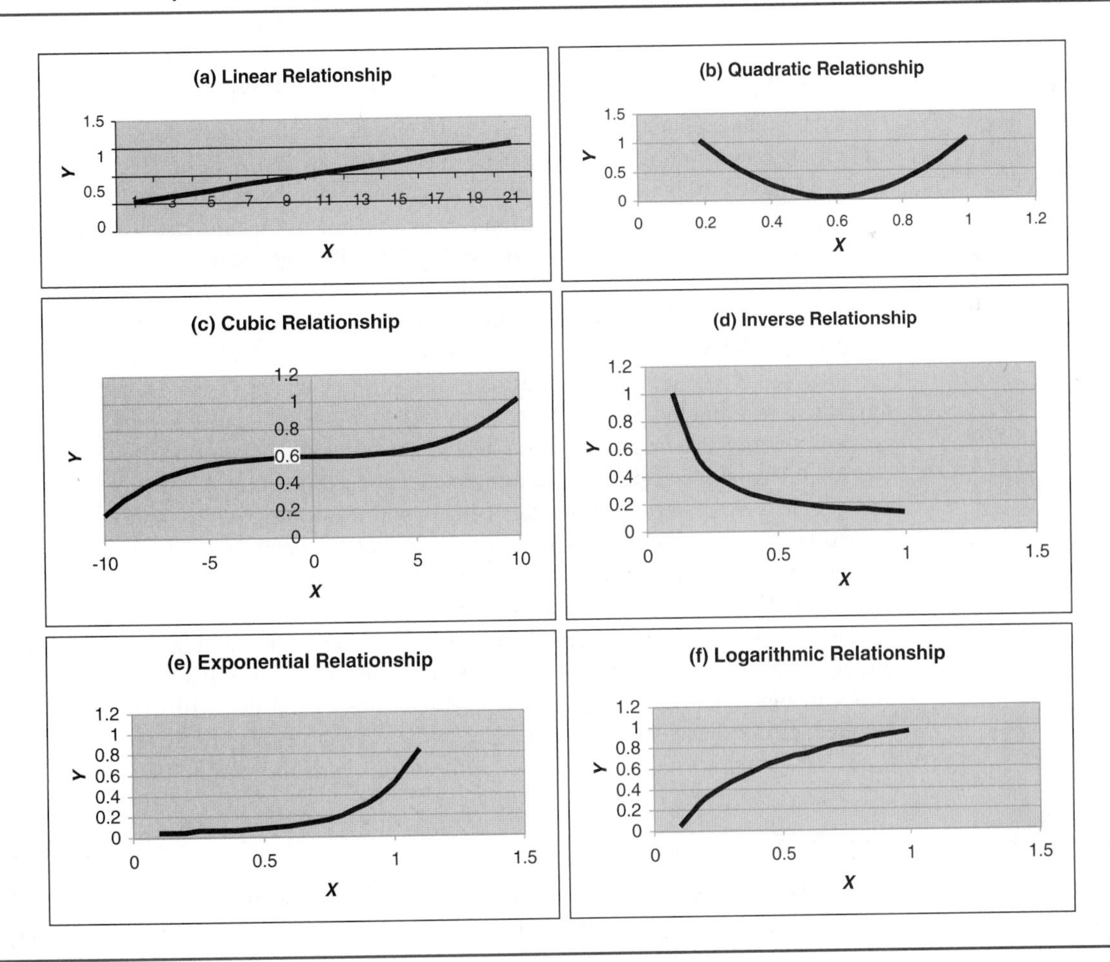

Panel *c* is indicative of a cubic relationship of the form $Y = X^3$. In this case a linear relationship can be obtained by taking the cube root of the *X* values (i.e., $\sqrt[3]{X} = X^{\frac{1}{3}}$). Panel *d* suggests an inverse relationship between the variables of the form $Y = 1/X$. In this case, a linear relationship can be obtained by taking the inverse of the *X* values (i.e., dividing 1 by the *X* values).

Panel *e* depicts an exponential relationship of the form $Y = e^X$. A linear relationship can be obtained in this case by using Excel's LN function to take the natural log of the *X* values. Finally, a logarithmic relationship is suggested in Panel F. In this case, a linear relationship can be obtained by raising the base *e* by the power of *X* (i.e., e^X). This can be easily accomplished using Excel's EXP function.

A couple of remarks are in order regarding transforming variables in a regression model. First, in addition to transforming the independent variables, it is also possible to transform the dependent variable. For example, if it is determined that several of the independent variables require the same transformation, the modeler may choose to simply transform the dependent variable rather than performing the same transformation on several of the independent variables. Second, note that the interpretation of the regression model is changed when the variables are transformed. To illustrate, suppose that in our house price example we determined that there was a quadratic relationship between house price and house size. If the subsequent regression model fit house price to the square root of house size, then the interpretation of the model parameter *b* would be altered. In this case, *b* would refer to the impact changes in the *square root* of house size would have on the price.

Step 3: Select the Variables to Include in the Model

Once the candidate independent variables have been specified, their correlations checked, and any necessary transformations performed, the next step is to determine specifically which variables to include in the regression model. Often computer packages use some type of *stepwise* procedure to determine which values to include in the final regression model. For example, with **backward elimination,** all independent variables are included in the model and the variables with the least predictive value are dropped one at a time with the model evaluated at each iteration. **Forward selection** works in exactly the opposite direction by selecting one new variable for inclusion at each iteration. One way that variables could be selected for inclusion is to determine which variable when added to the model will result in the greatest increase in R^2. Of course the opposite approach would be employed with backward elimination, where the variable that resulted in the smallest decrease in R^2 would be removed from the model at each iteration. Later in this chapter we address the issue of variable selection using hypothesis testing.

Although it is theoretically possible for R^2 not to change as additional independent variables are added to the model, it is impossible for it to decrease as additional variables are added. Therefore, in general, including additional independent variables will tend to increase R^2. The calculations are beyond our scope here, but we note that many analysts prefer to use a measure that adjusts R^2 for both the sample size and number of independent variables included in the model. This measure, called the **adjusted R^2**, is used to help reduce the chances that R^2 is inflated as more variables are added, and is included in Excel's Regression Output Report (see Adjusted *R* square in Exhibit 3.19).

Step 4: Analyze the Residuals

As was discussed previously in the section on simple linear regression, an analysis of the residuals is a useful way to validate that the assumptions of regression analysis are met. As noted earlier, two key assumptions are that the expected value of the error terms (residuals) equals zero and that they are normally distributed. Perhaps the easiest way to validate this assumption is to create a histogram for the residuals and note, in particular, if there tends to be a grouping around zero. Examination of the histogram can also indicate the extent to which the data are skewed (unsymmetrical) and whether outliers (extreme values) are present.

To investigate whether the assumption of constant variance is met, it is common to plot the residuals against the predicted values of Y. Also, plots for each independent variable against the residual can be developed. The interpretation of these plots is similar to our earlier discussion of the patterns exhibited in Exhibit 3.13.

Finally, the assumption that the error terms are independent is important because if they are correlated with one another (called *autocorrelation*), the regression model fit to the data based on the least squares method will tend to underestimate the values of the error terms. This creates problems later when we attempt to construct confidence intervals for predictions made on the basis of the model. One common approach to test the hypothesis that the residuals are not correlated with one another and are therefore independent is with the Durbin–Watson test statistic (see Albright et al. 1998) for autocorrelation.

3.5 Regression Hypothesis Tests

When running a regression to determine the relationship between two or more variables, we usually hope to conclude that there is a significant relationship (i.e., that one or more factors can help predict the dependent variable). We can test whether there is such a relationship by testing the hypothesis that the coefficient(s) of the independent variable(s) is (are) zero. If any of the coefficients are zero, then those independent variables have no relationship to (i.e., do not help predict) the value of the dependent variable. But if we can reject this hypothesis at some acceptable level of significance, we can feel more confident that we have indeed found a relationship.

For the overall regression, we can employ an ANOVA test, using a test statistic that compares the variance explained by the regression to the unexplained variance. This statistic again follows an F distribution and is interpreted as in Chapter 2—a high value provides evidence, depending on the p value, that the independent variable(s) is (are) related to the dependent variable. However, if there is more than a single independent variable, we will not know which variable it is that is significant. Hence, it is appropriate to then run individual t tests on each of the independent variables. The t tests take the same form: the null hypothesis is that the independent variable has no relationship to the dependent variable and thus its coefficient is zero. The results are interpreted just as before. If there is a single independent variable, then the p value of the F test and the p value of the t test will be identical.

To illustrate the use of hypothesis testing in conjunction with regression analysis, we return to the data obtained to predict the price of building a new house. These data are summarized in Exhibit 3.18. Briefly, the data set includes three potential independent variables: house size (in ft^2), the cost of the lot, and a dummy variable associated with whether the house has a daylight basement. The dependent variable is the price of the house.

The results of using Excel's Regression tool (Tools/Data Analysis. . ./Regression) are shown in Exhibit 3.19. The top portion of this report contains relevant statistics related to the regression model, including the multiple coefficient of determination, R^2. The middle of the report contains the ANOVA test for the overall model. In cell K12, we observe the relatively large F value of 71.9, which provides evidence that one or more of the independent variables is related to the dependent variable. Further verification of this

EXHIBIT 3.18 New Home Prices and Related Data

	A	B	C	D	E
1	Lot	X_1 - House	X_2 - Lot	X_3 - Daylight	Y - House
2	Number	Size (ft²)	Price	Basement	Price
3	145	2,620	$51,500	0	$266,500
4	144	2,635	$52,000	0	$266,900
5	119	3,019	$72,500	1	$364,500
6	136	3,049	$82,000	0	$384,900
7	7	3,141	$67,000	0	$389,900
8	114	3,141	$72,500	1	$399,900
9	97	3,264	$78,000	1	$439,000
10	90	3,319	$68,500	0	$405,000
11	108	3,403	$58,000	1	$414,500
12	200	3,578	$78,000	0	$442,000

EXHIBIT 3.19 Excel's Regression Output Report

	G	H	I	J	K	L
1	SUMMARY OUTPUT					
2						
3	*Regression Statistics*					
4	Multiple *R*	0.9864				
5	*R* square	0.9729				
6	Adjusted *R* square	0.9594				
7	Standard error	12651.9516				
8	Observations	10				
9						
10	ANOVA					
11		*df*	*SS*	*MS*	*F*	*Significance F*
12	Regression	3	34539597728	11513199243	71.92518346	0.0000
13	Residual	6	960431272.2	160071878.7		
14	Total	9	35500029000			
15						
16		*Coefficients*	*Standard Error*	*t Stat*	*P value*	
17	Intercept	-215717.12	43637.0313	-4.9434	0.0026	
18	*X* Variable 1	152.01	17.3556	8.7584	0.0001	
19	*X* Variable 2	1.66	0.4824	3.4359	0.0139	
20	*X* Variable 3	16295.87	8443.4409	1.9300	0.1018	
21						
22						
23						
24						
25						

Large *F* value suggests one or more independent variables has predictive value

Regression model parameters

is provided by the p value of almost zero in cell L12 (actually, it is larger than zero, but when rounded to four decimal places is displayed as zero). This indicates that there is almost no chance of obtaining a test statistic based on the F distribution as large as the one obtained if the null hypothesis that none of the independent variables has any predictive value regarding the dependent variable (i.e., their coefficients are all zero) were true.

The bottom of the report shown in Exhibit 3.19 provides the calculated parameters for the intercept and each of the independent variables (column H). In addition, this table provides the results of the individual t tests on each of the independent variables. For example, the first independent variable listed in row 18 corresponds to house size. According to the results shown, the null hypothesis that house size has no predictive value given that the other two variables have already been included in the regression model can be rejected. The calculated test statistic for this variable is 8.76 (cell J18) and there is almost no chance of getting a test statistic this large when the null hypothesis is true, as is indicated by the p value shown in cell K18. The third independent variable (the dummy variable) is more questionable. In this case, we see that there is a 10 percent probability of getting a test statistic of 1.93 when the null hypothesis that this variable has no predictive value given the other two variables are included in the model is true. This may not be strong enough evidence to reject the hypothesis that this variable has no predictive value, but it does make its contribution questionable. We also note that, while beyond the scope of this book, there are F tests that can be used to test the null hypothesis that a set of independent variables has no predictive value given the inclusion of another set of variables in the regression model.

3.6 Time Series Analysis

A **time series** is simply a set of values of some variable measured either at regular points in time or over sequential intervals of time. We measure stock closing prices at specific points in time and quarterly sales over specific intervals of time. If, for example, we recorded the number of books sold each month of the previous year at Amazon.com and kept those data points in the order in which they were recorded, the 12 numbers would constitute a 12-period time series. Time series data can be collected over very short intervals (such as hourly sales at a fast-food restaurant) or very long intervals (such as the census data collected every 10 years).

Components of a Time Series

We analyze a time series because we believe that knowledge of its past behavior might help us understand (and therefore help us predict) its behavior in the future, normally just the next period. In some instances, such as the stock market, this assumption may be unjustified, but in planning many operational activities, history does (to some extent, at least) repeat itself, and past tendencies continue. Our goal is to find a forecasting model that is easy to compute and use, responsive to changes in the data, and accurate in its predictions. To begin our discussion of time series analysis, let us consider the component parts of any time series.

To analyze time series data, it is often helpful to think of it as comprising four components:

1. Trend, T
2. Seasonal variation, S
3. Cyclical variation, C
4. Random variation, R

The *trend* is the long-run direction of the time series, including any constant amount of demand in the data. Exhibit 3.20 illustrates three fairly common trend lines showing changes in demand; a horizontal trend line would indicate a constant, unchanging level of demand.

EXHIBIT 3.20 Three Common Trend Patterns

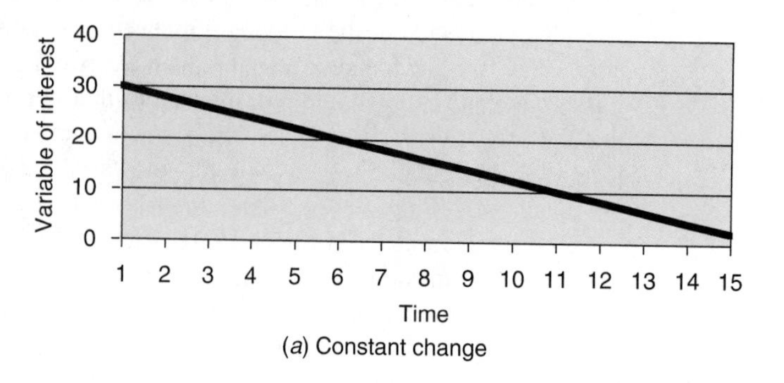

(*a*) Constant change

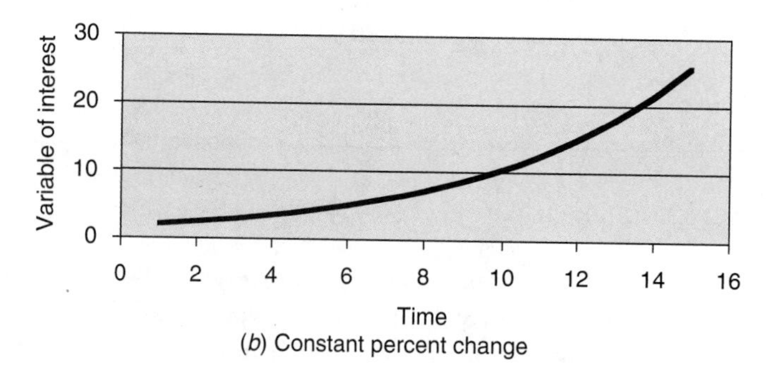

(*b*) Constant percent change

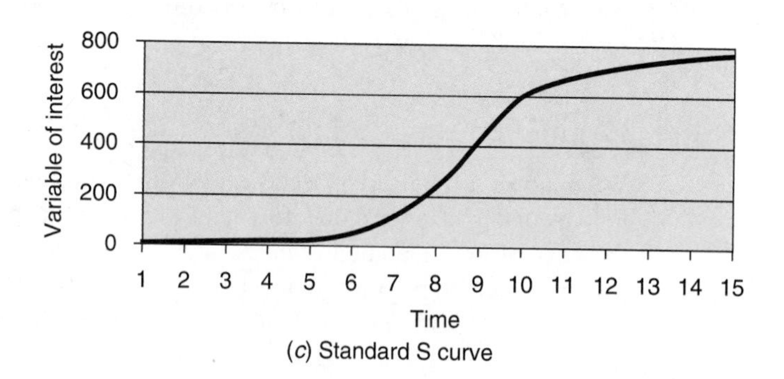

(*c*) Standard S curve

A straight-line or linear trend (showing a constant amount of change, as in Panel *a* of Exhibit 3.20) could be an accurate fit to the historical data over some limited range of time, even though it might provide a rather poor fit over an entire time series. Panel *b* in the figure illustrates the situation of a constant percentage change. Here, changes in a variable depend on the current size of the variable (rather than being constant each period, as in Exhibit 3.20*a*). The trend line shown in Panel *c* of Exhibit 3.20 resembles the life cycle or "stretched-S" growth curve that describes the demand many products and services experience over time.

Seasonal fluctuations are fairly regular fluctuations that repeat within one year's time, or whatever period encompasses the full set of seasonals. Seasonal fluctuations result primarily from nature, but they are also brought about by human behavior. Sales of heart-shaped boxes of candy and pumpkins are brought about by events that are controlled by humans. Snow tires and antifreeze enjoy a brisk demand during the winter months, whereas sales of golf balls and bikinis peak in the spring and summer months. Of course, seasonal demand often leads or lags behind the actual season. For example, the production season for meeting retailers' demand for Christmas goods is August through September. Also, seasonal variation in events need not be related to the seasons of the year. For example, fire alarms in New York City reach a "seasonal" peak at 7 P.M. and a seasonal low at 7 A.M. every day. And restaurants reach three seasonal demand peaks every day at 7:00 A.M., 12:30 P.M., and 8 P.M.

The *cycle* or *cyclical component* is obvious only in time series that span several sets of seasonals. A cycle can be defined as a long-term oscillation, or a swing of the data points about the trend line over a period of at least three complete sets of seasonals. National economic cycles of booms and depressions, as well as periods of war and peace, are examples of such cycles.

Cycles, particularly business cycles, are often difficult to explain. Economists have devoted considerable research and speculation to their causes. Identification of a cyclic pattern in a time series requires the analysis of a long period of data. For most decision-making situations, forecasting the cyclic component is not considered, because long-term data are typically unavailable to determine the cycle. In addition, cycles are not likely to repeat in similar amplitude and duration; hence, the assumption of repeating history does not hold.

Random variations are, as the name implies, without a specific assignable cause and without a pattern. Random variations are the fluctuations left in the time series after the trend, seasonality, and cyclical behaviors have been accounted for. Random fluctuations can sometimes be explained after the fact, such as an increase in the consumption of energy owing to abnormally harsh weather, but cannot be systematically predicted and, hence, are not included in time series models. In the treatment that follows, we will only consider the trend and seasonal components of time series models. See Makridakis and Wheelwright (1998) for more extensive models.

Time Series Models

The objective of time series analysis is to determine the magnitude of one or more of these components and to use that knowledge for the purpose of forecasting the next period. In the remainder of this section we will consider three models of time series analysis:

1. Moving averages (trend component of the time series)
2. Exponential smoothing (trend component of the time series)
3. Linear trend, multiplicative model (trend and seasonal components)

Moving Averages The **moving average** technique is one of the simplest ways to predict a trend. It generates the next period's forecast by averaging the actual demand for only the last n time periods (n is often in the range of 4 to 7). That is:

$$\text{Forecast} = \text{average of actual demand in past } n \text{ periods}$$

Any data older than n are thus ignored. Note, also, that the moving average values old data just the same as more recent data. The value of n is usually based on the expected seasonality in the data, such as four quarters or 12 months in a year; that is, n should encompass one full cycle of data. If n must be chosen arbitrarily, then it should be based on experimentation; that is, the value selected for n should be the one that works best for the available historical data.

Mathematically, a forecast using the moving average method is computed as

$$F_{t+1} = \frac{1}{n} \sum_{i=(t-n+1)}^{t} A_i$$

where

t = period number for the current period
F_{t+1} = forecast for the next period
A_i = actual observed value in period i
n = number of periods of demand to be included in the moving average (known as the "order" of the moving average)
$\sum_{i=(t-n+1)}^{t}$ = sum from $i = (t - n + 1)$ to t

To illustrate the use of the moving average, data was collected on Intel's monthly stock closing price as shown in Exhibit 3.21. In the spreadsheet shown, a four-period moving average was computed by entering the formula =AVERAGE(B4:B7) in cell C8 and copying it to cells C9:C28. The graph with the actual time series and moving average illustrates how a moving average smoothes out the fluctuations in the time series. The plot also demonstrates one of the weaknesses associated with using moving averages. Specifically, whenever there is an upward or downward trend in the data, a forecast based on the moving average approach will always lag the time series. Therefore, the moving average approach is most appropriate for situations where the decision maker would like to simply smooth out fluctuations around an assumed horizontal trend.

EXHIBIT 3.21 Four Period Moving Average of Intel's Monthly Stock Closing Price

	A	B	C
1			4 Period
2			Moving
3	Date	Closing Price	Average
4	Nov-97	38.813	
5	Dec-97	35.125	
6	Jan-98	40.500	
7	Feb-98	44.844	
8	Mar-98	39.032	39.82
9	Apr-98	40.407	39.88
10	May-98	35.719	41.20
11	Jun-98	37.063	40.00
12	Jul-98	42.219	38.06
13	Aug-98	35.594	38.85
14	Sep-98	42.875	37.65
15	Oct-98	44.594	39.44
16	Nov-98	53.813	41.32
17	Dec-98	59.282	44.22
18	Jan-99	70.469	50.14
19	Feb-99	59.969	57.04
20	Mar-99	59.438	60.88
21	Apr-99	61.188	62.29
22	May-99	54.063	62.77
23	Jun-99	59.500	58.66
24	Jul-99	69.000	58.55
25	Aug-99	82.188	60.94
26	Sep-99	74.313	66.19
27	Oct-99	77.438	71.25
28	**Forecast**		**75.73**
28			

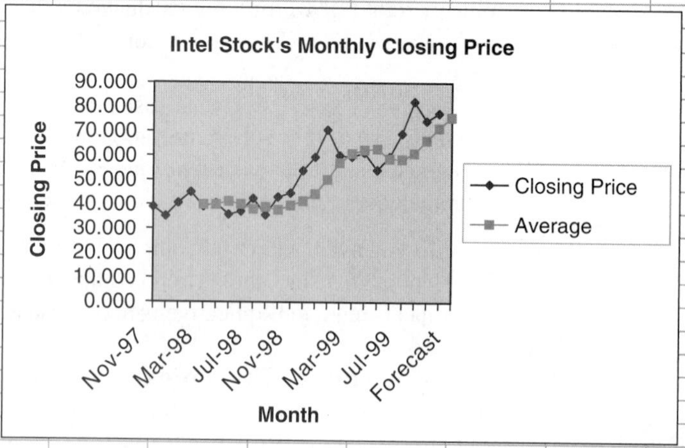

A refinement of the moving average approach is to vary the weights assigned to the values included in the average. Such an approach is called a **weighted moving average,** with the newer data typically weighted more heavily. The reason for weighting the newer data more heavily is that since it is more current, it is often considered to be more representative of the future. Referring to Intel's closing stock price, a weighted moving average could be constructed by weighting the fourth oldest observation in the average by 0.1, the third oldest by 0.2, the second oldest by 0.3, and the most recent observation by 0.4. Of course any combination of weights that summed to one could be used. Likewise, any number of periods could be included in the weighted moving average.

Time series analysis involves two inherent difficulties, and a compromise solution that addresses both must be sought. The first problem is producing as good a forecast as possible with the available data. Usually, this can be interpreted as using the most current data because those data are more representative of the current behavior of the time series. In this sense, we are looking for an approach that is responsive to recent changes in the data.

The second problem is to smooth the random behavior of the data. That is, we do not want a forecasting system that forecasts increases in demand simply because the last period's demand suddenly increased, nor do we want a system that indicates a downturn just because demand in the last period decreased. All time series data contain a certain amount of this erratic or random movement. It is impossible for a manager to predict this random movement of a time series, and it is folly to attempt it. The only reasonable conclusion is to avoid overreaction to a fluctuation that is simply random. The general interpretation of this objective is that several periods of data should be included in the forecast so as to "smooth" the random fluctuations that typically exist. Thus, we are also looking for an approach that is stable, even with erratic data.

Clearly, methods used to attain both responsiveness and stability will be somewhat contradictory. If we use the most recent data so as to be responsive, only a few periods will be included in the forecast; but if we want stability, large numbers of periods will be included. The only way to decide how many periods to include is to experiment with several different approaches and evaluate each on the basis of its ability to produce good forecasts and still smooth out random fluctuations.

Exponential Smoothing As just noted, we generally want to use the most current data and, at the same time, use enough observations of the time series to smooth out random fluctuations. One technique perfectly adapted to meeting these two objectives is **exponential smoothing.**

The computation of a demand forecast using exponential smoothing is carried out with the following equation:

New demand forecast = (α) current actual demand + ($1 - \alpha$) previous demand forecast

or

$$F_{t+1} = \alpha A_t + (1 - \alpha)\, F_t$$

where α is a smoothing constant that must be between zero and one, F_t is the exponential forecast for period t, and A_t is the actual demand in period t.

The smoothing constant α can be interpreted as the weight assigned to the last (i.e., the current) data point. The remainder of the weight ($1 - \alpha$) is applied to the last forecast. However, the last forecast was a function of the previous weighted data point and the forecast before that. To see this, note that the forecast in period t is calculated as

$$F_t = \alpha A_{t-1} + (1 - \alpha)F_{t-1}$$

Substituting the right-hand side in our original formula yields

$$F_{t+1} = \alpha A_t + (1 - \alpha)[\alpha A_{t-1} + (1 - \alpha)F_{t-1}]$$

Thus, the data point A_{t-1} receives a weight of $(1 - \alpha)\alpha$, which, of course, is less than α. Since this process is iterative, we see that exponential smoothing automatically applies a set of diminishing weights to each of the previous data points and is therefore a form of weighted averages. Exponential smoothing derives its name from the fact that the weights decline exponentially as the data points get older and older. In general, the weight of the *n*th most recent data point can be computed as follows:

$$\text{Weight of } n\text{th most recent data point in an exponential average} = \alpha(1 - \alpha)^{n-1}$$

Using this formula, the most recent data point, A_t, has a weight of $\alpha(1 - \alpha)^{1-1}$, or simply α. Similarly, the second most recent data point, A_{t-1}, would have a weight of $\alpha(1 - \alpha)^{2-1}$, or simply $\alpha(1 - \alpha)$. As a final example, the third most recent data point, A_{t-2} would have a weight of $\alpha(1 - \alpha)^{3-1}$ or $\alpha(1 - \alpha)^2$.

The higher the weight assigned to the current demand, the greater the influence this point has on the forecast. For example, if α is equal to 1, the demand forecast for the next period will be equal to the value of the current demand. The closer the value of α is to 0, the closer the forecast will be to the previous period's forecast for the current period. (Check these results by using the equation.)

Rearranging the terms of the original formula provides additional insights into exponential smoothing, as follows:

$$\begin{aligned} F_{t+1} &= \alpha A_t + (1 - \alpha)F_t \\ &= \alpha A_t + F_t - \alpha F_t \\ &= F_t + \alpha A_t - \alpha F_t \\ &= F_t + \alpha(A_t - F_t) \end{aligned}$$

In this formula, $A_t - F_t$ represents the forecast error made in period t. Thus, the formula shows that the new forecast developed for period $t + 1$ is equal to the old forecast plus some percentage of the error (since α is between 0 and 1). Notice that when the forecast in period t exceeds the actual demand in period t, we have a negative error term for period t and the new forecast will be reduced. On the other hand, when the forecast in period t is less than the actual demand in period t, the error term in period t is positive and the new forecast will be adjusted higher.

Our objective in exponential forecasting is to choose the value of α that results in the best forecasts. Forecasts that tend always to be too high or too low are said to be *biased*—positively if too high and negatively if too low. The value of α is critical in producing good forecasts, and if a large value of α is selected, the forecast will be very sensitive to the current demand value. With a large α, exponential smoothing will produce forecasts that react quickly to fluctuations in demand. This, however, is irritating to those who have to constantly change plans and activities on the basis of the latest forecasts. Conversely, a small value of α weights historical data more heavily than current demand and, therefore, will produce forecasts that do not react as quickly to changes in the data; that is, the forecasting model will be somewhat insensitive to fluctuations in the current data.

Generally speaking, larger values of α are used in situations in which the data exhibit low variability and can therefore be plotted as a rather smooth curve. On the other hand, a lower value of α should be used for data that exhibit a high degree of variability. Using a high value of α in a situation where the data exhibit a high degree of variability would result in a forecast that constantly overreacts to random changes in the most current demand.

As with *n*, the appropriate value of α is usually determined by trial and error; values typically lie in the range of 0.01 to 0.30. One method of selecting the best value is to try several values of α with the existing historical data (or a portion of the data) and choose the value of α that minimizes the average forecast errors. As you can probably imagine, spreadsheets can greatly speed the evaluation of potential smoothing constants and the determination of the best value of α. For example, the spreadsheet shown in Exhibit 3.22 forecasts the monthly closing price of Intel's stock using exponential smoothing. Various values of α can be easily investigated here by simply changing the number entered in cell B1. Also note that when exponential smoothing is used, a forecast value is needed for the very first period. Since a forecast value for the first period is typically not available, it is common to simply set $F_1 = A_1$.

Thus, the forecasts proceed as follows:

$$F_1 = A_1 = 38.813$$
$$F_2 = 0.2(38.813) + 0.8(38.813) = 38.813$$
$$F_3 = 0.2(35.125) + 0.8(38.813) = 38.08$$

and so on as shown in Exhibit 3.22.

Linear Trend, Multiplicative Model Exhibit 3.23 presents the quarterly number of visitors to a Web site providing medical information (http://www.Medfo.com). Demand is seen to be generally increasing, as is indicated by the linear trend line fit by Excel to the data (adding linear trend lines to Excel charts was discussed earlier in the chapter). Given the apparent quality of the fit between quarter number and the number of visitors, the webmaster has decided to try a linear trend time series model. The model parameters for the regression model with quarter number as the independent variable and ridership volume as the dependent variable were calculated in cells A19 and B19 using Excel's LINEST function. However, careful observation of the data reveals that the number of visitors is above average during the second and fourth quarters and below average during the first and third quarters, perhaps due to the weather-related illnesses.

EXHIBIT 3.22 Using Exponential Smoothing to Forecast the Monthly Closing Price of Intel's Stock

	A	B	C
1	Alpha	0.2	
2			Exponential
3	Date	Closing Price	Smoothing
4	Nov-97	38.813	38.81
5	Dec-97	35.125	38.81
6	Jan-98	40.500	38.08
7	Feb-98	44.844	38.56
8	Mar-98	39.032	39.82
9	Apr-98	40.407	39.66
10	May-98	35.719	39.81
11	Jun-98	37.063	38.99
12	Jul-98	42.219	38.61
13	Aug-98	35.594	39.33
14	Sep-98	42.875	38.58
15	Oct-98	44.594	39.44
16	Nov-98	53.813	40.47
17	Dec-98	59.282	43.14
18	Jan-99	70.469	46.37
19	Feb-99	59.969	51.19
20	Mar-99	59.438	52.94
21	Apr-99	61.188	54.24
22	May-99	54.063	55.63
23	Jun-99	59.500	55.32
24	Jul-99	69.000	56.15
25	Aug-99	82.188	58.72
26	Sep-99	74.313	63.42
27	Oct-99	77.438	65.60
28	Forecast		**67.96**
29			
30	*Key Formula*		
31	Cell C4	=B4	
32	Cell C5	=C4+(B$1*(B4-C4)) {copy to cells C6:C28}	

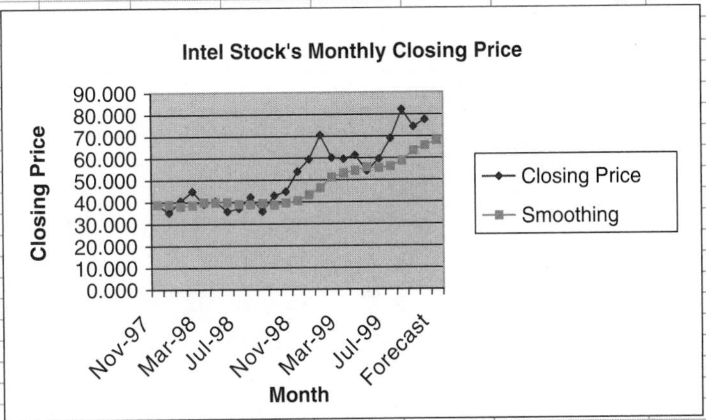

There are several versions of the linear trend time series model (e.g., there are additive and multiplicative versions) and also many different approaches to determining the components of these forecasting models. We will present one method for determining the two demand components of a simple multiplicative model. Conceptually, the model is presented as

$$\text{Forecast} = \text{trend component (or } T) \times \text{seasonal component (or } S)$$

In order to develop this model, we must first analyze the available historical data and attempt to break down the original data into trend and seasonal components.

As indicated earlier, a trend is a long-run direction of a series of data. In the medical Web site example, the trend in the number of visitors to the Web site appears to follow a straight line—that is, to be a trend with respect to time. In order to project this linear trend into the future, we first estimate the parameters of the trend line in exactly the same fashion that was discussed earlier in the chapter. Referring to Exhibit 3.23, we see that the trend line for the ridership volume is:

$$\text{Number of visitors}_X = 27{,}253 + 15{,}462X$$

where X represents the quarter.

As was noted earlier, and made even clearer in Exhibit 3.23, the data are above the trend line for all of the second and fourth quarters and below the trend line for all of the first and third quarters. Recognizing this distinct seasonal pattern in the data should allow us to estimate the amount of seasonal variation around the trend line (i.e., the seasonal component, S).

The trend line is the long-run direction of the data and does not include any seasonal variation. We can compute, for each available quarter of data, a measure of the "seasonality" in that quarter by dividing actual number of visitors by the computed value of the trend for that quarter. This method is known as the ratio-to-trend method. Using the notation developed thus far, we can write the seasonal component for any quarter X as

$$\frac{Y_X}{T_X}$$

EXHIBIT 3.23 Number of Visitors to Medfo.com

	A	B
1		Visitors to
2	Quarter	Web site
3	1	35,000
4	2	80,000
5	3	55,000
6	4	100,000
7	5	95,000
8	6	140,000
9	7	115,000
10	8	160,000
11	9	155,000
12	10	200,000
13	11	175,000
14	12	220,000
15	13	215,000
16	14	260,000
17		
18	*b*	*a*
19	15461.54	27252.75
20		

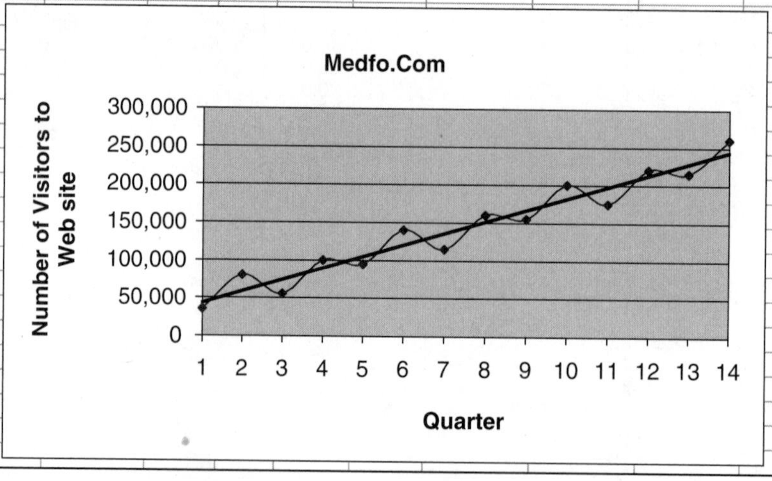

Medfo.Com

Number of Visitors to Web site: 0, 50,000, 100,000, 150,000, 200,000, 250,000, 300,000

Quarter: 1 2 3 4 5 6 7 8 9 10 11 12 13 14

where Y_X is the number of visitors to the Web site in quarter X and T_X is the trend estimate for quarter X. Excel's TREND function can be used to calculate the trend estimate for each quarter, as shown in Exhibit 3.24.

Consider the second and third quarters of the first year. The computed trend value for each of these two quarters is

$$T_2 = 27{,}253 + 15{,}462(2) = 58{,}176$$

and

$$T_3 = 27{,}253 + 15{,}462(3) = 73{,}637$$

The actual volumes (in thousands) in quarters 2 and 3 were

$$Y_2 = 80{,}000$$
$$Y_3 = 55{,}000$$

Dividing Y_2 by T_2 and Y_3 by T_3 gives us an indication of the seasonal pattern in each of these quarters.

$$\frac{Y_2}{T_2} = \frac{80,000}{58,176} = 1.38$$
$$\frac{Y_3}{T_3} = \frac{55,000}{73,637} = 0.75$$

Similar indices were calculated for all quarters in the spreadsheet shown in Exhibit 3.24.

In quarter 2 the actual number of visitors was 138 percent of the expected volume (i.e., the number of visitors predicted on the basis of a linear trend), but in quarter 3 the number of visitors was only 75 percent of that expected. Note that over the 14 periods of available data we have four observations of the number of visitors for first and second quarters and three observations of the number of visitors for third and fourth quarters. We can compute the average of each of these sets of quarterly data and use the averages as the seasonal components for our time series forecasting model, as shown in Exhibit 3.25.

EXHIBIT 3.24 Calculation of Quarterly Seasonal Factors

	A	B	C	D
1				Seasonal
2		Visitors to		Factor
3	Quarter	Web Site	T_X	(Y/T)
4	1	35,000	42714.29	0.82
5	2	80,000	58175.82	1.38
6	3	55,000	73637.36	0.75
7	4	100,000	89098.90	1.12
8	5	95,000	104560.44	0.91
9	6	140,000	120021.98	1.17
10	7	115,000	135483.52	0.85
11	8	160,000	150945.05	1.06
12	9	155,000	166406.59	0.93
13	10	200,000	181868.13	1.10
14	11	175,000	197329.67	0.89
15	12	220,000	212791.21	1.03
16	13	215,000	228252.75	0.94
17	14	260,000	243714.29	1.07

EXHIBIT 3.25 Calculating Seasonal Component (S) for Quarters 1 Through 4

	A	B	C	D	E
1	**Year**	**Quarter 1**	**Quarter 2**	**Quarter 3**	**Quarter 4**
2	**1**	0.82	1.38	0.75	1.12
3	**2**	0.91	1.17	0.85	1.06
4	**3**	0.93	1.10	0.89	1.03
5	**4**	0.94	1.07		
6	**Average**	**0.90**	**1.18**	**0.83**	**1.07**

Using both the trend component and the seasonal component, the webmaster now can forecast the number of visitors to the site for any quarter in the future. First, the trend value for the forecast quarter is computed and is, in turn, multiplied by the appropriate seasonal factor. For example, to forecast for the last quarter of the fourth year (quarter 16) and the first quarter of the fifth year (quarter 17), the webmaster would first compute the trend values.

$$T_{16} = 27{,}253 + 15{,}462(16) = 274{,}644$$
$$T_{17} = 27{,}253 + 15{,}462(17) = 290{,}106$$

Next, the forecast is computed by multiplying the trend value by the appropriate seasonal factor. For the fourth quarter S_4 is 1.07, so the forecast F is

$$F_{16} = 274{,}644 \times 1.07 = 293{,}869$$

The seasonal factor for the first quarter is 0.90; therefore, the forecast for quarter 17 is

$$F_{17} = 290{,}106 \times 0.90 = 261{,}095$$

These two forecasts correspond to the previous results for fourth and first quarters in that the fourth-quarter forecast is above the trend and the first-quarter forecast is below the trend. Seasonal indexes can be used in a similar way with exponential smoothing or moving averages. Again, simple ratios are calculated, averaged out, and then applied to the exponential smoothing or moving average forecasts.

3.7 Detailed Modeling Example

Dr. John Cash was recently appointed as the new director of the MBA program at a large midwestern public university. One of his primary charges from the dean is to determine what the school can do to increase the starting salaries of the graduates of the full-time MBA program. According to many faculty members, increasing the starting salaries would facilitate attracting better students and would further enhance the school's national reputation. In the remainder of this section, we illustrate in the broader modeling context how regression analysis can be used to address this issue.

Step 1: Opportunity/Problem Recognition

The opportunity/problem confronting Dr. Cash is to identify the factors that influence starting salaries of new MBA graduates so they may be more carefully considered in the future. Furthermore, it would also be useful to have some indication regarding the extent to which the identified factors affect starting salaries.

Step 2: Model Formulation

Dr. Cash thus decided to first simply investigate the statistical relationships between starting salaries and other factors that would likely affect these salaries. Simple graphs would indicate what types of relationships might be appropriate to test in a regression. In theory, a great many factors might be relevant to starting salaries, including the reputation of the school. However, Dr. Cash decided to first check the three factors shown in the influence diagram of Exhibit 3.26 that could be influenced by the school's admissions committee and that were easily obtainable.

Step 3: Data Collection

Dr. Cash thus hired a graduate assistant to collect data on the top-rated 50 peer institutions. In addition to collecting data on average starting salaries, Dr. Cash instructed his assistant to collect data for each school on the most recently graduated class's average undergraduate grade point average (GPA), GMAT score, and years of work experience prior to returning to graduate school. The assistant was able to find most of the information in recently published graduate school ratings and MBA program directories. On a few occasions the assistant needed to visit a particular school's Web site or call the school to obtain a particular piece of information. The graduate assistant was not able to obtain data for one of the schools. The data collected for the remaining 49 schools is shown in Exhibit 3.27.

Step 4: Analysis of the Model

In this section, we utilize the four-step process discussed earlier to illustrate the development and analysis of the regression model.

Step 1: Identify Candidate Independent Variables to Include in the Model The three candidate independent variables—undergraduate GPA, GMAT score, and years of

EXHIBIT 3.26 Factors Impacting Starting Salaries of MBA Graduates

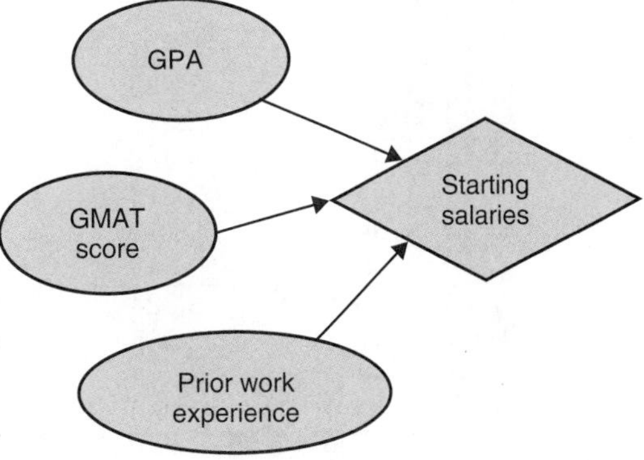

EXHIBIT 3.27 Data Collected on Top-Rated Peer Schools

	A	B	C	D	E
1		Undergraduate	GMAT	Years of	Starting
2	School	GPA	Score	Experience	Salary
3	1	3.307	628.41	3.41	$75,490
4	2	3.193	640.15	2.47	$86,947
5	3	3.315	638.87	2.94	$80,017
6	4	3.151	652.63	2.67	$88,369
7	5	3.350	663.38	3.44	$98,517
8	6	3.329	673.00	3.81	$96,988
9	7	3.350	628.45	2.26	$71,639
10	8	3.333	618.43	2.25	$65,979
11	9	3.280	623.73	2.69	$77,458
12	10	3.433	600.70	2.76	$64,940
13	11	3.427	683.76	4.13	$94,983
14	12	3.447	720.60	3.14	$105,724
15	13	3.233	630.71	4.13	$70,969
16	14	3.175	647.77	3.72	$92,484
17	15	3.274	629.88	3.49	$71,981
18	16	3.239	694.24	4.56	$95,034
19	17	3.259	612.48	2.62	$61,926
20	18	3.246	630.82	2.84	$80,046
21	19	3.409	634.55	1.68	$65,198
22	20	3.361	630.94	3.94	$79,987
23	21	3.265	618.59	3.40	$65,028
24	22	3.479	689.94	4.22	$99,993
25	23	3.290	686.67	3.93	$95,050
26	24	3.322	631.31	3.15	$65,953
27	25	3.245	632.64	2.10	$74,096
28	26	3.173	637.39	3.00	$63,022
29	27	3.221	652.91	2.98	$74,030
30	28	3.421	671.15	2.78	$99,962
31	29	3.150	614.34	3.59	$70,461
32	30	3.490	688.93	4.33	$104,959
33	31	3.536	682.53	4.06	$95,038
34	32	3.239	612.42	2.64	$65,005
35	33	3.269	659.95	3.58	$75,981
36	34	3.433	675.77	3.24	$95,028
37	35	3.365	684.56	5.09	$115,030
38	36	3.246	641.99	3.28	$70,002
39	37	3.175	619.38	3.52	$74,956
40	38	3.209	664.09	2.98	$67,061
41	39	3.431	634.26	2.52	$71,496
42	40	3.458	678.49	3.79	$106,002
43	41	3.217	636.39	2.47	$77,009
44	42	3.418	674.30	4.88	$89,938
45	43	3.074	622.76	3.02	$78,499
46	44	3.262	650.80	2.42	$76,962
47	45	3.144	640.71	3.30	$92,086
48	46	3.324	652.13	4.20	$74,965
49	47	3.308	646.11	2.96	$64,684
50	48	3.212	631.70	3.64	$62,005
51	49	3.580	684.33	4.73	$100,024

EXHIBIT 3.28 Pairwise Correlations Between Candidate Independent Variables

	A	B	C	D	E
54			**GMAT**	**Experience**	
55		**GPA**	0.52	0.31	
56		**GMAT**		0.56	
57					
58	**Key Formulas**				
59	Cell C55	=CORREL($B3:$B51,C3:C51) {copy to cell D55}			
60	Cell D56	=CORREL(C3:C51,D3:D51)			

experience—were initially identified based on the knowledge of the decision maker, Dr. Cash. Given that a candidate pool of independent variables has been identified and the data have been collected for these variables, the next task is to check the correlations between each of the variables. As shown in Exhibit 3.28, all the correlation coefficients are below 0.80, and therefore there does not appear to be a severe multicollinearity problem among the independent variables.

Step 2: Transform the Data Plots of each independent variable with the dependent variable were created to assess the validity of the linear relationship between the variables and to determine if any transformation of the variables was required. According to the plots shown in Exhibit 3.29, there does appear to be a linear relationship between each independent variable and the dependent variable. Therefore, no transformation of the data appears to be required.

Step 3: Select the Variables to Include in the Model Forward selection will be used to illustrate the process of determining which variables to include in the model. In the first iteration, three simple regression models with each independent variable included as the single predictor variable are developed, and the coefficient of determination is calculated. As shown in Exhibit 3.30, the independent variable "average GMAT score" explained considerably more of the variation in starting salaries than the other two independent variables ($R^2 = 0.684$). Therefore, GMAT is selected as the first variable to include in the model.

In the second iteration, we investigate the extent to which the model is improved by adding one of the remaining independent variables to the model. In our case, this requires investigating the effect of adding GPA and work experience to the model with GMAT already included. According to the results shown in Exhibit 3.30, adding either variable substantially improves the model, but prior work experience yields a slightly higher R^2. Therefore, the independent variable "work experience" would be added in the second iteration. Note that the R^2 values for multiple regression models can be calculated using the LINEST function or Excel's Tools/Data Analysis/Regression feature (see Excel's help screen or user manual for more information on these options), as explained earlier.

Finally, in the third iteration, we investigate whether the last remaining independent variable GPA should be added to the model. As shown in Exhibit 3.30, adding this third variable only increases R^2 by a tenth of a percent, or from 83.3 percent to 83.4 percent. Given a preference for what statistician's call *parsimonious* models (i.e., models without unnecessary variables), there appears to be no compelling reason to include GPA in the model.

EXHIBIT 3.29 Plots of Candidate Independent Variables with the Dependent Variable

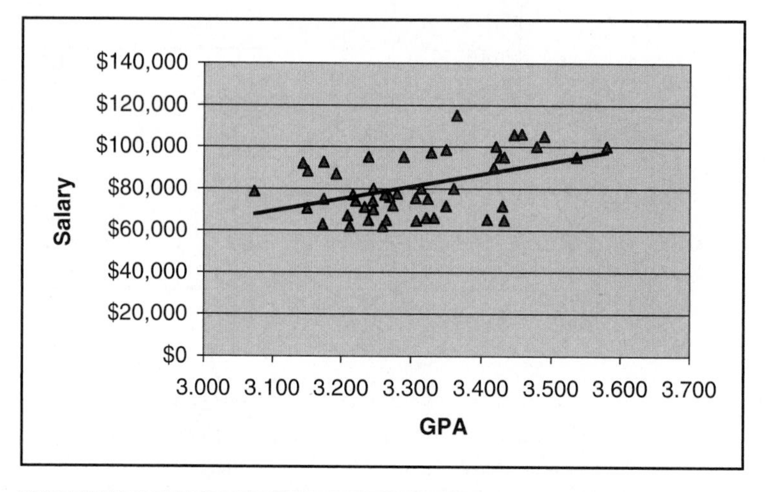

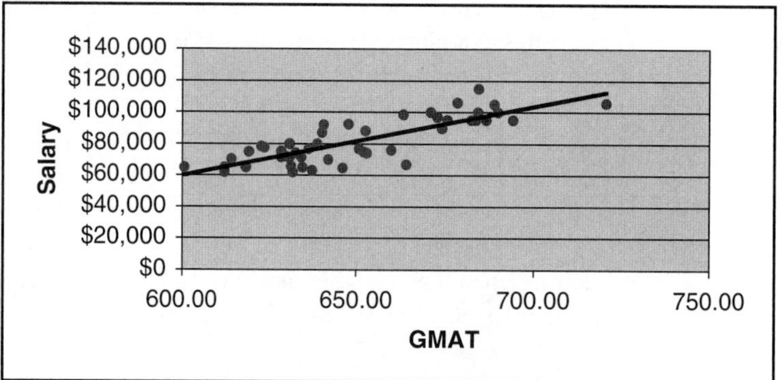

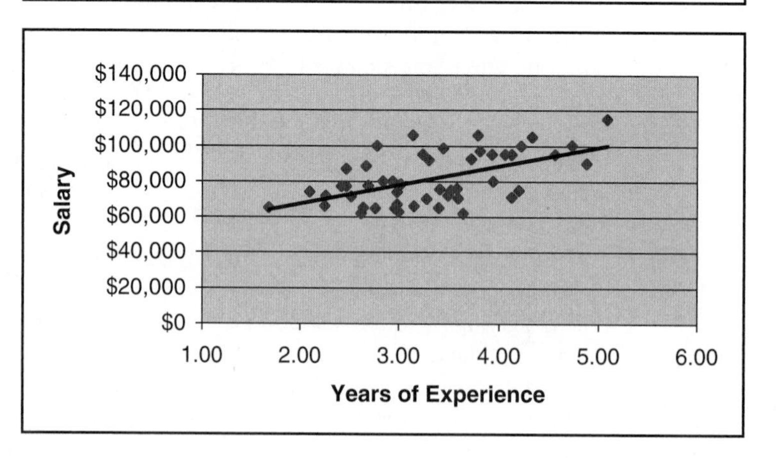

EXHIBIT 3.30
Summary of Forward Selection Results for Starting Salary Model

Iteration	Variable(s) Included in the Regression Model	R^2
1	GPA	21.1%
1	GMAT Score	68.4
1	Work Experience	31.6
2	GMAT and GPA	82.5
2	GMAT and Experience	83.3
3	GMAT, Experience, and GPA	83.4

Step 4: Analyze the Residuals The last step in regression model formulation is to perform an analysis of the residuals to confirm that the assumptions underlying regression analysis are met. To investigate the assumption that the error terms are normally distributed with a mean of zero, a histogram of the residuals was constructed, as shown in Exhibit 3.31. As can be seen, the histogram is approximately shaped like a normal distribution.

To further check on this problem and also to investigate whether the assumption of constant variance is met, the residuals were plotted against the predicted values of Y using the regression model. In addition, separate graphs of the residuals plotted against each independent variable were created. Inspection of these graphs shown in Exhibit 3.32 reveals no apparent systematic pattern, and therefore no indication that the assumption of constant variance is violated. If Dr. Cash was still concerned about the distribution of the residuals, he might consider using GPA instead of prior experience in the regression model, using both in the model, collecting additional data on other possible factors to include in the model, increasing the number of schools in the sample, or other such changes.

Based on the previous steps, a multiple regression model has been developed to provide insight into the factors that impact average starting salaries of new graduates of MBA programs. Analysis of the situation suggests that more than 83 percent of the variation in average starting salaries can be explained by two independent variables, namely, GMAT scores and prior work experience. It was further concluded that including undergraduate GPA as a third independent variable explained little of the variability in average starting salaries beyond what was already accounted for by GMAT scores and work experience.

The final regression model developed for this situation is

$$\text{Salary} = -183{,}670 + 394.3\ (\text{experience}) + 2{,}768.5\ (\text{GMAT score})$$

According to this model, each additional year of work experience increases starting salaries by \$394.3, while each one point increase in GMAT score results in a \$2,768.5 increase in starting salaries, more than six times as much.

Step 5: Implementation

Based on Dr. Cash's regression analysis, it appears that it might be worthwhile for the admissions committee to alter the profile of students recruited to enter the full-time MBA program. That is, placing greater emphasis on GMAT scores and prior work experience

EXHIBIT 3.31 Histogram of Residuals

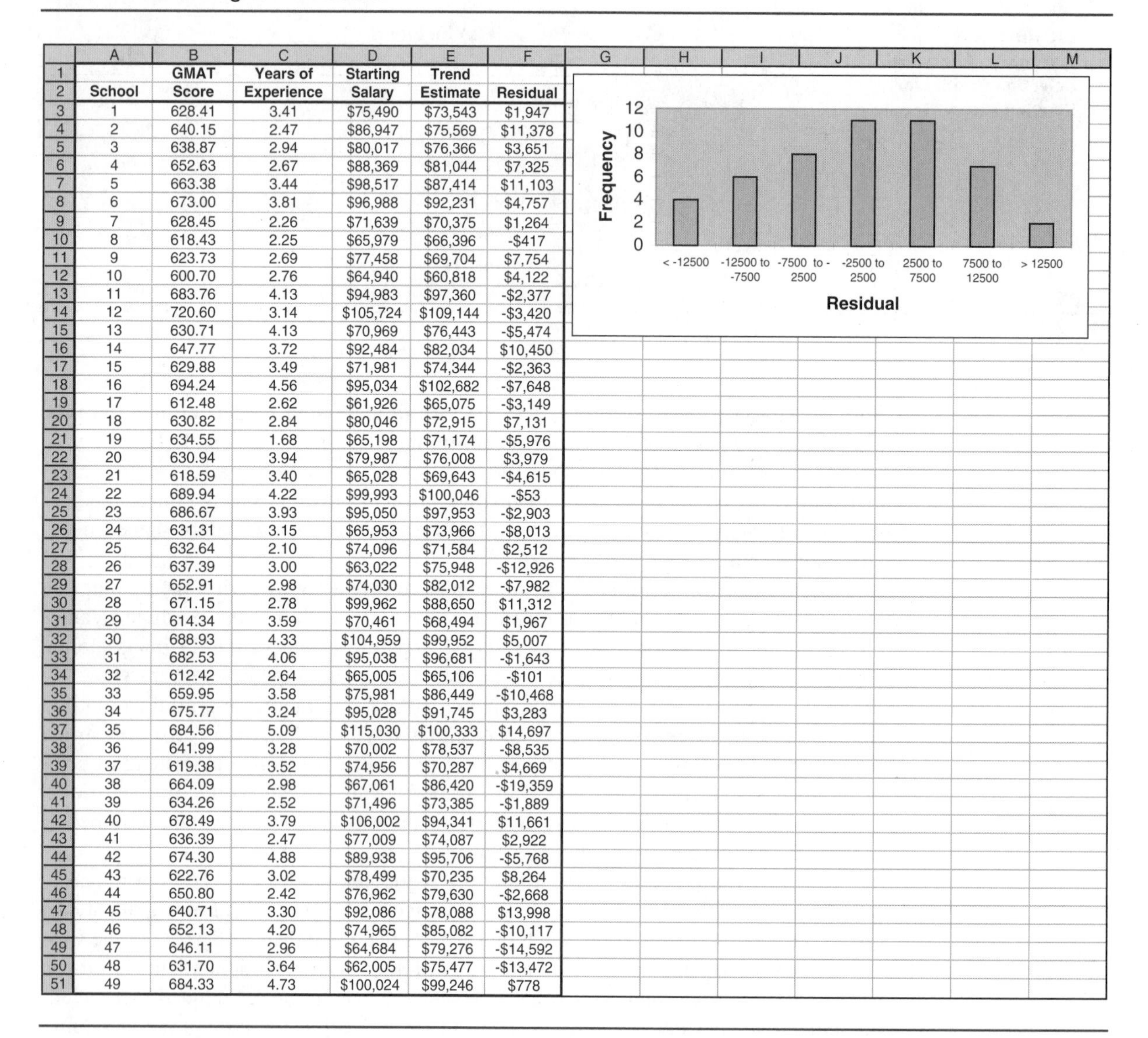

School	GMAT Score	Years of Experience	Starting Salary	Trend Estimate	Residual
1	628.41	3.41	$75,490	$73,543	$1,947
2	640.15	2.47	$86,947	$75,569	$11,378
3	638.87	2.94	$80,017	$76,366	$3,651
4	652.63	2.67	$88,369	$81,044	$7,325
5	663.38	3.44	$98,517	$87,414	$11,103
6	673.00	3.81	$96,988	$92,231	$4,757
7	628.45	2.26	$71,639	$70,375	$1,264
8	618.43	2.25	$65,979	$66,396	-$417
9	623.73	2.69	$77,458	$69,704	$7,754
10	600.70	2.76	$64,940	$60,818	$4,122
11	683.76	4.13	$94,983	$97,360	-$2,377
12	720.60	3.14	$105,724	$109,144	-$3,420
13	630.71	4.13	$70,969	$76,443	-$5,474
14	647.77	3.72	$92,484	$82,034	$10,450
15	629.88	3.49	$71,981	$74,344	-$2,363
16	694.24	4.56	$95,034	$102,682	-$7,648
17	612.48	2.62	$61,926	$65,075	-$3,149
18	630.82	2.84	$80,046	$72,915	$7,131
19	634.55	1.68	$65,198	$71,174	-$5,976
20	630.94	3.94	$79,987	$76,008	$3,979
21	618.59	3.40	$65,028	$69,643	-$4,615
22	689.94	4.22	$99,993	$100,046	-$53
23	686.67	3.93	$95,050	$97,953	-$2,903
24	631.31	3.15	$65,953	$73,966	-$8,013
25	632.64	2.10	$74,096	$71,584	$2,512
26	637.39	3.00	$63,022	$75,948	-$12,926
27	652.91	2.98	$74,030	$82,012	-$7,982
28	671.15	2.78	$99,962	$88,650	$11,312
29	614.34	3.59	$70,461	$68,494	$1,967
30	688.93	4.33	$104,959	$99,952	$5,007
31	682.53	4.06	$95,038	$96,681	-$1,643
32	612.42	2.64	$65,005	$65,106	-$101
33	659.95	3.58	$75,981	$86,449	-$10,468
34	675.77	3.24	$95,028	$91,745	$3,283
35	684.56	5.09	$115,030	$100,333	$14,697
36	641.99	3.28	$70,002	$78,537	-$8,535
37	619.38	3.52	$74,956	$70,287	$4,669
38	664.09	2.98	$67,061	$86,420	-$19,359
39	634.26	2.52	$71,496	$73,385	-$1,889
40	678.49	3.79	$106,002	$94,341	$11,661
41	636.39	2.47	$77,009	$74,087	$2,922
42	674.30	4.88	$89,938	$95,706	-$5,768
43	622.76	3.02	$78,499	$70,235	$8,264
44	650.80	2.42	$76,962	$79,630	-$2,668
45	640.71	3.30	$92,086	$78,088	$13,998
46	652.13	4.20	$74,965	$85,082	-$10,117
47	646.11	2.96	$64,684	$79,276	-$14,592
48	631.70	3.64	$62,005	$75,477	-$13,472
49	684.33	4.73	$100,024	$99,246	$778

and placing less emphasis on undergraduate grade point averages could have a positive impact on starting salaries. Further, since it is likely to be more difficult to raise average prior work experience by a year than it is to raise GMAT scores by a point, and since raising GMAT scores by a point increases starting salaries by six times as much as a one year increase in experience (according to Dr. Cash's model), it could be argued that the emphasis in the future should be on recruiting students with high GMAT scores. Exhibit 3.33 illustrates in the form of a memo how the results of this analysis might be communicated to the dean of the business school. The companion Web site for this book contains a sample PowerPoint presentation of the results of this analysis.

EXHIBIT 3.32 Residual Plots

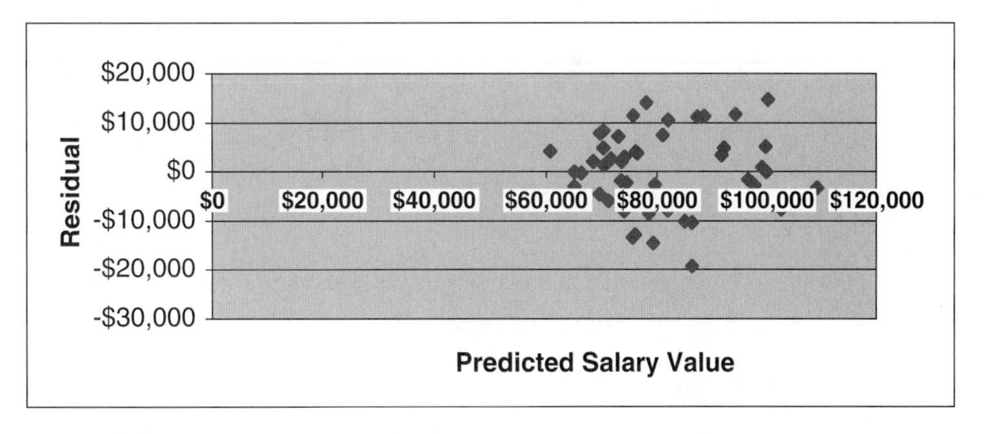

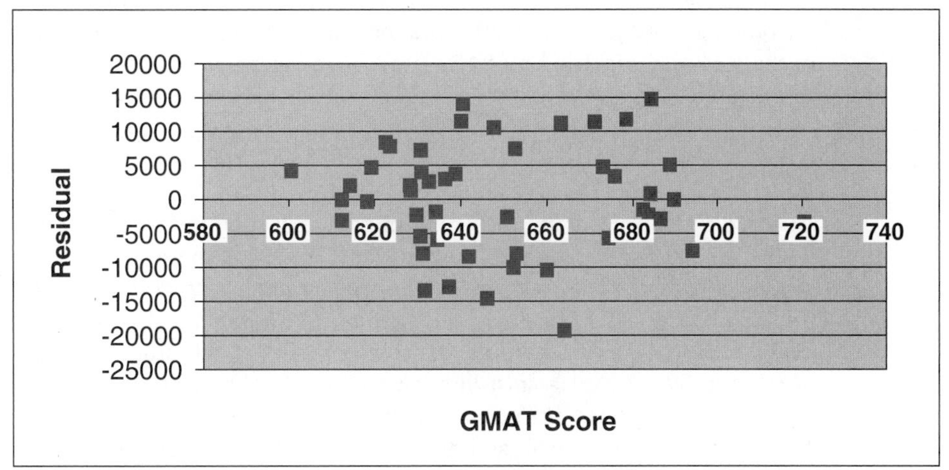

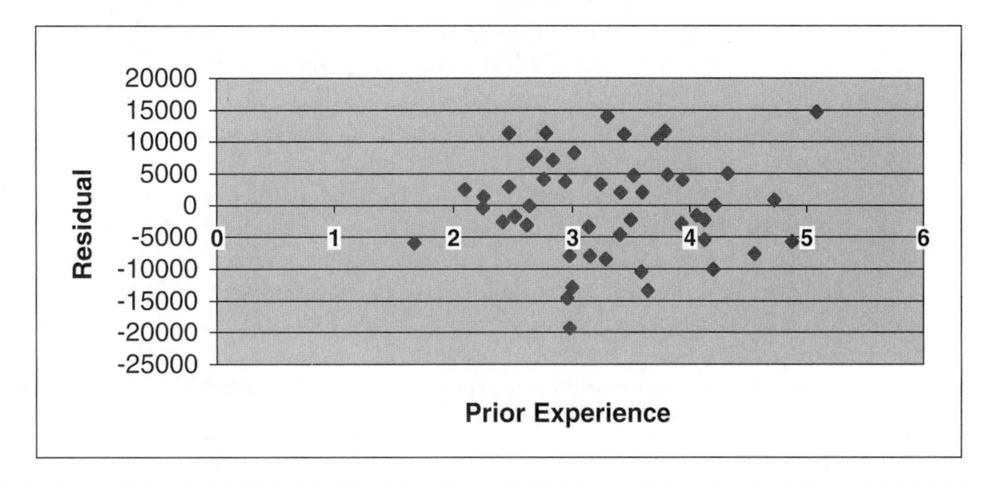

EXHIBIT 3.33 Memo Overviewing Results of Regression Analysis on Starting Salaries

MEMO

To: Dr. Cheryl Smith, Dean
From: Dr. John Cash, Director of MBA Program
CC: Faculty
Date: 3/26/02
RE: Preliminary results regarding factors that impact MBA graduate's starting salaries

Introduction

The purpose of this memo is to summarize the preliminary results I have obtained in my ongoing investigation of the factors that impact MBA starting salaries. The results of the study suggest that future MBA student recruiting efforts place more emphasis on student GMAT scores and then secondly on prior work experience.

Analysis

As a top 50 MBA program, the scope of data collection for this study was limited to the other MBA programs ranked in the top 50. For the most part, these peer schools are nearly alike in terms of the students they attract, their curricula, and the types of recruiters that visit their campuses. Unfortunately, data could not be obtained for one of the schools, hence the results of this study are based on 49 of the top 50 MBA programs. In addition to the average starting salaries of last year's graduating class, data was collected on the group's average undergraduate GPA, average GMAT score, and average years of prior work experience. These variables were chosen because they are frequently included in the published rankings of graduate business schools, and because of their use in making admission decisions. For the most part, the data were obtained from published rankings such as those done by *Business Week* and *U.S. News and World Reports.* On occasion we needed to visit a school's Web site or call the school to obtain a particular piece of information.

A forward selection procedure where the variable that results in the greatest improvement in the model is added iteratively was used to develop a regression model. In the first iteration, it was determined that average GMAT score accounted for over 68 percent of the variation in starting salaries and was therefore selected as the first variable to include in the model. In the second iteration, average prior work experience was added with the combination of these two variables explaining over 83 percent of the variation in starting salaries. Adding average undergraduate GPAs to the model resulted in almost no improvement in the model and therefore was not included in the regression model. The final regression model developed for MBA starting salaries is

Starting salary = −183,670 + 394.3 (prior experience) + 2,768.5 (GMAT score)

Thus, according to this model, each additional year of work experience increases starting salaries by $394.3 while each one point increase in the GMAT score results in a $2,768.5 increase in starting salaries.

Recommendations

These results suggest a shift in the profile of students recruited for our full-time program toward placing *greater emphasis* on GMAT scores, then secondly on prior work experience and finally *less emphasis* on undergraduate GPAs. The strategy of shifting our emphasis to GMAT scores is particularly appealing given that it is more difficult to recruit an entering class with more work experience than it is to recruit a class with higher GMAT scores. Furthermore, according to the regression model, a one point increase in GMAT scores increases average starting salaries by six times as much as a one year increase in average years of prior work experience.

Limitations

There are of course a number of limitations associated with this study. First, there are other potentially important variables that were not included such as student communication skills, computer literacy, quality of prior work experience, salaries prior to returning to graduate school, and so on. Second, it is important to recognize that the results of this study can only identify an apparent relationship between the variables investigated. It cannot be concluded on the basis of this study that there is a cause and effect relationship between GMAT scores or prior work experience, and average starting salaries. Finally, it is important to note that all data included in this analysis were based on averages and therefore it would be a misuse of the model to attempt to predict the average starting salary for a particular individual. Indeed, as a follow-up to this study it may be worthwhile to further investigate this issue with data collected at the individual level.

Based on the above discussion, the results of this study appear to have important implications regarding our MBA student recruitment strategy. As this is very much a work in progress, I certainly welcome any feedback you have regarding the study thus far.

QUESTIONS

1. Why are statistical models more commonly used descriptively instead of prescriptively?
2. How are the regression and time series statistical techniques in this chapter more akin to quantitative business modeling than the statistical techniques in the previous chapter?
3. The linear regression models discussed here all assume that the errors have no discernible pattern that can be predicted beforehand. But if we did see a predictable pattern among the errors, how would that alter our regression analysis?
4. Since the coefficient of determination is simply the square of the correlation coefficient and correlation never implies which variable is the cause and which is the effect, can we then infer that the coefficient of determination explains the variability in X due to Y? How then do we infer that X causes Y?
5. Distinguish between extrapolation and generalization in your own words.
6. Why don't we keep adding independent variables in multiple regression as long as the multiple coefficient of determination keeps increasing?
7. How might you evaluate a time series to see if there is any cyclic component in it? If you did identify a cyclic component, how could you use it to improve your forecast of the upcoming period?
8. What circumstances might lead a forecaster to weight older data more heavily than recent data?
9. Why is it suggested that a value of α between 0.01 and 0.30 be used in exponential forecasting? Why not larger values?
10. What would be the problem with including the ranking of the school in Dr. Cash's regression equation? (*Note:* School rankings are typically determined by student evaluations of their experience, deans' ratings of school quality, and employers' evaluations of each school, including their salary offers.)
11. Is there any harm in trying to manipulate independent variables such as GMAT scores in order to "improve" a dependent variable? (One easy way to improve class average GMAT scores is to not admit students who have the lowest 10 percent of the GMAT scores who would have previously been admitted.)

EXPERIENTIAL EXERCISE

Visit a Web site such as http://www.realtor.com or identify another source of house price data. Collect data on the prices of the houses as well as several variables that you imagine would impact the price. Collect this data for at least three independent variables and at least 50 houses. Also, restrict your data collection to a particular area such as a particular zip code, city, or subdivision. Next complete the following tasks:

1. Follow the four-step process explained in the chapter and develop a regression model for estimating the price of a house in the area you studied.
2. Collect data on five additional houses and plug the values of these houses' independent variables into your regression model. How good a job did the regression model you developed predict the prices of these additional houses?
3. List any potential limitations associated with your study.
4. Write a memo to a real estate professional overviewing the results of your study and explaining how the regression model you developed could be used. In the memo include a discussion of how the data were collected, how the collected data were analyzed, what key insights were obtained, and key limitations associated with this study.

MODELING EXERCISES

1. Given the following data, find the parameters of the simple linear regression line manually using the equations. Then predict period 8.

Period	1	2	3	4	5
Data	6	5	8	9	7

2. Forecast period 4 in the following data with exponential forecasting using two values of alpha: 0.1 and 0.3.

Period	1	2	3
Data	7	5	9

3. Given the following data:

Period	1	2	3	4	5	6
Data	24	22	26	23	25	19

   a. Forecast period 7 using a moving average of order 3.
   b. Repeat a (above) but use weights of 0.2, 0.3, and 0.5.
   c. Repeat a (above) using weights of 0.5, 0,3, and 0.2.
   d. Compare your results in a, b, and c.
   e. If the periods above were weeks, what order would you suggest using instead of 3? What if the periods were months?

4. Manually calculate the best linear regression to predict the number of new four-wheel drive SUVs sold at Lemon Auto Sales during 2003 when the number of snow-related accidents in the county the previous winter was 1200 and the following data were available concerning snow-related accidents.

	Year			
	1999	2000	2001	2002
Accidents, previous winter	870	940	1,010	1,020
Four-wheel SUVs sold	650	690	730	710

5. Determine the seasonals and trend equation for the following data.

Quarter	1	2	3	4	5	6	7	8
Sales	50	40	30	50	60	50	40	60

6. Manually determine the linear regression equation to predict the demand at a price of \$90, given the following data:

Demand	113	110	102	105
Price, \$	50	60	80	70

7. Consider the following data and develop a forward selection procedure for appropriate factors to include in a regression to predict *Y*.

Factor A	0	1	1	0	0	0	1	0
Factor B	3	4	6	5	2	4	6	6
Factor C	25	29	26	33	29	28	37	38
Y	332	348	302	355	339	348	371	366

8. Given the following data, use a trend line and seasonals to predict sales in summer, 2003.

	Quarter			
Year	Winter	Spring	Summer	Fall
1998	123	133	172	281
1999	155	189	205	286
2001	151	186	288	303
2002	178	225	272	296

9. The following equation has been estimated for monthly demand for a new wristwatch computer: $D = 16.7A + 12.3S - 5.4P - 2.1I$ where D is the demand in thousands of units, A is advertising in millions of dollars, S is average store shelf space in linear feet, P is the price in \$100, and I is the annual inflation rate in percent. If advertising is planned to be \$3 million, average store shelf space 4 linear feet, price \$500, and inflation is currently running 4 percent, what is demand expected to be? How sensitive is demand to a 10 percent increase in price? To an inflation increase to 6 percent? If management is considering investing another \$100,000 in this product, would it be better to put it into advertising or reducing the price? Do you think there might be a relationship between advertising and store shelf space? Between inflation and the price? If so, what effect might this have on demand?

10. Analyze the residuals in Problem 7 for normality, a mean of zero, constant variance across all three independent variables and the dependent variable, and independence.

11. Advertising expenditures and sales for the last 15 years in hundreds of millions of dollars were entered into the following spreadsheet.
   a. Using Excel or another software package, plot advertising expenditures against sales. Does there appear to be a linear relationship between these two variables?

b. Use your software package to add a linear regression line to the data you plotted. How good a fit is the linear regression line to the data?
c. How much of the variation in sales does the variable "advertising expenditures" explain?

	A	B	C
1	Year	Advertising	Sales
2	1	3.3	167.5
3	2	3.4	162.6
4	3	3.6	172.8
5	4	3.6	170.7
6	5	3.7	172.8
7	6	3.9	177.6
8	7	4.1	170.0
9	8	4.1	185.8
10	9	4.4	181.6
11	10	4.5	196.4
12	11	4.7	178.3
13	12	4.9	198.3
14	13	5.1	198.5
15	14	5.2	208.7
16	15	5.5	204.7

12. Referring to Exercise 11, use your software program to find the parameters of the simple linear regression model. Explain in plain English the interpretation of these parameters. What level of sales would you project for an advertising budget of $400,000,000?
13. Referring to Exercise 11:
    a. Using Excel or another software package, plot advertising expenditures in period t versus sales in period $t + 1$. Does there appear to be a linear relationship between the advertising expenditures and the lagged value of sales?
    b. Use your software package to add a linear regression line to the data plotted. How good a fit is the regression line to the data?
    c. How much of the variation in sales does the variable advertising expenditures in the prior year explain? Does advertising expenditures in the same year or advertising expenditures in the prior year provide a better predictor of sales?
14. Referring to Exercise 11, use your software package to calculate the regression line estimates for each value of advertising expenditures. Based on these estimates, calculate the residuals. Plot the residuals against the independent variable advertising expenditures. Are any violations of the assumptions associated with regression analysis apparent from the graph?
15. Data on the annual returns for 25 mutual funds were collected and entered into the following table. Data were also collected regarding the type of investments the fund focused on and the years of experience of the fund manager.
    a. Using Excel or other software package, plot the fund manager's years of experience versus the fund's annual return. Does there appear to be a linear relationship between these two variables?
    b. Use your software package to add a linear regression line to the plotted data and to calculate the parameters of the regression model. How good a fit is the linear regression model to the data?
    c. How much of the variation in a mutual fund's annual return does the variable years of experience explain?
    d. Calculate the regression line estimates for each value of years of experience, and then use these values to calculate the residuals. Create a histogram of the residuals and plot the residuals against the independent variable years of experience. Are any violations of the assumptions associated with regression analysis apparent from the graph?
    e. Based on your model, what would you estimate the annual return would be of a fund where the fund manager had 7 years of experience?

Type of Fund	Years of Experience	Annual Return
Stocks	4	16.7%
Stocks	2	11.4
Stocks	6	23.9
Bonds	9	27.5
Stocks	9	31.4
Bonds	10	30.7
Bonds	1	4.4
Stocks	5	19.8
Stocks	9	30.3

Bonds	2	10.0%
Bonds	3	9.2
Stocks	1	8.9
Bonds	1	6.2
Bonds	2	7.8
Stocks	3	14.4
Stocks	1	10.4
Bonds	4	13.4
Bonds	4	16.0
Stocks	6	22.7
Stocks	4	20.7
Stocks	4	19.6
Bonds	4	10.3
Stocks	9	32.2
Stocks	5	16.5
Bonds	5	12.7

16. Referring to Exercise 15:
    a. Develop a multiple regression model using both fund type and years of experience to predict annual return.
    b. How much of the variation in annual return is explained by these two independent variables?
    c. Calculate the parameters of the regression model and explain these parameters in terms that a typical manager would understand.
    d. Create a histogram of the residuals and plot the residuals against both independent variables. Are any violations of the assumptions associated with regression analysis apparent from the graph?
    e. Based on your model, what is your estimate for the annual return of a stock fund where the fund manager had 7 years of experience?
17. Would it be appropriate to use the model you developed in Exercise 16 to estimate the annual return of a fund that invested in foreign currencies, given that the fund manager had 5 years of experience? Why or why not? Would it be appropriate to use this model to estimate the annual return of a stock fund where the fund manager had 22 years of experience? Why or why not?
18. Data have been collected on the average daily number of visitors to several Web sites and the monthly advertising revenues these sites generate.
    a. Plot the average daily number of visitors against the advertising revenue. How would you characterize the relationship between these two variables?
    b. Develop a linear regression model. Use this model to estimate the advertising revenue for each data point and then calculate the residual for each data point. Based on a plot of the residuals against the independent variable number of visitors, what type of relationship is suggested between these variables?
    c. Transform advertising revenue based on your answer to part b. Would you conclude that the transformed values of advertising revenue and daily Web visitors are linearly related?

	A	B
1	**Visitors**	**Advertising**
2	554	$307,471
3	419	$183,852
4	227	$49,837
5	842	$710,387
6	992	$986,434
7	364	$135,298
8	213	$44,663
9	543	$296,788
10	174	$33,268
11	915	$833,297
12	247	$63,652
13	258	$67,498
14	425	$186,066
15	904	$810,966
16	354	$118,361
17	910	$827,364
18	959	$923,154
19	799	$636,544
20	748	$556,322
21	152	$30,471
22	878	$774,458
23	576	$328,460
24	230	$57,559
25	444	$196,410
26	250	$65,372

19. Given the following demand data:
    a. Plot the data using a spreadsheet. Then fit a linear trend line to the plotted data. What do the plot of the data and the trend line tell you?
    b. Use a spreadsheet to develop a linear trend, multiplicative model of the data, including seasonality, and forecast demand for each month of 2002.

	A	B	C	D
1			Demand	
2	Month	1999	2000	2001
3	Jan	27,000	33,750	33,750
4	Feb	18,000	27,000	49,500
5	Mar	40,500	31,500	40,500
6	Apr	29,250	40,500	40,500
7	May	31,500	33,750	36,000
8	Jun	40,500	40,500	45,000
9	Jul	38,250	45,000	63,000
10	Aug	45,000	49,500	63,000
11	Sep	56,250	58,500	45,000
12	Oct	63,000	63,000	67,500
13	Nov	40,500	45,000	49,500
14	Dec	40,500	49,500	63,000
15	Total	470,250	517,500	596,250

20. Referring to Exercise 19:
    a. Use exponential smoothing with an α of 0.025 and forecast the demand for January of 2002. Assume the forecast for January 1999 was 25,000.
    b. Forecast the demand for January 2002 using a four-period moving average.
    c. Plot the actual demand values against the forecast values created using the simple linear trend model, a linear trend model with seasonality, exponential smoothing, and moving averages all on the same graph for the period beginning in January 2000 and ending in December 2001. Based on a visual inspection of this graph, which forecasting technique appears to work best in this situation?
21. Referring to Exercise 11, test the hypothesis that advertising expenditures has predictive value regarding sales.
22. Referring to Exercise 15, test the hypothesis that the multiple regression model with both "type of fund" and "years of experience" included as independent variables has predictive value regarding the fund's annual return. Also test the hypotheses that these two variables independently provide predictive value, given the inclusion of the other variable in the regression model.

CASES

Resale Value of Long's Automobile

Stefani Long is contemplating the purchase of a new car. One of her primary concerns is how well the car will maintain its value. In particular, she is wondering how certain options affect a car's resale value.

One car Stefani is considering is a Chevrolet Camaro two-door coupe. She prefers a car with a five-speed manual transmission over one with an automatic transmission but would consider the automatic transmission if this significantly increased the car's resale value. Living in northern Maine, she is also somewhat indifferent about purchasing a car with air conditioning, but again, would consider adding this option if it significantly increased the car's resale value. Finally, Stefani would really like to indulge a bit and upgrade to a leather interior, but she is not clear on the value of doing this.

To analyze this situation, Stefani contacted an old high school friend that works at a used car lot. He agreed to provide her with data on all 50 two-door Camaro coupes that were sold this year. She entered this information into the spreadsheet shown in Exhibit A.

Questions

1. Develop a regression model to help Stefani assess the factors that affect a Camaro's resale value.
2. Explain in plain English to Stefani how to use the regression model and what its parameters mean.
3. How could your analysis be extended to compare how well different car models maintain their value?

Lewisville Crate Company

Lewisville Crate Company is a small, closely held corporation in Lewisville, North Carolina. Its stock is divided among three brothers, with the principal shareholder being the founding brother, Ben Thomas. Ben

EXHIBIT A Sales Data for Two-Door Camaro Coupes Sold This Year

	A	B	C	D	E	F
1	Model	Type of		Air	Leather	Sales
2	Year	Transmission	Mileage	Conditioning	Interior	Price
3	1996	Automatic	32757	Yes	Yes	$9,180
4	1992	Automatic	66416	Yes	Yes	$6,275
5	1997	Automatic	25908	No	Yes	$13,165
6	1996	Automatic	46828	Yes	No	$8,640
7	1993	Manual	56254	No	No	$6,175
8	1996	Manual	53503	No	No	$7,375
9	1995	Manual	66908	Yes	No	$8,460
10	1992	Manual	83896	No	Yes	$5,085
11	1996	Automatic	57577	Yes	No	$8,340
12	1996	Manual	40169	Yes	Yes	$8,365
13	1996	Manual	26294	Yes	No	$8,350
14	1997	Automatic	31881	No	No	$12,565
15	1998	Manual	18992	No	No	$13,440
16	1998	Manual	22705	Yes	No	$14,000
17	1994	Manual	65575	Yes	Yes	$7,735
18	1998	Manual	25116	No	Yes	$13,550
19	1997	Manual	27024	No	Yes	$12,400
20	1996	Automatic	58571	No	Yes	$8,180
21	1994	Manual	62544	No	Yes	$7,385
22	1994	Automatic	60940	Yes	No	$8,200
23	1995	Manual	55168	Yes	Yes	$9,200
24	1999	Manual	12013	Yes	Yes	$16,275
25	1997	Manual	29137	Yes	Yes	$12,935
26	1999	Manual	8027	Yes	No	$16,110
27	1996	Manual	40611	Yes	Yes	$8,365
28	1993	Automatic	69978	No	Yes	$6,810
29	1994	Automatic	44223	No	No	$8,125
30	1994	Automatic	63642	No	Yes	$8,050
31	1999	Manual	9853	No	Yes	$15,840
32	1997	Automatic	36186	Yes	No	$12,875
33	1993	Automatic	80214	Yes	Yes	$6,795
34	1996	Automatic	33055	Yes	No	$8,915
35	1999	Automatic	9744	No	No	$16,110
36	1996	Automatic	31639	No	Yes	$8,780
37	1993	Automatic	87653	No	No	$6,050
38	1997	Manual	32057	No	No	$11,625
39	1994	Automatic	72450	Yes	No	$7,850
40	1993	Manual	74562	Yes	Yes	$6,420
41	1997	Automatic	29717	No	No	$12,565
42	1996	Manual	45446	No	Yes	$7,865
43	1997	Manual	30905	Yes	Yes	$12,935
44	1993	Manual	89408	Yes	No	$5,735
45	1995	Manual	30294	Yes	Yes	$9,650
46	1998	Automatic	15826	No	No	$14,315
47	1999	Automatic	11518	No	Yes	$16,350
48	1998	Manual	14995	No	No	$13,515
49	1995	Manual	40832	Yes	No	$9,135
50	1998	Manual	18819	No	No	$13,440
51	1994	Manual	75796	Yes	Yes	$7,360
52	1995	Automatic	47903	Yes	No	$9,760

formed the company 20 years ago when he resigned as an accountant for a large manufacturer of corrugated boxes. Lewisville Crate supplies wooden crates to a variety of exporters.

Ben attributes his success to the fact that he can serve the five-state area he considers "his territory" better than any of his large competitors. He recognizes the danger of becoming too dependent upon any one client and has enforced the policy that no single customer can account for more than 20 percent of sales. Two of the exporters account for 20 percent of sales each, and hence are limited in their purchases. Ben has persuaded the purchasing agents of these two companies to add other suppliers, since this alternative supply protects them against problems Lewisville Crate might have—shipping delays, raw material shortages, or labor problems.

Lewisville Crate currently has more than 550 customers with orders ranging from a low of 100 crates to blanket orders for 5,000 crates per year. Crates are produced in 8 standard sizes.

Lewisville Crate competes on the basis of offering quick delivery times to its customers. Providing such high levels of service, however, requires tight inventory control and close production scheduling. So far, Ben has always forecast demand and prepared production schedules on the basis of experience, but because of the ever-growing number of accounts and changes in personnel in customer purchasing departments, the accuracy of his forecasts has been rapidly declining. The number of backorders is on the increase, late orders are more common, and inventory levels of finished crates are also on the increase. A second warehouse has recently been leased because of overcrowding in the main warehouse. Plans are to shift some of the slower-moving crates to the leased space.

There has always been an increase in demand for crates just before the holiday season, when customers begin stocking up. This seasonality in demand has always substantially increased the difficulty of making a reliable forecast.

Ben thinks that it is now important to develop an improved forecasting method to help smooth production and warehousing volume. He has compiled the demand data shown below.

Questions

1. Develop a forecasting method for Lewisville Crate and forecast total demand for 2001.
2. Should Ben's experience with the market be factored into the forecast? How?

	A	B	C	D	E	F
1		Sales (in number of crates)				
2	**Month**	**1995**	**1996**	**1997**	**1999**	**2000**
3	Jan	20400	13600	20400	25500	25500
4	Feb	13600	23800	13600	20400	37400
5	Mar	17000	30600	30600	23800	30600
6	Apr	30600	25500	22100	30600	30600
7	May	23800	27200	23800	25500	27200
8	Jun	17000	30600	30600	30600	34000
9	Jul	27200	23800	28900	34000	47600
10	Aug	30600	47600	34000	37400	47600
11	Sep	34000	37400	42500	44200	34000
12	Oct	45900	45900	47600	47600	51000
13	Nov	40800	44200	30600	34000	37400
14	Dec	30600	17000	30600	37400	47600

BIBLIOGRAPHY

Albright, S. C., W. L. Winston, and C. Zappe. *Data Analysis and Decision Making with Microsoft® Excel.* Pacific Grove, CA: Brooks/Cole, 1999.

Anderson, D. R., D. J. Sweeney, and T. A. Williams. *Statistics for Business and Economics,* 8th ed. Cincinnati, OH: South-Western, 2002.

Georgoff, D. M., and R. G. Murdick. "Manager's Guide to Forecasting." *Harvard Business Review,* 64 (Jan.–Feb. 1986): 110–120.

Makridakis, S., and S. C. Wheelwright. *Forecasting: Methods and Applications,* 3rd ed. New York: Wiley, 1998.

Mason, R. D., D. A. Lind, and W. G. Marchal. *Statistical Techniques in Business and Economics,* 10th ed. Boston: Irwin/McGraw-Hill, 1999.

Montgomery, D., and E. Peck. *Introduction to Linear Regression Analysis,* 3rd ed. New York: Wiley, 2002.

Newbold, P., and T. Bos. *Introductory Business Forecasting,* 2nd ed. Cincinnati, OH: South-Western Publishing Co., 1994.

Sanders, N. R., and K. B. Manrodt. "Forecasting Practices in U.S. Corporations: Survey Results." *Interfaces,* 24 (March–April 1994): 92–100.

Siegel, A. F. *Practical Business Statistics,* 4th ed. Boston: Irwin/McGraw-Hill, 2000.

4 Optimization and Mathematical Programming

INTRODUCTION

Congratulations again on your job offer from that top consulting firm. Unfortunately, you can forget about a nice, leisurely orientation. You will be needed immediately to help collect data on a project involving the development of a production planning model for a small electronics manufacturing services company.

After doing a little research, you learn that many electronic manufacturing services (EMS) companies are prospering due to the recent move toward outsourcing. More specifically, EMS companies provide a variety of services related to the design, production, distribution, and after-sale servicing of medical, telecommunication, and computer products for original equipment manufacturers (OEMs). OEMs outsource these services for a variety of reasons: to lower production costs, to gain access to advanced manufacturing technologies, and to better utilize their own manufacturing assets. Two leading electronic manufacturing service providers include Solectron (http://www.solectron.com) and SCI Systems (http://www.sci.com).

Startron, a relatively young EMS provider, has contacted your consulting firm to help it better plan its production. Startron produces three special niche products for its clients: desktop computers, server computers, and notebook computers.

Initial discussions with management indicate that currently the demand for all three of the products produced by Startron exceeds available production capacity.

Thus, it becomes clear that the major problem facing Startron is how many orders of each product it should accept from its clients in order to maximize overall profit.

Having defined the problem, your specific assignment is to help determine what information is needed to solve this problem and to then collect this information. Since the objective is to maximize overall profit, you will need information related to the profitability of each product. You thus make a note to schedule an appointment with the controller of Startron to obtain this data.

Your first priority, however, is to get a better handle on Startron's operations. Fortunately, the plant manager is available to give you a tour of the factory. During the plant tour, you discover that Startron's operations are divided into four departments: assembly, software installation, testing, and packaging. You also learn that the plant operates only one shift per day, that factory employees work 8 hours per day, and that overtime is never used because of its high costs. On the plant tour you inquire into the total number of factory workers and how they are allocated across the four departments. You also ask how many labor hours each type of product requires from each department. The plant manager is able to tell you off the top of her head exactly how many workers there are in each department, and she promises to get back to you in a day or so regarding the labor hours.

Because Startron's operations are labor intensive, the plant manager mentions that as a rule the company insists on producing a minimum quantity of each product, even though other products might be more profitable. The plant manager explains that over the long term it is more important to Startron to maintain the skills to produce a variety of products than to maximize its short-term profitability. When pressed further on the issue, the plant manager suggests that each type of product should make up at least 20 percent of the production on a unit basis.

Exhibit 4.1 contains a summary of the data collected based on your discussions with the controller, plant manager, and other key employees.

With the data collected, the project team now turns its attention to modeling this situation. Because of the complexity associated with this situation, the team agrees that it would be useful to develop an influence diagram to help them better understand the situation, rather than jumping right in and trying to develop a formal mathematical model. The influence diagram developed by the team is shown in Exhibit 4.2.

According to the influence diagram shown in Exhibit 4.2, the primary factor that influences Startron's profits is the number of each type of computer to produce each day. This factor also corresponds to the decision variables. The influence diagram further indicates that the decision regarding how many of each computer type to produce depends on three sets of parameters that cannot be changed, at least in the short term: minimum production quantities set by management, unit profitability, and available

EXHIBIT 4.1
Summary of Data Collected from Startron

	Computer Model			Number of Workers
	Desktop	**Server**	**Notebook**	
Profit per unit	$75	$145	$125	
Assembly time (hrs)	0.5	0.75	1.5	12
Software installation (hrs)	0.25	0.4	0.3	3
Testing (hrs)	1	1.5	1	20
Packaging (hrs)	0.1	0.1	0.2	2

EXHIBIT 4.2 Influence Diagram for Startron's Production Planning Problem

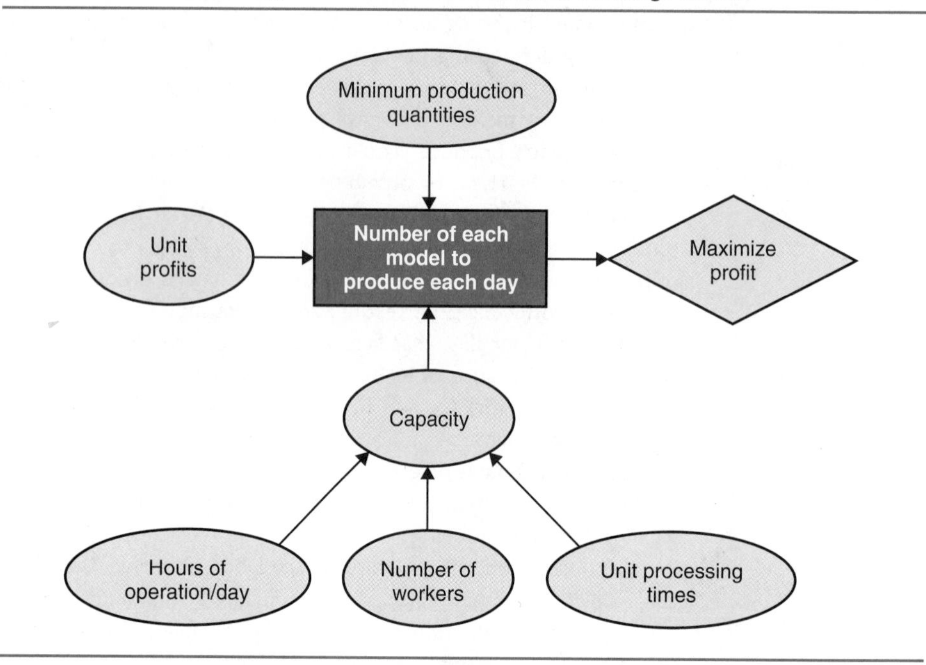

capacity. We further observe that available capacity is influenced by the parameters: number of workers available, the hours the plant is operated each day, and unit processing times.

A couple of comments related to the influence diagram are in order. First, if the situation facing Startron were to change such that demand no longer exceeded capacity, the influence diagram would likely need to be modified along the lines shown in Exhibit 4.3. Second, a number of the elements considered to be parameters in Exhibit 4.2 would become decision variables over a longer planning horizon. For example, the number of workers, hours of operation time per day, and possibly even unit processing times could all be changed if management so desired.

By clearly identifying the objective, decision variables, parameters, and key relationships among these components, the influence diagram shown in Exhibit 4.2 can be used to guide the development of a formal mathematical model. A mathematical model is needed in this situation because there are a massive number of combinations of feasible solutions. The model will help identify the best of all these possible solutions. As the rectangular box indicates in Exhibit 4.2, the decision variables or factors that management has control over and must make a decision about are the production quantities for each product type. The team agrees to let DT represent the number of desktop computers to be produced each day. Likewise, SV and NB are defined to be the number of server computers and notebook computers to be produced each day, respectively.

Next, the team turns its attention to Startron's goal, shown in the diamond at the right of Exhibit 4.2. Can Startron's goal of profit maximization be formulated in a mathematical model? The answer is yes. For example, the daily profit earned from producing desktops is equal to the \$75/unit profit for desktops times the number of desktops produced or $75DT$. The same logic is used to determine the daily profits from producing server computers and notebook computers. As is shown below, summing the profits of all three products provides Startron's total daily profit:

$$\text{Maximize profit} = 75DT + 145SV + 125NB$$

EXHIBIT 4.3 Modified Influence Diagram Including Demand

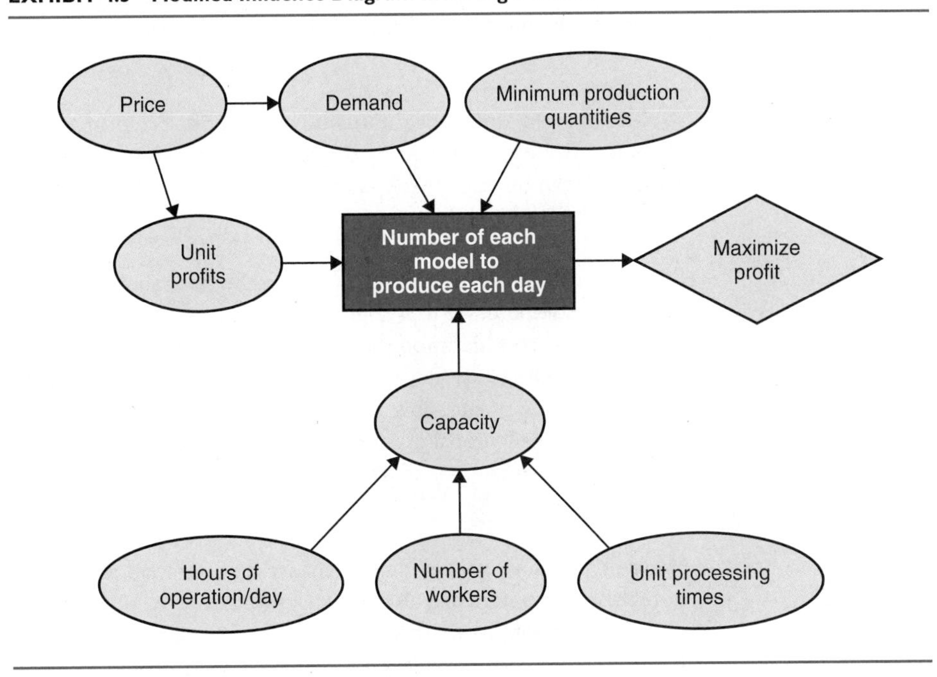

In addition to Startron's objective, the model must incorporate all relevant constraints and limitations, as shown in the influence diagram. Referring to Exhibit 4.2, we observe that one key limitation is available capacity, which further depends on the number of workers, the hours the plant operates, and unit processing times. Rechecking Exhibit 4.1, there are a total of 12 assembly workers available each day. Since each assembly worker is available 8 hours each day, a total of 96 hours (12 workers × 8 hours/worker) of assembly hours are available each day. Therefore, in determining the daily production quantities of the three products, it is important to ensure that no more than 96 hours of assembly time are required. In addition to providing the number of workers available in each department, Exhibit 4.1 also lists the labor processing times for each computer model in each department. For example, we see that each desktop computer requires 0.50 hour of assembly time. Likewise, each server computer and notebook computer requires 0.75 hour and 1.50 hours of assembly time, respectively. To ensure that the available assembly time is not exceeded, we could add a constraint to the model as follows:

$$0.5DT + 0.75SV + 1.5NB \leq 96$$

In English, this formula says that the hours needed to assemble desktop computers plus the hours needed to assemble server computers plus the hours needed to assemble notebook computers must be less than or equal to the 96 hours that are available each day. Labor constraints for available software installation time, testing time, and packing time would be developed in a similar fashion. The influence diagram shown in Exhibit 4.2 indicates that these capacity constraints incorporate all three parameters that were identified in the diagram as affecting capacity.

Referring to the top of the influence diagram, we observe that another important constraint is that each type of computer should account for at least 20 percent of the production on a per-unit basis. The total units produced is simply the sum of the three decision variables: *DT, SV, NB*. One way to express the constraint that desktop computers should account for at least 20 percent of the total units produced is as follows:

$$DT \geq 0.2\ (DT + SV + NB)$$

In developing mathematical models, many modelers prefer having all the decision variables on the left-hand side of the inequality or equality. Thus, the terms in the above constraint can be rearranged as follows:

$$0.8DT - 0.2SV - 0.2NB \geq 0$$

Similar constraints are needed to ensure the minimum number of server computers and notebook computers are produced.

One final consideration that is not explicitly identified in the influence diagram is that it is not possible to produce a negative number of computers. To ensure that our model does not consider infeasible solutions where a negative number of computers are produced, we add additional constraints. For example, the constraint below ensures that producing a negative number of desktop computers is not suggested by solving the model:

$$DT \geq 0$$

Similar constraints are needed to ensure that negative numbers of server computers and notebook computers are also not suggested.

The complete mathematical model your project team developed to help Startron better plan its production is listed in Exhibit 4.4. In reviewing this mathematical model, pay particular attention to the way each element identified in the influence diagram has been incorporated in the complete model. Thus, we see that the influence diagram can be used to help validate a mathematical model in addition to helping guide its development.

Before trying to solve Startron's dilemma, we need additional information related to formulating optimization models, constructing appropriate spreadsheets, interpreting the results, solving the model in Excel, and conducting sensitivity analyses. We will then return to Startron's situation. Following this, we discuss some special integer programming situations and complete the chapter with a detailed example.

EXHIBIT 4.4
Formulation of the Mathematical Model

Let DT = the number of desktop computers to produce each day
SV = the number of server computers to produce each day
NB = the number of notebook computers to produce each day

Maximize profit = $75DT + 145SV + 125NB$

Subject to:

$0.5DT + 0.75SV + 1.5NB \leq 96$ (available assembly hours)

$0.25DT + 0.4SV + 0.3NB \leq 24$ (available software installation hours)

$1DT + 1.5SV + 1NB \leq 160$ (available testing hours)

$0.1DT + 0.1SV + 0.2NB \leq 16$ (available packaging hours)

$0.8DT - 0.2SV - 0.2NB \geq 0$ (minimum number of desktop computers that must be produced)

$-0.2DT + 0.8SV - 0.2NB \geq 0$ (minimum number of servers that must be produced)

$-0.2DT - 0.2SV + 0.8NB \geq 0$ (minimum number of notebook computers that must be produced)

$DT \geq 0$ (cannot produce a negative number of desktop computers)

$SV \geq 0$ (cannot produce a negative number of servers)

$NB \geq 0$ (cannot produce a negative number of notebook computers)

4.1 The Modeling Process for Optimization Studies

Optimization is perhaps the most widely taught of all topics in the field of management science. Optimization is prescriptive because it determines or prescribes the "best" possible action for a manager to take, under the assumptions of the model being studied. In the Startron example, the solution to the model will prescribe exactly how many of each type of computer should be produced each day in order to maximize profits. Descriptive models on the other hand simply portray or depict a situation.

It is important to note, however, that the model assumptions made for optimization studies are usually somewhat severe compared to the variety of conditions that exist in the real world of managers, organizations, competition, and governmental regulation. Thus, what we determine from our quantitative models to be "best" is only true in a very restricted sense, and the modeler needs to be aware of this when presenting the results of the study to management. Ideally, the modeler would be able to translate the study results into the real world of the manager and discuss the tradeoffs affecting the manager's real-world decision.

Next, we describe the optimization process in more detail and then the topics covered in this chapter. Once the reader has a basic understanding of the optimization process, we describe the modeling process that is involved when modeling a situation for optimization studies.

Optimization

Optimization is the name for a family of tools designed to help solve managerial problems in which the decision maker must allocate scarce (or limited) resources among various activities to optimize a measurable goal. For example, distributing machine time (the resource) among various products (the activities) is a typical allocation problem. Allocation problems usually display the following characteristics:

1. A limited quantity of economic resources (such as labor, capital, machines, or raw materials) is available for allocation.
2. The resources are used in the production of products or services.
3. There are two or more ways in which the resources can be used. Each is called a **solution.** (Usually the number of ways is very large, or even infinite.)
4. Each activity (product or service) in which the resources are used yields a return (or reward) in terms of the stated goal.
5. The allocation is usually restricted by several limitations and requirements called **constraints.**

Optimization also entails a number of assumptions, as follows:

1. Returns from different allocations can be compared; that is, they can be measured by a common unit (such as dollars).
2. All data are known with certainty.
3. The resources are to be used in the most economical manner.

In the case of linear programming, the most common form of optimization and the main topic of this chapter, two additional assumptions are made:

4. The return from any allocation is independent of other allocations.
5. The total return is the sum of the returns yielded by the different activities.

The allocation problem can generally be stated this way: Find the way of allocating a set of limited resources to various activities so the total reward is maximized. Allocation problems usually have many possible alternative solutions. Depending on the underlying assumptions, the number of solutions can be either infinite or finite. Also, different solutions yield different rewards. Of the available solutions, one (sometimes more) is the *best,* in the sense that the degree of goal attainment associated with it is the highest (i.e., total reward is maximized). This is called the *optimal* solution.

Many problems in organizations are related to the allocation of resources (money, people, time, power, space, equipment). The reasons are limited resources, various ways of allocation, the difficulty of measuring the contribution of the allocation to the goals, and disagreement concerning the importance of the results. Optimization provides a relatively unbiased approach to this allocation problem.

The uses of optimization, especially linear programming, are so numerous and varied that many common software programs, such as spreadsheets, have built in this functionality. A glance at professional or trade journals reveals many applications of optimization. The following examples illustrate the breadth of optimization, both in terms of the types of organizations that use it and its applicability to a wide variety of managerial decisions.

- Welch's, a large processor of grapes, needed a way to calculate recipes for its products based on raw material availability and factory capacity. To accomplish this, a linear programming model was formulated. This linear programming model is solved each day using spreadsheet optimization. Welch estimates that it saved between $130,000 and $170,000 using this model during its first year (Schuster and Allen 1998).
- Minnesota's Nutrition Coordinating Center developed an optimization model to estimate the nutrient content of commercial food products. A key benefit of using optimization is its ability to estimate nutrient values four times faster than conventional methods with no loss in accuracy. Based on this success, a decision support system using linear programming was developed to maintain a food composition database (Westrich, Altmann, and Potthoff 1998).
- Headquartered in Germany, AgrEvo is in the crop-protection and plant-biotechnology industry. A key challenge facing the firm is the selection of research and development projects for funding. To aid in this process, a spreadsheet optimization model was developed. Use of this model provided management with many important insights. For example, the model indicated that a resource that was perceived to be the limiting factor in completing projects actually had excess capacity. Indeed, further analysis of the model suggested that all projects could be funded by simply reallocating resources. In total, this allowed the organization to retain additional projects worth tens of millions of dollars (Sonntag and Grossman 1999).
- Like other gas utilities, the Peoples Gas, Light and Coke Company's profits depend on gas supply and storage contracts. One factor that complicates planning in this industry is that demand for natural gas depends on difficult-to-forecast weather conditions. To address this problem, the Peoples Gas, Light and Coke Company configures its supply portfolio based on the use of an optimization model that considers alternative weather scenarios. Using this model, the company has been able to save in excess of $50 million per year by restructuring its supply portfolios (Knowles and Wirick 1998).

The Modeling Process

The modeling process when conducting an optimization study is identical to that described in Chapter 1. However, the modeler should be aware of special considerations in optimization studies that are different from those of descriptive studies such as statistical analyses.

Step 1: Opportunity/Problem Recognition As opposed to descriptive models, which are commonly used to identify and better understand opportunities or problems, optimization is commonly used to find new ways to conserve resources, identify more productive use of equipment or labor, reduce overall plant and capital investments, and so on. That is, optimization studies typically search for the *best* way of doing something, often something currently being done but in a way that is thought to perhaps be less effective or efficient than might be possible.

Thus, optimization studies are typically more action-directed and may come later, in a larger overall study of a problem or opportunity. Descriptive studies may have uncovered an opportunity, or given management a better understanding of a situation, but now it is time to determine how to best exploit this opportunity or solve this problem.

Step 2: Model Formulation The issue of model formulation will be handled thoroughly in the remainder of this chapter. In overview, however, we note that a systems approach is taken which involves establishing a "boundary" for the opportunity under study, and an initial influence diagram is then proposed. After this, management is commonly consulted and some data are collected to better understand the process. Then the influence diagram is usually modified to reflect this better understanding and some more data are collected. This iterative process continues until everyone is satisfied that the model formulation adequately represents the real situation.

Step 3: Data Collection As noted, we advise the modeler to stay in constant touch with the manager throughout the modeling process in an optimization study, not only because the data may change but also because the opportunity itself may change. Moreover, an error due to changing data or conditions when trying to gain additional understanding about a problem or opportunity is usually less serious than a similar error when making a definite recommendation to take a specific action because that action was determined to be optimal.

Another reason the data collection is more critical in optimization than with other quantitative modeling studies is because the only allowance made for variability and chance is through *sensitivity analysis,* where the effect of possible changes in the variables is examined. Hence, optimization tends to relate more closely to decision making under certainty rather than risk or uncertainty. That is, optimization assumes that the data are known, a dangerous assumption even under the best of circumstances. The data collection process thus calls for constant vigilance.

Finally, an aid to data collection is the sensitivity analysis conducted during the model analysis step because this process indicates what data are particularly sensitive to changes and errors, and hence must be collected accurately. Thus, the data collection step is not only iterative with the model formulation step but also with the model analysis step, described next.

Step 4: Model Analysis The issue of model analysis will be detailed in the remainder of this chapter. It basically consists of invoking an algorithmic analysis process to yield the "optimum" outputs. Optimization usually employs a static model of the situation and employs mathematical programming, which includes an objective function and usually a large number of decision variables and constraints. The mathematical program generates and evaluates alternative solutions to the situation, improving the outcome with each iteration of the program until a "best" or optimal outcome is achieved. However, sensitivity analysis is easily conducted with optimization models, and this often proves to be valuable in the implementation step of the study, as well as in the data collection step.

Step 5: Implementation As noted, optimization studies result in one "best" answer, so the issue of choosing a course of action would appear to be a minor issue. However, a sensitivity analysis is typically also conducted to check the impact on the results of changes or errors in the data. In addition, the use of sensitivity analysis allows the managers to ask "what-if" questions in case the "certain" data turn out to not be as certain as the modeler was led to believe. In the example at the beginning of the chapter, sensitivity analysis could, for example, be undertaken to investigate the impact of removing the minimum production requirements for each product. Other types of sensitivity analyses that could be undertaken include investigating the impact of changes in the profitability of the various computer models and studying the effects of changes in the availability of the various resources. In addition, conducting the sensitivity analysis with the managers by itself facilitates the implementation step through the communication and instantaneous feedback between the modeler and these other agents in the process.

Following a managerial decision about the situation and its implementation, it is wise for the modeler to stay in touch with the situation to make sure that the predicted results are achieved. Unfortunately, it is more commonly the case that all the projected benefits described in the study are not attained because of countervailing impacts, human resistance, unexpected forces arising to oppose the actions taken, and so on. And even if the predicted results are achieved, it may be at the expense of some other part of the organization. That is, the opportunity/problem "solution" may have been a suboptimal action that made things worse elsewhere in the organization, either to the overall detriment of the organization or, more commonly, to the advantage of the organization but not as great as was predicted.

Structure of the Chapter

In this chapter, we cover a number of optimization topics. We start with the basic but most common and useful category, called **linear programming.** Linear programming deals with allocation problems in which the goal or objective and all the requirements imposed on the opportunity are expressed as *linear functions* (see Appendix A). We cover the basic concepts and operation of linear programming in detail through both text and graphics. We then describe how to use Microsoft Excel to solve linear programs and illustrate this approach with an example, including how to interpret the results. Next we address an extension of linear programming called *integer* programming, which can also be implemented through Excel. When the requirement that some or all of the decision variables must be integers (whole numbers, such as 12 people) is added to a linear programming problem, it becomes one of integer linear programming. Last, we look at a special case of integer programming where one or more variables can assume only the values of 0 or 1 (e.g., on and off, or purchase and do not purchase).

4.2 Linear Programming

Sean and his wife, Caitlin, were enjoying a barbecued salmon dinner on their patio, their first unhurried meal in days. Final tests had just been successfully completed on the new ergonomic suspension system for OfficeComfort Corporation's desk chairs that Sean had developed, and the outlook was encouraging. "They'll be installing the system on two models for initial sales next week," he was telling Caitlin. "With a larger work force, more machine time, and a bigger marketing budget, I'm sure we could sell a lot of these new chairs. I wonder how many of each model we should produce?"

Caitlin was thinking about her chemistry problem at the Diskote Company. A new, expensive coating for computer hard drives had been developed, and the production manager had asked Caitlin to see if she could find a combination of two new ingredients, code-named Alpha and Beta, that would result in an equivalent granularity and density but at less cost than the original ingredients. She felt confident she could.

Caitlin did not realize that her problem, a typical *blending* problem, was in many ways equivalent to Sean's, a typical *output-mix* problem.

Linear programming (LP) is one of the best known tools of management science. The most general statement of its objective is that it is used to determine an optimal allocation of an organization's limited resources among competing demands. Many managerial problems can be considered allocation problems. Two of the most common are the output-mix problem faced by Sean and the blending problem faced by Caitlin. But there are many others as well, including the scheduling of facilities, the assignment of resources to tasks, transportation between multiple points, and even dietary planning. Linear programming deals with a special class of allocation problems; namely, those in which all the mathematical functions in the model are linear.

The Output-Mix Problem

In an **output-mix problem,** there are two or more *outputs* (i.e., products or services), such as computer models or patients, competing for limited resources, such as limited production capacity or time with the doctor. The problem is to find out *which outputs* to include in the production plan *and in what quantities* these should be produced or delivered (output mix) in order to maximize profit, market share, or some other goal. In the example at the beginning of the chapter the objective was to determine how many desktop, server, and notebook computers to produce given the limitation of available worker hours in several departments.

Although a solution to an output-mix problem does specify the quantities to be produced, what it tells more generally, in effect, is *how to allocate* scarce resources. This is because the technology of production is given and once a decision has been made on the outputs and quantities to produce, a determination has actually been made of what resources to use, where, and in what quantities.

The Blending Problem

Blending problems involve the determination of the *best blend* of available ingredients to form a certain quantity of an output under strict specifications. The best blend means the least-cost blend of the required inputs. Blending problems are especially important in the process industries such as petroleum, chemicals, and food.

The blending problem is similar to the output-mix problem. However, the objectives usually differ. In the output-mix problem, the profit derived from selling the products made from the given amount of resources is to be maximized. In blending, the cost of certain ingredients is to be minimized, while adhering to given specifications. Therefore, a blending problem is also considered a problem of allocating resources in the best manner. The experiential exercise at the end of the chapter deals with a special case of the blending problem sometimes referred to as the *diet problem.*

Formulating the Linear Programming Model

Every linear programming (LP) model is composed of the following six components:

Decision Variables The **decision variables** in LP depend on the type of LP model being developed. For example, the decision variables can be the quantities of the resources to be allocated, or the number of units to be produced. The manager is searching for the value of these variables (usually denoted by x_1, x_2, . . . or x, y, and z) that will provide an optimal solution (e.g., the highest profit or lowest cost) to the problem.

Objective Function This is a mathematical statement of the desired goal. It is expressed as a linear function showing the relationship between the decision variables and the goal (or objective) under consideration. The **objective function** is a measure of goal attainment. With LP we seek to find the maximum level of a desired goal such as total profit or market share, or the minimum level of some undesirable outcome such as total cost. LP is a type of optimization model because it attempts to either maximize or minimize the value of the objective function.

Profit or Cost Coefficients The coefficients of the variables in the objective function (e.g., 75, 145, and 125 in the example at the beginning of the chapter) express the *rate* at which the value of the objective function increases or decreases by adding to the solution one unit of a given decision variable. For example, each additional desktop computer Startron produces contributes an additional $75 to profits. Although we consider these coefficients to be fixed, this is generally true only within a certain range of the variable due to other effects such as economies of scale.

Constraints The manager is searching for the values of the decision variables that will maximize (or minimize) the value of the objective function. Such a process is usually subject to several uncontrollable restrictions, requirements, or regulations that are called constraints. The optimization (i.e., the maximization or minimization) is performed subject to a set of constraints. Therefore, LP can be defined as a **constrained optimization problem.** These constraints are expressed in the form of linear inequalities (or, sometimes, equalities). They reflect the fact that resources are limited or that other limiting factors must be considered. In the Startron example there were limited resources in four departments, as well as limitations on the minimum number of computers that should be produced of each computer model.

Constraint Coefficients The numbers in front of the decision variables in the constraints are called the **constraint coefficients.** They indicate the rate at which a given resource is depleted or utilized. They typically appear on the left-hand side of the constraints. In the Startron example at the beginning of the chapter the constraint coefficients for the assembly time constraint were 0.5, 0.75, and 1.5. The 0.5 coefficient indicates that each desktop computer produced uses up half-an-hour of the total available assembly hours.

Right-Hand-Side Constants The capacities (or availability) of the various resources, usually expressed as some upper or lower limit, are given on the *right-hand side* of the constraints. The **right-hand-side constants** can also express minimum requirements. In the example at the beginning of the chapter, the right-hand-side constant of the assembly constraint was 96, indicating that the maximum amount of assembly time that can be used on any given day is 96 hours.

Output-Mix and Blending Problems: Two Examples

Let us return to the output-mix and blending problems and present both with the relevant data. For the purpose of clarity, our focus here will be on the development of the actual LP model and not on the entire modeling process. An example later in this chapter will illustrate LP in the broader modeling context.

EXAMPLE

The Output-Mix (Maximization) Problem

The two models of desk chairs Sean developed for the OfficeComfort Corporation are a leather and a fabric model. The company is in the market to maximize profit. The U.S. profit realized from exports is $300 for the leather model and $250 for the fabric model. Obviously, the more chairs produced and sold, the better. The trouble is that certain limitations prevent OfficeComfort Corporation from producing and selling thousands of chairs daily:

1. Only 40 hours of labor each day are available in the production department (a labor constraint).
2. Only 45 hours of machine time a day (a machining constraint) are available.
3. The company cannot sell more than 12 chairs of the leather model each day (a marketing constraint).

OfficeComfort's problem is to determine *how many chairs of each model to produce each day so that the total profit will be as large as possible.*

In terms of the overall modeling process, we have now clearly defined the opportunity/problem confronting OfficeComfort and have collected the relevant data. Therefore, the focus of the ensuing discussion will be on *formulating* the model (step 2 in the modeling process) and *solving/analyzing* the model (step 4 in the modeling process).

Formulation of the Output-Mix Problem

We describe the formulation of the problem in terms of the decision variables, the objective function, and the constraints. In this case, the *decision variables* are

$$x_1 = \text{Number of leather models to be produced daily}$$
$$x_2 = \text{Number of fabric models to be produced daily}$$

The *objective function* is the equation that reflects the desired goal—in this case, to sell the combination of chairs that result in the greatest profit. The daily profit realized from selling leather chairs is $300x_1$ (i.e., the profit per unit times the number of units). Similarly for the fabric chair a profit of $250x_2$ will be realized each day. The total profit, z, is, therefore, $300x_1 + 250x_2$. This total profit is called the *value of the objective function.* OfficeComfort wishes to maximize the value of the objective function z. This goal is tempered by four sets of *constraints:*

1. **Labor Constraint** There are 40 hours of labor available each day. Each leather model requires 2 hours of labor, whereas each fabric model requires only 1 hour. This situation can be expressed as

Demand for labor			Supply
Total labor for leather model		Total labor for fabric model	Total labor available
$2x_1$	$+$	$1x_2$	$\leq$ 40

Note that the less-than-or-equal-to sign ($\leq$) is used to represent this linear inequality. That is, 40 hours is the *available* capacity, which does not necessarily have to be used in full.

2. **Machine Time Constraint** There are up to 45 machine-hours available per day. Machine processing time for one unit of the leather model is 1 hour and for one unit of the fabric model, 3 hours. This limitation can be expressed as

$$1x_1 + 3x_2 \leq 45$$

3. **Marketing Constraint** It is only possible to sell *up to* 12 of the leather model each day. This can be expressed as

$$1x_1 \leq 12$$

4. **Nonnegativity Constraint** Finally, it is impossible to produce a negative number of chairs; that is, both x_1 and x_2 must be nonnegative (zero or positive). This constraint is expressed as

$$x_1 \geq 0, x_2 \geq 0$$

In summary, the problem is to find the best daily production plan so that the total profit will be maximized. The problem can now be written in standard form, as follows: Find x_1 and x_2 that maximize *z*, the value of the objective function, subject to constraints.

$$\begin{aligned}
&\text{Maximize } z = 300x_1 + 250x_2 \\
&\text{Subject to:} \\
&2x_1 + 1x_2 \leq 40 \text{ (labor constraint)} \\
&1x_1 + 3x_2 \leq 45 \text{ (machine time constraint)} \\
&1x_1 + 0x_2 \leq 12 \text{ (marketing constraint)} \\
&x_1, x_2 \geq 0 \quad \text{(nonnegativity constraints)}
\end{aligned}$$

Note: Because production continues day after day, it is not necessary to complete all chairs at the end of the day; that is, a fractional number of chairs is permissible (e.g., 5.73 chairs). However, had it been necessary to complete all chairs at the end of the day, additional constraints limiting x_1 and x_2 to whole numbers could have been added. Such an addition changes the problem to one of *integer programming,* a topic we cover a bit later in the chapter.

We shall return to the solution of this problem later.

EXAMPLE

The Blending (Minimization) Problem

Let us now consider the technical aspects of Caitlin's problem at Diskote. In preparing coatings, it is required that the coating have a granularity rating of at least 300 units and a density rating of at least 250 units. Granularity and density levels are determined by two ingredients: Alpha and Beta. Both Alpha and Beta contribute equally to the granularity rating, one ounce (dry weight) of either producing one degree of granularity in one drum of coating. However, the density is controlled entirely by the amount of Alpha, one ounce of it producing three units of density in one drum of coating. The cost of Alpha is 45 cents per ounce, and the cost of Beta is 12 cents per ounce. Assuming that the objective is to minimize the cost of the ingredients, then the problem is to find the quantity of Alpha and Beta to be included in the preparation of each drum of coating. The problem is shown graphically in Exhibit 4.5.

EXHIBIT 4.5 The Blending Problem

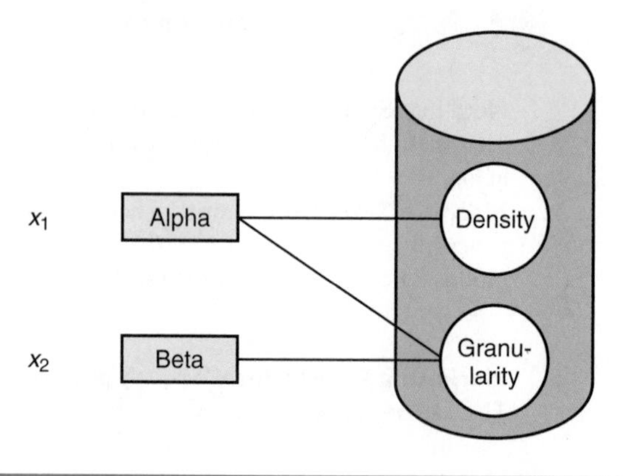

Formulation of the Blending Problem

The problem is formulated for only one drum of coating. The reason is that an optimal answer for one drum of coating will remain optimal for any number of drums as long as the relationships are linear. The total quantity of coating to be produced is, of course, larger than one drum, and it is determined mainly by the demand and the manufacturing technology.

The *decision variables* are:

x_1 = Quantity of Alpha to be included, in ounces, in each drum of coating
x_2 = Quantity of Beta to be included, in ounces, in each drum of coating

The cost of Alpha is 45 cents per ounce, and because x_1 ounces are going to be used in each drum, then the cost per drum is $45x_1$. Similarly, for Beta the cost is $12x_2$. The total cost is, therefore, $45x_1 + 12x_2$, and as our objective function, it is to be *minimized* subject to the following constraints:

1. To provide a granularity rating of at least 300 units in each drum. Because each ounce of Alpha or Beta increases the granularity by one unit, the following relationship exists:

$$\underbrace{1x_1}_{\text{Supplied by Alpha}} + \underbrace{1x_1}_{\text{Supplied by Beta}} \geq \underbrace{300}_{\text{Demand}}$$

2. To provide a density level of at least 250 units. The effect of Alpha (alone) on density can similarly be written as

$$\underbrace{3x_1}_{\text{Supplied by Alpha}} + \underbrace{0x_2}_{\text{Supplied by Beta}} \geq \underbrace{250}_{\text{Demand}}$$

In summary, the blending problem is formulated as follows. Find x_1 and x_2 so as to

Minimize $z = 45x_1 + 12x_2$
Subject to:
$1x_1 + 1x_2 \geq 300$ (granularity specification)
$3x_1 + 0x_2 \geq 250$ (density specification)
$x_1, x_2 \geq 0$ (nonnegativity constraints)

The blending problem is going to be solved later in the chapter. At this time, the reader may note that the blending problem is very similar in structure to the output-mix problem. One difference, however, is that here we wish to *minimize* the value of the objective function, whereas in the output-mix problem we were *maximizing* the objective function.

In our example, all the output-mix constraints are of the less-than-or-equal to ($\leq$) type, and all the blending problem constraints are of the greater-than-or-equal-to ($\geq$) type. In reality, when we deal with larger problems, we find all types of constraints, $\leq$, $\geq$, and = in both types of problems, as will be demonstrated later.

The General LP Model

The general LP model can be presented in the following mathematical terms. Let

a_{ij} = Constraint coefficients	c_j = Objective function coefficients
b_i = Right-hand-side constants	x_j = Decision variables
i = Row number	j = Column number

Maximize (or minimize) $z = c_1x_1 + c_2x_2 + c_3x_3 + \cdots + c_jx_j + \cdots + c_nx_n$

Subject to the linear constraints:

$$
\begin{aligned}
a_{11}x_1 + a_{12}x_2 + \cdots + a_{1n}x_n &\le b_1 \\
a_{21}x_1 + a_{22}x_2 + \cdots + a_{2n}x_n &\le b_2 \\
\cdots\cdots\cdots\cdots\cdots\cdots\cdots \\
a_{i1}x_1 + a_{i2}x_2 + \cdots + a_{in}x_n &\le b_i \\
\cdots\cdots\cdots\cdots\cdots\cdots\cdots \\
a_{m1}x_1 + a_{m2}x_2 + \cdots + a_{mn}x_n &\le b_m
\end{aligned}
$$

and the nonnegativity constraints:

$$x_1 \ge 0, x_2 \ge 0, \ldots, x_n \ge 0$$

The blending problem discussed earlier is used to illustrate the components and terminology associated with LP.

Find x_1 and x_2 that will minimize the value of the linear objective function:

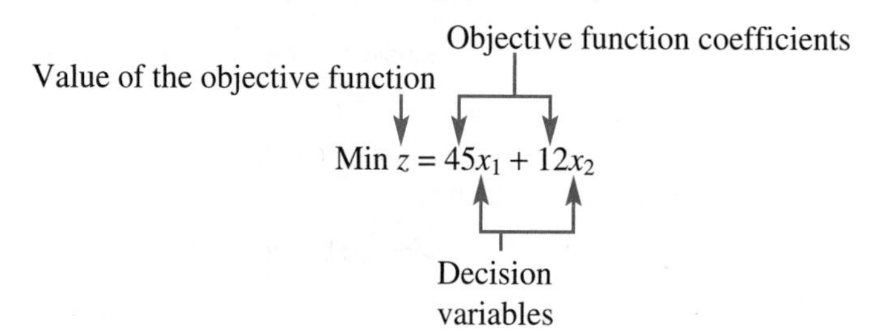

Subject to the linear constraints:

$$
\begin{aligned}
1x_1 + \quad 1x_2 &\ge 300 \\
3x_1 + \quad 0x_2 &\ge 250
\end{aligned}
$$

Constraint coefficients

Right-hand sides

and subject to the nonnegativity of the decision variables: $x_1 \ge 0$; $x_2 \ge 0$.

Advantages, Assumptions, and Solution Methods

The advantages and limitations of LP are described next. The general solution methodology is also overviewed, in order to introduce the more detailed descriptions in the remainder of the chapter.

Advantages of Linear Programming LP can be used to solve allocation-type problems that are common and extremely important in organizations. LP not only provides the optimal solution, but does so in an efficient manner. Further, it provides additional information concerning the value of the resources that are allocated.

Specifically, the advantages of LP are as follows:

1. Finds an optimal solution(s)
2. Determines the solution quickly through computer packages
3. Finds solutions to a wide variety of situations that can be formulated with LP, as was discussed earlier in this chapter
4. Handles situations involving a large number of variables and constraints
5. Finds solutions to situations with a very large or infinite number of possible alternatives
6. Allows a fast sensitivity analysis

Assumptions of Linear Programming The applicability of LP is limited by several assumptions. As in all mathematical models, assumptions are made for reducing the complex real-world problem into a simplified form that can be more readily analyzed. The major assumptions are:

Certainty. It is assumed that all the data in the LP are known with certainty. In problems under risk, the expected value of the input data can be considered as a constant, thus enabling the treatment of risky situations by LP. Then a sensitivity analysis can be conducted to explore other possible values of the data.

Linear Objective Function. It is assumed that the objective function is linear. This means that per unit cost, price, and profit are assumed to be unaffected by changes in production methods or quantities produced or sold. This assumption is usually reasonable within a limited range. However, beyond this range the assumption may not be reasonable. For example, some organizations offer price breaks for large orders. As another example, a firm's cost structure would likely change if it is required to add additional capacity. Some of these kinds of situations can be handled with integer programming, as discussed later.

Linear Constraints. It is also assumed that the constraints are linear. The linearity assumptions in LP for both constraints and the objective function are reflected in three properties: additivity, independence, and proportionality.

1. **Additivity** It is assumed that the total utilization of each resource is determined by adding together that portion of the resource required for the production of each of the various products or activities. For example, in the Startron example at the beginning of the chapter, the total assembly time was found by adding the amount of assembly time required for the desktop, server, and notebook computers. The assumption of additivity also means that the effectiveness of the joint performance of activities, always equals the sum of the effectiveness resulting from the individual performance of these activities.
2. **Independence** Complete independence of coefficients is assumed, both among activities and among resources. For example, the price of one product has no effect on the price of another, or the time to perform one activity is independent of the time for other activities.
3. **Proportionality** The requirement that the objective function and constraints must be linear is a proportionality requirement. This means that the amount of resources used, and the resulting value of the objective function, will be proportional to the value of the decision variables.

Nonnegativity. Negative activity levels (or negative production) are not permissible. It is required, therefore, that all decision variables take nonnegative values.

Divisibility. Variables can, in general, be classified as continuous or discrete. Continuous variables are subject to *measurement* (e.g., weight, temperature), whereas discrete variables are those that can be *counted:* 1, 2, 3, . . . In LP, it is assumed that the unknown variables x_1, x_2, . . . , are continuous, that is, they can take any fractional value (divisibility assumption). If the variables are restricted to whole numbers and thus are indivisible, a problem in integer programming exists. As will be discussed later, extra constraints can be added to ensure that the optimal solution contains only whole numbers for all integer decision variables.

Solving Linear Programs A set of decision variables, each having a value, is called a *solution.* For example, $x_1 = 5$ and $x_2 = 0$ is a solution for the OfficeComfort Corporation problem. Proposed solutions to an LP problem that *satisfy all the constraints* are called *feasible solutions.* The collection of all feasible solutions is called the feasible solution space or area. Any proposed solution that violates one or more of the constraints is termed *infeasible.*

Once a problem has been formulated, one of several available methods of solution can be applied. Normally, this is done with the aid of a computer. For our purposes, two of the solution methods have a significant value. They are (1) the **graphical method,** whose main purpose is to illustrate the concepts involved in the solution process; and (2) the general, computationally powerful **simplex method** used in Excel and other software packages.

Distribution Problems: Transportation, Transshipment, Assignment

Distribution problems are a special type of linear programming problem but can still be solved as such. For manual solutions, there are special solution procedures, but they will not be considered here. However, it is worthwhile to be able to recognize the special structure of these distribution problems.

The Transportation Problem The **transportation problem** deals with *shipments* (or what can be considered to be equivalent to shipments) from one set of sources to a number of destinations. Typically:

- A limited supply of items, such as cement or oil, is available at certain sources, such as factories or refineries.
- There is a demand for the items at several destinations such as warehouses, distribution centers, or stores.
- The quantities of supply at each source and the demands at each destination are known.
- The per-unit costs of transporting the commodity from each source to each destination are known. They are usually based on the distance between the two points, or they may be negotiated.
- No shipments occur between the sources or between the destinations.
- The objective is to determine how many units should be shipped from each source to each destination so that all demands are satisfied (if possible) at the minimum total shipping cost.

Many problems other than those involving shipments between points can be interpreted as equivalent to a transportation problem (see Turban and Meredith 1994). For example, multiple plants may be available to make multiple types of clothing that are required in certain amounts every month. Here, the plants may be considered as the sources with limited production supplies, the clothing types as the destinations with the amounts of each needed as being the demands, and the production costs for each item at the various plants being the shipment costs. Similarly, the best crops to grow in multiple farming regions can be considered as a transportation problem, with a certain supply of acres available in each region, a certain acreage of production demand for each crop in the economy, and different costs for growing each crop in each region. And even multiperiod production scheduling problems can be considered to be transportation problems, as when there are different methods (e.g., regular time, overtime, using past inventory, subcontracting) for a supplier to produce items demanded in each period (month or quarter).

The Transshipment Problem The **transshipment problem** relaxes the requirement that no shipments are allowed between the sources and between the destinations. However, an extra constraint must be added to the formulation specifying that there is a material balance at every point. That is, the amount shipped in plus the amount produced must equal the amount shipped out plus the amount consumed for every supply point and every demand point.

The Assignment Problem The **assignment problem** involves finding the best one-to-one match for each of a given (finite) number of "candidates" to a number of "positions." Typical situations include assigning workers to machines, teachers to classes, houses to potential buyers, and so on. Different benefits or costs are involved in each match, and the goal is to maximize the total reward or minimize the total cost. The assignment problem is a different type of distribution problem in that it requires integer solutions of the zero–one form for the variables, that is, each supply-destination amount is either 1 or 0, designating an assignment or not. The solution of integer programming problems will be covered in Section 4.6, where the special case of zero–one problems is also addressed.

4.3 Analysis of the Model by the Graphical Method

The graphical method is used mainly to illustrate certain characteristics of LP problems and to help in understanding and interpreting the solution provided by spreadsheets. The only case where it has a practical value is in the solution of small problems with two decision variables and only a few constraints, or problems with two constraints and only a few decision variables.

Example 1: A Maximization Problem

In order to illustrate the graphical method, let us return to the output-mix problem discussed previously:

$$\text{Maximize } z = 300x_1 + 250x_2$$

Subject to:

$$2x_1 + 1x_2 \leq 40 \text{ (labor constraint)}$$
$$1x_1 + 3x_2 \leq 45 \text{ (machining constraint)}$$
$$1x_1 + 0x_2 \leq 12 \text{ (marketing constraint)}$$
$$x_1, x_2 \geq 0 \qquad \text{(nonnegativity constraints)}$$

The plotting of algebraic equations and inequalities in terms of geometric lines and curves was initiated by the French philosopher Descartes in the seventeenth century. Two straight lines intersecting at right angles are used as a reference, and points are located by giving two coordinates (distances from each of the lines). The plane formed by the two lines is divided into four regions called quadrants. An example of such a system is shown in Exhibit 4.6. We use x_1 and x_2 to designate the lines. Sometimes x and y are used instead.

Because the nonnegativity constraints appear in all LP problems, we are limited to the first quadrant only ($x_1 \geq 0$, $x_2 \geq 0$). The graphical solution procedure consists of two phases: (a) graphing the feasible area and (b) identifying the optimal solution.

Phase 1: Graphing the Feasible Area The feasible area is established through graphing all of the inequalities and equations that describe the constraints.

Graphing the First (Labor) Constraint. This constraint is expressed as $2x_1 + 1x_2 \leq 40$. The steps in drawing the constraint are as follows:

Step 1 An inequality of the type *less than or equal to* has two parts. We will first consider only the equality part of the constraint: $2x_1 + 1x_2 = 40$. Because an equation can be

EXHIBIT 4.6 Feasible and Infeasible Quadrants on the Two-Dimensional Grid

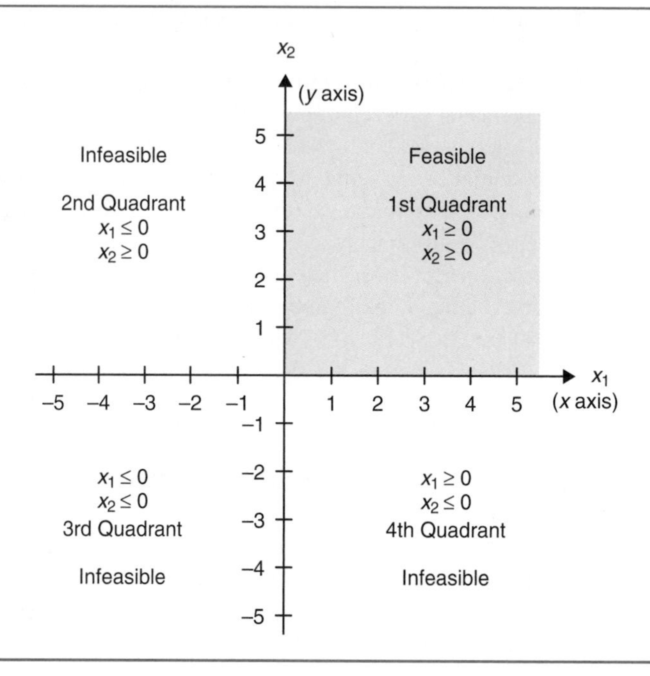

shown graphically as a straight line, it is sufficient to find the coordinates of two points to graph the entire equation as a line. To do so, first set x_1 to zero and then solve the equation for x_2. We get: $2(0) + x_2 = 40$. Thus, $x_2 = 40$. This yields a point (0, 40), which is shown as point A in Exhibit 4.7. This solution means that if only fabric chairs are produced, it will be at a rate of 40 per day. To find a second point, we set x_2 to zero:

$$2x_1 + 0 = 40$$

Solving this, we get $x_1 = 20$. This is shown as point B (20, 0) in Exhibit 4.7.

Step 2 Joining points A and B by a straight line is a representation of the equation $2x_1 + x_2 = 40$. However, it is not just the equation that is of interest, but also the inequality $2x_1 + x_2 < 40$. This inequality is represented by an *area* below and to the bottom left of the equality (Exhibit 4.7). Because the original constraint was of the $\leq$ type, the area of interest includes both the line and the shaded area below it.

This area represents all possible combinations of leather and fabric chairs that use 40 or less hours of labor, and therefore we call it a *feasible* area.

Step 3 *How to find the side of the constraint on which the feasible area is located.* Sometimes it is not obvious which side of the line represents the feasible area. Of the several methods that can he used, the following is the simplest. Take the point with coordinates (0, 0) and check if it is feasible (i.e., does not violate the constraint). If the point is feasible, then the entire area is feasible. For example, point O in Exhibit 4.7, which is inside the area, has coordinates of (0, 0). If we substitute these coordinates into the constraint $2x_1 + 1x_2 \leq 40$, we get $2(0) + 1(0) = 0$. Because 0 is smaller than 40 and still within our constraint that x_1, x_2 must be greater than or equal to zero, the point is in the feasible area. Point $U = (20, 10)$, on the other hand, is not feasible, because $2(20) + 1(10) = 50$.

The inequality area in Exhibit 4.7 would normally include negative values of x_1 and x_2, except for the fact that in LP, negative values are excluded by the nonnegativity constraint $x_1 \geq 0$ and $x_2 \geq 0$. The end result is a feasible area inside (and including the boundary of) the triangle OAB in Exhibit 4.7 (shaded).

EXHIBIT 4.7 The Labor Constraint

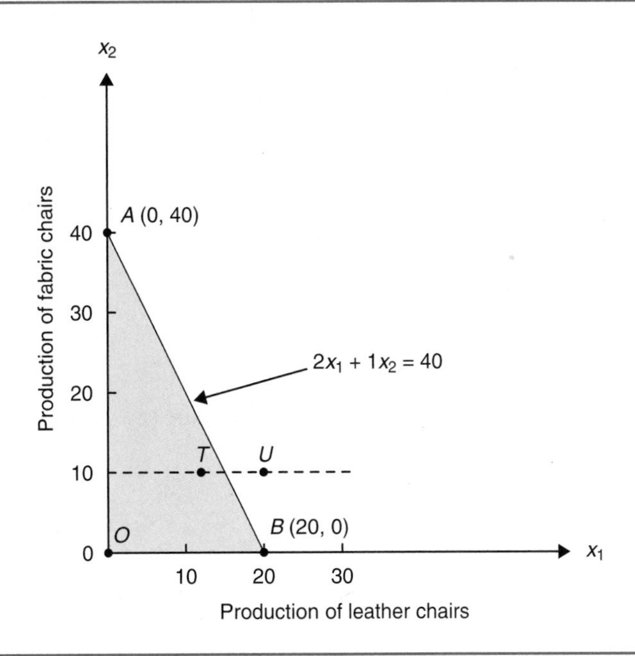

Graphing the Second (Machine Time) Constraint. Similarly, the shaded area OCD in Exhibit 4.8*a* represents the area of feasible solutions for the machining constraint $1x_1 + 3x_2 \leq 45$.

Graphing the Third (Marketing) Constraint. Next, the equality part $x_1 = 12$ of the marketing constraint $x_1 \leq 12$ is plotted as a straight line, vertically from point E and parallel to the x_2 axis (Exhibit 4.8*b*). Again, the feasible area is left of, and including, the line.

EXHIBIT 4.8 The Second and Third Constraints

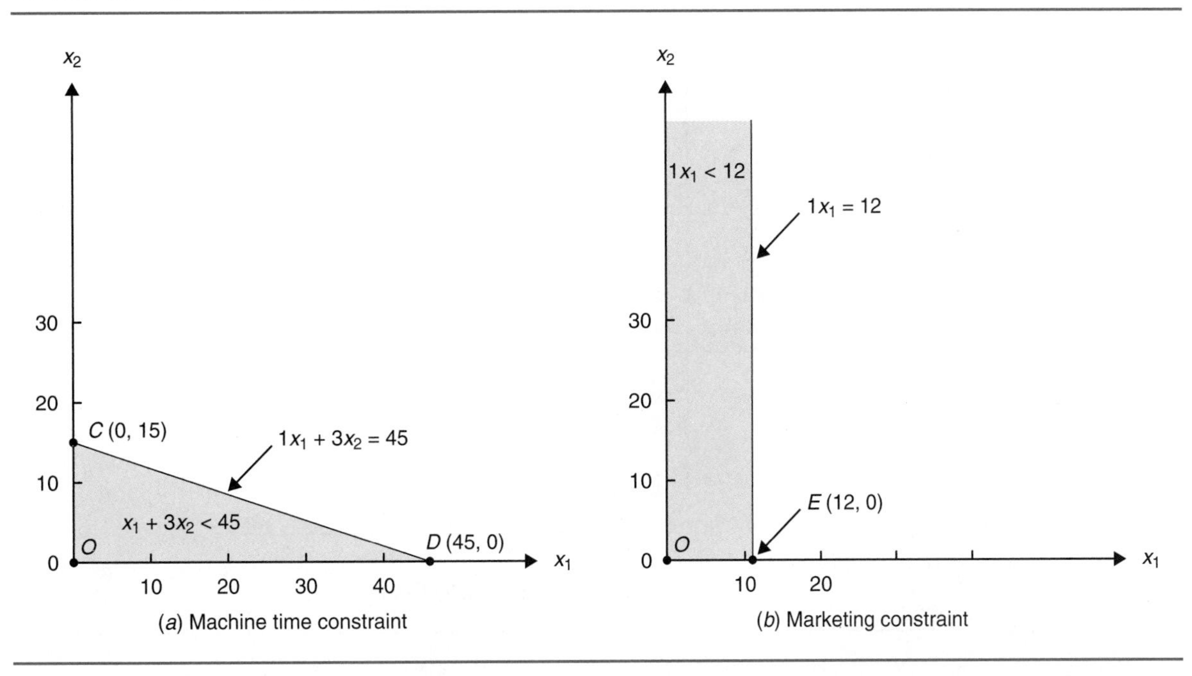

Combining the Constraints. Once all the constraints have been drawn, they can be put on one graph. As a matter of practicality, they can be built on one graph from the beginning. We used the graphs of Exhibits 4.7 and 4.8 for instructional purposes only. The combination of all constraints is shown in Exhibit 4.9. The shaded area *OCGE* is the area that satisfies all constraints simultaneously. Therefore, any solution in this area (Exhibit 4.10) is *feasible* with respect to *all* the constraints.

EXHIBIT 4.9 All Three Constraints Simultaneously

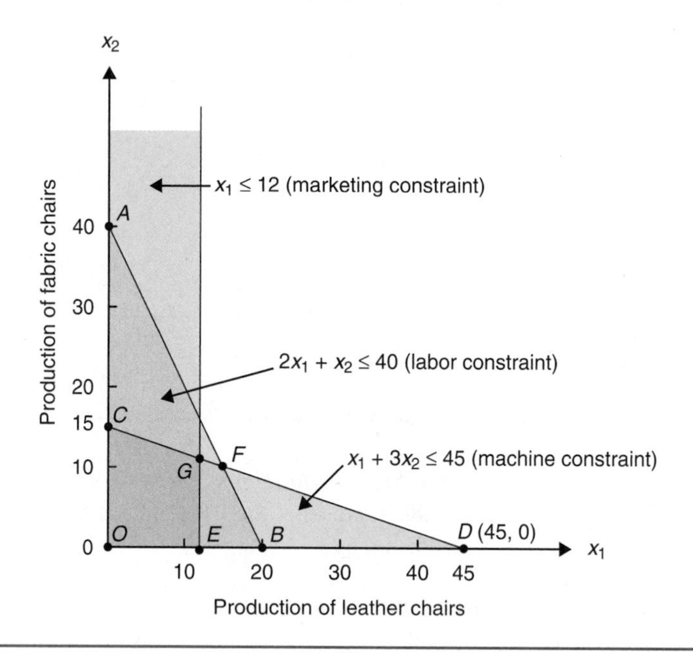

EXHIBIT 4.10 The Feasible Area

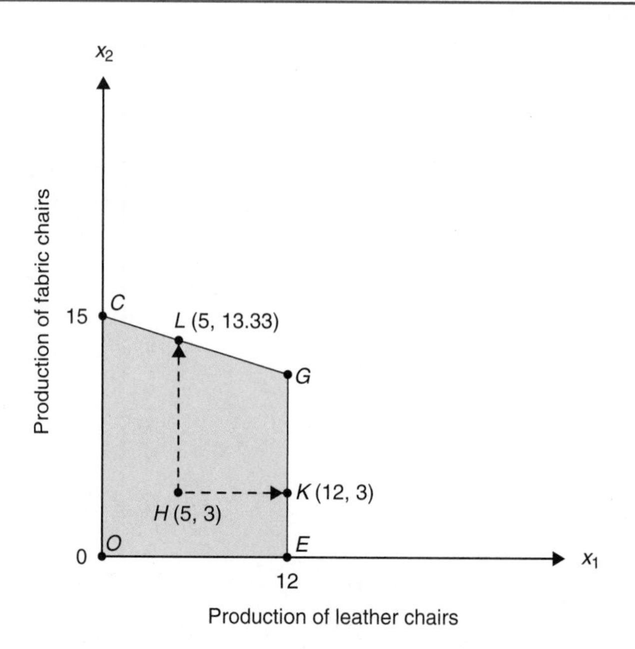

Phase 2: Identifying an Optimal Solution Any point in the shaded area *OCGE* of Exhibit 4.10 and its boundary is a feasible solution. Because there are an infinite number of points in this area, there are an *infinite number of feasible solutions* for this problem. To find an optimal one, it is necessary to identify a solution (point) in the feasible area that maximizes the profit (objective) function.

How can this task be accomplished? Let us examine a feasible solution inside the feasible area; say, point *H*. This point is a solution plan that calls for five leather chairs ($x_1 = 5$) and three fabric chairs ($x_2 = 3$). This solution is not optimal, because production of x_1 can be increased from point *H* to point *K* (12 units of x_1 and 3 of x_2), resulting in a higher profit. Notice that such an increase may continue only until we are limited by a constraint. Similarly, point *L*, where more x_2 is produced than at *H*, will result in a higher profit than point *H*.

The reason that both points *K* and *L* are better than point *H* is that both yield a higher profit, because the coefficients of the objective function are positive (more units, more profit). It can, however, be shown that no matter what sign the coefficients of the objective function have, there will always be at least one point somewhere *on a boundary* of the feasible area that is *superior* to any point inside the area. Therefore, an optimal solution to a linear programming problem can *never* be inside the feasible area but must be on a boundary. As we will see, the optimal solution always occurs at a corner point. Therefore, an efficient search for an optimal solution need only consider corner points (*O, C, G,* and *E* in our case). Two methods can be used for such a search.

Method 1: Enumeration of All Corner Points. In this method, the solution values at all the corner points are compared. This is done by determining the coordinates of all corner points and then computing and comparing the values of the objective function at these points. Let us demonstrate: For points *O, C,* and *E,* the exact coordinates can be read directly from Exhibit 4.10, as given in Exhibit 4.11.

For point *G,* the approximate coordinates can be read from the graph, or exact coordinates can be computed as follows: Point *G* is at the intersection of two straight lines. Therefore, the value at point *G* (as given by x_1 and x_2) must be the same for the two intersecting lines. Such a value can be found by simultaneously solving the two intersecting linear equations, one of which describes the marketing constraint and the other the machine time constraint. In this case:

$$\text{Equation 1: } 1x_1 = 12 \quad \text{(marketing constraint)}$$
$$\text{Equation 2: } 1x_1 + 3x_2 = 45 \text{ (machine constraint)}$$

The solution is simple in this situation, because the value of one decision variable, x_1, is already known to be 12. This value is introduced into Equation 2, which can then be solved. We get

$$1x_1 + 3x_2 = 1(12) + 3x_2 = 45, \text{ or } 3x_2 = 33, \text{ or } x_2 = 11$$

Therefore, the coordinates of point *G* are (12, 11).

EXHIBIT 4.11
Coordinates of the Corner Points

Point	x_1	x_2
O	0	0
C	0	15
E	12	0
G	?	?

Now the profits at each corner point can be calculated, using the objective function $300x_1 + 250x_2$. The results are shown in Exhibit 4.12.

The point that yields the greatest profit is point *G*. The optimal solution, therefore, is to produce 12 of the leather model and 11 of the fabric model. The total profit is $6,350.

It is important not to confuse the optimal solution with the *value* of the optimal solution. The optimal solution in this example is $x_1 = 12$ and $x_2 = 11$. The value of the optimal solution is $6,350. The distinction between the optimal solution and the value of the optimal solution is important to understanding sensitivity analysis, discussed later in this chapter.

The process of comparing profits at all corner points may be very lengthy, because in larger linear programming problems many corners exist. The graphical method can also be used to evaluate the corners. This method is presented next.

Method 2: The Use of Isoprofit (Constant Profit) Lines. According to this procedure, the optimal solution is found by using the *slope* of the objective function as a guide. Let us illustrate.

In examining solution *H* (5, 3) in Exhibit 4.10, it was indicated that better solutions are available, such as *K* (12, 3). The question is: In what direction should one move to find better solutions? The answer to this question is found from the objective function. If the objective function (a profit function in our example) is graphed and the direction of increasing profit is identified, then when one starts moving the profit function in that direction, the profit will continue to increase as it is moved. The limit to the increase is reached when the function touches the farthest point(s) on the boundary of the feasible area (if the feasible area is bounded; the unbounded situation will be discussed in a later section). This point is then an optimal solution. The **graphical method** accomplishes this task in a systematic fashion.

Profit functions (such as $z = 300x_1 + 250x_2$) describe an infinite number of equations that depend on the value of z and, therefore, cannot be presented as a single line. To overcome this difficulty, a family of linear equations, called **isoprofit** lines (or **isocost** lines, in the minimization case) is constructed. An *isoprofit* line is a collection of points, each of which designates a solution with the *same* profit. By assigning various values to z, we get different profit lines. Graphically, such a family can be plotted as many lines *parallel* to each other (see Exhibit 4.13).

To start, we pick a value for z. Any arbitrary profit figure will work. However, it is simple to pick a profit number that gives an integer as an answer to x_1 when we set $x_2 = 0$, and vice versa. A good choice is to use a number that is divided easily by the coefficients of both variables. For example, 1,500/300 = 5, and 1,500/250 = 6. Thus, let us assume that $z = 1{,}500$. Then, the profit equation $300x_1 + 250x_2 = 1{,}500$ can be drawn as a straight line exactly in the same manner as the equality constraints were drawn. This line intersects the x_1 axis at point *N* in Exhibit 4.13 (where $x_2 = 0$ and $x_1 = 5$) and the x_2 axis at

EXHIBIT 4.12
Profit Values at Corner Points

Point (Corner)	Solution Coordinates	Total Profit $300x_1 + 250x_2 = z$
O	(0, 0)	300 (0) + 250 (0) = 0
C	(0, 15)	300 (0) + 250(15) = $3,750
E	(12, 0)	300(12) + 250 (0) = $3,600
G	(12, 11)	300(12) + 250(11) = $6,350 ← *Maximum*

point M (where $x_1 = 0$ and $x_2 = 6$). Point N tells us how many units of product x_1 *alone* are required to produce a profit of \$1,500. Because the profit contribution per unit of product x_1 is \$300, the answer is $1,500 \div 300 = 5$ units of x_1. Similarly, because the per unit profit contribution of x_2 is \$250, it takes 6 units of x_2, if only x_2 is produced (as indicated by point M), to produce a profit of \$1,500. All points on the isoprofit line MN are within the shaded area, representing feasible solutions, giving a total profit of \$1,500.

In a similar fashion, other isoprofit lines could be drawn, yielding different levels of profits. For example, line PQ in Exhibit 4.13 represents a \$3,000 isoprofit level, and the line RS represents a \$6,000 isoprofit level. An examination of lines MN, PQ, and RS shows that they are *parallel* to each other. Note, also, that profit levels increase as the isoprofit lines get *farther away* from the origin, or point (0, 0).

A little reflection will show that an even higher profit can be achieved if additional isoprofit lines can be drawn farther away from the origin. The question is: Where is the farthest *isoprofit* line? Let us see how this question can be answered.

The feasible area is reproduced from Exhibit 4.10 as Exhibit 4.14. Superimposed on the figure is the isoprofit line MN. Now, if one starts building parallel isoprofit lines in the northeast direction (see the heavy arrow), these lines will represent higher and higher profits. We shall now show how to build such parallel lines, and how to find the farthest possible isoprofit line.

To build parallel lines to MN, take a ruler and a 90-degree triangle. Place one edge of the right angle along the MN and set the ruler against the other right-angle side. This makes the ruler perpendicular to line MN. To find the farthest isoprofit line, slide the triangle along the ruler to the northeast until the point where it is about to leave the feasible area $OCGE$.

EXHIBIT 4.13 Family of Isoprofit Equations $z = 300x_1 + 250x_2$

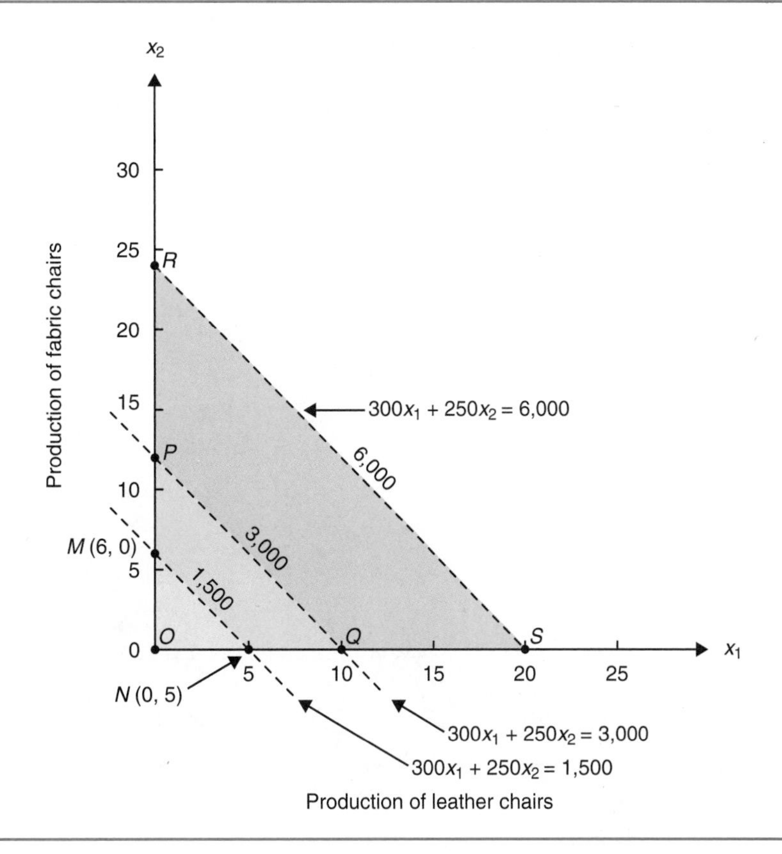

In our example, the farthest isoprofit line (VW) hits point G, which is declared the *optimal solution.* The coordinates of this point were previously computed as 12 and 11. The value of the objective function at that point is $300(12) + 250(11) = 6{,}350$. Therefore, line VW designates the isoprofit line $z = 6{,}350$.

When employing this method, one of four cases may be expected. First, it can be shown that an optimal solution must be at a corner (called a vertex or *extreme* point) where two constraints intersect. Thus, the farthest isoprofit line may intersect only one corner point, providing a single optimal solution. Second, the farthest isoprofit line may coincide with one of the boundary lines of the feasible area. Then *at least* two optimal solutions must be on two adjoining vertices, and the others will be on the boundary connecting them. (This second case is discussed in more detail later.) Third, the isoprofit line may possibly go on without limit from the constraints, providing an *unbounded* solution. This usually indicates that an error has been made in formulating the model. Last, there may be no feasible solutions at all, due to the strictness of the constraints. This situation is indeed possible, and indicates to the modeler that some of the constraints will need to be relaxed in order to allow a solution to the opportunity or problem.

Note that if the optimal values of x_1 and x_2 are introduced into the inequality constraints, the solution will fully utilize only the second and the third resources in this case. That is, the labor constraint is not fully utilized (*binding*). This can be clearly seen in Exhibit 4.9 where the labor constraint does not intersect the optimal point.

Example 2: A Minimization Problem

The graphical method can be used for minimization problems in a manner similar to the maximization case. To illustrate, let us examine the blending problem discussed earlier.

EXHIBIT 4.14 Moving to an Optimal Solution

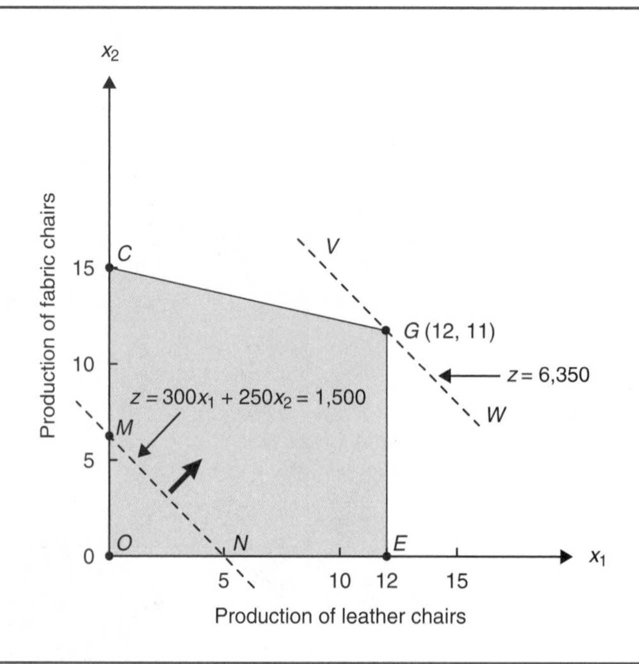

EXHIBIT 4.15 The Blending Problem

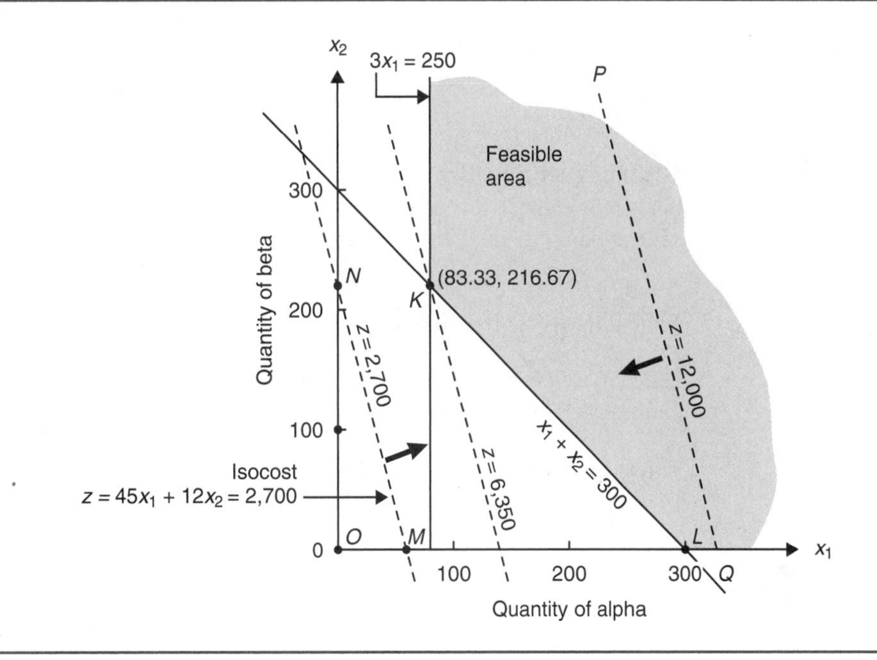

The problem is reproduced below:

$$\text{Minimize } z = 45x_1 + 12x_2$$
$$\text{Subject to:}$$
$$1x_1 + 1x_2 \geq 300$$
$$3x_1 + 0x_2 \geq 250$$
$$x_1, x_2 \geq 0$$

To begin with, the inequality constraints are considered equations. They are drawn in Exhibit 4.15. Because the inequalities are of the greater-than-or-equal-to type, the feasible area is formed by considering the area to the *upper right* side of each equation (away from the origin, the shaded area in Exhibit 4.15). Next, a family of lines that represents various levels of the objective function is drawn (broken lines in Exhibit 4.15). These lines, in the minimization case, are isocost lines.

In the diagram, a value of $z = 2{,}700$ is arbitrarily selected. The isocost equation is

$$45x_1 + 12x_2 = 2{,}700$$

This isocost line is shown as line *MN* in Exhibit 4.15, note that the line *does not* intersect the feasible solution space. Therefore, this isocost line must be moved toward the feasible area until it first intersects a point (or points) in this region (point *K*). The coordinates of point *K* can be read from the graph, or they can be computed as the intersection of the two linear equations. The solution for point *K* is

$$x_1 = 83.33$$
$$x_2 = 216.67$$

This means that for every drum of the coating, the following raw materials should be used: 83.33 ounces of Alpha at a cost of 45 cents × 83.33 = 3,750 cents; and

216.67 ounces of Beta at a cost of 12 cents × 216.67 = 2,600 cents. The total cost is, therefore, 3,750 cents + 2,600 cents = 6,350 cents. If the isocost line is constructed inside the feasible area, like line *PQ* in Exhibit 4.15, then the search moves in the direction of the arrow, toward the origin. Because the objective is cost minimization, we search for the *lowest* level isocost that is still feasible.

Graphing an Equality Thus far, we have graphed only inequalities. If a constraint appears as an equality, then the feasible region of solutions is not an area anymore but a *line segment* directly on the line describing the equality constraint.

Utilization of the Resources—Slack and Surplus Variables

Next, we introduce the concept of slack and surplus variables that convert inequalities to equalities, which are more easily handled in our solution procedure.

Slack Variables The optimal solution for the output-mix problem calls for $x_1 = 12$ and $x_2 = 11$. Let us examine now what will happen to the constraints if the optimal solution is used.

1. **The Labor Constraint** This constraint is expressed as

$$\overbrace{2x_1 + 1x_2}^{\text{Demand for labor}} \leq \overbrace{40}^{\text{Supply of labor}}$$

Introducing the optimal values, we get

$$2(12) + 1(11) = 35 \text{ hours}$$

That is, the demand for labor is 35 hours, whereas the supply is 40. Therefore, there would be 40 – 35 = 5 hours of unused labor potential. This unused supply is called *slack.* The slack must take only nonnegative values. Designating this **slack variable** by s_1, (where the 1 designates the first constraint), the labor constraint can be rewritten as

$$2x_1 + 1x_2 + s_1 = 40$$

transforming the constraint into an equality. The solution for s_1 is

$$s_1 = 40 - (2x_1 + 1x_2) = 40 - 2(12) - 1(11) = 5$$

2. **The Machine Constraint** Similarly, the machine slack variable can be inserted into the constraint as

$$x_1 + 3x_2 + s_2 = 45$$

Substituting the values of the optimal solution and solving for s_2:

$$s_2 = 45 - (1x_1 + 3x_2) = 45 - 12 - 33 = 0$$

In this case, the machine time is *fully utilized* (no slack exists). A fully utilized constraint is also called a *binding constraint,* and it constitutes a bottleneck in the organization's operations.

3. **The Marketing Constraint** Similarly:

$$1x_1 + s_3 = 12$$

Solving:

$$s_3 = 12 - 1x_1 = 12 - 12 = 0$$

The implication again is that there is no slack; the constraint is being exactly met.

Note: A glance at Exhibits 4.9 and 4.14 will show that a constraint with zero slack intersects the optimal solution point. A constraint with positive slack does not intersect the point of the optimal solution.

Surplus Variables Any greater-than-or-equal-to constraint has a surplus variable. For example, assume that a fourth constraint is added:

$$2x_1 + 3x_2 \geq 50$$

Let the optimal solution be

$$x_1 = 12; x_2 = 11$$

Introducing these values into the left side of the inequality, we get

$$2\,(12) + 3\,(11) = 57$$

The value of the left side of the inequality is larger than the value of the right side. The difference, 7 in this case, is called called *surplus.* It indicates how much the requirements of the right-hand side are exceeded.

The fourth constraint is now written with a **surplus variable** that can either be positive or zero. It is written as

$$2x_1 + 3x_2 - s_4 = 50$$

Introducing the values of the decision variables and solving for s_4:

$$s_4 = (2x_1 + 3x_2) - 50 = 2(12) + 3(11) - 50 = 7$$

As another example, take the first constraint of the blending problem:

$$1x_1 + 1x_2 - s_1 = 300$$
$$s_1 = 1x_1 + 1x_2 - 300 = 83.33 + 216.67 - 300 = 0$$

The value of the surplus is zero, which indicates that the minimum specification is just being met.

Note: An LP inequality can be multiplied or divided by a constant without changing the optimal solution. For example, the inequality $3x_1 + 6x_2 \geq 15$ can be rewritten $x_1 + 2x_2 \geq 5$. However, if this is done, the magnitude of the surplus (or slack) variable will be changed. Therefore, one *should not* change the original constraints unless it is absolutely necessary (e.g., for a computer solution).

Summary Each less-than-or-equal-to constraint has a slack variable whose value is either positive or zero. Each greater-than-or-equal-to constraint has a surplus variable whose value is either positive or zero. Each equality constraint has neither slack nor surplus. Any constraints that intersect at the point of the optimal solution have zero slack or surplus.

Special Situations

Case 1: Unbounded Solutions Graphically, in this case, the feasible solution space extends indefinitely, creating an unbounded problem (see Exhibit 4.16 for a maximization case). The result is that the optimal solution is infinite. Such a result usually indicates that an error was made, the problem was misstated, or an incorrect assumption was made.

Case 2: No Feasible Solution In some cases, there will not be a feasible solution to the problem. Graphically, no solution space that simultaneously satisfies all constraints exists (see Exhibit 4.17).

EXHIBIT 4.16 An Unbounded Maximization Problem

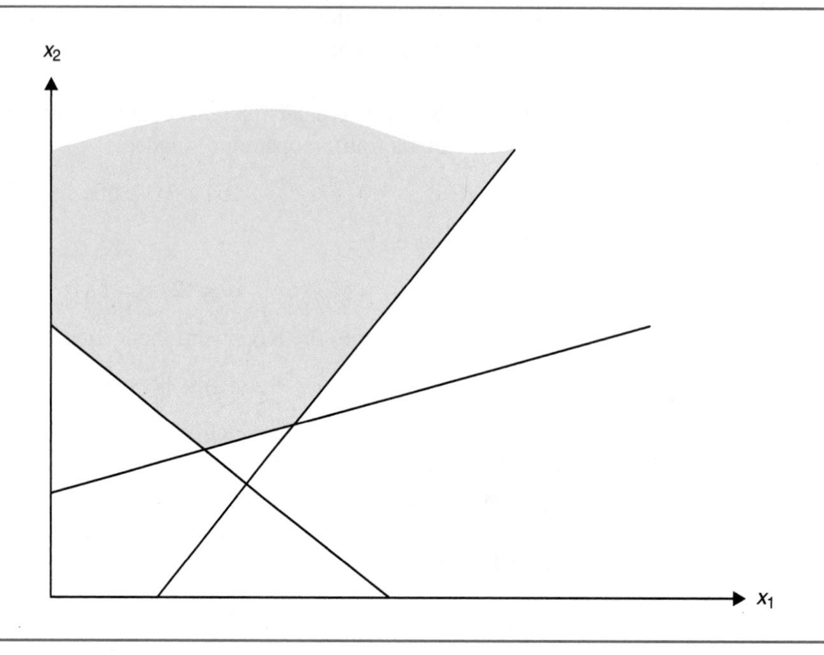

EXHIBIT 4.17 No Feasible Solution Space (Shaded Regions Are Infeasible)

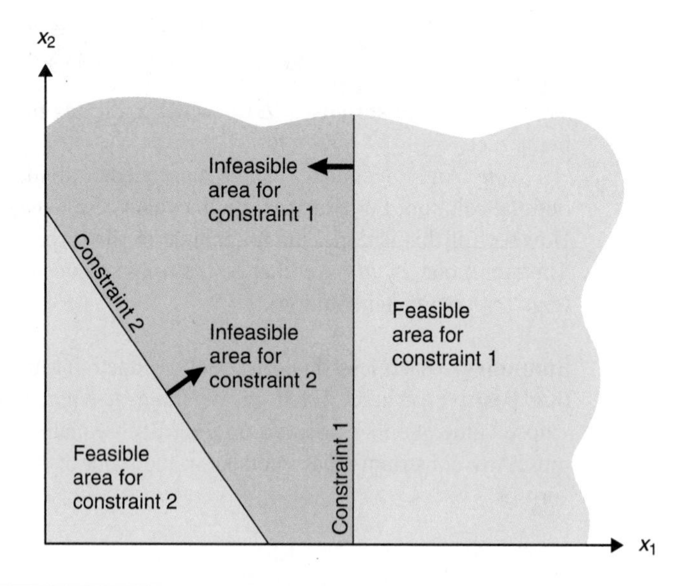

Case 3: Multiple Optimal Solutions This solution is shown, for a maximization case, in Exhibit 4.18. The objective function, when moved as far as possible from the origin to the right, does not touch a corner point but rather a line segment between two points (*P* and *Q*). This happens when the isoprofit line is parallel to an "active" (binding) constraint (2). In such a case, there are optimal solutions at corner *P*, at corner *Q*, and at an infinite number of points on the line segment *PQ*.

EXHIBIT 4.18 Multiple Solutions Case

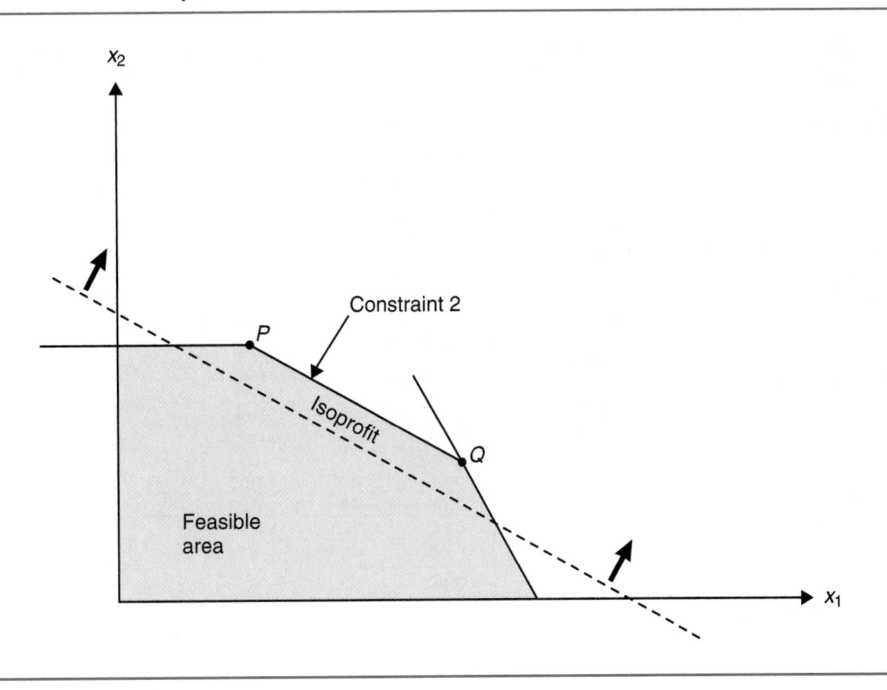

4.4 Solving Linear Programming Models with Excel

To illustrate the use of spreadsheets for solving linear programming models, the data for OfficeComfort were entered into the spreadsheet shown in Exhibit 4.19. This spreadsheet has been organized into three sections. These sections are the "Model Parameters" section at the top of the spreadsheet, an area for the values of the "Decision Variables" in the middle of the spreadsheet, and the "Model Outputs" section at the bottom of the spreadsheet. As you may recall, OfficeComfort must decide how many leather and fabric desk chairs to produce each day, subject to available labor capacity, machine capacity, and market demand.

The formulas that were entered in the "Model Outputs" section are shown in the top right of the spreadsheet. Cells B17:B20 use the SUMPRODUCT function. As was discussed in Chapter 1, this function multiplies the values in two rows and then returns the sum of these products. The remaining formulas are straightforward.

There are two ways the spreadsheet shown in Exhibit 4.19 can be used to find the optimal values for the decision variables. One approach would be to use trial and error by systematically entering values in cells B12 and C12 and observing the values in the "Model Outputs" section. Given that there are an infinite number of possible values for the decision variables, a better approach is to use the spreadsheet's built-in optimization capabilities to find the optimal solution.

Using Excel's Solver

Although solving a linear programming model with a spreadsheet is similar to solving the model using the graphical approach presented earlier, the terminology is a little different. For example, in Excel a formula corresponding to the objective function is entered in a

EXHIBIT 4.19 Production Planning Spreadsheet for OfficeComfort

	A	B	C	D	E	F	G	H	I
1	**Office Comfort Production Planning Model**								
2									
3	**Model Parameters**								
4		**Leather**	**Fabric**	**Limit**	***Key Formulas***				
5	Unit profit ($)	300	250		B17	=SUMPRODUCT(B$12:C$12,B5:C5)			
6	Labor time (hrs/unit)	2	1	40		{Copy B17 to Cells B18:B20}			
7	Machine processing time (hrs/unit)	1	3	45	B21	=B12			
8	Demand for 20"	1	0	12	B22	=C12			
9					C18	=D6 {Copy C18 to Cells C19:C20}			
10	**Decision Variables**				D18	=C18-B18 {Copy Cell D18 to D19:D20}			
11	Quantity of leather and fabric chairs to produce	x_1	x_2		D21	=B21-C21 {Copy Cell D21 to D22}			
12		0	0						
13									
14	**Model Outputs**								
15	Profit earned			**Slack/**					
16		**Value**	**RHS**	**Surplus**					
17	Objective function	$0.00							
18	Labor hours	0.00	40	40.00					
19	Machine hours	0.00	45	45.00					
20	Marketing (demand)	0.00	12	12.00					
21	Nonnegativity x_1	0.00	0	0					
22	Nonnegativity x_2	0.00	0	0					

cell in the spreadsheet and is referred to as the **target cell.** Likewise, the cells that contain the values of the decision variables are referred to as the **changing cells.** Referring to Exhibit 4.19, the target cell (objective function) is cell B17, where a formula that calculates profit was entered; and the changing cells (value of the decision variables) are cells B12 and C12.

Excel's built-in optimization routine is called Solver. To access Excel's Solver, select Solver from the Tools menu at the top of the screen. If Solver is not displayed after selecting Tools, click on Add-Ins in the Tools drop down menu list to access the Add-Ins dialog box. Check the Solver Add-In check box and then click on the OK button. See Excel's Help menu under Solver-Installing for more information. After selecting Solver, the Solver Parameters dialog box is displayed as shown in Exhibit 4.20.

In Exhibit 4.20 the box labeled Set Target Cell: refers to the cell we wish to maximize or minimize. In OfficeComfort's case the objective is to maximize profits. Thus, in the Set Target Cell box we specify cell B17 by first clicking inside this box with the mouse and then either entering the cell reference directly in the box or clicking on cell B17 with the mouse. Since our objective is to find values for our decision variables that provide the maximum amount of profit, we next select the Max radio button. The other two radio buttons are used for solving minimization problems or to select values for the decision variables such that a specific value of the objective function is found (e.g., to find values for x_1 and x_2 that provide a profit of exactly $5,000).

As mentioned earlier, the By Changing Cells: are the cells Excel's Solver is permitted to change to arrive at the maximum value of profit. In OfficeComfort's case, there are two variables that can be changed: the number of leather chairs to produce each day (cell B12) and the number of fabric chairs to produce each day (cell C12). To enter this information, click on the box directly below the label By Changing Cells: and then enter the cell range B12:C12, using either the keyboard or the mouse. You can have Excel propose which cells contain your decision variables by clicking on the Guess button, but this is not generally recommended.

The next step is to enter the constraints of the LP model. To enter constraints, click on the Add button in the Subject to Constraints: section of the Solver Parameters dialog

EXHIBIT 4.20 Excel's Solver Parameters Dialog Box

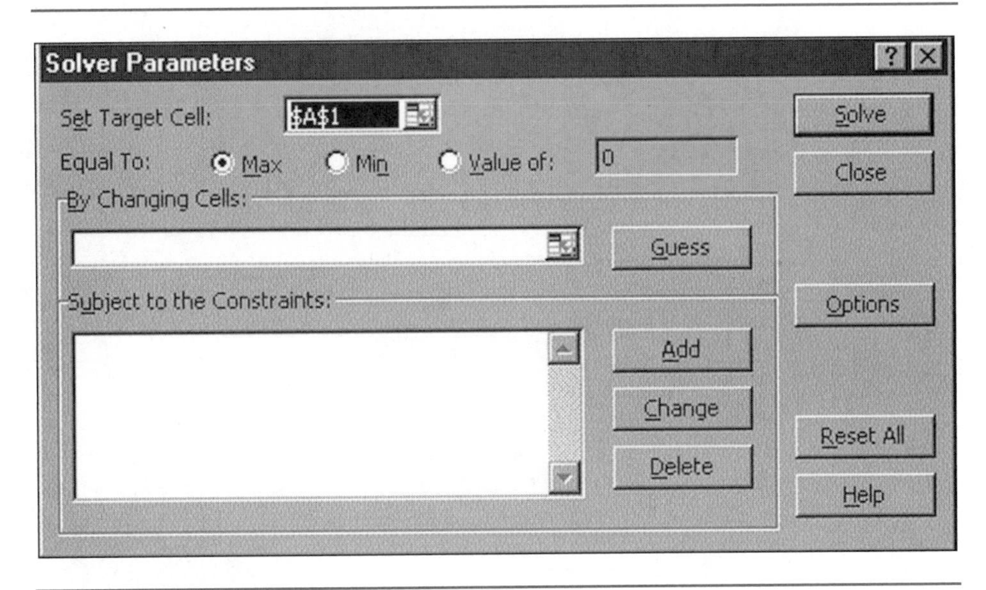

box. After we click on this button, the Add Constraint dialog box is displayed, as is shown in Exhibit 4.21.

The first constraint in our LP model is that the demand for total labor hours must be less than or equal to 40 hours. In the spreadsheet shown in Exhibit 4.19, this constraint is equivalent to specifying that cell B18 must be less than or equal to C18. This information is entered into the Add Constraint dialog box, as shown in Exhibit 4.22. Note that entering a cell

EXHIBIT 4.21 Excel's Add Constraint Dialog Box

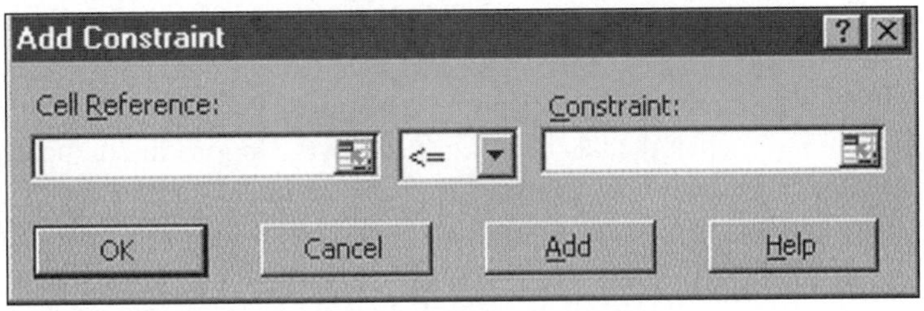

EXHIBIT 4.22 Entering Labor Constraint

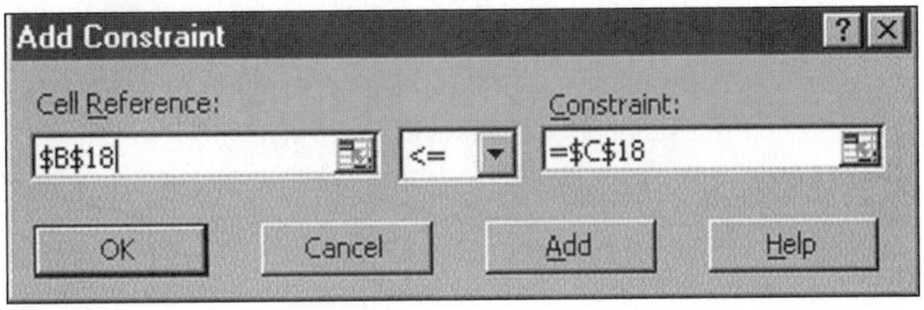

EXHIBIT 4.23 Specifying that Model Is Linear in Excel's Solver Options Dialog Box

Solver Options

Max Time: 100 seconds
Iterations: 100
Precision: 0.000001
Tolerance: 5 %
Convergence: 0.001
OK
Cancel
Load Model...
Save Model...
Help
Assume Linear Model
Assume Non-Negative
Use Automatic Scaling
Show Iteration Results
Estimates: Tangent, Quadratic
Derivatives: Forward, Central
Search: Newton, Conjugate

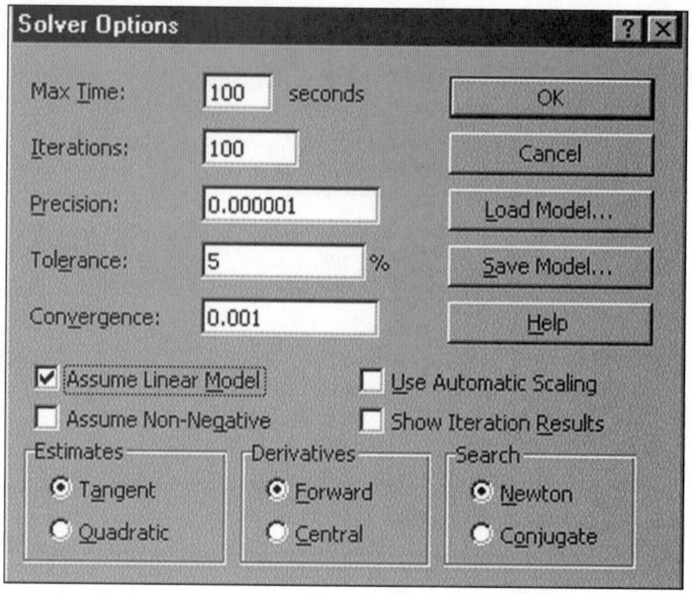

reference for the right-hand side constant (i.e., entering C18 instead of 40) makes it easier to make changes to the LP model and is considered good modeling practice.

Once the information has been entered in the Add Constraint dialog box as shown in Exhibit 4.22, click the Add button. A new blank Add Constraint dialog box similar to Exhibit 4.21 is displayed. The remaining constraints are entered in a similar fashion. When the last constraint has been entered, select the OK button in the Add Constraint dialog box to redisplay the Solver Parameters dialog box. Note that once a constraint has been entered it can be easily modified or deleted by first selecting the constraint in the Subject to the Constraints: box and then selecting the Change or Delete button, respectively.

Next, since we are dealing with a linear programming model, select the Options button and then select the Assume Linear Model check box as shown in Exhibit 4.23. Then click the OK button in the Solver Options dialog box. Also notice that in the Solver Options dialog box you can specify that all the decision variables are non-negative by selecting the check box directly below the Assume Linear Model check box as an alternative to adding non-negativity constraints in the Add Constraint dialog box.

Once all this information has been entered, as shown in Exhibit 4.24, select the Solve button in the upper right-hand corner of the Solver Parameters dialog box to find the optimal solution to the linear programming model. When Excel finds the optimal solution, the Solver Results dialog box is displayed, as shown in Exhibit 4.25. Select the OK button to keep the solution found by Excel. According to the results shown in Exhibit 4.26, Excel found the same optimal solution that we found earlier with the graphical method. More specifically, the optimal solution shown in Exhibit 4.26 calls for producing 12 of the leather chairs each day (cell B12) and 11 of the fabric chairs (cell C12). Producing this number of chairs will result in a daily profit of $6,350 (cell B17).

According to the optimal solution, 35 hours of labor are needed each day, leaving 5 hours of slack each day. All the machine hours and the demand for the leather model is fully met, leaving no slack for either of these constraints. Finally, notice that in the Solver Results dialog box shown in Exhibit 4.25 that Excel provides a variety of reports, including a sensitivity report, which will be discussed a bit later in this chapter.

EXHIBIT 4.24 Excel's Solver Parameters Dialog Box for OfficeComfort

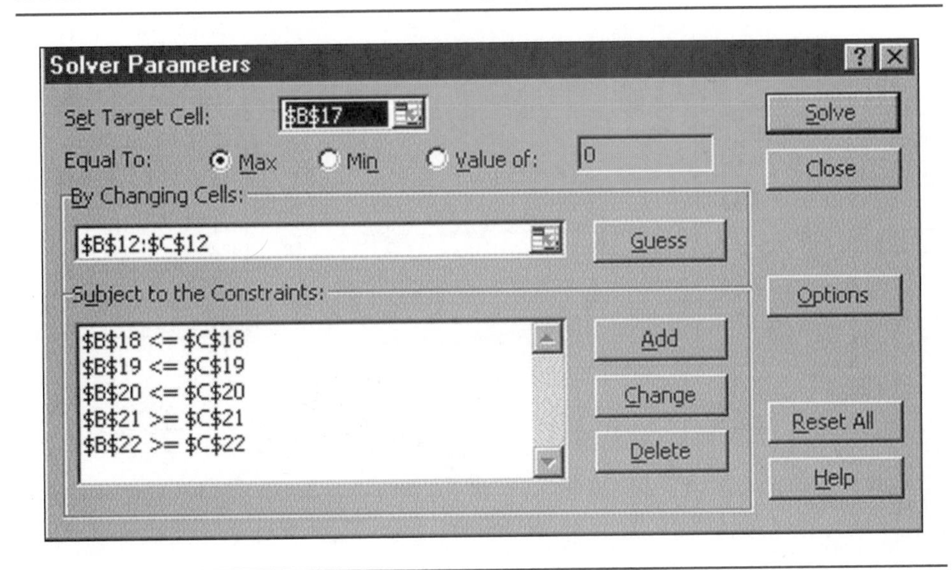

EXHIBIT 4.25 Excel's Solver Results Dialog Box

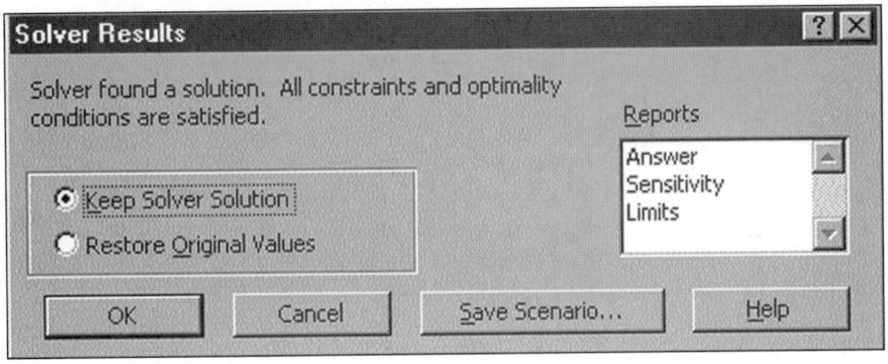

Solver Results

Solver found a solution. All constraints and optimality conditions are satisfied.

Reports: Answer, Sensitivity, Limits

Keep Solver Solution

Restore Original Values

OK | Cancel | Save Scenario... | Help

Solving Large Problems

The approach for entering an optimization model into a spreadsheet with separate sections for the model parameters, decision variables, and model outputs offers the advantages of being logical and easily understood. Unfortunately, for larger and more realistic-sized problems, entering the formulas in the "Model Outputs" section can become quite cumbersome. It should be noted that there are a wide variety of software packages available on the market in addition to Excel for solving LP problems, especially large problems. The magazines *OR/MS Today* and *Datamation* frequently include timely reviews of these packages.

In this section we demonstrate an approach that is better suited to solving larger optimization problems with spreadsheets. To illustrate this approach, consider the model shown below, which identifies the lowest-cost aggregate production schedule for the next eight planning periods (Shafer 1991). Although this model contains 64 decision variables and 41 constraints, it is still considered to be a relatively small model. It is, however, of sufficient size to illustrate our approach for solving more realistically sized optimization models.

EXHIBIT 4.26 Optimal Solution Found by Excel's Solver

	A	B	C	D	E	F	G	H	I
1	**OfficeComfort Production Planning Model**								
2									
3	**Model Parameters**								
4		**Leather**	**Fabric**	**Limit**	***Key Formulas***				
5	Unit profit ($)	300	250		B17	=SUMPRODUCT(B$12:C$12,B5:C5)			
6	Labor time (hrs/unit)	2	1	40		{Copy B17 to Cells B18:B20}			
7	Machine processing time (hrs/unit)	1	3	45	B21	=B12			
8	Demand for 20"	1	0	12	B22	=C12			
9					C18	=D6 {Copy C18 to Cells C19:C20}			
10	**Decision Variables**				D18	=C18-B18 {Copy Cell D18 to D19:D20}			
11	Optimal solution showing quantity of leather and fabric chairs to produce	x_1	x_2		D21	=B21-C21 {Copy Cell D21 to D22}			
12		12	11						
13									
14	**Model Outputs**								
15	Profit earned based on optimal solution			**Slack/**					
16		**Value**	**RHS**	**Surplus**					
17	Objective function	$6,350.00							
18	Labor hours	35.00	40	5.00					
19	Machine hours	45.00	45	0.00					
20	Marketing (demand)	12.00	12	0.00					
21	Nonnegativity x_1	12.00	0	12					
22	Nonnegativity x_2	11.00	0	11					

$$\text{Min cost} = 7(I_1+I_2+I_3+I_4+I_5+I_6+I_7+I_8)+9(B_1+B_2+B_3+B_4+B_5+B_6+B_7+B_8)$$
$$+15(R_1+R_2+R_3+R_4+R_5+R_6+R_7+R_8)+17(O_1+O_2+O_3+O_4+O_5+O_6+O_7+O_8)$$
$$+22(S_1+S_2+S_3+S_4+S_5+S_6+S_7+S_8)+2(H_1+H_2+H_3+H_4+H_5+H_6+H_7+H_8)$$
$$+2(R_1+R_2+R_3+R_4+R_5+R_6+R_7+R_8)$$

Subject to:

$$R_1+O_1+S_1+0-B_0-I_1+B_1 = 50 \quad (1)$$
$$R_2+O_2+S_2+I_1-B_1-I_2+B_2 = 55 \quad (2)$$
$$R_3+O_3+S_3+I_2-B_2-I_3+B_3 = 69 \quad (3)$$
$$R_4+O_4+S_4+I_3-B_3-I_4+B_4 = 170 \quad (4)$$
$$R_5+O_5+S_5+I_4-B_4-I_5+B_5 = 150 \quad (5)$$
$$R_6+O_6+S_6+I_5-B_5-I_6+B_6 = 80 \quad (6)$$
$$R_7+O_7+S_7+I_6-B_6-I_7+B_7 = 90 \quad (7)$$
$$R_8+O_8+S_8+I_7-B_7-I_8+B_8 = 75 \quad (8)$$

$$R_1-5W_1=0 \quad (9) \qquad W_1 \le 20 \quad (17) \qquad O_1-0.2R_1 \le 0 \quad (25)$$
$$R_2-5W_2=0 \quad (10) \qquad W_2 \le 20 \quad (18) \qquad O_2-0.2R_2 \le 0 \quad (26)$$
$$R_3-5W_3=0 \quad (11) \qquad W_3 \le 20 \quad (19) \qquad O_3-0.2R_3 \le 0 \quad (27)$$
$$R_4-5W_4=0 \quad (12) \qquad W_4 \le 20 \quad (20) \qquad O_4-0.2R_4 \le 0 \quad (28)$$
$$R_5-5W_5=0 \quad (13) \qquad W_5 \le 20 \quad (21) \qquad O_5-0.2R_5 \le 0 \quad (29)$$
$$R_6-5W_6=0 \quad (14) \qquad W_6 \le 20 \quad (22) \qquad O_6-0.2R_6 \le 0 \quad (30)$$
$$R_7-5W_7=0 \quad (15) \qquad W_7 \le 20 \quad (23) \qquad O_7-0.2R_7 \le 0 \quad (31)$$
$$R_8-5W_8=0 \quad (16) \qquad W_8 \le 20 \quad (24) \qquad O_8-0.2R_8 \le 0 \quad (32)$$

$$W_1-H_1+F_1=15 \quad (33)$$
$$W_2-W_1-H_2+F_2=0 \quad (34)$$
$$W_3-W_2-H_3+F_3=0 \quad (35)$$
$$W_4-W_3-H_4+F_4=0 \quad (36)$$
$$W_5-W_4-H_5+F_5=0 \quad (37)$$
$$W_6-W_5-H_6+F_6=0 \quad (38)$$
$$W_7-W_6-H_7+F_7=0 \quad (39)$$
$$W_8-W_7-H_8+F_8=0 \quad (40)$$
$$B_8=0 \quad (41)$$

The approach presented here for solving large optimization models involves creating a spreadsheet with three major sections. In the top section, a table is created with rows for the objective function and each constraint and a column for each decision variable. Then, the coefficients of the decision variables are entered in the cells of the table. This table for our sample problem spans the range A1:BO44 in Exhibit 4.27. Notice that the last two columns of the table (i.e., columns BN and BO) were added to keep track of the constraint type and the right-hand sides.

The middle section of the spreadsheet (i.e., cells B46:BM47) consists of two rows and is used to keep track of the values of the decision variables. The top row in this range simply contains labels identifying the decision variables while the bottom row contains the actual values of the decision variables. In setting up the spreadsheet, the labels in row 46 were entered by simply copying the labels previously entered in row 2. You need not enter anything in the cells in row 47 as Excel's Solver will determine the values for these cells. Referring to Exhibit 4.27 we see, for example, that the decision variable I_3 was assigned a value of 31 by Solver.

The final section of the spreadsheet used to solve the problem was entered in the range A49:D91 and is shown in Exhibit 4.28. This section is similar to our initial approach in that it contains formulas corresponding to the objective function and constraints. In the first column of this section, labels to identify the rows in the optimization model were entered by simply copying the labels entered earlier in the cells A3:A44. Column B contains formulas for the objective function and the left-hand sides of the constraints. For example, in cell B50, the formula

=SUMPRODUCT(B$47:BM$47,B3:BM3)

was entered to calculate the value of the objective function based on the values of the decision variables entered in cells B47:BM47. This formula can then be copied to cells B51:B91 to calculate the left-hand sides of the constraints based on the values of the decision variables entered in cells B47:BM47. In comparing this approach to our initial approach, you can now see that this approach provides a much easier way to enter the formulas for the objective function and constraints. The values in cells BN4:BO44 were then copied to cells C51:D91.

Using Solver to find an optimal solution is now relatively straightforward. The target cell is cell B50 and the range for the changing cells is B47:BM47. To simplify entering the 41 constraints, we can take advantage of a shorthand approach that spreadsheets make available. To illustrate, we could enter the first two constraints as follows:

B51 = D51
B52 = D52

A quicker way to enter these two constraints is to capitalize on a spreadsheet's ability to handle ranges and therefore combine these two constraints into one constraint as follows:

B51:B52 = D51:D52

A quick scan of column C in Exhibit 4.28 reveals that the first 16 constraints are all of the equal-to type. Therefore, rather than entering 16 individual constraints, we can enter one combined constraint as follows:

B51:B66 = D51:D66

In a similar fashion we observe that the next 16 constraints (i.e., 17 to 32) are all of the less-than-or-equal-to type. Therefore, these 16 constraints can be combined and entered as:

B67:B82 ≤ D67:D82

EXHIBIT 4.27 Creating a Table of Coefficients to Solve Large Optimization Models

	A	B	C	D	E	F	G	H	I	J	K	L	M	N	O	P	Q	R	S	T	U	V	W	X	Y	Z	AA	AB	AC
1																									Decision Variables				
2	Row	I_1	I_2	I_3	I_4	I_5	I_6	I_7	I_8	B_1	B_2	B_3	B_4	B_5	B_6	B_7	B_8	R_1	R_2	R_3	R_4	R_5	R_6	R_7	R_8	O_1	O_2	O_3	O_4
3	Obj Funct	7	7	7	7	7	7	7	7	9	9	9	9	9	9	9	9	15	15	15	15	15	15	15	15	17	17	17	17
4	(1)	-1								1								1								1			
5	(2)	1	-1							-1	1								1								1		
6	(3)		1	-1							-1	1								1								1	
7	(4)			1	-1							-1	1								1								1
8	(5)				1	-1							-1	1								1							
9	(6)					1	-1							-1	1								1						
10	(7)						1	-1							-1	1								1					
11	(8)							1	-1							-1	1								1				
12	(9)																	1											
13	(10)																		1										
14	(11)																			1									
15	(12)																				1								
16	(13)																					1							
17	(14)																						1						
18	(15)																							1					
19	(16)																								1				
20	(17)																												
21	(18)																												
22	(19)																												
23	(20)																												
24	(21)																												
25	(22)																												
26	(23)																												
27	(24)																												
28	(25)																	-0.2								1			
29	(26)																		-0.2								1		
30	(27)																			-0.2								1	
31	(28)																				-0.2								1
32	(29)																					-0.2							
33	(30)																						-0.2						
34	(31)																							-0.2					
35	(32)																								-0.2				
36	(33)																												
37	(34)																												
38	(35)																												
39	(36)																												
40	(37)																												
41	(38)																												
42	(39)																												
43	(40)																												
44	(41)																1												
45																													
46	Decision	I_1	I_2	I_3	I_4	I_5	I_6	I_7	I_8	B_1	B_2	B_3	B_4	B_5	B_6	B_7	B_8	R_1	R_2	R_3	R_4	R_5	R_6	R_7	R_8	O_1	O_2	O_3	O_4
47	Variables	0	0	31	0	0	0	0	0	0	0	0	0	0	0	0	0	50	55	100	100	100	80	90	75	0	0	0	20

Finally, the last nine constraints are of the equal-to type and can be entered as

B83:B91 = D83:D91

Notice that by capitalizing on the spreadsheet's ability to use ranges we were able to enter the 41 constraints by entering only 3 constraints. Had we done some additional up-front planning and grouped all the equal-to constraints together and all the less-than-or-equal-to constraints together, we could have actually entered all 41 constraints by entering only 2 constraints!

Now all that remains is to actually solve the model. Before solving the model, select the Options button in the Solver Parameters dialog box and click on the Assume Linear Model and the Assume Non-Negative check boxes to tell Excel the model is a linear programming model and that all the decision variables must greater-than-or-equal-to zero, respectively. The Solver Parameters dialog box is shown in Exhibit 4.29.

EXHIBIT 4.27 ***continued***

	AD	AE	AF	AG	AH	AI	AJ	AK	AL	AM	AN	AO	AP	AQ	AR	AS	AT	AU	AV	AW	AX	AY	AZ	BA	BB	BC	BD	BE	BF	BG	BH	BI	BJ	BK	BL	BM	BN	BO
1																																					Type of	
2	O_5	O_6	O_7	O_8	S_1	S_2	S_3	S_4	S_5	S_6	S_7	S_8	H_1	H_2	H_3	H_4	H_5	H_6	H_7	H_8	F_1	F_2	F_3	F_4	F_5	F_6	F_7	F_8	W_1	W_2	W_3	W_4	W_5	W_6	W_7	W_8	Constraint	RHS
3	17	17	17	17	22	22	22	22	22	22	22	22	2	2	2	2	2	2	2	2	2	2	2	2	2	2	2	2	0	0	0	0	0	0	0	0		
4					1																																=	50
5						1																															=	55
6							1																														=	69
7								1																													=	170
8	1								1																												=	150
9		1								1																											=	80
10			1								1																										=	90
11				1								1																									=	75
12																													-5								=	0
13																														-5							=	0
14																															-5						=	0
15																																-5					=	0
16																																	-5				=	0
17																																		-5			=	0
18																																			-5		=	0
19																																				-5	=	0
20																													1								≤	20
21																														1							≤	20
22																															1						≤	20
23																																1					≤	20
24																																	1				≤	20
25																																		1			≤	20
26																																			1		≤	20
27																																				1	≤	20
28																																					≤	0
29																																					≤	0
30																																					≤	0
31																																					≤	0
32	1																																				≤	0
33		1																																			≤	0
34			1																																		≤	0
35				1																																	≤	0
36													-1								1								1								=	15
37														-1								1							-1	1							=	0
38															-1								1							-1	1						=	0
39																-1								1							-1	1					=	0
40																	-1								1							-1	1				=	0
41																		-1								1							-1	1			=	0
42																			-1								1							-1	1		=	0
43																				-1								1							-1	1	=	0
44																																					=	0
45																																						
46	O_5	O_6	O_7	O_8	S_1	S_2	S_3	S_4	S_5	S_6	S_7	S_8	H_1	H_2	H_3	H_4	H_5	H_6	H_7	H_8	F_1	F_2	F_3	F_4	F_5	F_6	F_7	F_8	W_1	W_2	W_3	W_4	W_5	W_6	W_7	W_8		
47	20	0	0	0	0	0	0	19	30	0	0	0	0	1	9	0	0	0	2	0	5	0	0	0	0	4	0	3	10	11	20	20	20	16	18	15		

Back to Startron's Dilemma

Having defined Startron's problem, collected the data, and formulated the model, the next step is to solve and analyze the model. The spreadsheet shown in Exhibit 4.30 was developed in Excel, and Excel's built-in Solver was used to find the optimal (i.e., highest profit) production plan for Startron. Note in the current model it was assumed that producing a fraction of a unit was possible (i.e., units could be started on one day and finished on the following day).

The spreadsheet to model Startron's production planning problem is divided into three sections. At the top of the spreadsheet are the model parameters, or the inputs to the model that are *not* under the control of the decision maker. In the middle of the spreadsheet (cells

EXHIBIT 4.28 Bottom Section of Spreadsheet to Solve Large Model

	A	B	C	D
49	Model	Value	Type	RHS
50	Obj funct	11773		
51	(1)	50	=	50
52	(2)	55	=	55
53	(3)	69	=	69
54	(4)	170	=	170
55	(5)	150	=	150
56	(6)	80	=	80
57	(7)	90	=	90
58	(8)	75	=	75
59	(9)	0	=	0
60	(10)	0	=	0
61	(11)	0	=	0
62	(12)	0	=	0
63	(13)	0	=	0
64	(14)	0	=	0
65	(15)	0	=	0
66	(16)	0	=	0
67	(17)	10	≤	20
68	(18)	11	≤	20
69	(19)	20	≤	20
70	(20)	20	≤	20
71	(21)	20	≤	20
72	(22)	16	≤	20
73	(23)	18	≤	20
74	(24)	15	≤	20
75	(25)	10	≤	0
76	(26)	11	≤	0
77	(27)	20	≤	0
78	(28)	0	≤	0
79	(29)	0	≤	0
80	(30)	-16	≤	0
81	(31)	-18	≤	0
82	(32)	-15	≤	0
83	(33)	15	=	15
84	(34)	0	=	0
85	(35)	0	=	0
86	(36)	0	=	0
87	(37)	0	=	0
88	(38)	0	=	0
89	(39)	0	=	0
90	(40)	0	=	0
91	(41)	0	=	0

B15:D15) are the decision variables. Finally, in the bottom of the spreadsheet are all the model outputs. Included in the "Model Outputs" section is a formula that calculates the value of the objective function or daily profit (cell B20) based on the values of the decision variables (cells B15:D15). In column B of rows 21–27, formulas were entered that calculate the values of the left-hand sides of the constraints based on the values of the decision variables entered in cells B15:D15. For example, cell B25 is *not* the minimum number of desktop computers but, rather, the value of the constraint equation in the

EXHIBIT 4.29 Solver Parameters Dialog Box for Large-Model Example

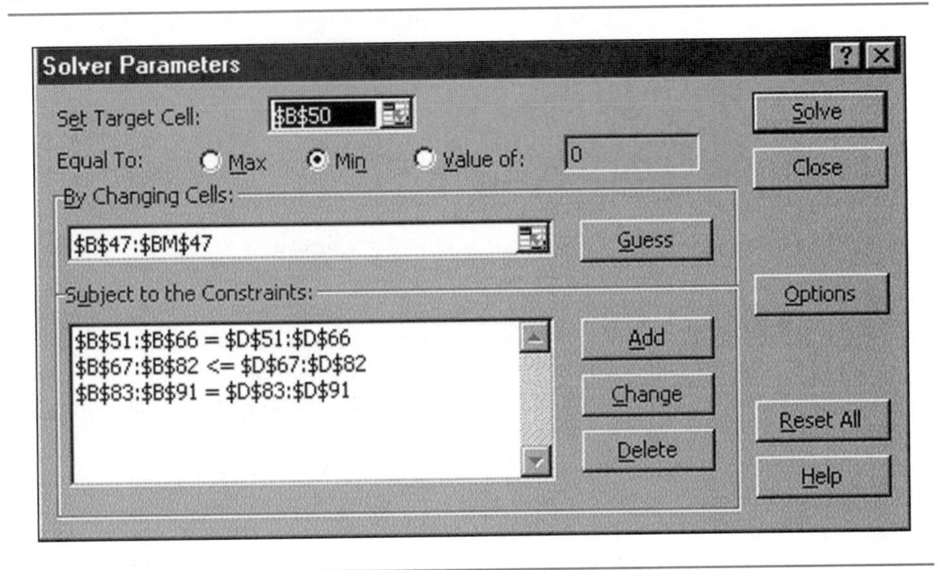

EXHIBIT 4.30 Excel Spreadsheet Developed to Solve and Analyze Startron's Production Planning Problem

	A	B	C	D	E	F
1	**Startron Production Planning Model**					
2						
3	**Model Parameters**					
4					**Number**	**Available**
5		**Desktop**	**Server**	**Notebook**	**of Workers**	**Hours**
6	Profit	$75	$145	$125		
7	Assembly time (hrs)	0.5	0.75	1.5	12	96
8	Software installation (hrs)	0.25	0.4	0.3	3	24
9	Testing (hrs)	1	1.5	1	20	160
10	Packaging (hrs)	0.1	0.1	0.2	2	16
11						
12						
13	**Decision Variables**					
14		**DT**	**SV**	**NB**		
15		15.5	15.5	46.5		
16						
17	**Model Outputs**					
18				**Slack/**		
19		**Value**	**RHS**	**Surplus**		
20	Objective function	$9,212.90				
21	Assembly time (hrs)	89.03	96.00	6.97		
22	Software installation (hrs)	24.00	24.00	0.00		
23	Testing (hrs)	85.16	160.00	74.84		
24	Packaging (hrs)	12.39	16.00	3.61		
25	Min # of desktops	0.00	0.00	0.00		
26	Min # servers	0.00	0.00	0.00		
27	Min # of notebooks	30.97	0.00	30.97		
28	DT $\geq$ 0	15.48	0.00	15.48		
29	SV $\geq$ 0	15.48	0.00	15.48		
30	NB $\geq$ 0	46.45	0.00	46.45		

Callouts: Optimal daily production quantities (row 15); Daily profit (cell B20)

mathematical model. In rows 21–27 of column C the actual right-hand sides of the constraints are entered. Finally, in rows 21–27 of column D formulas were entered to calculate the values of the slack and surplus variables for each constraint.

According to the results shown in Exhibit 4.30, 15.5 desktops should be produced each day. In effect, this means following a pattern of completing 15 desktops one day and 16 desktops the next day. Likewise, 15.5 servers and 46.5 notebook computers should also be produced each day. If Startron follows this production plan it will earn an average $9,212.90 in profit each day.

The slack and surplus variables show the amounts below or above each of the constraints. After reviewing the values of the slack and surplus variables, it can be seen that there is unused capacity in the assembly, testing, and packaging departments. For example, the assembly department has 6.97 hours of unused capacity each day while the testing department has 74.84 unused hours each day. The only department currently using all of its capacity is the software installation department, which is using all 24 hours of its available time and therefore has zero slack. Further analysis of the results indicates that the plan calls for producing the minimum number of desktops and servers, while 30.97 more notebook computers are produced beyond the minimum requirement.

After looking over these results, you wonder how much it is costing Startron to require that it produce at least 20 percent of each computer type. To answer this question, we can simply eliminate these constraints and solve the model again. Exhibit 4.31 contains the results of the model solved without the minimum production quantity constraints included. Without these constraints, the new optimal solution calls for producing 0 desktops, 19.2 servers, and 54.4 notebook computers, for an average daily profit of $9,584.

EXHIBIT 4.31 Sensitivity of Optimal Solution to Minimum Production Quantity Constraints

	A	B	C	D	E	F	G
1	**Startron Production Planning Model**						
2							
3	**Model Parameters**						
4					**Number**	**Available**	
5		**Desktop**	**Server**	**Notebook**	**of Workers**	**Hours**	
6	Profit	$75	$145	$125			
7	Assembly time (hrs)	0.5	0.75	1.5	12	96	
8	Software installation (hrs)	0.25	0.4	0.3	3	24	
9	Testing (hrs)	1	1.5	1	20	160	
10	Packaging (hrs)	0.1	0.1	0.2	2	16	
11							
12					Optimal daily production quantities with no minimum production quantity constraints		
13	**Decision Variables**						
14		**DT**	**SV**	**NB**			
15		0.0	19.2	54.4			
16							
17	**Model Outputs**						
18	Increased daily profit			**Slack/**			
19		**Value**	**RHS**	**Surplus**			
20	Objective function	$9,584.00					
21	Assembly time (hrs)	96.00	96.00	0.00			
22	Software installation (hrs)	24.00	24.00	0.00			
23	Testing (hrs)	83.20	160.00	76.80			
24	Packaging (hrs)	12.80	16.00	3.20			
25	Min # of desktops	-14.72	0.00	-14.72			
26	Min # servers	4.48	0.00	4.48			
27	Min # of notebooks	39.68	0.00	39.68			
28	DT ≥ 0	0.00	0.00	0.00			
29	SV ≥ 0	19.20	0.00	19.20			
30	NB ≥ 0	54.40	0.00	54.40			

Comparing this solution with our original solution indicates that Startron could earn an additional $371.10/day ($9,584 – $9,212.90) in profit if it eliminated the requirement that all models represent at least 20 percent of the unit volume.

There are a wide variety of other types of "what-if" or *sensitivity analyses* that can be performed. For example, it can be determined how much the unit profits can be changed for each computer model before the current optimal solution changes. Also, for bottleneck resources such as software installation, it can be determined how much the profit would increase if additional units of this resource were available. These are the issues we address next.

4.5 Sensitivity ("What-If") Analysis

The optimal solution to an LP model is based on a set of assumptions and on the estimation of future data such as prices and costs. In a deterministic model, there is no provision for risk or uncertainty. Therefore, it is important for management to know *what will happen* to the optimal solution *if* changes occur in the input data on which the LP model is based. The name *sensitivity analysis* derives from the fact that an analysis is made to find out how sensitive the optimal solution is to changes in the input data. Recall that the optimal solution refers specifically to the optimal values of the decision variables. Because sensitivity analysis is done *after* the optimal solution is found, it is often called *postoptimality analysis.*

Why a Sensitivity Analysis?

One of the major purposes of sensitivity analysis is to provide the decision maker with additional information to adapt to changing conditions. For example, the availability of resources in an actual situation may change from day to day. Sensitivity analysis permits one to investigate the impact of these changes without constantly having to resolve the problem. Sensitivity analysis also provides the decision maker with information as to which parameter estimates are the most critical in the formulation.

There are two approaches for conducting sensitivity analysis.

1. **A Trial-and-Error Approach** According to this approach, one may change the input data, hence forming a new model. Then the problem is resolved and the results are compared with that of the original model. This process can be repeated for all desired changes, either singly or in combination. The deficiency of this approach is that it may become very lengthy because there are typically large numbers of possible changes in the data. Another possible deficiency is that resolving a problem may be expensive or time consuming for large-scale problems.
2. **Use of an Analytical Approach** When an analytical approach is used, there is no need to completely resolve the LP model each time a change is made. Furthermore, information such as the "permissible range of change," which is directly provided by the analytical approach, can otherwise only be provided after an extremely lengthy experimentation period. The analytical approach presented here finds the effects on the optimal solution of each of the following changes, assuming one change is made at a time, in the input data:
   - Changes in the coefficients in the objective function
   - Changes to the right-hand sides of the constraints
   - Changes in the coefficients of the constraints
   - Adding or deleting a constraint

We will illustrate here only the first two: changes in the coefficients of the objective function and in the right-hand sides of the constraints. Changes in the coefficients of the constraints are similar to changes in the coefficients of the objective function. And adding or deleting a constraint is similar to changing the right-hand sides in the sense that a constraint can become or stop being binding.

Sensitivity Analysis: Objective Function

Changes in the coefficients of the objective function are often related to *output pricing* decisions. Assuming that the unit cost of producing a given output remains the same, two important managerial questions can be answered by the analysis:

- When does a price (or profit) *decrease* of an output justify discontinuing or reducing the quantity of the output?
- How much of a price (or profit) *increase* in an output justifies increasing the quantity of the output in the optimal solution?

Graphical Explanation of the Changes in the Coefficients Let us reproduce the output-mix problem:

$$\begin{aligned}&\text{Maximize } z = 300x_1 + 250x_2\\ &\text{Subject to:}\\ &2x_1 + 1x_2 \leq 40\\ &1x_1 + 3x_2 \leq 45\\ &1x_1 + 0x_2 \leq 12\\ &\quad x_1, x_2 \geq 0\end{aligned}$$

The decision as to whether the optimal solution should include a particular product, and how much, depends on its relative contribution to profit, as expressed in the objective function. In graphical terms, such a decision depends on the *slope* of the objective function. Let us examine the optimal graphical solution to the output-mix problem (Exhibit 4.32). The existing objective function (line *KL*) yields a solution at point *G* (12 units of x_1 and 11 units of x_2). However, if we change the slope of the objective function so it is parallel to line *MN,* then the optimal solution will be at point *C* (produce 15 of x_2).

Recall from Chapter 3 that the equation for a line is often expressed as $y = mx + b$. When expressed in this fashion, m represents the slope of the line and b represents where the line intersects the y-axis. In our discussions here, we have used x_1 instead of x and x_2 in place of y. Generally speaking, the objective function has been expressed as follows:

$$\text{Maximize (or minimize) } z = c_1x_1 + c_2x_2$$

where c_1 and c_2 are the coefficients for x_1 and x_2 respectively. Alternatively, had we used x and y instead of x_1 and x_2, we could have expressed the objective function as:

$$\text{Maximize (or minimize) } z = c_1x + c_2y$$

Rearranging the terms so that the objective function is in the slope–intercept format is easy:

$$\begin{aligned}z &= c_1x_1 + c_2x_2\\ z - c_2x_2 &= c_1x_1\\ -c_2x_2 &= c_1x_1 - z\\ x_2 &= (-c_1/c_2)x_1 + z/c_2\end{aligned}$$

Based on this, we see that the slope of the objective function is $-c_1/c_2$ and the y intercept is z/c_2. Therefore, the slope of the objective function can change if the coefficient of x_1 changes, the coefficient of x_2 changes, or the ratio between the coefficients of x_1 and x_2 changes.

EXHIBIT 4.32 Solution to the Product Mix Problem

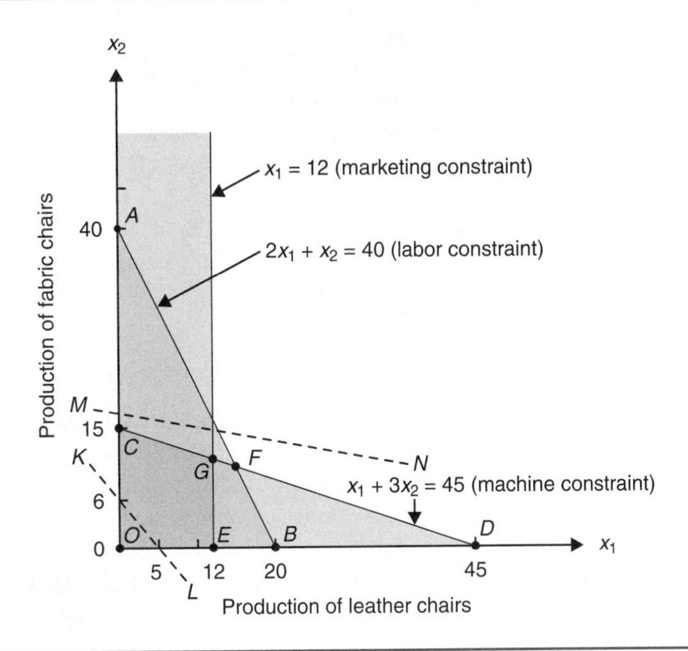

A change in solution from point G to point C in Exhibit 4.32 will be achieved when the coefficient of x_2 is more than three times that of x_1. The reason for this is that with a ratio of exactly one to three, the objective function will be parallel to the machine constraint *CD,* with both points C and G being optimal. In other words, the objective function would have the same slope as the machine constraint.

Limits on the Coefficients Consider Exhibit 4.32. The current solution involves both leather chairs (x_1) and fabric chairs (x_2). Assume that the profit contribution of x_1 remains 300, whereas the profit contribution of x_2 increases. This increase will change the slope of line *KL* toward that of *MN*. When the profit contribution reaches 900 (exactly 3 times 300), the line will be parallel to *CD* and both points C and G will be optimal solutions. Thus, 900 is the upper limit of the coefficient of x_2 (i.e., c_2). If the profit contribution of fabric chairs is *more* than 900, the solution will be at point C.

Similarly, the solution may move from point G to point E (produce only leather chairs). For this to happen, assuming that the profit contribution of x_1 remains 300, the profit contribution of x_2 would have to decrease *below* zero. If the profit contribution of x_2 is *exactly* zero, then both points G and E will be optimal solutions. Thus, the upper limit on the coefficient of x_2 (i.e., c_2) is 900, and the lower limit is zero. As long as c_2 is between zero and 900, our optimal solution of producing 12 leather chairs and 11 fabric chairs remains optimal. Note, however, that while the optimal solution does not change by changing c_2 within this range, the value of the objective function will change. To illustrate, if the profit of the fabric chair increases to \$350, while the profit of the leather chair is held constant at \$300, the optimal solution would still be to produce 12 leather chairs and 11 fabric chairs. However, the daily profit would increase from \$6,350 to $300 \times 12 + 350 \times 11 = \$7,450$.

Let us consider the variable x_1. If the coefficient of x_2 remains 250, then the *lower* limit of c_1 is 83.33 (i.e., 1/3 of 250). If the profit contribution of leather chairs becomes less than 83.33, then the optimal solution will be point C and we will not produce any of the leather chairs.

The upper limit on c_1 is the amount required to move the solution from point G to E. This can happen only if c_2 is negative. Because c_2 is held constant at 300, no matter what the value of c_1 is, point G will be superior to E. Thus, the upper limit of c_1 is ∞.

Note: The example so far has dealt only with products that are in the solution. However, assume that the optimal solution was at point C to begin with (produce fabric chairs only). Again, by figuring the appropriate slope, it would have been possible to calculate the necessary increase in c_1 that would result in the inclusion of leather chairs in the optimal solution.

The graphical analysis is limited, of course, to two decision variables. A similar analysis is possible for any number of variables. We discuss how to obtain this information from Excel later in this chapter.

Sensitivity Analysis: Right-Hand Sides

The right-hand side (RHS) constants of the constraints, b_i, express the capacities or limitations of the resources, or make explicit certain requirements. Management may be interested in finding out the effect of changes in these values of b_i on the optimal solution. Such changes might or might not affect the optimal solution and might or might not affect the value of the objective function. Graphically, a change in any b_i represents a movement of the constraint, *parallel to itself,* toward or away from the origin O.

As an example consider the following LP model:

Maximize $z = 3x_1 + 4x_2$
Subject to:
(1) $3x_1 + 5x_2 \leq 15$
(2) $2x_1 + 1x_2 \leq 8$
(3) $0x_1 + 1x_2 \leq 2$
(4) $x_1, x_2 \geq 0$

The graphical solution of the problem is shown in Exhibit 4.33, where point C is the optimal solution. Moving a binding constraint such as (1) or (2) will change the location of the optimal solution. For example, if the capacity of constraint (2) is increased from 8 to 9 (line ST), then the optimal solution moves to point G. On the other hand, slightly moving constraint (3) will not affect the optimal solution at all (because the constraint is not binding). However, moving it farther down may change the location of the optimal solution. Management may be interested in finding the range of such changes. For example, moving constraint (2) more to the right will result in a situation where the optimal solution is at point F (5 units of x_1, 0 of x_2). This will happen when the capacity of constraint (2) is 10; that is, 10 is the upper limit on constraint (2).

Sensitivity Analysis with Excel

Most real-life problems are too large or complex for a graphical solution. Therefore, they are solved with the aid of computer software. Dozens of specialized computer programs are available. In this section we overview the interpretation of the sensitivity analysis information provided by Excel.

Earlier we demonstrated how OfficeComfort's output-mix LP model could be solved with Excel. As you may recall from our earlier discussion, once Excel finds the optimal solution to an LP model, the Solver Results dialog box is displayed (refer back to Exhibit 4.25). To display the sensitivity analysis information, click on Sensitivity in the Reports section at the far right of this dialog box. Excel's sensitivity report for OfficeComfort is shown in Exhibit 4.34.

EXHIBIT 4.33 Changes in Capacity

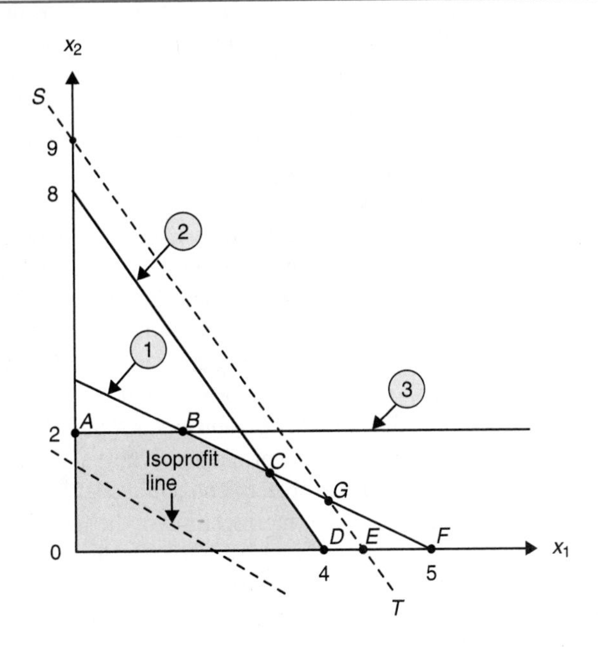

EXHIBIT 4.34 Sensitivity Report for OfficeComfort

	A	B	C	D	E	F	G
1	Microsoft Excel 9.0 Sensitivity Report						
2	Worksheet: [OfficeComfort chapter 4.xls]Sheet1						
3	Report Created: 6/5/01 1:27:22 PM						
4							
5							
6	**Adjustable cells**						
7			**Final**	**Reduced**	**Objective**	**Allowable**	**Allowable**
8	**Cell**	**Name**	**Value**	**Cost**	**Coefficient**	**Increase**	**Decrease**
9	\$B\$12	x_1	12	0	300	1E+30	216.6666667
10	\$C\$12	x_2	11	0	250	650	250
11							
12	**Constraints**						
13			**Final**	**Shadow**	**Constraint**	**Allowable**	**Allowable**
14	**Cell**	**Name**	**Value**	**Price**	**R.H. Side**	**Increase**	**Decrease**
15	\$B\$18	Labor hours value	35.00	0.00	40	1E+30	5
16	\$B\$19	Machine hours value	45.00	83.33	45	15	33
17	\$B\$22	Nonnegativity x_2 value	11.00	0.00	0	11	1E+30
18	\$B\$21	Nonnegativity x_1 value	12.00	0.00	0	12	1E+30
19	\$B\$20	Marketing (demand) value	12.00	216.67	12	3	12

As shown in Exhibit 4.34, Excel's sensitivity report is divided into two sections. The top section provides sensitivity analysis information for the objective function coefficients. The bottom section provides sensitivity analysis information for the right-hand side constants.

Referring to the top portion of the sensitivity report, the first column labeled "Cell" lists the cells in the spreadsheet associated with each decision variable. You may want to refer back to Exhibit 4.19 to verify that the decision variable x_1 was assigned to cell B12 and that x_2 was assigned to cell C12. The "Name" column provides a label for the cell by concatenating the text, if any, directly to the left of the cell (none exists in Exhibit 4.19) with the text entered as a column heading (x_1 and x_2 in Exhibit 4.19). The "Final Value" column lists the optimal values for each decision variable. In the case of OfficeComfort the optimal solution was to produce 12 leather chairs (x_1) and 11 fabric chairs (x_2) each day. For decision variables that have a zero final value, the "Reduced Cost" column indicates how much that variable's objective function coefficient would have to improve in order for the variable to be included in the optimal solution and not have a final value of zero. Multiple optimal solutions exist when a given variable has both a final value and reduced cost equal to zero. The "Objective Coefficient" column lists the objective function coefficient for each decision variable. You may recall that the profit for each leather chair produced was $300, and the profit for each fabric chair was $250. Finally, the last two columns tell us how much each objective function coefficient can be increased and decreased so that the current solution remains optimal. (The value 1E+30 in the table means a value of 1 followed by 30 zeros; i.e., ∞). Again, note that the sensitivity analysis information provided in this report is based on the assumption that only one change is made at a time and that all other parameters and inputs are held constant. Also recall that while the optimal solution may not change, the value of the objective function will most likely change if one of the objective function coefficients is changed.

To illustrate the interpretation of the information contained in the top part of the sensitivity report, consider decision variable x_2 (i.e., the number of fabric chairs to produce each day). According to the table shown in Exhibit 4.34, the optimal solution calls for producing 11 fabric chairs each day (from the "Final Value" column). Since the final value was greater than zero, we do not need to consider the "Reduced Cost" column. The profit obtained from producing each chair is $250 (from the "Objective Coefficient" column). As long as the unit profit for fabric chairs does not increase by more than $650 or decrease by $250, the current optimal solution will remain the same (from the "Allowable Increase" and "Allowable Decrease" columns, respectively). Stated another way, as long as the unit profit on fabric chairs is between $0 (250 – 250) and $900 (250 + 650) the optimal solution will be to produce 12 leather chairs and 11 fabric chairs. Again, as was demonstrated earlier, while the optimal solution will remain the same, the value of the objective function will change. Also, note that we obtained exactly the same result using the graphical method.

You may be wondering what the value of this sensitivity information is, given that the value of the objective function must be recalculated. The value is that as a manager at OfficeComfort you know within what range the profit per unit can change without requiring you to resolve the LP model. For example, if it turns out that the unit profit for each fabric chair is only $150, the current production plan remains optimal. On the other hand, if the unit profit from each fabric increases above $900, the current solution would not be the optimal solution and the LP model would need to be resolved.

The bottom portion of Excel's sensitivity report provides sensitivity analysis information for the right-hand side constants. Each row in this table corresponds to one of the constraints in the LP model. The first two columns labeled "Cell" and "Name" are the same as was discussed for the top table. According to the table, the label "Labor Hours" was entered in the cell directly to the left of cell B18 and the column heading "Value" was entered in column B. This is easily verified by referring back to Exhibit 4.19.

Continuing to move to the right, the next column is labeled "Final Value." The "Final Value" column calculates the value of the left-hand side of the constraint based on the optimal values of the decision variables.

The "Shadow Price" column is perhaps the most useful information shown in this part of Exhibit 4.34. More specifically, the "Shadow Price" column tells you what will happen to the objective function if a unit change is made to a particular right-hand-side constraint constant. In general, the interpretation of the information provided in the "Shadow Price" column depends on the type of objective function and on the type of constraint. Exhibit 4.35 summarizes how to interpret Excel's shadow price information. Note that while the "Shadow Price" provides insight into the impact a change in a particular constraint's right-hand side will have on the value of the optimal solution, it does not provide insight into how the values of the decision variables will change.

The "Constraint R.H. Side" column simply lists the right-hand side constants for each constraint. Finally, the last two columns provide the range over which each right-hand constant can be changed. As long as a particular right-hand side constant is within this range, the shadow prices listed are valid. If a change is made beyond this range, the LP model must be resolved.

To illustrate the interpretation consider the labor hour constraint listed in the first row of this table. According to the table, the solution calls for using 35 hours of labor (from "Final Value" column). Since the right-hand side constant or upper limit for this constraint was 40 hours (from "Constraint R.H. Side" column), there are 5 unused labor hours. The shadow price for the labor hour constraint is 0.00. This tells us that increasing the right-hand side constant from 40 hours to 41 hours would have no impact on the value of our optimal solution. This makes sense intuitively. Specifically, since we already have 5 extra hours of labor, adding more labor hours would only increase the number of unused labor hours and would not improve our solution. In general, we note that constraints with slack or surplus (i.e., nonbinding constraints) will have a shadow price of zero. Also note that we can increase the available labor hours by a very large number (essentially, infinity) or decrease it by 5 hours, and the shadow price will still be zero. Again this makes sense. We have already discussed that since there are already unused hours of labor available, adding more hours only adds to the surplus hours and will not change the value of the objective function. On the other hand, decreasing the labor hours will not affect the value of the objective function as long as they are not reduced by more than 5 hours (the amount of slack this constraint has). If the amount of labor hours is decreased by more than 5 hours, a new solution would have to be found because the current solution depends on having available 35 hours of labor.

As an example of a binding constraint, consider the machine hours constraint. Referring to Exhibit 4.34, the solution calls for using 45 machine hours ("Final Value" column), and there 45 hours of machine time available ("Constraint R.H. Side" column). Therefore, this constraint has no slack and is a binding constraint. The shadow price for this constraint is \$83.33. This means that for each extra hour of machine time, the value of the objective function would increase by \$83.33. Likewise, for each hour that machine time is decreased, the value of the objective function would decrease by \$83.33. This shadow price is valid as long as the available machine hours is 45 – 33 = 12 to 45 + 15 = 60 hours. For example, if the available machine hours could be increased from 45 to 50, the value of the objective function would increase by \$416.65 (5 extra hours × \$83.33). On the other hand, if the machine hours could be increased to 70 hours, the model would have to be resolved to determine the impact this change would have.

EXHIBIT 4.35
Interpreting Excel's Shadow Prices: The Shadow Price Indicates How Much a Unit Increase (Decrease) in the RHS Constant Will ______

	Type of Constraint	
Type of Objective Function	**Less-Than-or-Equal-to**	**Greater-Than-or-Equal-to**
Maximize	Increase (decrease) the objective function.	Decrease (increase) the objective function.
Minimize	Decrease (increase) the objective function.	Increase (decrease) the objective function.

As this example illustrates, shadow price information allows decision makers to assess the impact that changes in the levels of the resources will have on the value of the objective function without having to resolve the model. In OfficeComfort's case, we have determined that each additional hour of machine time would increase daily profits by \$83.33. The managerial implication of this is that OfficeComfort could increase its profits if it were able to acquire additional machine hours for less than \$83.33/hour.

4.6 Integer Programming

The previous sections introduced the reader to the topic of linear programming. In this section we continue our discussion of optimization and consider integer programming where some or all of the decision variables must be an integral number of units: 14 nurses, three trucks, and so on. Adding this integrality restriction to an LP problem increases the computational burden tremendously. Next, we describe and illustrate the integer programming problem and then solve it with Excel. Then a special type of integer programming model is discussed called the **zero–one formulation.** In the zero–one formulation, there are one or more variables that can assume only one of two values, zero or one.

Overview of Integer Programming

Integer programming is important not only because it allows us to solve practical problems with indivisibility requirements, but also because it can be used as a computational tool in the solution of several complicated problems that cannot otherwise be formulated (or cannot otherwise be solved effectively). For example, many nonlinear as well as combinatorial problems can be reduced to integer programming models. (A combinatorial problem consists of finding, from among a finite set of feasible solutions—usually very large—a solution that optimizes the value of the objective function.)

There are many examples of organizations that use integer programming models. We provide two here:

- Valley Metal Container (VMC) located in Golden, Colorado, produces over 4 billion aluminum cans per year and is the largest aluminum can facility in the world. The company, which is actually a joint venture between the Coors Brewing Company and American National Can, developed an integer programming model with a spreadsheet user interface to help plan the weekly production schedule. VMC management estimates that the model has helped reduce direct costs by more than \$150,000 annually (Katok and Ott 2000).
- Grantham, Mayo, Van Otterloo and Company LLC (GMO) is an investment management firm with approximately \$26 billion in assets under management. The firm uses integer programming to construct portfolios that match the performance of target portfolios, but with fewer individual stocks and transactions. The firm believes that the benefits of this approach include a 40 to 60 percent reduction in the number of individual stocks owned, a \$4 million annual reduction in trading costs, and an improved trading process (Bertsimas, Darnell, and Soucy 1999).

Integer programming is different from LP in several ways, as follows.

Constraints Adding the indivisibility requirement results in additional constraints. This means that the optimal integer solution will be either inferior (the usual case) or, at best, as good as the optimal noninteger solution. Clearly, the solution to an optimization problem cannot be improved by adding additional constraints or limitations. Thus, there is often a cost attached to imposing the indivisibility requirement.

Complexity In contrast to LP, there is no simple solution procedure for integer programming models. One approach, called the *Gomory Cutting Plane Method* (see Turban and Meredith 1994), keeps adding constraints to the initial LP problem until an integer solution is found, but it is a complex method. There are also a variety of "enumeration" techniques such as *complete enumeration* (checking all feasible integer solutions), *graphical examination* (for small problems consisting of either two variables or two constraints), or *intelligent search methods* like *branch and bound* (see Turban and Meredith 1994). In the branch and bound approach, the problem is successively divided into subproblems for which bounds are determined. If the bounds are worse than an existing feasible solution, those subproblems are discarded and the remaining subproblems are considered. The process is continued until no further subdivision is possible and the optimal solution has thus been determined. Since there are no simple solution procedures, one may have to settle for near-optimal rather than optimizing techniques, particularly in cases where there are a large number of integer decision variables. One approach is to round the LP solution to the nearest integer variable values that do not violate the constraints. Another is to use this rounded solution and then further check all feasible integer solutions adjacent to it.

Finite Solutions With LP there is either one optimal solution or an infinite number of them. With integer programming, there is either one or a finite number of optimal solutions. However, this finite number could still be astronomical.

Computer Options Although many general purpose computer programs including Excel can solve models with integer decision variables, the sensitivity analyses generated by these general programs do not usually have any meaning.

Integer Programming Models Integer programming models can be classified into the following three categories:

1. **All-Integer Model** This model adds a requirement that all decision variables (x_1, x_2, etc.) must be integers. This situation is illustrated in the next example.
2. **Mixed Integer Model** In this model, there is a requirement that some, but not all, of the decision variables must be integers. This situation can be handled similarly to the all-integer approach illustrated in the next example.
3. **Zero–One Integer Model** In this model there are sets of decision variables whose value is limited to either zero or one. These variables are important because a set of 0–1 values can represent "yes" and "no" situations such as to undertake a project, invest in a certain stock, and so on. This situation is illustrated in the next subsection.

EXAMPLE

Southern General Hospital

Jack Kalini, the administrator of Southern General Hospital, just returned from a meeting of the hospital board. He was somewhat disappointed because the board only approved funds for a minor expansion, $210,000, to be used for not more than 10 hospital beds. He wanted to make the most of this authorization; that is, to get as much additional income as possible to improve the hospital's cash-flow situation. After consulting with his staff, it was determined that there were two places where additional beds could be added—the maternity ward or the cardiac ward. Each additional bed in the maternity ward would generate $20 of profit per day, versus $25 in the cardiac ward. For a moment, he thought of placing all the additional beds in the cardiac ward; however, he remembered that the board insisted on at least two additional beds for the maternity ward. The expansion would cost \$24,600 per bed in the cardiac ward and only \$20,000 per bed in the maternity ward. The problem/opportunity was how many beds to add to each ward.

Having defined the problem/opportunity and collected the necessary data, the next step is to formulate the model. Let us begin by defining the decision variables as follows:

x_1 = Number of beds to add to the maternity ward
x_2 = Number of beds to add to the cardiac ward

Given these decision variables, we can now formulate the model as follows:

Maximize $z = 20x_1 + 25x_2$
Subject to:

$1x_1 + 1x_2 \leq 10$	(total bed limitation)
$\$20{,}000x_1 + \$24{,}600x_2 \leq \$210{,}000$	(budget constraint)
$1x_1 \geq 2$	(at least two beds in maternity ward)
$x_1, x_2 \geq 0$	(nonnegativity constraints)

The spreadsheet shown in Exhibit 4.36 was created to solve this model. To obtain the optimal solution that is shown, cell B17 was specified as the Target Cell, the Max radio button was selected, and the range B12:C12 was entered for the Changing Cells. The constraints were entered as discussed earlier and are listed in the Solver Parameters dialog box shown in Exhibit 4.37. You might notice that equations were not entered for the nonnegativity constraints in the spreadsheet shown in Exhibit 4.36, nor are they listed in Exhibit 4.37. The reason for this is that the Excel shortcut for entering the nonnegativity constraints was used by selecting Assume Non-Negative in the Solver Options dialog box (see Exhibit 4.38).

According to the optimal solution, Southern General should acquire two maternity beds and 6.91 cardiac beds. Doing this will increase profits by \$212.76 per day. According to the solution, Southern General Hospital would completely utilize its budget and exactly meet the minimum requirement for maternity beds.

One serious limitation with the current optimal solution is that for practical purposes it is not feasible. As you probably already recognized, it is generally not possible to purchase 6.91 beds.

To alleviate this problem, we can add additional constraints that require our decision variables to be integers. In our case, we want to restrict cells B12:C12 to be integers. Doing this in Excel is

EXHIBIT 4.36 Optimal Number of Beds for Southern General Hospital to Acquire

	A	B	C	D	E	F	G
1	**Southern General Hospital**						
2							
3	**Model Parameters**						
4		**Maternity**	**Cardiac**	**Limit**		***Key Formulas***	
5	Added profit/day per new bed	\$20	\$25			B17	=SUMPRODUCT(B\$12:C\$12,B5:C5)
6	Maximum new beds			10		B18	=B12+C12
7	Cost of new beds	\$20,000	\$24,600	\$210,000		B19	=SUMPRODUCT(B\$12:C\$12,B7:C7)
8	Minimum new maternity beds			2		B20	=B12
9						C18	=D6 (Copy to Cells C19:C20)
10	**Decision Variables**					D18	=C18-B18 (Copy to Cell D19)
11		x_1	x_2			D20	=B20-C20
12		2.00	6.91				
13							
14	**Model Outputs**						
15				**Slack/**			
16		**Value**	**RHS**	**Surplus**			
17	Objective function	\$212.76					
18	Maximum new beds	8.91	10.00	1.09			
19	Budget constraint	210,000.00	210,000.00	0.00			
20	Minimum new maternity beds	2.00	2.00	0.00			

Value of objective function

Optimal solution

similar to adding any other constraint. First, while in the Solver Parameters dialog box, select the Add button. In the Add Constraint dialog box, enter the cell address or range of cells corresponding to the decision variables that must be integers. In the middle box, select the int option. *Note:* When the int option is selected, integer is automatically entered into the last box. Exhibit 4.39 shows the Solver Parameters dialog box with the added integer constraint and Exhibit 4.40 shows the optimal solution obtained with the added integer constraint.

As expected, adding new constraints did not improve the value of the objective function. More specifically, adding the integer constraint reduced the value of the objective function by $2.76 (212.76 – 210.00). The new optimal solution calls for purchasing three maternity beds and six cardiac

EXHIBIT 4.37 Solver Parameters Dialog Box for Southern General Hospital

Solver Parameters
Set Target Cell: $B$17
Equal To: Max Min Value of: 0
By Changing Cells:
$B$12:$C$12
Guess
Subject to the Constraints:
$B$18 <= $C$18
$B$19 <= $C$19
$B$20 >= $C$20
Add
Change
Delete
Solve
Close
Options
Reset All
Help

EXHIBIT 4.38 Solver Options Dialog Box for Southern General Hospital

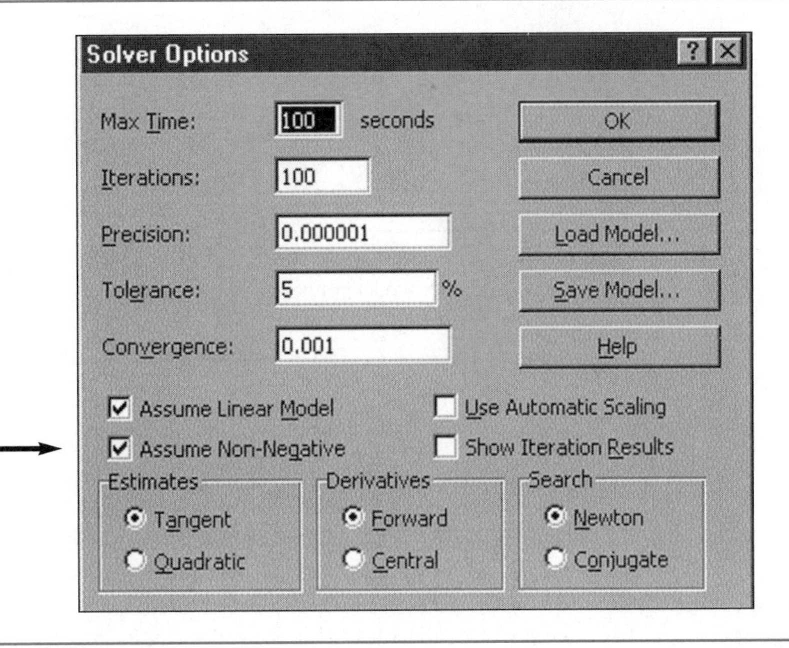

EXHIBIT 4.39 New Solver Parameters Dialog Box with Integer Constraint Added

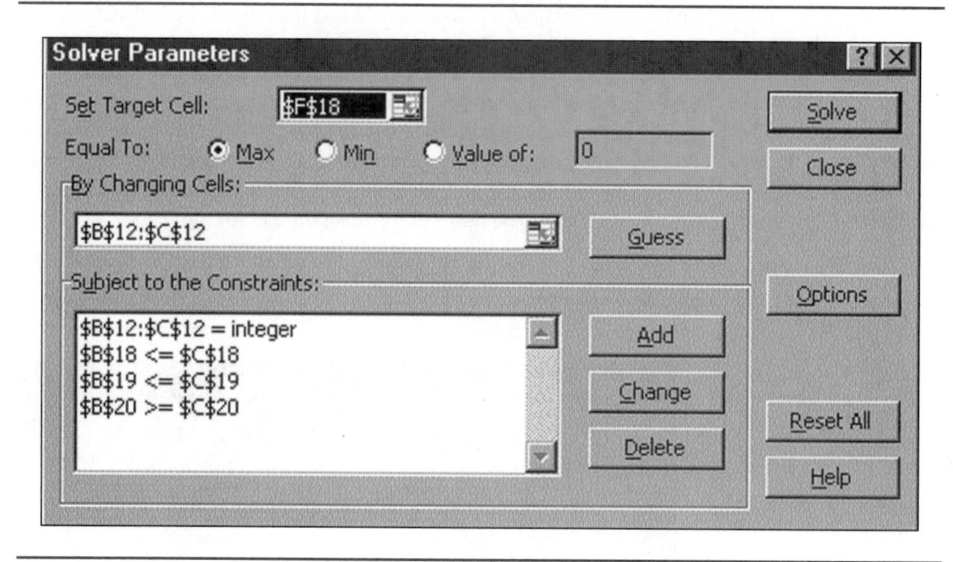

EXHIBIT 4.40 Optimal Solution with Integer Requirement

	A	B	C	D	E	F	G	H	I	J
1	**Southern General Hospital**									
2										
3	**Model Parameters**									
4		**Maternity**	**Cardiac**	**Limit**		***Key Formulas***				
5	Added profit/day per new bed	$20	$25			B17	=SUMPRODUCT(B$12:C$12,B5:C5)			
6	Maximum new beds			10		B18	=B12+C12			
7	Cost of new beds	$20,000	$24,600	$210,000		B19	=SUMPRODUCT(B$12:C$12,B7:C7)			
8	Minimum new maternity beds			2		B20	=B12			
9						C18	=D6 (Copy to Cells C19:C20)			
10	**Decision Variables**					D18	=C18-B18 (Copy to Cell D19)			
11		x_1	x_2			D20	=B20-C20			
12		3.00	6.00							
13										
14	**Model Outputs**									
15				**Slack/**						
16		**Value**	**RHS**	**Surplus**						
17	Objective function	$210.00								
18	Maximum new beds	9.00	10.00	1.00						
19	Budget constraint	207,600.00	210,000.00	2400.00						
20	Minimum new maternity beds	3.00	2.00	1.00						

New value of objective function

Optimal solution with integer requirement

Generally more slack

beds. Clearly, we would not have arrived at the new optimal solution by simply rounding the original solution. Also note that all the slack and surplus variables are positive in the new optimal solution.

The Zero–One Model

A special and important application of integer programming is the case where the value of the decision variables is limited to two "logical" values. For example, a variable may be either yes or no, or match or not. These are symbolized by the values 0 and 1 and known as zero–one or binary variables. This restriction is very common in many managerial situations and greatly extends the power of mathematical programming for managerial purposes. As we demonstrate in the following example, Excel can solve these models as well.

EXAMPLE

The Fixed-Charge Situation

Many real-life problems involve a combination of fixed and variable costs. The fixed costs are incurred only if certain projects are undertaken or a certain capacity level is exceeded. A common example is machine scheduling.

Eastland Corporation is planning to produce at least 900 printed circuit boards. Three production lines are available with the setup costs, unit processing (variable) costs, and capacities given in Exhibit 4.41. Eastland's objective is to determine which lines to use in order to minimize the total costs.

Having defined the problem/opportunity and collected the necessary data, our next step is to formulate the model. In terms of decision variables, we define x_A, x_B, and x_C to represent the quantities of printed circuit boards produced on lines A, B, and C, respectively; and d_A, d_B, d_C to indicate whether a line is to be used (1) or not (0). Using these decision variables, the model can be formulated as

$$\text{Minimize } z = \underbrace{(850d_A + 150d_B + 520d_C)}_{\text{Fixed cost}} + \underbrace{(20x_A + 58x_B + 36x_C)}_{\text{Variable cost}}$$

Subject to:

$$x_A + x_B + x_C \geq 900$$
$$x_A \leq 500d_A$$
$$x_B \leq 700d_B$$
$$x_C \leq 620d_C$$

and

$$d_A, d_B, d_C = 0, 1$$
$$x_A, x_B, x_C \geq 0$$

Notice in our formulation that we capture the setup cost through the capacity constraint. For example, in line A, instead of just writing the capacity constraint as "production must be less than or equal to 500 units," we include the setup variable d_A to force x_A to zero if line A is *not* set up ($d_A = 0$). (*Note:* This problem is actually a mixed-integer programming problem, because only *some* of the variables are restricted to 0 or 1 values.)

The spreadsheet shown in Exhibit 4.42 was developed to solve this model. When solving optimization models with Excel it is desirable to have all the decision variables on the left-hand side. Therefore, the maximum capacity constraints for the three lines required a slight transformation. For example, the maximum capacity constraint for line A was transformed as follows:

$$x_A - 500d_A \leq 0$$

The maximum capacity constraints for lines B and C were transformed in a similar fashion.

The only other new aspect of the model developed for Eastland is that it utilizes zero–one, or binary variables. Specifying that a variable is zero–one in Excel is similar to specifying that the variable is integer. First, while in the Solver Parameters dialog box, select the Add button. In the Add Constraint dialog box, enter the cell address or range of cells corresponding to the decision variables that must be zero–one. In the middle box, select the bin option. *Note:* When the bin option is selected, binary is automatically entered into the last box. Exhibit 4.43 shows the Solver

EXHIBIT 4.41
Data for Eastland Corporation

Line	Setup Cost	Unit Processing Cost	Maximum Capacity
A	$850	$20	500
B	150	58	700
C	520	36	620

EXHIBIT 4.42 Optimal Solution for Eastland Corporation

	A	B	C	D	E	F	G	H
1	**Eastland Corporation**							
2								
3	**Model Parameters**							
4		**Line A**	**Line B**	**Line C**	**Limit**		***Key Formulas***	
5	Setup cost	$850	$150	$520			B18	=SUMPRODUCT(B12:D12,B6:D6)+
6	Unit processing cost	$20	$58	$36				SUMPRODUCT(B13:D13,B5:D5)
7	Capacity	500	700	620			B19	=SUM(B12:D12)
8	Minimum production				900		B20	=B12-(B7*B13)
9							B21	=C12-(C7*C13)
10	**Decision Variables**						B22	=D12-(D7*D13)
11		**Line A**	**Line B**	**Line C**			C19	=E8
12	*x*	500.00	0.00	400.00			C20:C22	0
13	*d*	1.00	0.00	1.00			D19	=B19-C19
14							D20	=C20-B20 (Copy to Cells D21:D22)
15	**Model Outputs**							
16				**Slack/**				
17		**Value**	**RHS**	**Surplus**				
18	Objective function	$25,770.00						
19	Minimum production	900.00	900.00	0.00				
20	Capacity of Line A	0.00	0.00	0.00				
21	Capacity of Line B	0.00	0.00	0.00				
22	Capacity of Line C	-220.00	0.00	220.00				

Value of objective function

Optimal solution

Excess capacity on Line C

EXHIBIT 4.43 Parameter Dialog Box for Eastland Corporation

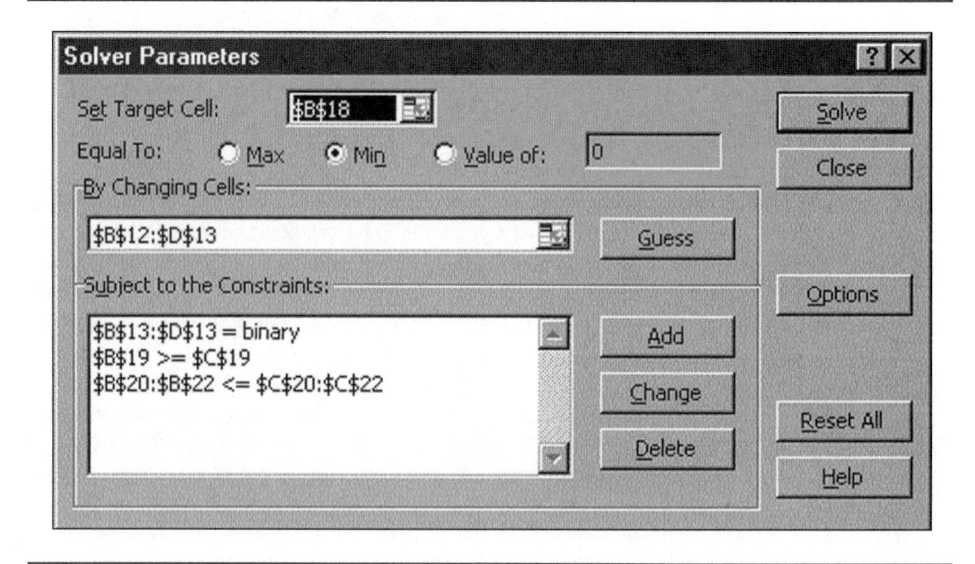

Parameters dialog box with the added binary constraint. After reviewing Exhibits 4.42 and 4.43 you may notice that the nonnegativity constraints appear to be missing. As discussed earlier, you can select the Options button in the Solver Parameters dialog box, select the Assume Non-Negative check box and then click on the OK button.

Also, as discussed earlier, spreadsheets provide a shorthand for entering constraints that can be a significant time saver. Notice in Exhibit 4.43 that the constraint specifying that d_A, d_B, and d_C (cells B13:D13) are zero–one (binary) variables was entered as one constraint using the range that contained all the zero–one variables. Entering a range of cells is faster but equivalent to entering a new constraint for each decision variable one at a time. Thus, the constraint B13:D13 = binary is equivalent to the three constraints B13 = binary, C13 = binary, and D13 = binary.

In a similar fashion, this shorthand approach was used for entering the capacity constraints for each line. Referring to Exhibit 4.42, note that the capacity constraints for the three lines are contained in rows 20–22, respectively. Also observe that the left-hand side of the three constraints is always calculated in column B and the right-hand side constant is contained in column C of these rows. Since these constraints are all of the less-than-or-equal type, we can enter all three constraints simultaneously with Excel as B20:B22 ≤ C20:C22. This is equivalent to entering the following three constraints: B20 ≤ C20, B21 ≤ C21, and B22 ≤ C22. Using this shortcut permits entering moderately large models in a relatively short amount of time. To facilitate using this shortcut, it is helpful to group the constraints together by their type (i.e., ≤, =, ≥) and the decision variables by their type (i.e., regular continuous variables, integer, binary). Doing this makes it easier to specify the appropriate ranges. Finally, range names can be used to further simplify the process of entering constraints.

Referring to Exhibit 4.42, we see that the optimal solution for Eastland is to produce 500 printed circuit boards on line A and 400 on line C. Although line B has the lowest setup (fixed) cost, its relatively high unit processing cost made utilizing this line unattractive, given the volume of output that was needed. From that slack/surplus column we observe that line A is at capacity while line C could produce an additional 220 printed circuit boards. Also, Eastland exactly meets the minimum requirement of 900 units.

4.7 Detailed Modeling Example

Up to this point, the focus of this chapter has been primarily on the model formulation and analysis stages of the modeling process. In this section, we illustrate (though briefly) the use of optimization in the broader modeling context.

Step 1: Opportunity/Problem Recognition

Palm General Hospital's chief administrator is concerned about the increased cost of staffing its emergency department. Salaries in this department have been unexpectedly increasing, with a sudden scarcity of medically qualified personnel for this critical function. However, the budget for this and the coming years did not anticipate this increase, and thus, the resources to handle this problem are unavailable. The chief administrator wishes to staff the emergency room (ER) to ensure complete and competent coverage, of course, but at the lowest possible cost. She senses that its current scheduling process may be missing some efficiencies or other opportunities to make some savings in the scheduling of their personnel.

Step 2: Model Formulation

To guide the development of the formal mathematical model, the influence diagram shown in Exhibit 4.44 was developed. According to the diagram, our decision is to determine how many nurses to schedule per shift in order to minimize total nurse staffing cost. Patient demand, patient processing times, and the number of nurses available are identified as factors that will influence the number of nurses scheduled during each shift.

Step 3: Data Collection

The sign-in sheets and other records were collected from the emergency department for the last three months. A 12-week analysis was undertaken to determine the average demand level the ER experiences throughout the day. For now, it was assumed there would

EXHIBIT 4.44 Influence Diagram for Palm General Hospital

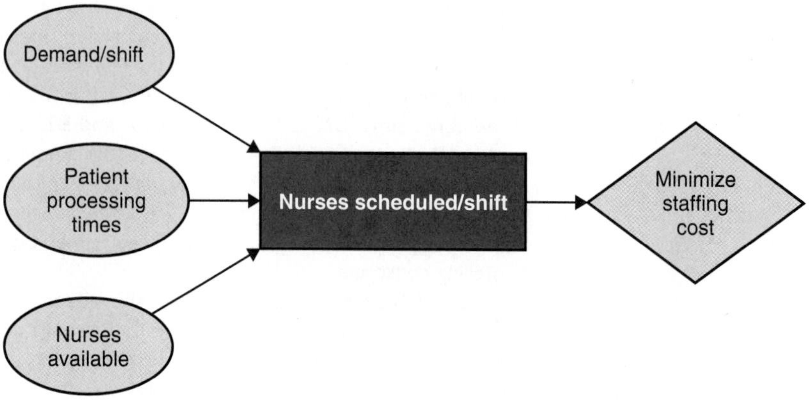

EXHIBIT 4.45 Sign-In Sheet Summary

Time Period	Average Number of Cases
7 A.M.–11 A.M.	80
11 A.M.–3 P.M.	60
3 P.M.–7 P.M.	120
7 P.M.–11 P.M.	60
11 P.M.–3 A.M.	40
3 A.M.–7 A.M.	20

be no increase or decrease in demand rates in the near future. If it appeared that savings were possible by improved personnel scheduling, then a time series analysis (Chapter 3) would be cost justified to determine a more accurate demand forecast.

For the purpose of this analysis, each day was divided into six periods. The average number of emergency cases arriving by time period is summarized in Exhibit 4.45.

Further analysis of the data indicates that one emergency room nurse can handle 2.5 cases each hour. The ER nurses work in shifts of 8 hours. The normal shifts are: 7 A.M.–3 P.M.; 3 P.M.–11 P.M.; and 11 P.M.–7 A.M. During the course of the study it was noted that, as an exception to these normal shifts, three nurses would be willing to start their shift at 11 A.M. and two indicated a willingness to start their shift at 7 P.M. Finally, it was determined that all nurses in the ER department had approximately the same experience and earned equivalent salaries.

The minimum number of nurses needed for each time period can be calculated based on the rate at which nurses can handle cases and the number of cases per time period. For example, between 7 A.M. and 11 A.M. there is a need to process 20 cases/hour (80 cases ÷ 4 hours). Based on the nurses' ability to handle 2.5 cases per hour, there is thus a need for 8 nurses (20 ÷ 2.5) between 7 A.M. and 11 A.M. The minimum number of nurses for the other periods is shown in Exhibit 4.46.

Based on the influence diagram, the problem is to find the minimum number of nurses to meet the demand, assuming they all receive the same wages. We can define the decision variables as follows:

x_1 = Number of nurses starting at 7 A.M.
x_2 = Number of nurses starting at 11 A.M.

EXHIBIT 4.46 Nurse Staffing Requirements

Time period	7 A.M.–11 A.M.	11 A.M.–3 P.M.	3 P.M.–7 P.M.	7 P.M.–11 P.M.	11 P.M.–3 A.M.	3 A.M.–7 A.M.
Required nurses	8	6	12	6	4	2

x_3 = Number of nurses starting at 3 P.M.
x_4 = Number of nurses starting at 7 P.M.
x_5 = Number of nurses starting at 11 P.M.

Note: a decision variable was not defined for nurses to start at 3 A.M. since this is not the start of a regular shift and no nurses indicated a willingness to begin their shifts at this time.

Based on these decision variables, an optimization model can be formulated as follows:

Minimize $z = x_1 + x_2 + x_3 + x_4 + x_5$
Subject to:
$x_1 \geq 8$ (for the 7 A.M.–11 A.M. time period)
$x_1 + x_2 \geq 6$ (for 11 A.M.–3 P.M.)
$x_2 + x_3 \geq 12$ (for 3 P.M.–7 P.M.)
$x_3 + x_4 \geq 6$ (for 7 P.M.–11 P.M.)
$x_4 + x_5 \geq 4$ (for 11 P.M.–3 A.M.)
$x_5 \geq 2$ (for 3 A.M.–7 A.M.)
$x_2 \leq 3$ (3 nurses willing to start at 11 A.M.)
$x_4 \leq 2$ (2 nurses willing to start at 7 P.M.)
$x_1, x_2, x_3, x_4, x_5 \geq 0$

Step 4: Analysis of the Model

The optimization model formulated with the data collected above was entered into an Excel spreadsheet and is shown in Exhibit 4.47. Most of the formulas that needed to be entered in the "Model Outputs" section of the spreadsheet were unique and therefore limited the ability to copy formulas across cells. However, as shown in Exhibit 4.47, the formulas needed were quite straightforward and not very complex. One primary advantage of the model shown in Exhibit 4.47 is its ease of entry in Excel. As shown in Exhibit 4.48, entering the optimization model in Excel required specifying a cell for the Target Cell, one range for the Changing Cells and only two constraints by exploiting Excel's ability to accept ranges of cells for the left-hand and right-hand sides of constraints. Also, the Assume Linear Model and Assume Non-Negative check boxes were selected in the Solver Options dialog box. Also observe that the optimal solution contained only integer values without the extra integer constraint. Had this not been the case, we could have added the integer constraint and resolved the problem. In general, it is a good idea to leave the integer constraint out to begin with because general LP models can be solved much faster than integer programming models.

According to the optimal solution, a total of 24 nurses will be needed. Eight nurses would begin their shift at 7 A.M., three at 11 A.M., nine at 3 P.M., two at 7 P.M. and two at 11 P.M. The only time periods that have extra nurses are the 11 A.M.–3 P.M. and the 7 P.M.–11 P.M. time periods, each with an extra five nurses. All the other time periods are staffed with the minimum required nurses. Also, all nurses that volunteered to start their shifts at 11 A.M. and 7 P.M. were scheduled to do so.

The sensitivity report generated by Excel for Palm General Hospital is shown in Exhibit 4.49. You may recall that it was assumed that all the nurses were paid the same salary.

EXHIBIT 4.47 Spreadsheet Model and Optimal Solution for Palm General Hospital

	A	B	C	D	E	F	G
1	**Palm General Hospital**						
2							
3	**Model Parameters**						
4		**7 - 11 A.M.**	**11 - 3 P.M.**	**3 - 7 P.M.**	**7 - 11 P.M.**	**11 - 3 A.M.**	**3 - 7 A.M.**
5	Required nurses	8	6	12	6	4	2
6	Nurses available to start		3		2		
7							
8	**Decision Variables**						
9		x_1	x_2	x_3	x_4	x_5	
10	Number of nurses to start each shift	8	3	9	2	2	
11							
12	**Model Outputs** Total nurses needed each day						
13				**Slack/**			
14		**Value**	**RHS**	**Surplus**			
15	Objective function	24.00					
16	Nurses needed 7 - 11 A.M.	8.00	8	0.00			
17	Nurses needed 11 - 3 P.M.	11.00	6	5.00			
18	Nurses needed 3 - 7 P.M.	12.00	12	0.00			
19	Nurses needed 7 - 11 P.M.	11.00	6	5.00			
20	Nurses needed 11 - 3 A.M.	4.00	4	0.00			
21	Nurses needed 3 - 7 A.M.	2.00	2	0.00			
22	Nurses available to start at 11 A.M.	3.00	3	0.00			
23	Nurses available to start at 7 P.M.	2.00	2	0.00			
24							
25		*Key Formulas*					
26		B15	=SUM(B10:F10)				
27		B16	=B10				
28		B17	=B10+C10				
29		B18	=C10+D10				
30		B19	=D10+E10				
31		B20	=E10+F10				
32		B21	=F10				
33		B22	=C10				
34		B23	=E10				
35		C16	=B5				
36		C17	=C5				
37		C18	=D5				
38		C19	=E5				
39		C20	=F5				
40		C21	=G5				
41		C22	=C6				
42		C23	=E6				
43		D16	=B16-C16 (Copy to Cells D17:D21)				
44		D22	=C22-B22 (Copy to Cell D23)				

EXHIBIT 4.48 Solver Parameters Dialog Box for Palm General Hospital

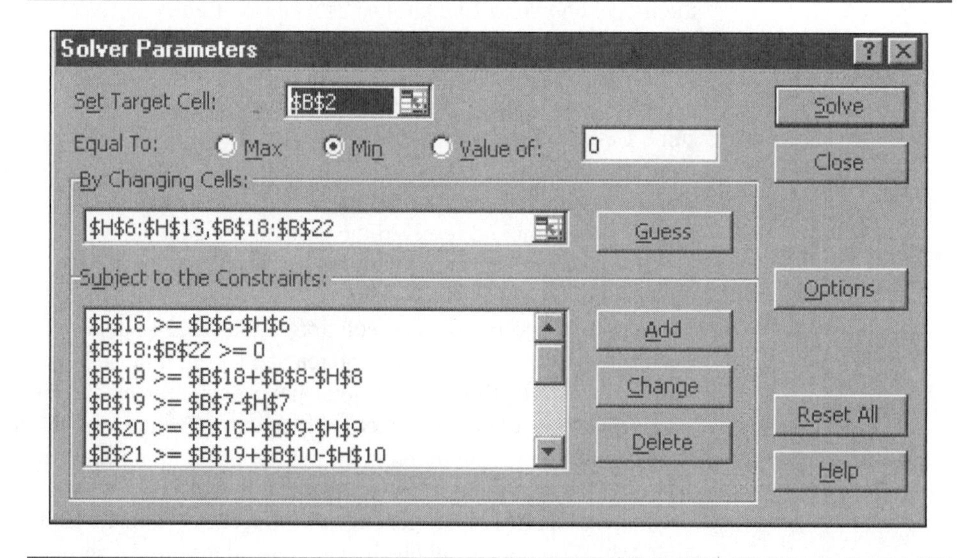

Therefore, all the coefficients in the objective function are one. Referring to Exhibit 4.49, we observe that the Final Value for x_1 (the number of nurses to start their shift at 7 A.M.) is eight. According to the last two columns in the top table shown in Exhibit 4.49, as long as the objective function coefficient for x_1 remains between zero and infinity, the solution shown in Exhibit 4.47 will remain optimal. One interpretation of this result is that if the cost of staffing the shift beginning at 7 A.M. were to change relative to the staffing costs of the other shifts, the optimal solution would remain unchanged. The situation is quite different for x_2. If the cost of staffing the shift beginning at 11 A.M. were to increase at all relative to the costs of staffing the other shifts, the current solution would no longer be the optimal solution and the model would have to be resolved.

Referring to the bottom table in Exhibit 4.49, we observe that if we could reduce the nurses needed during 7 A.M.–11 A.M., the total number of nurses needed would be reduced by one, from 24 to 23 nurses. Likewise, if the number of nurses required during this shift were to increase to 9, the total number of required nurses would increase to 25 nurses. This result is valid as long as the required number of nurses during the 7 A.M.–11 A.M. period remains between three and infinity. Therefore, based on this range, we know that if the number of required nurses during the 7 A.M.–11 A.M period were to increase to 100, 92 (100 – 8) additional nurses would be needed. Likewise, if the number of nurses required

EXHIBIT 4.49 Sensitivity Report for Palm General Hospital

Adjustable Cells

Cell	Name	Final Value	Reduced Cost	Objective Coeefficient	Allowable Increase	Allowable Decrease
$B$10	x_1	8	0	1	1E+30	1
$C$10	x_2	3	0	1	0	1E+30
$D$10	x_3	9	0	1	1E+30	0
$E$10	x_4	2	0	1	0	1E+30
$F$10	x_5	2	0	1	1E+30	0

Constraints

Cell	Name	Final Value	Shadow Price	Constraint R.H. Side	Allowable Increasse	Allowable Decrease
$B$16	Nurses needed 7–11 A.M. value	8.00	1.00	8	1E+30	5
$B$17	Nurses needed 11–3 P.M. value	11.00	0.00	6	5	1E+30
$B$18	Nurses needed 3–7 P.M. value	12.00	1.00	12	1E+30	5
$B$19	Nurses needed 7–11 P.M. value	11.00	0.00	6	5	1E+30
$B$20	Nurses needed 11–3 A.M. value	4.00	1.00	4	1E+30	0
$B$21	Nurses needed 3–7 A.M. value	2.00	0.00	2	0	1E+30
$B$22	Nurses available to start at 11 A.M. value	3.00	0.00	3	5	3
$B$23	Nurses available to start at 7 P.M. value	2.00	0.00	2	0	2

during this period dropped to 5, then the total number of nurses required would be reduced by 3, to 21 nurses. On the other hand, if the number of required nurses during the 7 A.M.–11 A.M period were to decrease to 2 nurses, we would have to resolve the model because a reduction of six nurses is outside the range that the Shadow Price is valid.

As a final example, observe that if the number of nurses required during the 11 A.M.–3 P.M. period were to increase by one nurse, this would have no impact on the optimal solution. This makes sense because there are 5 extra nurses scheduled for this time period (8 that start at 7 A.M. + 3 that start at 11 A.M. – 6 required nurses). Changing the number of required nurses for the 11 A.M.–3 P.M. time period will have no effect on the value of the objective function as long as the number required nurses for this time period does not increase by more than 5 nurses. If the number of required nurses increased by more than 5, all available slack would be used up and the model would have to be resolved.

Step 5: Implementation

Now that we have analyzed the situation facing Palm General Hospital, the final remaining task is to implement a solution. The first step is to select a course of action—we, of course, choose the solution shown in the previous subsection. Next is to communicate it to the chief administrator. If administration accepts the recommended solution, the next step is to apply this solution to the problem. If administration is not satisfied with the recommended action, then we must reconsider the model and derive new solutions, based on administration's new instructions. Following the final decision and application of the solution, we must then monitor the results to make sure we achieve the benefits we anticipated and haven't initiated any new problems elsewhere in the system.

Thus, our next step is to communicate our recommendation to the chief administrator. For practical purposes, this is a vitally important task because no matter how well the preceding steps were executed, if the product of this process is not effectively communicated to key decision makers it might as well have not been done. Exhibit 4.50 illustrates in the form of a memo how the results of our analysis might be communicated to hospital administrators. The companion Web site for this book contains a sample PowerPoint presentation of the results of our analysis. (*Note:* If acceptable multiple optimal solutions had been identified, those should also be communicated to hospital administration.)

There are a number of reasons why we may need to reconsider the results of our solution. One reason may be that the decision makers don't like the solution for some reason. In the case of Palm General Hospital, the administrators may think that the staffing levels vary too much, for example. Another reason the model may need to be reconsidered is that important constraints may have been omitted. For example, in the case of Palm General, perhaps we overlooked the fact that at least two head nurses need to be available at all times of the day. As a final example, the model may need to be reconsidered because one or more of the assumptions were not correct. In the case of the model developed for Palm General, it was assumed that all nurses earned the same salary. If it were later determined that this assumption was not valid, the model would need to be modified and resolved.

EXHIBIT 4.50 Sample Memo to Hospital Administrators of Palm General Hospital

MEMO

To: Dr. Joann Galt, Chief Administrator
From: Brianna Regan, Senior Business Analyst
CC: Dr. George Samuelson, Chief of Staff
Date: 3/28/02
Subject: Nurse Staffing of ER

I have concluded my analysis of the nurse staffing needs of the ER. A summary of my recommended schedule is contained in Table A. The proposed schedule requires a total of 24 nurses.

A couple of notes about this table: First, the total number of nurses available was calculated based on the assumption that nurses work a consecutive eight-hour shift from the time they start. The number of required nurses was determined on the basis of a review of the ER records over the previous 12 weeks, where it was concluded that nurses can handle, on average, 2.5 cases per hour.

An optimization model was developed to determine the minimum number of nurses required to meet expected ER demand. The model incorporated information about the demand per time period, patient processing times, and the number of available nurses. The schedule calls for exactly meeting the number of required nurses in four of the six time periods. In the other two time periods, there are five extra nurses available. All three nurses that volunteered to start at 11 A.M. and the two nurses that volunteered to start at 7 P.M. have been scheduled to start at these times, respectively. It is important to note, however, that the schedule presented here requires the smallest number of total nurses while meeting the required number of nurses in each of the six time periods.

I have also conducted an extensive follow-up analysis. I have learned that if there were to be an increase in the staffing cost for the 7 A.M.–11 A.M. time period, it would not change the proposed schedule. On the other hand, if the cost of staffing

TABLE A Recommended Nurse Schedule for the ER

Time Period	Number of Nurses that Start Their Shift	Total Number of Nurses Available Based on Start Time	Required Number of Nurses	Extra Nurses
7 A.M.–11 A.M.	8	8	8	0
11 A.M.–3 P.M.	3	11	6	5
3 P.M.–7 P.M.	9	12	12	0
7 P.M.–11 P.M.	2	11	6	5
11 P.M.–3 A.M.	2	4	4	0
3 A.M.–7 A.M.	0	2	2	0

(continued)

EXHIBIT 4.50 Continued

the 11 A.M.–3 P.M. or 7 P.M.–11 P.M. time periods were to increase relative to the costs of staffing the other time periods (perhaps due to incentive wages or overtime), a new schedule would need to be developed. Likewise, if the staffing cost of the 3 P.M.–7 P.M. or the 11 P.M.–3 A.M. time periods were to increase relative to the other time periods, a new schedule would need to be developed.

Furthermore, according to my follow-up analysis I have learned that any change in the nurse requirements for the 7 A.M.–11 A.M., 3 P.M.–7 P.M., and 11 P.M.–3 A.M. time periods will have a proportional change on the total nurses needed. For example, if the number of required nurses were reduced by two in any of these time periods, the number of total nurses would also be reduced by two. This relationship is valid as long as the nurse requirements are not reduced by more than five nurses for the 7 A.M.–11 A.M. and 3 P.M.–7 P.M. time periods. It is also important to point out that all the results discussed regarding the follow-up study assume that only one change is being made and that all other parameters are held constant.

There are several limitations of this analysis that should be pointed out. First, the nurse requirements were based on a three-week study conducted in July. It is quite possible that there is some seasonality in the demand for our ER services. If this is the case, the model would need to be modified and resolved for different demand patterns. Second, the model was based on the assumption that nurses are equally paid. If this assumption is not valid because there are shift premiums or the nurses are paid based on their experience, the model again would need to be modified and resolved. Finally, it was assumed that nurses could only start their shift during one of the three normal schedules—7 A.M.–3 P.M., 3 P.M.–11 P.M., and 11 P.M.–7 A.M.—unless they volunteered otherwise. A change in this policy could have an impact on the proposed schedule.

QUESTIONS

1. Would an optimization model be classified as deterministic or probabilistic? Descriptive or prescriptive? Why?
2. What are the advantages and limitations of LP?
3. How does the graph for a minimization problem differ from that of a maximization problem?
4. Describe the difference between slack and surplus variables. What kind of constraints does each occur in? What would an equality constraint have?
5. In what cases might a slack or surplus variable have a coefficient that is not equal to zero in the objective function?
6. Briefly describe and contrast situations where one optimal solution exists, alternate optimal solutions exist, the problem is infeasible, and the problem is unbounded.
7. Graphically, what would a sensitivity analysis of the coefficients of the constraints look like? What would adding or deleting a constraint look like?
8. Excel doesn't provide a built-in function for the sensitivity analysis of the constraint coefficients. If you were interested in doing this, how might you proceed?

9. Why isn't the best integer solution to a problem just the rounded up or down solution to the noninteger problem?
10. How does the zero–one situation differ from the integer situation? If you wanted to solve a zero–one problem using an integer programming package, how might you formulate the problem to do this?
11. How can an influence diagram be used to help formulate an optimization model?
12. What is the difference between the optimal solution and the value of the optimal solution?
13. Briefly list and describe the six components of a linear programming model.

EXPERIENTIAL EXERCISE

In the book *Enter the Zone* (Sears 1995), Dr. Barry Sears proposes a new diet. Specifically, according to this diet, 40 percent of your calories should come from carbohydrates, 30 percent from protein, and 30 percent from fat. Note that each gram of carbohydrate and each gram of protein have four calories, and each gram of fat has nine calories.

Your task in this assignment is to draw an influence diagram and formulate an optimization model that, when solved, creates a one-week meal plan (including snacks) that minimizes your total food costs. At a minimum, your optimization model should restrict the following:

- The total amount of carbohydrates consumed (e.g., to be 35 to 45 percent of total calories).
- The total amount of protein consumed (e.g., to be 25 to 35 percent of total calories)
- The total amount of fat consumed (e.g., to be 25 to 35 percent of total calories)
- The total number of calories consumed. (As a guideline for males, this may be 2200 to 2500 calories per day. This will vary based on your activity level and overall weight goals).
- The total amount of saturated fat consumed (e.g., to be less than 50 percent of fat calories consumed)

There are other constraints you may want to include in your model. Some examples might include the following:

- Restrictions on how many times you eat a particular item (e.g., no more than three apples per week)
- Constraints to ensure adequate levels of important vitamins and minerals are consumed (e.g., at least 1,000 milligrams of Vitamin C each day)
- Constraints to ensure that adequate levels of fiber are consumed
- Constraints that specify exactly what is eaten at each meal day by day, rather than determining in aggregate what should be eaten over the course of the week

The Web site accompanying this book (http://meredith.swcollege.com) has a list of links that you might find helpful in completing this exercise.

MODELING EXERCISES

1. Nitron Corp. of Canada produces two products. Unit profit for product A is $60; for product B, $50. Each must pass through two machines, P and Q. Product A requires 10 minutes on machine P and 8 minutes on machine Q. Product B requires 20 minutes on machine P and 5 minutes on machine Q. Machine P is available 200 minutes a day, whereas machine Q is available 80 minutes a day. The company must produce at least two units of product A and five of product B each day. Units that are not completed in a given day are finished the next day; that is, a portion of a product can be produced in the daily plan. What is the most profitable daily production plan?
   a. Develop an influence diagram for Nitron.
   b. Formulate the optimization model.
   c. Solve graphically (specifically show all constraints and the objective function as well as the optimal solution).
   d. How should the resources be allocated?
   e. Solve using Excel. How does your solution obtained using Excel compare with the one you obtained graphically?

2. Given:
   Maximize $z = 5x_1 + 5x_2$
   Subject to:
   (1) $1x_1 + 2x_2 \leq 30$
   (2) $1x_1 + 1x_2 \leq 19$
   (3) $8x_1 + 3x_2 \leq 120$
   a. Solve the problem graphically.
   b. If the problem has more than one optimal solution, explain why this is so and identify the two corner-point optimal solutions.
   c. Find the slacks, or surpluses, or both, for all constraints.
3. Given a set of constraints:
   (1) $10x_1 + 8x_2 \leq 120$
   (2) $5x_1 + 5x_2 \geq 30$
   (3) $1x_1 \geq 2$
   (4) $1x_2 \leq 10$
   (5) $1x_2 \geq x_1$
   a. Graphically display the feasible area.
   b. Compute the coordinates of all feasible corner (intersection) points.
   c. If the objective function is: maximize $z = 5.5x_1 + 3x_2$, find the value of the objective function at all corner points. Which corner point is the best one?
   d. Graphically draw the slope of the objective function. Confirm the findings of part c.
   e. Find the slacks/surpluses on all constraints.
   f. Given: minimize $z = 5.5x_1 + 3x_2$, what will the optimal solution be now?
4. A knitting machine can produce 1,000 pairs of pants or 3,000 shirts (or a combination of the two) each day. The finishing department can handle either 1,500 pairs of pants or 2,000 shirts (or a combination of the two) each day. The marketing department requires that at least 400 pants be produced each day. The company's stated objective is profit maximization.
   a. If the profit from a pair of pants is $4 and that derived from a shirt is $1.50, how many of each type should be produced? Solve graphically.
   b. Assume that the profit from a shirt increases to $2. What should be the minimum profit derived from selling a pair of pants that will justify production of pants only (i.e., $X_{\text{shirts}} = 0$)?
   c. Examine the solutions to a and b and interpret the difference between them. Can this interpretation be generalized to all LP graphical solutions?
   d. Verify your answers to parts a–c using Excel and its Sensitivity Report.
5. The owner of Black Angus Ranch of Australia is trying to determine the correct mix of two types of beef feed, A and B, which cost 50 cents and 75 cents per pound, respectively. Five essential ingredients are contained in the feed, as shown in the following table, which also indicates the minimum daily requirements of each ingredient:

	Percent per Pound of Feed		
Ingredient	**Feed A**	**Feed B**	**Minimum Daily Requirements (Pounds)**
1	20	25	30
2	30	10	50
3	0	30	20
4	24	15	60
5	10	20	40

   Find the least-cost daily blend for the ranch. That is, how many pounds of feed A and feed B will be included in the mix? (Solve graphically.)
6. Given the two LP graphical solutions as follows; the optimal solution is marked by x. In each case, find
   a. Which constraints are redundant.
   b. Which constraints will have a slack.
   c. Which constraints will have a surplus.

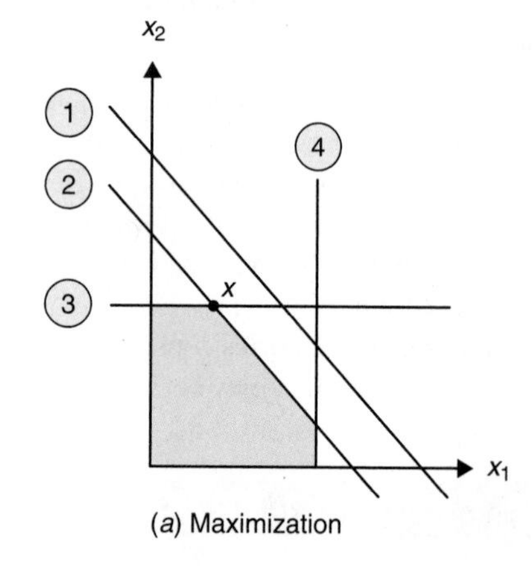

(a) Maximization

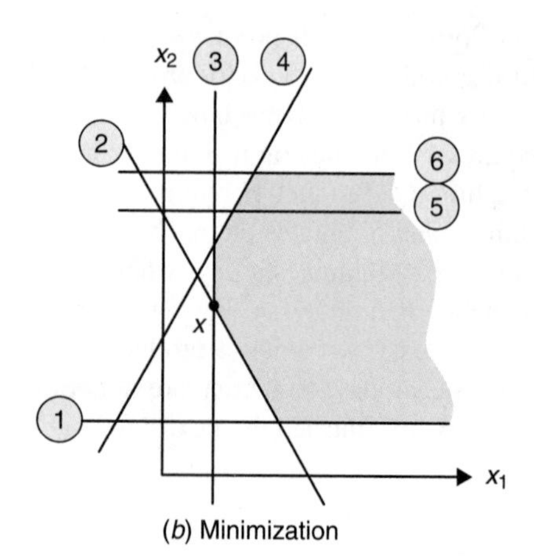

(b) Minimization

7. Given an LP problem:
   Minimize $z = 6x_1 + 14x_2$
   Subject to:
   $4x_1 + 7x_2 = 56$
   $4.5x_1 + 2.5x_2 \leq 45$
   $x_2 \geq 3$
   $x_1 \leq 12$
   a. Draw the feasible area.
   b. Compute the value of all intersecting feasible points.
   c. Solve graphically.
8. For each of the following constraints and proposed solutions find
   a. Whether the solution is feasible.
   b. The value of the slack or surplus.

   (1) $2x_1 + 2.5x_2 - 0.5x_3 \geq 40$; solution (2, 4, 2)
   (2) $5x_1 + 3x_2 + x_3 \leq 60$; solution (10, 2, 4)
   (3) $2x_1 + x_2 \geq 3$; solution (4, 3)

9. In 1717, Edward Teach, better known as "Blackbeard," sailed his ship into the harbor of Arecibo in the West Indies and attacked and looted the town. His plunder consisted of 29,000 pounds of gold (worth 350 shillings per troy ounce), 21,000 pounds of silver (worth 25 shillings/oz.), and 9,000 pounds of gems and jewelry (worth 200 shillings/oz). Blackbeard's ship includes three holds: the forward hold has a capacity of 11,000 pounds or 900 cubic feet, the center hold has a capacity of 15,000 pounds or 1,400 cubic feet, and the aft hold has a capacity of 8,000 pounds, or 500 cubic feet. Draw an influence diagram to help determine how much of each type of plunder Blackbeard should put in each hold to maximize his haul. *Note:* There are 12 troy ounces in each pound, and gold occupies a quarter of a cubic foot per pound when stored for shipping, silver takes a third of a cubic foot per pound, and the gems/ jewelry two-thirds of a cubic foot per pound.
10. COM Food Corporation specializes in menu preparation for British restaurants, institutions, and individual families. The major idea of the service is to provide an adequate (and tasteful) diet at a minimal cost. Given below is a *simplified* diet problem. The minimum daily nutrient requirements for a certain group of adults are

Calories	2,860
Protein	80 grams
Iron	15 milligrams
Niacin	20 milligrams
Vitamin A	20,000 units

A menu, when recommended by COM Food, must supply *at least* these minimum daily requirements. In this simplified problem, we shall assume that COM Food prepares one menu only. (In actual cases, a different menu would be prepared for every day of the week, or even every day of the month.) The following table lists foods and prices (pounds sterling per pound, £) offered to the customers.

Food	Price per Pound (£)
Beef	1.00
Butter	0.79
Bread	0.26
Carrots	0.15
Halibut	0.80
Eggs	0.34
Cheese	1.10

The quantities of calories, protein, iron, niacin and vitamin A included in each 100 grams of the above foods are given below:

Food	Calories	Protein (Grams)	Iron (mg)	Niacin (mg)	Vitamin A (Units)
A Beef	309	26.0	3.1	4.1	0
B Butter	716	0.6	0.0	1.0	3,300
C Bread	276	8.5	0.6	0.9	0
D Carrots	42	1.2	0.6	0.4	12,000
E Halibut	182	26.2	0.8	10.5	0
F Eggs	162	12.8	2.7	0.3	1,140
G Cheese	368	21.5	0.5	0.4	1,240

Find the least-cost food mix of the daily menu that satisfies minimum daily requirements.

a. Draw an influence diagram and set up the problem as a linear program.
b. Solve the problem: Give the quantities of the foods to be included and the total daily cost.
c. In what nutritional elements will you have a surplus? In what quantities?
d. The customers are unhappy with the proposed menu solution because it does not include beef and includes too much bread and eggs, so the following requirements are imposed:
   (1) The menu must include at least 100 grams of beef.
   (2) The menu should not include more than 500 grams of bread and 200 grams of eggs.
   Resolve the problem and find the optimal menu now. What is the additional cost, and in what nutritional elements will there be a surplus? (How much?)

11. A pharmaceutical firm produces two grades of mouthwash for the retail market. The firm must supply 600 gallons of green mouthwash and 200 gallons of blue mouthwash a day. It uses two vats whose capacities are 25 (Vat A) and 50 (Vat B) gallons to produce the mouthwashes. It takes an hour to produce a batch in Vat A, costing $30/hour, and 1.5 hours per batch in Vat B, which costs $35/hour. The company works a two-shift day (16 hours). Green mouthwash sells for $5/gallon and must include at least 45 percent of liquid "p" (which costs $2/gallon) and not more than 25 percent of specially distilled water (cost of $0.50/gallon). Blue mouthwash sells for $4/gallon and must include at least 25 percent of liquid "p" and not more than 50 percent water. The remaining ingredients consist of special fillers that cost $1/gallon. Draw an influence diagram and find the best daily schedule to maximize profits, considering the batching requirements.
12. Western Swiss Machine shop makes deluxe and regular skis on a weekly schedule for the area skiing enthusiasts. They have a contract with Apple Hill to supply 18 regular pairs of skis per week. They also sell both regular and deluxe skis to local sporting goods stores. A deluxe pair of skis requires 40 minutes for roughing and 20 minutes for finishing, whereas a regular pair of skis requires 20 minutes for roughing and 26.6667 minutes for finishing. With only 1,000 minutes for roughing and 800 minutes of finishing time available per week and a profit realization of 4 SFr (Swiss francs) and 3 SFr for the deluxe and regular, respectively, what weekly mix of the two types of skis should be produced to meet the contract's requirement and maximize profits?
    a. Draw the influence diagram and formulate an LP.
    b. Solve with Excel.
13. The Swiss Construction Company is building roads on the side of the Alps. It is necessary to use explosives to blow up the underground boulders to make the surface level. There are three ingredients (A, B, C) in the explosive used. It is known that at least 10 grams of the explosive must be used to get results. If more than 20 grams are used, the explosion will be too damaging. Also, for an explosion, at least 1/4 gram of ingredient C must be used for every gram of ingredient A, and at least 1 gram of ingredient B must be used for every gram of ingredient C. The costs of ingredients A, B, and C are 6 Swiss francs (SFr), 18 SFr, and 20 SFr per gram, respectively. Draw an influence diagram to find the least-cost explosive mix necessary to produce a safe explosion.
14. Westcan Corporation is considering producing five different types of special computers that yield the following unit profits:

Type	A	B	C	D	E
Net profit ($000)	16	8	11	6	10

The company has $1 million capital to invest in production and a capability of 10,000 working days. The capital and labor requirements for each type of computer are given below.

Type	Required Capital per Unit ($000)	Required per Unit Working Days
A	20	200
B	15	120
C	16	150
D	10	80
E	14	100

Find the best production plan by considering two different influence diagrams:
    a. If the company's objective is *profit maximization.*
    b. If the company's objective is to produce the *maximum number of total units* (of all types combined).
15. Glades Discount Store is opening a new department with a storage area of 10,000 square feet. Management considers four products for display:
Product A: Costs $55, sells for $80, and requires 24 square feet per unit for storage.
Product B: Costs $100, sells for $130, and requires 20 square feet per unit for storage.
Product C: Costs $200, sells for $295, and requires 36 square feet per unit for storage.
Product D: Costs $300, sells for $399, and requires 50 square feet per unit for storage.
At least 10 units of each product must be displayed. The company's objective is profit maximization. Draw an influence diagram and find out how many units of each product should be on display if the company has $600,000 available for purchasing the products.
16. Paint Fair Company advertises its weekly sales in newspapers, television, and radio. Each dollar spent on advertising in newspapers is estimated to reach an exposure of 12 buying customers, each

dollar in TV reaches an exposure of 15 buying customers, and each dollar in radio reaches an exposure of 10 buying customers. The company has an agreement with all three media services that it will spend not less than 20 percent of its total money in each medium. Further, it is agreed that the combined newspaper and television budget will not be larger than three times the radio budget. The company has decided to spend no more than $17,000 on advertising. Construct an influence diagram. How much should the company budget for each medium if it is interested in reaching as many buying customers as possible?

17. Venezuela-Grand Hospital has recently modernized one of its operating rooms so that it now contains the very latest in a variety of equipment. Only two types of operations are performed. The hospital desires to schedule as many operations as possible in this room, which can handle up to 12 operations of type A, or 30 operations of type B (or any linear combination) per day. Each A-type operation requires 4 pints of blood, and each B-type operation requires 5.5 pints of blood. The hospital has a blood inventory of 100 pints. At least 7 type A operations and no more than 20 type B operations should be performed. Find the best possible schedule.
    a. Draw the influence diagram.
    b. Set up the problem as an integer linear program.
    c. Solve.

18. Products A, B, and C are to be made on three machines. Net profits per unit of A, B, and C, respectively, are $21, $26, and $22. Each product is processed by three different machines. Processing time per unit of production and data on machine availability are shown in the table below.

Machine	Product A	Product B	Product C	Machine Availability (Minutes per 2-Week Period)
1	273	221	374	9,282
2	273	442	187	9,282
3	91	182	159	4,732

Assuming no set-up requirements, draw the influence diagram and find the best product mix (i.e., what products should be produced and in what quantities) if the production schedule must meet all-integer constraints (i.e., no fractions of products can be produced).

19. The research department of ABC is selecting projects for the next two years. Seven proposed projects are to be evaluated. The yearly cost of each project in labor-hours (LH) required and the available labor-hours are given in the following table. Also, the expected profits (discounted to time zero) are given. The research department wants to maximize its profits. Find the projects they should select.
    a. Construct the influence diagram.
    b. Set up as a mixed-integer programming problem.
    c. Solve.

Project	Labor-Hours Required: 1st Year	Labor-Hours Required: 2nd Year	Discounted Expected Profits ($1000s)
Available LH/year	10,000	12,000	
A	1,000	4,000	120
B	1,200	2,000	100
C	1,800	1,600	80
D	2,000	2,400	140
E	1,200	1,800	100
F	2,600	2,000	160
G	2,200	2,200	140

20. Four different processes are available for producing paint. The processing cost of each gallon in any of the four available processes, the maximum capacity of each process, and the set-up costs are given in the following table. Assume that a daily demand of 35,000 gallons must be supplied. Find the best processing schedule (minimize total costs). Base your solution on an elapsed time of one day.

Process	Set-up Cost, Dollars	Processing Cost, Cents per Gallon	Maximum Capacity, Gallons
A	500	6	20,000
B	600	5	15,000
C	1,000	4	40,000
D	600	3	25,000

a. Construct an influence diagram to help you formulate the problem as an integer-programming problem.
b. Which processes should be used, and to what extent, in order to minimize the total cost?

21. A commodities wholesaler maintains a 400-ton capacity warehouse to buy, sell, and store cocoa beans. Cocoa purchased in one month cannot be sold until the first day of the following month. The wholesaler's current stock is 200 tons of beans, and the cost of storage is $30/ton/month. The wholesaler believes that the buy and sell prices for cocoa beans for the next four months will be about July 900 (buy) and 1100 (sell), August 800/1200, September 1000/900, October 1100/1200. Draw an influence diagram to help formulate a buying and selling plan for the wholesaler to maximize her profits. Solve.
22. A French market research firm wishes to conduct home visit interviews according to the quotas specified in the following table:

Type of Household	Desired No. of Responding Calls (Quota)
Single person	50
Married, no children	100
Married, with children	150

Because not all persons are at home at the time of the visit and not all persons cooperate, there is only a certain "probability of response" to the home visits (calls). This probability is shown in the following table.

Type of Household	Probability of Response to Calls		
	Morning	Afternoon	Evening
Single person	0.1	0.1	0.5
Married, no children	0.5	0.4	0.7
Married, with children	0.75	0.6	0.9

Thus, the following requirements exist:

a. The total number (responding and nonresponding) of planned morning calls must not exceed the total number of afternoon calls.
b. The total number of responding evening calls must be at least 20 percent and no more than 30 percent of the total number of all responding calls.
c. An evening call costs twice as much as a morning or an afternoon call.

Draw an influence diagram to help decide how the calls should be distributed among the three types of households (by the three times during the day) such that the desired quotas are fulfilled at minimum cost. Construct an influence diagram and formulate the model but do not solve.

23. The ABC Corporation produces two types of paint, I and II, which it sells at $2.20 and $1.80 per gallon, respectively. Three different raw materials can be used to make the paint. Raw material 1 costs $1 per gallon and contains 70 percent ingredient A, 20 percent ingredient B, and 10 percent ingredient C. Raw material 2 costs $1.50 per gallon and contains 30 percent, 40 percent, and 30 percent of ingredients A, B, and C, respectively. Raw material 3 costs $0.80 per gallon and contains 50 percent ingredient A and 50 percent ingredient C. The product specification of paint type I calls for at least 40 percent each of ingredients A and C. Paint type II requires at least 30 percent of ingredient B, no more than 50 percent of ingredient C, and exactly 10 percent of A.
    a. Management would like to know the relative amount of raw materials (in percent) to be used in regular production so as to maximize total profit. Construct an influence diagram and formulate the situation, but do not solve.
    b. Separate the problem into two independent problems and explain why this is possible in this case.
    c. Formulate the same problem with the assumption that the company cannot use raw material 3 at more than twice the rate of raw material 2.
    d. How would the problem change if 5 percent of ingredient A is wasted in the production process?
24. Sunoil sells two types of Gasohol: regular and premium. Gasohol is prepared by mixing gasoline and alcohol. Sunoil can buy up to 100,000 barrels per week of gasoline at $40 per barrel and up to 12,000 barrels per week of alcohol at $47 per barrel. Regular Gasohol is made by blending nine portions of gasoline with one portion of alcohol, whereas premium Gasohol is made of 87 percent gasoline and 13 percent alcohol. Each barrel is equivalent to 42 gallons. A gallon of regular Gasohol sells for $1.35 and a gallon of premium Gasohol sells for $1.43. The company must provide at least 600,000 gallons of premium and 1 million gallons of regular each week. In addition, the amount of regular Gasohol produced must be at least twice the amount of premium produced.

    Draw an influence diagram to help find the most profitable production plan. *Note:* This problem can be formulated with either two or four decision variables. Find:
    a. The quantities of gasoline and alcohol used.
    b. The quantities of regular and premium produced.
    c. The total cost.
    d. The total profit.

CASES

The Daphne Jewelry Company*

The Daphne Jewelry Company markets the bulk of its products through seven salespersons operating in seven separate sales territories (on a one-to-one basis). This is due to the fact that other area salespersons use Daphne's products only as a supplement to some other distributor's line. The seven salespersons follow just the opposite practice, using the products of other manufacturers to augment the Daphne line. For this reason the firm sets periodic sales quotas only for the seven salespersons.

The firm sells nine product lines. These are listed in Table A, where each column shows how $1 in sales in each of the seven territories is distributed among the various product lines. For example, the 0.07 coefficient for the first product, belts, indicates that on average, 7 cents of every dollar's worth of merchandise sold in territory 1 is generated by belts. These distributions were found to be quite stable over time, regardless of the size of the account.

The market potential of each sales territory for the next planning period is presented in Table B. This is

*Case developed by Dr. Malcolm Golden, University of Miami, and Dr. Alan Parker, Florida International University.

TABLE A Dollar Distribution Value of Sales for Nine Products in Seven Territories

	Sales Territory						
Product Lines	**No. 1**	**No. 2**	**No. 3**	**No. 4**	**No. 5**	**No. 6**	**No. 7**
1. Belts	0.07	0.02	0.01	0.15	0.18	0.15	0.00
2. Buckles	0.05	0.00	0.00	0.10	0.10	0.07	0.00
3. Package goods	0.20	0.35	0.30	0.25	0.25	0.25	0.50
4. Necklaces	0.07	0.07	0.07	0.10	0.15	0.10	0.03
5. Earrings	0.15	0.15	0.15	0.15	0.15	0.15	0.15
6. Bracelets	0.10	0.20	0.10	0.10	0.10	0.10	0.05
7. Gold stone	0.18	0.10	0.17	0.10	0.05	0.05	0.12
8. Hematite	0.15	0.08	0.17	0.02	0.02	0.10	0.12
9. Job turquoise	0.03	0.03	0.03	0.03	0.00	0.03	0.03
Total	1.00	1.00	1.00	1.00	1.00	1.00	1.00

TABLE B Daphne's Market Potential in Each of Seven Selling Areas

Sales Territory	Market Potential (Maximum)
No. 1	$ 225,000
No. 2	135,000
No. 3	150,000
No. 4	100,000
No. 5	210,000
No. 6	80,000
No. 7	250,000
Total	$1,150,000

TABLE C Material Costs and Sales Commission for the Nine Product Lines

Product Lines	Cost of $1 in Merchandise	Sales Commission on $1 in Merchandise
1–6 inclusive	$0.50	$0.15
7–9 inclusive	0.67	0.10

TABLE D Product Line Capacity

Product Line	Product Line Capacity
1	$ 70,000
2	20,000
3	210,000
4	70,000
5	150,000
6	100,000
7	150,000
8	150,000
9	30,000

Daphne's estimate of "potential" demand for its products next year for each of the seven territories at the present level of advertising. These demand forecasts were based on past sales records, information gathered from trade associations and governmental agencies, as well as independent forecasts made by consulting firms that specialize in economic analysis of trade areas.

The cost of a dollar's worth of merchandise required in the production of each product line, together with the corresponding sales commission paid on each dollar of sales, is depicted in Table C. The production capacity of the nine production lines is given in Table D.

During recent years, the company's sales and profits have been growing very slowly. Last year, the company netted about $200,000 on sales of about $650,000. Mr. Brown, the president of Daphne Jewelry, was not pleased with the results. He felt there was a large quantity of unutilized production capacity as well as market potential. Mrs. Grant, the vice president of marketing, disagreed with Mr. Brown's assessment. She felt that the company was at or near optimal operating conditions, and that very little could be done within the framework of the existing conditions.

Last Monday, the president called the executive management team together and requested proposals for improving the situation. The vice president for marketing suggested an increase in the marketing efforts, especially in territories 2, 4, and 6, where current market potential is lowest. The vice president for production suggested increasing production of those product lines that yield the highest return. The controller suggested dropping the least profitable products or territories, or both. The president was reluctant to accept any of these suggestions, because both the market potential and the production capacity were underutilized. Furthermore, the specific marketing plan proposed violated the production capabilities, and the proposed increase in certain product lines violated the marketing capabilities.

The president finally decided to call in a management scientist, who was asked to prepare a report to include the following items:

a. Evaluate the existing situation; determine if the company is indeed close to optimal operating conditions.
b. Analyze the marketing and production proposals brought forth by the vice presidents and the controller.
c. Submit other proposals; determine their feasibility and profitability.
d. Analyze the pricing and commission policies; submit recommendations.

Hensley Valve Corp. (A)

It was Monday morning and the weekly meeting of the executive committee was in full swing. There were two primary items on the agenda, and both directly affected Henley's profit margin.

Agenda Item A: The Proposed Tax Increase on Diesel Fuel

Agenda Item B: Record Interest Rates

"Gus, as regional manager in our largest selling region, what will be the result of our raising prices to offset this possible increase in our trucking costs?"

"Well J. B., it certainly won't help our sales effort. Valve JBH-1 is only marginally profitable now, but takes twice the time to sell as our more profitable JBH-2. I'm afraid a price increase might wipe out the viability of our JBH-1 valves altogether."

"That's what I was afraid of, too. Pat, how about the effect of that cancellation of the NC machine we were hoping would help our productivity? I know you were counting on that to increase our output rate, but with the last two big jumps in the prime rate we simply can't afford it at this time."

"Yes, I realize that, J.B. I certainly hadn't expected the prime rate to go quite this high. Basically, our output rates will remain limited, especially on old line 3, which produces the JBH-1 and -2 valves. Given the limited floor space and equipment, and working three shifts on this line, we can still produce at most 600 JBH-1s or 100 JBH-2s a week, or any combination in between. I wondered if it would be worthwhile to have Tim Moran in our controller's office look at the interacting effect of all these changes? It seems that because so many things are happening at once, it may be best to totally change our product mix as well as our prices."

"I agree, Pat. It seems appropriate to undertake a complete contingency analysis of what we should do given any specific change in the market or combination of changes. I'll work up a memo to Tim this afternoon."

MEMO

To: Tim Moran, Controller's Assistant
From: J. B. Hensley, President
Subject: Reanalysis of Product Line

Please undertake a review of our JBH-1 and -2 valves for the next meeting of the executive committee on Monday morning. For this purpose you may assume their profitability to be $10 and $40 each, respectively. We have figured that a JBH-1 takes, on average, 4 hours to sell and a -2 takes 2 hours to sell. Sales has at most 1,000 hours a week available. Check with Pat Johnson for production figures on line 3. Items we would specifically like to know include:

Given our limited capacities, how many of each valve should we currently be producing and selling to maximize our profits?

What is an extra hour of sales time worth?

What is an increase in the capacity of line 3 worth?

What will be the effect on the solution of improving the JBH-1 marketing effort so that it only takes 2.5 hours to sell a unit?

Please add any other information you find relevant. Thank you.

Hensley Valve Corp. (B)

How does the all-integer solution compare to the general LP solution? What does this suggest to you about simply rounding the solution of the general LP model to obtain an integer solution?

BIBLIOGRAPHY

Bertsimas, D., and R. M. Freund. *Data, Models, and Decision: The Fundamentals of Management Science.* Cincinnati, OH: South-Western, 2000.

Bertsimas, D., C. Darnell, and R. Soucy. Portfolio Construction Through Mixed-Integer Programming at Grantham, Mayo, Van Otterloo and Company. *Interfaces,* 29:1, 1999, 49–66.

Camm, J. D., and J. R. Evans. *Management Science and Decision Technology,* Cincinnati, OH: South-Western, 2000.

Fylstra, D., L. Lasdon, J. Watson, and A. Warren. Design and Use of the Microsoft Solver. *Interfaces,* 28:5, 1998, 29–55.

Katok, E., and D. Ott. Using Mixed-Integer Programming to Reduce Label Changes in the Coors Aluminum Can Plant. *Interfaces,* 30:2, 2000, 1–12.

Knowles, T. W., and J. P. Wirick, Jr. The Peoples Gas Light and Coke Company Plans Gas Supply. *Interfaces,* 28:5, 1998, 1–12.

Nemhauser, G. L., and L. A. Wolsey. *Integer and Combinatorial Optimization.* 2nd ed. New York: Wiley, 1999.

Ragsdale, C. T. *Spreadsheet Modeling and Decision Analysis.* 3rd ed. Cincinnati, OH: South-Western, 2001.

Sears, B. *Enter the Zone.* New York: Harper Collins, 1995.

Sonntag, C., and T. A. Grossman, Jr. End-user Modeling Improves R&D Management at AgroEvo Canada, Inc. *Interfaces,* 29:5, 1999, 132–142.

Shafer, S. M. A Spreadsheet Approach to Aggregate Scheduling. *Production and Inventory Management Journal,* 32:4, 1991, 4–10.

Turban, E., and J. R. Meredith. *Fundamentals of Management Science.* 6th ed. Burr Ridge, IL: Irwin, 1994.

Wallace, S. W. Decision Making Under Uncertainty: Is Sensitivity Analysis of Any Use? *Operations Research,* 48:1, 2000, 20–25.

Westrich, B. J., M. A. Altmann, and S. J. Potthoff. Minnesota's Nutrition Coordinating Center Uses Mathematical Optimization to Estimate Food Nutrient Values. *Interfaces,* 28:5, 1998, 86–99.

Williams, H. *Model Building in Mathematical Programming,* 4th ed., New York: Wiley, 1999.

Winston, W. L. and S. C. Albright. *Practical Management Science,* Pacific Grove, CA: Duxbury, 2001.

Wolsey, L. A. *Integer Programming,* New York: Wiley, 1998.

Decision Analysis

5

INTRODUCTION

Irrational exuberance! This is precisely the way Federal Reserve Chairman Alan Greenspan described the skyrocketing stock market at the end of the 1990s. Take the example of Amazon.com. Between September 1999 and September 2000, Amazon's stock price fluctuated between $28 per share to $113 per share (it was selling at around $15 a share in May 2001, following the huge tech sell-off that began in early 2000). From September 1999 to September 2000, Amazon's market value (or market capitalization) fluctuated between $10 billion and $40 billion, based on the assumption that Amazon had approximately 356 million shares outstanding during this period.[1] In other words, Amazon's value changed by $30 billion (with a b) in a single year! Further, given that this is just one of numerous examples that could have been cited, it would certainly appear difficult to make a case that the markets are behaving rationally.

However, in a recent article published in *The McKinsey Quarterly*,[2] the authors combine traditional discounted cash flow analysis with decision theory to demonstrate that the huge gyrations many dot-com stocks have exhibited may indeed be the result of rational decision making. To demonstrate their approach, the authors generated a variety of possible scenarios for Amazon.com by varying the parameters shown in the influence diagram in Exhibit 5.1. The impact these changes had on the value of the discounted cash flows and thereby Amazon's market capitalization

EXHIBIT 5.1 Influence Diagram of Factors Driving Market Value of Amazon.com

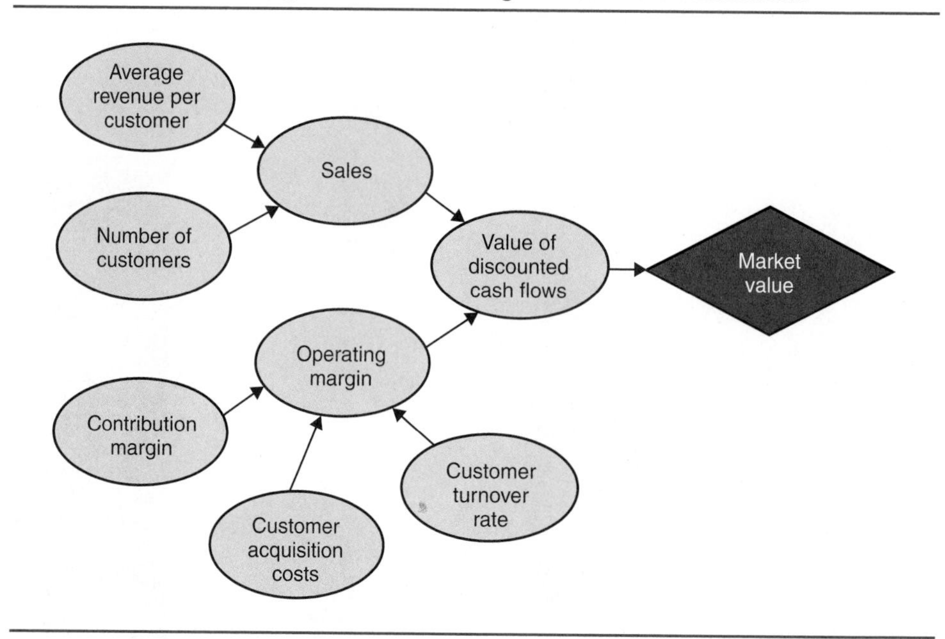

was then calculated. Next, the probabilities for each scenario were estimated and the probability-weighted average value across the scenarios calculated. (*Note:* the process of calculating probability-weighted averages is discussed later in this chapter.) The article clearly demonstrated that changes in the subjective probability estimates for the various scenarios had a significant impact on Amazon's market value. Further, given the extremely dynamic nature of Web-based businesses, it is not hard to imagine why individual investors and professional analysts alike would tend to change their subjective estimates of the likelihood of alternative scenarios from one day to the next.

Returning to *The McKinsey Quarterly* article, the authors identified four scenarios (labeled A through D) and calculated the discounted cash flows for each scenario as $73 billion, $37 billion, $15 billion, and $3 billion, respectively. The authors then showed that if Scenario A was expected to have no chance of occurring, Scenario B a 25 percent chance, Scenario C 35 percent, and Scenario D 40 percent, then Amazon's market value based on a probability-weighted average of the discounted cash flows would be $16 billion. Revising the probabilities of Scenarios A through D to 5 percent, 35 percent, 35 percent, and 25 percent, respectively resulted in a market value of $23 billion. As a further example, if investors became even more optimistic and changed the probabilities to 10 percent, 50 percent, 35 percent, and 5 percent for Scenarios A through D, respectively, Amazon's market value increased to $32 billion. This example clearly demonstrates that simple and perhaps even rationale adjustments to the probabilities of likely scenarios can help explain widely fluctuating stock prices. Furthermore, one would expect that the frequency and magnitude of changes in subjective probability estimates would increase in proportion to the amount of uncertainty inherent in the environment.

5.1 The Modeling Process for Decision Analysis Studies

Decision analysis is a combination of a descriptive and prescriptive business modeling approach. Although it simply depicts the expected result of a situation facing a decision maker when alternative situations, managerial actions, and outcomes are compared, the best action can be determined, at least among those actions contemplated. (*Note:* In contrast, optimization does not require the specification of the alternative actions; it simply finds the best among the perhaps infinite number of possibilities.)

Whether to invest in Amazon.com is a typical managerial decision problem. Investors, the decision makers in this case, must make a *choice* among several courses of action. They must choose not only how much to invest in a particular stock, but also the timing of such investments. The difficulty is that there is *uncertainty* with respect to what is going to happen in the future. No matter which choice the investors make, they are going to assume some *risk* that the future they had hoped for is not going to materialize. What is the degree of risk that they are assuming? How does it relate to the available alternatives? Can the risk be reduced? Can the risk be eliminated? **Decision analysis** (or sometimes, *decision theory*), a quantitative modeling procedure applied specifically to managerial decision making, attempts to answer such questions.

The Modeling Process

The modeling process for decision analysis follows the process described in Chapter 1 quite closely. Yet, there are a few special points worth noting.

Step 1: Opportunity/Problem Recognition The opportunity/problem recognition step is normally quite direct, with management facing a complex decision situation. This may be due to a recent change or problem arising that wasn't expected, an evolving situation where a decision was anticipated, or a sudden opportunity that has arisen and management is considering taking advantage of.

Step 2: Model Formulation Model formulation for decision analysis is based on relatively simple tabular arrangements of the data, or equally understandable graphical "tree" structures. Of all the quantitative modeling techniques, decision analysis may be the easiest to understand because it primarily involves a somewhat graphical depiction of the situation and the calculation of simple expected values. The only complexity arises when Bayesian statistics is invoked for the imperfect information case. Most of the chapter will be concerned with formulating and analyzing various types of decision models.

Step 3: Data Collection The data collection is usually straightforward and not unlike our previous descriptions concerning data collection. However, we need to discuss two aspects to data collection for decision analysis. One is that in decision analysis, we are not usually working in a certainty situation, as is assumed in optimization studies. That is, we only have probabilistic information on the situations—the possible future states—we are considering, and thus the outcomes are uncertain. And in other situations we may not even have these probabilities, thereby requiring guidelines to follow when the likelihood of the various future situations is completely unknown.

The second issue is when additional information regarding the possible outcomes is desired. In this case, an outside service to obtain this information may be needed, incurring additional cost. Decision analysis can determine what this additional information should be worth, even if it isn't perfectly credible.

Step 4: Model Analysis Model analysis is relatively straightforward unless more sophisticated statistical models involving conditional probabilities are required. As noted earlier, model formulation and analysis will be covered in detail in the remainder of the chapter.

Step 5: Implementation Implementation of decision analysis studies is not usually a serious difficulty since management is generally closely involved in formulating the alternative actions that may be taken, the possible future states of nature, and possibly even the possible outcomes that may occur. In reality, most such decisions are repetitive and must be reconsidered in a later time period anyway, so the issue of a *permanent* solution is not a factor. For example, investors may decide to wait some period of time, say six months, and then reconsider whether to invest in Amazon.com. Or if they decide to invest now, to some degree, they may want to revisit the decision in six months to decide whether to sell out, hold, or increase the amount invested. Reconsideration of the issue keeps the modeler in relatively continuous touch with management, further aiding successful implementation.

It is worth noting that the situation being analyzed by the modeler typically only represents a portion of the actual situation. The real situation may involve personnel issues, regulatory issues, and a host of other issues that don't have a direct cost or profit impact but are still important to the organization—possibly even more important. What managers may really want is a baseline set of values telling them approximately what each alternative is going to cost them if they don't choose the cost or profit "optimum" alternative. For example, if going with the second alternative is expected to cost the organization $100,000 but increase the morale of the employees and improve the quality of their performance, management might decide that the cost is worth the benefit obtained. However, if the cost was half a million, management might decide the cost was too high.

Structure of the Chapter

We start our discussion with a more illustrative example of the decision analysis situation to explain the various elements of decision analysis such as the states of nature and the payoffs, as well as the common decision table used to evaluate the decision situation and come to a conclusion. There are three different kinds of decision situations managers may face—certainty, risk, and uncertainty—and we describe each of these and the approaches used to address them. For help in visualizing a decision situation, and particularly for multiperiod, sequential decisions, the decision tree has become a well-known and managerially favored decision tool, and we describe it in detail. Last, we show how to determine the value of obtaining additional information, including imperfect information such as might be found by a marketing research firm, when making a managerial decision.

5.2 The Decision Analysis Situation

Decision analysis is a numerical, enumerative technique usually concerned with a static, one-shot decision situation. However, it can be extended to dynamic situations involving sequential decisions over multiple time periods. Also, it can take either an optimizing, prescriptive orientation (by identifying the *best* of all the alternatives) or a descriptive orientation (by simply identifying the possible outcomes, and the likelihood of each, of making certain decisions). Next, we pose a simple, single-period opportunity to illustrate the various tools of decision analysis. For consistency, we will maintain a financial context.

Mary's Dilemma

"The Dow Jones Industrial Average lost 437 points today," the computer screen calmly proclaimed. Mary Golden, vice president of Friendly Trust Company, read the story on the screen over and over again. Analysts were citing fears of inflation and climbing interest rates as the main reasons behind the market weakness. Mary moaned to herself, "This is the sixth time in the last four months that the Dow has lost more than 400 points. How am I supposed to make decisions in this environment?" She was still in shock when she scrolled down to another disappointing message: "Nation's Bank raised its prime rate by 1/4 of 1 percent." This was too much for one day.

Mary was in charge of Friendly Trust's investment department. She had just been authorized to invest a large sum of money in one (and only one) of three alternatives: corporate bonds, common stocks, or certificates of deposit (time deposits).

Friendly Trust's objective was to maximize the yield on the investment over a one-year period. The problem was that the economic situation seemed to be uncertain, and no one was able to predict the movements of the stock or bond markets. It was obvious to Mary that the yields (in percent of return on investment) depended on the future state of the economy. Therefore, she consulted the firm's economic research department. The researchers were not sure what the state of the economy would be after one year, but they told Mary they expected the economy to be in one of three possible states: solid growth, stagnation, or inflation. When asked for the likelihood of each condition, the researchers estimated a 50 percent chance for solid growth, a 30 percent chance for stagnation, and a 20 percent chance for inflation.

Mary examined the relationship between the yields on the possible investments and the state of the economy and noted the following trends:

1. If there is solid growth in the economy, bonds will yield 12 percent, stocks 15 percent, and time deposits 6.5 percent.
2. If stagnation prevails, bonds will yield 6 percent, stocks 3 percent, and time deposits 6.5 percent.
3. If inflation prevails, bonds will yield 3 percent, the value of stocks will drop 2 percent, and time deposits will yield 6.5 percent.

Mary examined all the above information and realized that the investment decision would not be simple at all.

The Structure of Decision Tables

The quantitative data of many decision situations can be arranged in a standardized tabular form known as a **decision table** (or a *payoff table*). The objective of doing so is to enable a systematic analysis of the problem. Although not all decision situations are explicitly amenable to a tabular presentation, many concepts used in decision tables are common to all decision situations.

Decision tables typically contain four elements:

1. The alternative courses of action or options
2. The states of nature or scenarios
3. The probabilities of the states of nature
4. The payoffs

The Alternative Courses of Action Decision making, by definition, involves two or more options or **alternative courses of action,** also called *strategies.* One, *and only one,* of these alternatives must be selected. The alternative courses of action are designated as $a_1, a_2, \ldots, a_n$ (sometimes $d_1, d_2, \ldots, d_n$), where n is the number of available alternatives that *may* be either finite or infinite. For example, the decision to select a textbook for a particular class may involve numerous but finite alternatives. If, however, one were producing beer, the quantity of water to add to the mix may include, at least in theory, an

EXHIBIT 5.2 Decision Table (Payoffs in Percentage Yield)

	A	B	C	D
1		States of Nature		
2		Solid Growth (s_1)	Stagnation (s_2)	Inflation (s_3)
3	Probabilities	0.5	0.3	0.2
4	a_1 Bonds	12	6	3
5	a_2 Stocks	15	3	-2
6	a_3 Time deposits (CDs)	6.5	6.5	6.5

Decision variables → (rows 4–6); Uncontrollable ← (row 2); Payoffs ← (rows 4–6)

infinite number of combinations. For example, one can add 2.1 gallons, 2.11 gallons, 2.111 gallons, and so on. In this text, the alternatives are listed on the left side of the table, one to a row. For example, see Exhibit 5.2, which is based on Mary's situation.

In most operating circumstances, not all possible alternatives are considered, but only those within a limited range. For example, in beer making, rules of good brewing establish that the water added ought to be within the range of, say, 7 to 7.5 gallons to a 10-gallon barrel. This is the range of *feasible* solutions. It still leaves an infinite number of points between 7 and 7.5, but one might decide to structure the alternatives in only tenths of gallons, thus reducing the number to only five (7.1, 7.2, 7.3, 7.4, 7.5).

Decision tables are used when the number of alternatives is *finite* and usually small (e.g., less than a few hundred). In the investment problem, there are three alternatives. The ability to generate alternatives depends on the creativity and imagination of the manager. A creative manager sees more alternatives than does a conservative one. For example, if Mary wants to consider investing in both stocks and bonds simultaneously, she can create more alternatives (e.g., a_4: 50 percent stocks, 50 percent bonds; or a_5: 30 percent stocks, 70 percent bonds).

The States of Nature At the top of the table, the possible **states of nature** (also called *events, possible futures,* or *scenarios*) are listed. They are generally labeled s_1, s_2, . . . , s_m. A state of nature can be a state of the economy (e.g., inflation), a weather condition, a political development, or other situation that the decision maker cannot control. In the investment example, three states of nature are the possible states of the economy: solid growth, stagnation, and inflation. The states of nature are usually not determined by the action of a single individual or an organization. They are basically the result of an "act of God," or the result of many forces pushing in various directions.

Frequently, it is useful to identify the most relevant states of nature through some technique such as *scenario analysis.* That is, there may be many, many possible states of nature, some of which may not have a serious impact on the decision, and others of which could be quite serious. In scenario analysis, various knowledgeable parties are interviewed—stakeholders, long-time managers, people experienced with similar situations—to determine the most relevant states of nature to the decision. Some of these states may be important on the downside and others on the upside. Such states need not necessarily come from people—the project or decision plan, the history of the industry, or other sources may lead the modeler to some conclusions about the most relevant states of nature to consider in the study.

The Probabilities of the States of Nature What is the likelihood of these states of nature occurring? Whenever it is possible to answer this question in terms of explicit chances (or probabilities), the information is recorded at the top of the table. The probabilities are given either in percent or in percentage fractions; for example, 50 percent = 0.5.

Because it is assumed that one *and only one* of the given states of nature will occur in the future, then the sum of the probabilities must always be one. This is expressed as

$$p_1 + p_2 + \cdot \cdot \cdot + p_m = 1$$

or

$$\sum_{j=1}^{m} p_j = 1$$

where p_1 = probability of s_1 occurring, p_2 = probability of s_2 occurring, and so on. In Mary's example, $p_1 = 0.5$, $p_2 = 0.3$, and $p_3 = 0.2$. The subscript m designates the fact that m states of nature are considered.

The Payoffs The **payoff** (or the *outcome*) associated with a certain alternative and a specific state is given in that cell within the body of the table located at the intersection of the alternative in question (given by a row) and the specific state of nature (given by a column). The payoff is designated by o_{ij} where i indicates the row and j the column. For example, in Exhibit 5.2, if the decision maker selects alternative a_1 and future s_2 occurs, then the outcome of the decision is predicted as a payoff (yield) of 6 percent. The payoffs can be thought of as *conditional* because a specific payoff results from a specific state of nature occurring but only after a certain alternative course of action has been taken. An important point to remember is that the payoff is measured within a *specified* period (e.g., after one year). This period is sometimes called the **decision horizon.** Alternatively, a payoff can be the *present value* of several payoffs realized at different times in the future as was the case in the Amazon.com example at the beginning of the chapter.

Payoffs can be measured in terms of money, market share, or other measures. The payoffs considered in most decisions are monetary. However, other consequences also will occur and should be taken into account once the quantitative answer is derived.

Exhibit 5.3 shows the general structure of decision tables. In terms of the elements of mathematical models, the decision table can be described in this way:

Decision Table	Mathematical Model
Alternative courses of action	= Independent decision variables
States of nature	= Uncontrollable parameters
Probabilities of the states of nature	= Uncontrollable parameters
Payoffs	= Expected results (dependent variables)

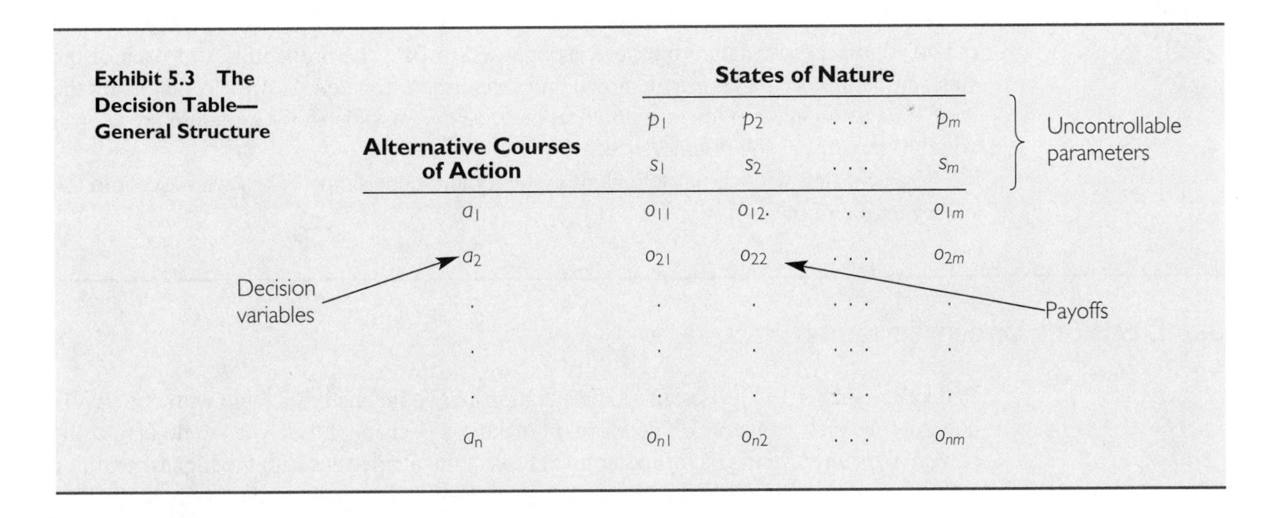
Exhibit 5.3 The Decision Table—General Structure

Alternative Courses of Action	States of Nature			
	p_1	p_2	. . .	p_m
	s_1	s_2	. . .	s_m
a_1	o_{11}	o_{12}	. . .	o_{1m}
a_2	o_{21}	o_{22}	. . .	o_{2m}
.	.	.	. . .	.
.	.	.	. . .	.
.	.	.	. . .	.
a_n	o_{n1}	o_{n2}	. . .	o_{nm}

EXHIBIT 5.4 The Zones of Decision Making

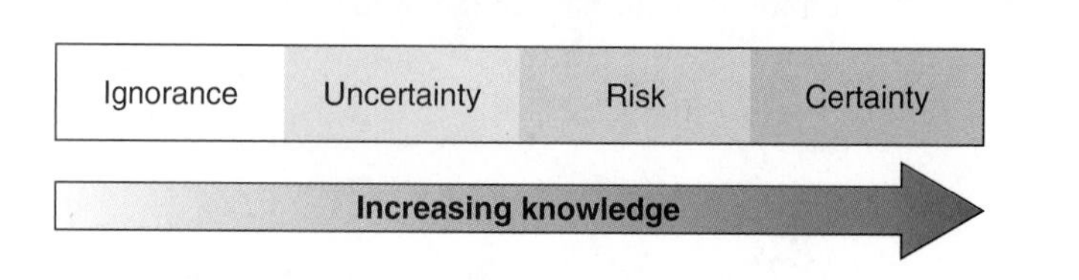

Classification of Decision Situations

Decision situations are frequently classified on the basis of what the decision maker knows or believes about the situation. It is customary to divide the degree of knowledge into four categories, as shown in Exhibit 5.4. Complete knowledge, or *certainty,* is on the far left and complete ignorance is on the far right. Between the two are *risk* and *uncertainty.* Decision analysis considers the three categories of decision making under certainty, risk, and uncertainty.

Certainty When making decisions under **certainty,** it is assumed that complete information is available so that the decision maker knows exactly what the outcome of each course of action will be. Such situations are termed *deterministic* and are approximated most often when the time horizon is very short. A substantial amount of knowledge and understanding of the behavior of the system under consideration is required to assume a state of certainty. The decision table corresponding to certainty is composed of a single state of nature column since it is assumed that only one state of nature will occur. Thus, there is only one possible payoff for each decision alternative.

Risk A decision under **risk,** also known as a *probabilistic decision,* involves two or more possible payoffs, depending on which state of nature occurs. In decision under risk, we assume that the chance of occurrence of each of the states of nature is known or can be estimated. A classic example is gambling at a casino, where the chances of winning at roulette or blackjack are known or can be calculated. In placing a bet, the gambler knows the long-run probability of winning the bet and thus can assess the degree of risk being assumed. With risk, less information is available than under certainty because it is not definitely known which outcome will occur.

Uncertainty Decision making under **uncertainty** also involves two or more possible payoffs and states of nature but here, as opposed to risk, the probabilities of each of the states of nature occurring are unknown and cannot be estimated. However, this is not the case of decision making under ignorance, because the possible states of nature are *known,* which is not the case with ignorance.

We deal first with the situations of certainty and uncertainty. Then we move into the murky region of risk.

5.3 Decisions Under Certainty

In decision making under certainty, the situation can be mapped as a table with one payoff column (one state of nature). Therefore, in making a decision, all one has to do is to compare all the entries in the payoff column and select the alternative with the highest profit or lowest cost. In executing such a comparison, one distinguishes between two cases:

- When the number of alternatives is relatively small, an approach known as complete enumeration is used.
- When the number of alternatives is large or even infinite, a selective search for the best solution is conducted with the aid of mathematical models.

Complete Enumeration

Complete enumeration means examining every payoff, one at a time, comparing the payoffs to each other (e.g., in pairs), and discarding inferior solutions. The process continues until *all* payoffs are examined.

EXAMPLE

Assignment of Employees to Machines

A maintenance crew of three machinists is to be assigned to the repair of three machines, on a one-to-one basis, in a manner that minimizes repair time. Based on historical data, the supervisor knows the repair time for each machine, depending on the machinist-machine assignment. The repair times, in hours, are shown in Exhibit 5.5. (For example, if Jack works on machine A, it takes him 3 hours to fix it; the repair time is the payoff in this case.)

Solution. All alternative assignments, with their resultant total repair times, are shown in Exhibit 5.6. Note that the decision table has only one column of payoffs (one state of nature). By comparing the total repair times for *all* possibilities, it is found that alternative a_5 is the best, because the total repair time is the smallest. Larger assignment problems have an extremely large number of possible solutions and are therefore solved by algorithms rather than by enumeration.

EXHIBIT 5.5 Repair Times

	A	B	C	D
1			Machine	
2	**Machinist**	**A**	**B**	**C**
3	Jack	3	7	4
4	Gene	4	6	6
5	Melanie	3	8	5

EXHIBIT 5.6 Assignment Payoffs

Alternatives	Total Payoff (Total Repair Time)		
a_1 Jack—A, Gene—B, Melanie—C	3 + 6 + 5 =	14	
a_2 Jack—A, Gene—C, Melanie—B	3 + 6 + 8 =	17	
a_3 Jack—B, Gene—A, Melanie—C	7 + 4 + 5 =	16	
a_4 Jack—B, Gene—C, Melanie—A	7 + 6 + 3 =	16	
a_5 Jack—C, Gene—B, Melanie—A	4 + 6 + 3 =	13	← Best
a_6 Jack—C, Gene—A, Melanie—B	4 + 4 + 8 =	16	

Computation with Analytical Models

Although complete enumeration is an effective approach in many situations, especially when using computers, there are two cases where it either doesn't work at all or works very poorly. The first is the case of an *infinite number of alternatives.* Managerial problems such as allocation of infinitely divisible resources or blending liquid materials are examples of such situations. To cope with these problems, models such as linear programming have been developed.

The second case involves problems with a finite but very large, sometimes astronomical, number of alternatives, such as determining the best locations for a group of warehouses across Europe. Such problems are frequently referred to as **combinatorial problems.** Several scheduling and sequencing problems are of this nature. In these cases, it is possible to enumerate all the alternatives, but it may take years to do so, even with the aid of high-speed computers. Special models such as integer programming (Chapter 4) and others have been developed for these kinds of situations.

5.4 Decisions Under Uncertainty

In the condition of uncertainty, the decision maker recognizes different potential states of nature but cannot confidently estimate the probabilities of their occurrence. This is an undesirable but often unavoidable situation. It may occur when one faces a completely new phenomenon (e.g., the fall of the Soviet Union) or when a completely new product, process, or state of nature is under consideration. In many such situations, even prominent experts cannot agree on the chances of the various states of nature occurring. In the limiting case of uncertainty, called *ignorance,* even the possible states of nature are not known; we will not consider this case here.

As an example of decision making under uncertainty, consider the following situation. The Palm Tree Hotel is considering the construction of an additional wing. Management is evaluating the possibility of adding 30, 40, or 50 rooms. The success of the addition depends on a combination of local government legislation and competition in the field. Four states of nature are being considered. They are shown, together with the anticipated payoffs (in percent of yearly return on investment), in Exhibit 5.7. Management cannot agree on the probabilities of each of the states of nature occurring. The problem is to find how many rooms to build in order to maximize the return on investment.

EXHIBIT 5.7 Payoff Table (Percent of Return on Investment)

	A	B	C	D	E	F
1		States of Nature				
2		Positive	Positive	No	No	
3		Legislation	Legislation	Legislation	Legislation	
4		and Low	and Strong	and Low	and Strong	
5		Competition	Competition	Competition	Competition	
6	Alternatives	s_1	s_2	s_3	s_4	Average
7	a_1 30 rooms	10	5	4	-2	4.25
8	a_2 40 rooms	17	10	1	-10	4.5
9	a_3 50 rooms	24	15	-3	-20	4
10						
11	Cell F7	=AVERAGE(B7:E7) Copy to cells F8:F9				

Currently, decision theory does not provide a single best criterion for selecting an alternative under conditions of uncertainty. Instead, there are a number of different criteria, each with its benefits and limitations. The choice among these is determined by organizational policy, the attitude of the decision maker toward risk, or both.

Five criteria of choice are presented in this section:

1. Equal probabilities (Laplace) criterion
2. Pessimism (maximin or minimax) criterion
3. Optimism (maximax or minimin) criterion
4. Coefficient of optimism (Hurwicz) criterion
5. Regret (Savage) criterion

Equal Probabilities (Laplace) Criterion

The user of the **equal probabilities (Laplace) criterion** assumes that all states of nature are *equally likely* to occur. Thus, equal probabilities are assigned to each state of nature. The expected values are then computed and the alternative with the highest expected payoff is selected.

Using the example given in Exhibit 5.7, probabilities of 0.25 are assigned to each of the four states of nature. The expected payoffs are then calculated as follows (shown in Exhibit 5.7).

$$E(a_1) = 0.25 \times 10 + 0.25 \times 5 + 0.25 \times 4 + 0.25 \times (-2)$$
$$= 0.25\,(10 + 5 + 4 - 2) = 4.25$$
$$E(a_2) = 0.25\,(17 + 10 + 1 - 10) = 4.5 \text{ (largest expected yield)}$$
$$E(a_3) = 0.25\,(24 + 15 - 3 - 20) = 4$$

Thus, the best alternative is a_2, with an expected payoff of 4.5. The major argument against this criterion is that there is absolutely no reason to assume the probabilities are all equal. Such an assumption may be as erroneous as assuming any other values. It is equivalent to flipping a coin to make a decision.

Pessimism (Maximin or Minimax) Criterion

The user of this criterion is completely pessimistic because he or she assumes that the worst will happen, no matter which alternative is selected. To provide protection under the **pessimism criterion,** the decision maker should select the alternative that will give as large a payoff as possible under the worst conditions (best of the worsts).

Let us reproduce Exhibit 5.7 as Exhibit 5.8 (payoffs in percent yield). Assume that the decision maker selects a_1; then the *worst* that can happen is a loss of 2 percent when s_4 occurs. Similarly, the worst for a_2 is –10, and for a_3 is –20 (the lowest number in the

EXHIBIT 5.8 Pessimistic Approach in the Case of Profit

	A	B	C	D	E	F	G
1		States of Nature				Worst	Best of Worst
2	Alternatives	s_1	s_2	s_3	s_4	(Minimum)	(Maximum of Minimums)
3	a_1 30 rooms	10	5	4	-2	-2	-2
4	a_2 40 rooms	17	10	1	-10	-10	
5	a_3 50 rooms	24	15	-3	-20	-20	
6							
7	Cell F3	=MIN(B3:E3) Copy to cells F4:F5					
8	Cell G3	=IF(F3=MAX($F$3:$F$5),F3," ") Copy to cells G4:G5					

EXHIBIT 5.9 Profits Under Two Alternatives

	A	B	C	D	E	F
1		States of Nature			Worst	Best of Worst
2	Alternatives	s_1	s_2	s_3	(Minimum)	(Maximum of Minimums)
3	a_1	40,000	20,000	500	500	
4	a_2	550	520	510	510	510
5						
6	Cell E3	=MIN(B3:D3) Copy to cell E4				
7	Cell F3	=IF(E3=MAX($E$3:$E$4),E3," ") Copy to cell F4				

row is selected in the case of maximization). This information is captured in column F, labeled Worst in the spreadsheet shown in Exhibit 5.8. From this column, the best entry is then selected (–2 in the example). The decision maker maximizes the minimum payoffs, and therefore this criterion is called *maximin.* (In the case of cost minimization, the decision maker will minimize the maximum possible costs; that is, the decision maker will *minimax.*) The use of this criterion will guarantee the decision maker that the loss will not exceed 2, even in the *worst possible case.*

One drawback of this criterion (which is also a drawback of *all* the remaining criteria) is that the decision is based on only a *small portion* of the available information. Thus, valuable information is completely disregarded, leading to poor choices. This is shown in Exhibit 5.9, which illustrates a deliberately exaggerated maximization case. According to the criterion of pessimism, a_2 should be selected. The decision is based on the "minimum" column, which includes only one entry from each row. The rest of the data are ignored. In reality, most decision makers would pay attention to the remaining information and, consequently, select a_1. The pessimistic decision maker acts in a *superconservative* manner, paying attention only to the risks and completely ignoring the opportunities.

Optimism (Maximax or Minimin) Criterion

An optimistic decision maker assumes that the very *best* outcome will occur and selects the alternative with the best possible payoff. Thus, the decision maker would use the **optimism criterion.** To do so, the decision maker searches for the best possible payoff for each alternative. This can be easily accomplished with spreadsheets using the formulas shown in Exhibit 5.10. The alternative with the best payoff in column F is then identified (best of bests) and shown in column G of Exhibit 5.10. If the data were costs, then the optimistic decision maker would select as best the *lowest* cost payoff for each alternative and then select the *lowest* of these lowests. Such an approach is labeled **minimin.**

EXHIBIT 5.10 Maximax Choice

	A	B	C	D	E	F	G
1		States of Nature					
2	Alternatives	s_1	s_2	s_3	s_4	Best	Best of Bests
3	a_1 30 rooms	10	5	4	-2	10	
4	a_2 40 rooms	17	10	1	-10	17	
5	a_3 50 rooms	24	15	-3	-20	24	24
6							
7	Cell F3	=MAX(B3:E3) Copy to cells F4:F5					
8	Cell G3	=IF(F3=MAX($F$3:$F$5),F3," ") Copy to cells G4:G5					

EXHIBIT 5.11
Profits Under Two Alternatives

	s_1	s_2	s_3	**Best**	**Best of Best**
a_1	50,010	0	0	50,010	**50,010**
a_2	50,000	50,000	50,000	50,000	

Reproducing the data of Exhibit 5.7 in Exhibit 5.10, the "best" column is created. According to the **maximax** criterion, alternative a_3 would be selected. Notice again that no consideration is given to most of the available information; only the highest payoff is considered. Thus, an optimistic decision maker is a gambler who *disregards* the risks and looks forward only to the opportunities.

Similar to the pessimism criterion, we can construct an exaggerated case, Exhibit 5.11, to illustrate the danger of ignoring the rest of the data. In this decision situation, most decision makers would choose alternative a_2 for a guaranteed 50,000.

Coefficient of Optimism (Hurwicz) Criterion

Most decision makers are neither completely optimistic nor completely pessimistic. Therefore, it was suggested by Hurwicz that a degree of optimism labeled alpha, α, be measured on a 0 to 1 scale (0 = completely pessimistic, 1 = completely optimistic). The **coefficient of optimism (Hurwicz) criterion** suggests that the best alternative is the one with the highest (in maximization) weighted value, where the weighted value, *WV*, for each alternative (each row in the decision table) is expressed by

$$(WV)_i = \alpha\ [\text{best } o_{ij}] + (1 - \alpha)\ [\text{worst } o_{ij}]$$

where o_{ij} is the payoff. Then the best $(WV)_i$ is selected.

Examining Exhibit 5.10 with α given as 0.7 we get

$$WV(a_1) = (0.7 \times 10) + (1 - 0.7) \times -2) \;\; = 6.4$$
$$WV(a_2) = (0.7 \times 17) + (1 - 0.7) \times -10) = 8.9$$
$$WV(a_3) = (0.7 \times 24) + (1 - 0.7) \times -20) = 10.8 \text{ (maximum)}$$

Thus, alternative a_3 is the best. (*Note:* In the case of minimization (such as with costs), where the best is the lowest, select the alternative with the *lowest WV*.)

The major difficulty in applying this criterion is the assessment of alpha. Note that use is made of more information than in minimax, yet only the two extreme payoffs are considered—the remaining information is ignored.

As an alternative to assessing the exact value of alpha, another approach is to determine the range of values of alpha under which each alternative would be preferred. Using this approach, the decision maker need only specify the range in which alpha is expected to fall as opposed to trying to specify the exact value of alpha. One way to investigate this issue is by creating a *data table.* More specifically, data tables facilitate the investigation of how changes in one variable impact one or more formulas. (*Note:* Later we consider how data tables can be used to investigate the impact that two variables will have on a single formula.) In this case, we are interested in investigating how the variable alpha affects our choice of the three alternatives.

To begin this analysis, we create a spreadsheet that mimics our analysis to this point, shown in Exhibit 5.12. Our goal is to determine how changes in alpha (entered in cell B2) will impact the weighted value (*WV*) for each alternative. To illustrate the use of data tables, we will first modify the spreadsheet shown in Exhibit 5.12 to analyze the impact changes in alpha have on a_1. Then we will extend the spreadsheet to analyze the impact

EXHIBIT 5.12 Choosing an Alternative Based on the Coefficient of Optimism Criterion

	A	B	C	D	E	F	G
1		**Best o_{ij}**	**Worst o_{ij}**				
2	Weights	0.7	0.3				
3	a_1	10	-2	6.4			
4	a_2	17	-10	8.9			
5	a_3	24	-20	10.8			

=SUMPRODUCT(B$2:C$2,B3:C3)
Copy to cells D4:D5

that changes in alpha have on a_2 and a_3. Finally, we will enhance the spreadsheet to automatically select the best alternative for each value of alpha and use this information to identify the ranges of alpha for which each alternative is best.

Excel supports both one- and two-input data tables. In this example, the one-input data table is appropriate because we are interested in investigating only one variable, alpha. Regardless of the type of data table used, a strict format is required. More specifically, and as is illustrated in Exhibit 5.13, creating a data table requires the specification

EXHIBIT 5.13 Format of a One-Input Data Table with Input Values of the Variable Entered in Column B

	A	B	C	D	E	F	G	H
1		**Best o_{ij}**	**Worst o_{ij}**					
2	Weights	0.7	0.3					
3	a_1	10	-2	6.4				
4	a_2	17	-10	8.9				
5	a_3	24	-20	10.8				
6								
7								
8			**a_1**					
9			6.4					
10		0.00	-2.0					
11		0.05	-1.4					
12		0.10	-0.8					
13		0.15	-0.2					
14		0.20	0.4					
15		0.25	1.0					
16		0.30	1.6					
17		0.35	2.2					
18		0.40	2.8					
19		0.45	3.4					
20		0.50	4.0					
21		0.55	4.6					
22		0.60	5.2					
23		0.65	5.8					
24		0.70	6.4					
25		0.75	7.0					
26		0.80	7.6					
27		0.85	8.2					
28		0.90	8.8					
29		0.95	9.4					
30		1.00	10.0					

=SUMPRODUCT(B2:C2,B3:C3)
This formula calculates *WV* for alternative 1 based on the input values of alpha entered in B10:B30.

Input values of alpha to be investigated. These values are substituted into cell B2.

Results of substituting values listed in input range into cell B2.

EXHIBIT 5.14 Specifying Input Cell for One-Input Data Table When Input Values Entered in Column Format

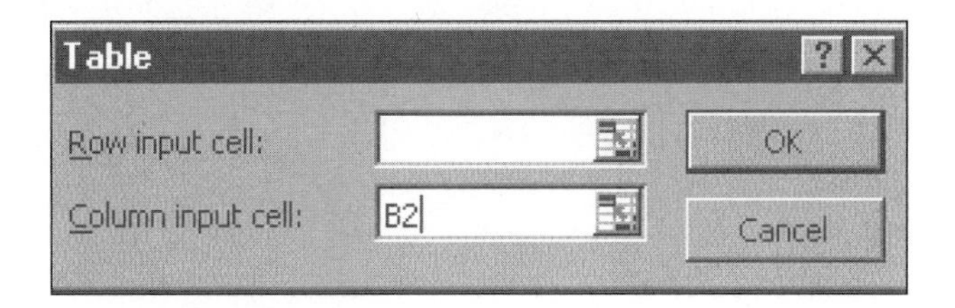

of three elements. The first element corresponds to a range containing values for the variable we wish to investigate. In this example, we have chosen to investigate the values of α from zero to one in increments of 0.05 (B10:B30). The second element is the formula that depends on the variable of interest. In our example, we are interested in calculating the *WV* for alternative 1 based on the values of alpha entered in cells B10:B30. Hence, the formula to calculate this was entered in cell C9. The last thing we must do is tell Excel we wish to construct a data table such that the values entered in cells B10:B30 are substituted for alpha (cell B2) in the formula that was entered in cell C9. To accomplish this, first highlight the range B9:C30. Next, select Data from Excel's main menu and then select Table. . . from the drop-down menu that appears. Finally, we tell Excel that the input variable should be substituted into cell B2, as shown in Exhibit 5.14.

A couple of notes are in order. First, when creating a one-input data table, Excel requires that the input values be entered in the first column (beginning in the second row) of the specified range (or the first row if the input values were entered across a single row instead of down a single column). Second, Excel requires that the formula related to the result of interest be entered in the first row of the second column.

Modifying the spreadsheet to analyze the impact of the alternative values of alpha on alternatives 2 and 3 is straightforward. First, we enter formulas that calculate the *WV* for alternatives 2 and 3 in cells D9 and E9, respectively (see Exhibit 5.15). Next we create the data table as before, with the only difference being that we highlight the range B9:E30 prior to selecting Data/Table. . .Finally, we enter formulas in column F to identify the best alternative for each value of alpha. According to the results shown in Exhibit 5.15, alternative one should be selected if management believes that alpha will be 0.5 or less. Likewise, if management believes alpha will be at least 0.6, then alternative 3 should be chosen. Otherwise, alternative 2 should be selected. Note that the results shown in the data table will be automatically updated should new values for alpha be entered in column B. For example, the values 0.51, 0.52, 0.53, and 0.54 could be entered to more precisely determine the range under which alternatives 1 and 2 are preferred. You might try and verify that alternative 1 is preferred for alphas of 0.53 and less.

The structure for a two-input decision table is slightly different. Specifically, the formula that depends on the two variables of interest is entered in the first row and first column of the data table. Then the range of values for one of the variables of interest is entered beginning in the second row of the first column, while the range of values for the second variable of interest is entered across the first row beginning in the second column. As an illustration, imagine you are interested in purchasing a car. Your best guesses are that the car will cost around $12,000, that your annual interest rate will be 8 percent and that you will finance it for 4 years. A data table to investigate alternative loan amount–interest rate combinations is shown in Exhibit 5.16. The Excel PMT (payment) function used has the following syntax: PMT (interest rate, term of loan, loan amount). Note that in the formula entered in cell B5, the annual interest rate was

EXHIBIT 5.15 Final Spreadsheet Analyzing Impact of Alpha

	A	B	C	D	E	F	G	H	I	J
1		Best o_{ij}	Worst o_{ij}							
2	Weights	0.7	0.3							
3	a_1	10	-2	6.4						
4	a_2	17	-10	8.9						
5	a_3	24	-20	10.8						
6										
7										
8			a_1	a_2	a_3	Best				
9			6.4	8.9	10.8					
10		0.00	-2.0	-10.0	-20.0	a_1				
11		0.05	-1.4	-8.7	-17.8	a_1				
12		0.10	-0.8	-7.3	-15.6	a_1				
13		0.15	-0.2	-6.0	-13.4	a_1				
14		0.20	0.4	-4.6	-11.2	a_1				
15		0.25	1.0	-3.3	-9.0	a_1				
16		0.30	1.6	-1.9	-6.8	a_1				
17		0.35	2.2	-0.6	-4.6	a_1				
18		0.40	2.8	0.8	-2.4	a_1				
19		0.45	3.4	2.2	-0.2	a_1				
20		0.50	4.0	3.5	2.0	a_1				
21		0.55	4.6	4.9	4.2	a_2				
22		0.60	5.2	6.2	6.4	a_3				
23		0.65	5.8	7.6	8.6	a_3				
24		0.70	6.4	8.9	10.8	a_3				
25		0.75	7.0	10.3	13.0	a_3				
26		0.80	7.6	11.6	15.2	a_3				
27		0.85	8.2	13.0	17.4	a_3				
28		0.90	8.8	14.3	19.6	a_3				
29		0.95	9.4	15.7	21.8	a_3				
30		1.00	10.0	17.0	24.0	a_3				

=IF(C10=MAX(C10:E10),$C$8,IF(D10=MAX(C10:E10),$D$8,$E$8))
Copy to cells F11:F30

EXHIBIT 5.16 Constructing a Two-Input Data Table

	A	B	C	D	E	F	G	H	I
1	Annual interest rate	8.00%							
2	Term (years)	4							
3	Amount of loan	$12,000							
4									
5		(292.96)	0.0700	0.0725	0.0750	0.0775	0.0800	0.0825	0.0850
6		10,000.00	(244.13)	(244.13)	(244.13)	(244.13)	(244.13)	(244.13)	(244.13)
7		10,500.00	(256.34)	(256.34)	(256.34)	(256.34)	(256.34)	(256.34)	(256.34)
8		11,000.00	(268.54)	(268.54)	(268.54)	(268.54)	(268.54)	(268.54)	(268.54)
9		11,500.00	(280.75)	(280.75)	(280.75)	(280.75)	(280.75)	(280.75)	(280.75)
10		12,000.00	(292.96)	(292.96)	(292.96)	(292.96)	(292.96)	(292.96)	(292.96)
11		12,500.00	(305.16)	(305.16)	(305.16)	(305.16)	(305.16)	(305.16)	(305.16)
12		13,000.00	(317.37)	(317.37)	(317.37)	(317.37)	(317.37)	(317.37)	(317.37)
13		13,500.00	(329.57)	(329.57)	(329.57)	(329.57)	(329.57)	(329.57)	(329.57)
14		14,000.00	(341.78)	(341.78)	(341.78)	(341.78)	(341.78)	(341.78)	(341.78)
15		14,500.00	(353.99)	(353.99)	(353.99)	(353.99)	(353.99)	(353.99)	(353.99)
16		15,000.00	(366.19)	(366.19)	(366.19)	(366.19)	(366.19)	(366.19)	(366.19)
17									
18									
19									
20									
21									
22									

=PMT(B1/12,B2*12,B3)

These values of annual interest rates are substituted into cell B1 in the formula entered in cell B5.

These values of loan amounts are substituted in for cell B3 in the formula entered in cell B5.

Table

Row input cell: B1

Column input cell: B3

OK

Cancel

Highlight the range B5:I16. Enter values into the **Table** dialog box that are displayed after selecting Data/Table...

EXHIBIT 5.17
Regret Table for Exhibit 5.7

Alternatives \ States of Nature	s_1	s_2	s_3	s_4	Largest Regret (Worst)	
a_1	24 – 10 = 14	15 – 5 = 10	4 – 4 = 0	–2 – (–2) = 0	14	
a_2	24 – 17 = 7	15 – 10 = 5	4 – 1 = 3	–2 – (–10) = 8	8	←*Minimum*
a_3	24 – 24 = 0	15 – 15 = 0	4 – (–3) = 7	–2 – (–20) = 18	18	

divided by 12 to convert it to an approximate monthly rate, while the term in years (cell B2) was multiplied by 12 to convert it to months. In using the PMT function, it is important that the interest rate and term of loan be in the same time units. Finally, observe that the PMT function returns the actual monthly payment values as negative numbers. This is because in performing these types of analyses, cash flows that you receive are assumed to have positive values while cash flows that you must pay are considered to have negative values.

Regret (Savage) Criterion

Savage argued that people frequently act to minimize their anticipated average (expected) *regret,* the relative loss resulting from selecting an alternative, given that a particular state of nature occurred, as compared with the *best* alternative that could have been selected for that particular outcome. For example, under the **regret (Savage) criterion,** if the management of Palm Tree decides to add 30 rooms (a_1) and both positive legislation and strong competition occur, they will *regret* not having added 50 rooms (a_3) which would have been the best alternative for that outcome. Because they will make 5 percent return versus 15 percent that they could have made with 50 rooms, their regret is 10 percent. Exhibit 5.17 shows the complete regret data. Within each payoff column, each payoff is subtracted from the largest payoff in the column. Once the table is constructed, the *largest* regret is found and entered in a new column on the right. Then, the alternative with the *smallest* regret in this column is chosen. In this example, the lowest regret is for alternative a_2. This selection guarantees that regardless of what happens, Palm Tree management will never have a regret larger than 8.

This procedure works for any regret table, whether derived from profit or loss data. Regret is always bad and is to be avoided or minimized. (*Note:* The value of regret, by definition, can never be negative.) However, as with the optimism and pessimism criteria, since this criterion focuses on only the worst possible outcome and ignores the other data, an exaggerated case could have been constructed illustrating the danger of relying on this criterion for all situations.

5.5 Decisions Under Risk

Decision situations in which the chance (or probability) of occurrence of each state of nature is known (or can be estimated) are defined as *decisions made under risk.* In such cases, the decision maker can assess the degree of risk that he or she is taking in terms of probability distributions. For example, a decision not to purchase fire insurance means that the decision maker takes a chance of perhaps 1 in 10,000 of losing property in a fire.

Probabilities in decision making should be viewed as a means of expressing the decision maker's judgment about an uncertain future. We all use phrases such as:

"There is a pretty good chance that interest rates will decline."
"It is likely to rain today."
"It's not likely that mortgage rates will come down."
"We are not sure how our competitor will react."

Probabilities constitute a language similar to these statements but are more precise (although not necessarily more *accurate*). By definition, a probability is the relative frequency of an event occurring when a situation is repeated many times under identical circumstances. But probabilities in managerial language are also used for one-time decision situations. For example, instead of saying that our new product has a pretty good chance of succeeding, we say that there is an 85 percent chance that the product will be a success.

Objective and Subjective Probabilities

There are two approaches to the assessment of the probabilities of the states of nature: objective and subjective. **Objective** or **empirical probabilities** can be derived based on historical occurrences, experimentation, or sometimes with statistical formulas. For example, the probability of a "head" in the toss of a coin is computed statistically as one half because only two states of nature (a head or a tail) can occur; and assuming the coin is fair, there is an equal chance for each (i.e., 50 percent). Thus, the use of objective probabilities narrows the scope for judgment.

Unfortunately, the use of objective probabilities requires several crucial assumptions that restrict its use:

1. Objective probabilities are usually based on observation of past events, experimentation, or both. Therefore, in using objective probabilities for decision making, it must be assumed that future conditions will follow the same pattern as past conditions (or that a clear trend for future events has been established).
2. A necessary result of the first assumption is that the process observed must be *stable* (i.e., the probabilities do not change over time).
3. It must be assumed that if a sample was observed for determining past behavior, it was large enough (statistically) and representative of the process under study.

The **subjective probability** approach measures the degree of belief in the likelihood of the future occurrence of a given outcome. Thus, such probabilities are a subjective appraisal of the nature of reality, in contrast to objective probabilities, which must be actual, countable, observable fact. Subjective probabilities are made by an individual(s) possessing experience with the phenomena involved. Subjective probabilities are used in cases where objective probabilities cannot be obtained. The major problem with subjective probabilities is that different decision makers may give different estimates of the probabilities, and even change their estimates as a result of psychological, emotional, or other behavioral factors. Also be aware of the fact that objective probabilities are not intrinsically more accurate than subjective probabilities.

(*Note:* With either objective or subjective probabilities, we incorporate *judgment* into the decision process. For example, in objective probabilities, we must judge which historical data are appropriate.)

Solution Procedures to Decision Making Under Risk

One solution approach to decision making under risk is the use of *expected value* (EV) as a criterion of choice. Alternatively, the *expected opportunity loss* (EOL) criterion may be used.

EXHIBIT 5.18 Yields (in Percent) of Investment Alternatives

	A	B	C	D	E
1		States of Nature			Expected
2		Solid Growth (s_1)	Stagnation (s_2)	Inflation (s_3)	Value
3	Probabilities →	0.5	0.3	0.2	Percent
4	a_1 Bonds	12	6	3	8.4
5	a_2 Stocks	15	3	-2	8.0
6	a_3 Time deposits (CDs)	6.5	6.5	6.5	6.5
7					
8	Cell E4	=SUMPRODUCT($B$3:$D$3,B4:D4) Copy to cells E5:E6			

Both lead to exactly the same result. Other criteria are the *most probable state of nature* criterion and some of those used for decision making under uncertainty, as just described.

As an example, reconsider Mary's dilemma, described earlier. Mary was considering investing in one of the following alternatives: a_1: corporate bonds, a_2: common stocks, or a_3: time deposits. Suppose that Friendly Trust wants to invest in only *one alternative,* with a declared objective of maximizing the yield over a one-year period. The yield (in percent of return) will depend on the state of the economy, which can be solid growth, stagnation, or inflation. The estimated yields under each alternative and state of the economy are shown in Exhibit 5.18. The problem is to find the best investment alternative.

Solution Procedure A: Expected Payoff Criterion This approach prescribes that the decision maker select the alternative with the best expected (average) payoff. This alternative should be selected each time the decision maker confronts the investment situation. Over the long run, the average yearly yield will be the same as the expected payoff. *The expected payoff of an alternative is the sum of all possible payoffs of that alternative, weighted by the probabilities of those payoffs occurring.*

Finding the Expected Payoff. The expected payoffs are computed, for each alternative, one at a time, as follows.

Step 1 Multiply each payoff by its corresponding probability. For example, for Alternative a_1 of Exhibit 5.18: 12×0.5, 6×0.3, and 3×0.2.

Step 2 Sum the results of the multiplication of Step 1; the total is the expected payoff.

In mathematical terms, let:

a_i = alternative i
s_j = state of nature j
p_j = probability that state of nature s_j *will occur*
o_{ij} = payoff resulting from the selection of alternative a_i when s_j occurs

Then, the expected value $E(a_i)$ is

$$E(a_i) = p_1 o_{i1} + p_2 o_{i2} + \cdots = \sum_{j=1}^{m} p_j o_{ij}$$

For alternatives a_i in Exhibit 5.18, the results are

$$E(a_1) = (12 \times 0.5) + (6 \times 0.3) + (3 \times 0.2) = 8.4 \text{ percent}$$
$$E(a_2) = (15 \times 0.5) + (3 \times 0.3) - (2 \times 0.2) = 8.0 \text{ percent}$$
$$E(a_3) = (6.5 \times 0.5) + (6.5 \times 0.3) + (6.5 \times 0.2) = 6.5 \text{ percent}$$

Selecting an Alternative. Once the expected payoffs for all alternatives are determined, only the newly formed column of expected payoffs need be considered (see the right-hand column of Exhibit 5.18). This is the dependent (or result) variable of the decision model. If the problem is one of maximization, then the *highest* expected payoff is searched for, using complete enumeration. In minimization, the alternative with *the lowest* expected payoff is sought. In the investment (maximization) example, alternative a_1 has the highest expected yield, and therefore investing in corporate bonds is recommended. The meaning of this choice is that the decision maker should invest in a_1, and only a_1, each time that he or she is confronted with the decision. Furthermore, the average yearly yield, over the long run, per decision, is 8.4 percent. *Note:* The use of this criterion when the payoffs are expressed in dollars is called the *expected monetary value,* or the EMV criterion.

Solution Procedure B: Expected Opportunity Loss (EOL) or Regret Criterion The basic idea of this criterion is that people frequently act to minimize their *regret* or *opportunity loss,* as defined earlier. Exhibit 5.19 shows the complete opportunity loss data. Once the table is constructed, the *expected regret* is computed by the same method used in computing expected payoffs, and the alternative with the *smallest expected regret* is selected. Note also that alternative a_1 was selected under both the expected opportunity loss criterion and the expected payoff criterion. This is *not* a coincidence. Both will *always* lead to the *same choice* because the mathematical operations are basically the same. The difference is only philosophical. These are merely two different explanations of why people make certain choices.

Solution Procedure C: Most Probable State of Nature Criterion This criterion prescribes that as the decision maker confronts the various possible states of nature in a decision under risk, he or she ignores all but the *most probable* state. By doing so, the decision maker changes the situation to a decision under *assumed certainty.* Some nonrepetitive decisions are treated with this principle. The following description illustrates this criterion.

Consider the investment decision of Exhibit 5.18. According to the most-probable-state-of-nature criterion, the decision maker assumes that solid growth will occur because it has the largest chance of occurring. The other states of nature are ignored. Therefore, Mary Golden will select a_2 because it will give her the largest yield (15 percent) in the event that solid growth occurs. However, if the decision maker is wrong, she may end with a yield of 3, or even lose 2. If this is a repeating situation, she will only make 8 (the expected value) over the long run.

One problem with this criterion is that it ignores the magnitude of some of the payoffs. We illustrate this potential problem with the following fictitious, but feasible, example. According to the most probable state of nature criterion, a decision maker would choose a_1 over a_2 in the following data no matter what the profits are under the s_2 column. In reality, a_2 would almost always be selected because of the large potential payoff.

EXHIBIT 5.19 Opportunity Loss Table (Percent Yield) of Exhibit 5.18

States of Nature	0.5	0.3	0.2	← Probabilities
Alternatives	Solid Growth	Stagnation	Inflation	Expected Regret
a_1	15 − 12 = 3	6.5 − 6 = 0.5	6.5 − 3 = 3.5	(3 × 0.5) + (0.5 × 0.3) + (3.5 × 0.2) = 2.35 ← Smallest
a_2	15 − 15 = 0	6.5 − 3 = 3.5	6.5 − (−2) = 8.5	(0 × 0.5) + (3.5 × 0.3) + (8.5 × 0.2) = 2.75
a_3	15 − 6.5 = 8.5	6.5 − 6.5 = 0	6.5 − 6.5 = 0	(8.5 × 0.5) + (0 × 0.3) + (0 × 0.2) = 4.25

Probability	0.6	0.4
State	s_1	s_2
Alternative a_1	10	100
Alternative a_2	9	500

Solution Procedure D: Uncertainty Criteria Earlier in this chapter we introduced five criteria of choice for the case of uncertainty. These could be used for the case of risk, as well. However, with the exception of nonrepetitive decisions, it is unlikely that these criteria will be used when actual probabilities of occurrence are available because the criteria possess too many deficiencies and do not include consideration of the additional probability information.

Note: The most probable state of nature criterion ignores the *relative magnitude* of some of the outcomes, whereas the uncertainty criteria ignore the *likelihood* of each of the events occurring. The expected value criterion combines *both* the size of the consequences and their likelihood, and is therefore a superior criterion for most situations.

Solution Procedure E: Simulation There is one other approach that is more involved but much more powerful—conducting a decision analysis through simulation, generally known as **risk analysis.** (An example is given in the "Detailed Modeling Example" in Section 5.9, and simulation is covered in detail in Chapter 7.) Risk analysis involves constructing a mathematical model of a situation on a computer and then running the model using what is called a **Monte Carlo simulation** to see what the outcomes will be under various circumstances. In a Monte Carlo simulation, the model is run many, many times, starting from a different point each time, based on the probabilities of the variables. Equations in the model are then used to construct a statistical distribution of the outcomes of interest, such as costs and times. Thus, the different states of nature are simulated according to their likelihood of occurring, just as in the real world. In this way, the model can simulate what would have likely happened if a particular managerial action had been taken. As many managerial actions as can be devised can be simulated to see what the results might have been. It is common in risk analysis to plot the outcome (e.g., the resulting cost) of each major managerial alternative against the likelihood of attaining that cost, and then comparing the final *risk profiles* of the alternatives. If one alternative's cost is always better at every probability than another, we say it *dominates* the other alternative. However, sometimes the profiles intersect and the choice of a good alternative is more complicated.

This approach was used to simulate the moving of a computer to a new location. A simulation run of 2000 trials was made, simulating various failures and variations in cost and time for each of three methods of moving the computer. A cost–probability distribution was then constructed, shown in Exhibit 5.20, to help identify the lowest-cost alternative, as

EXHIBIT 5.20 Risk Profiles for Three Alternatives

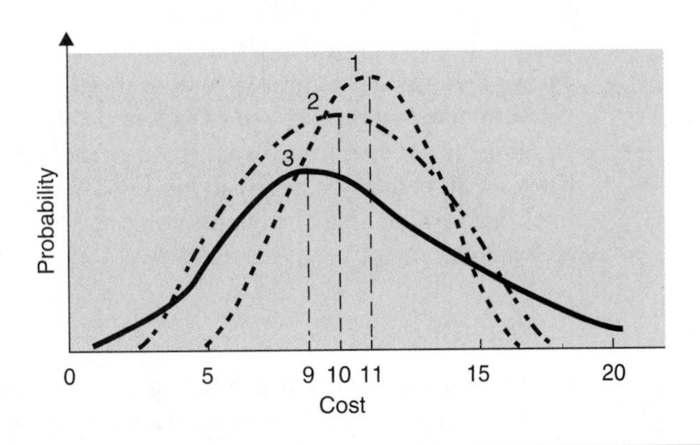

well as the alternative with the lowest risk of incurring a high cost. As seen in the exhibit, alternative 3 has the lowest expected cost (9) but also has the highest likelihood for a cost of 20 or more, as well as the highest likelihood for a cost of about 2 or less. Alternative 1 has the highest expected cost but the smallest variance and thus, risk. As seen in the exhibit, there is virtually no chance with alternative 1 of the cost exceeding 16, whereas there is a small chance with alternative 2 and a higher chance with alternative 3. Of course, with such a small variance in alternative 1, there is also virtually no chance of the cost ever being *less* than 4, but there is a small chance with alternatives 2 and 3.

Notes on Implementation

Nonrepetitive Decisions The expected value approach is based on achieving the best payoff over the long run. For example, if Mary looks at Exhibit 5.18 and selects alternative a_1, she will make either a 12, 6, or 3 percent yield each time the decision is made. Only over the *long run* will these yields average out to 8.4 percent. The question may thus be asked: Is there any justification in using an expected value approach for a one-shot, **nonrepetitive decision?**

There is at least one case in which the answer to this question is clearly yes. If a company is making several one-shot decisions whose payoff is more or less of the same magnitude, then the overall impact of using an expected value approach is similar to that of a repetitive decision. That is, repeating one decision 30 times, or making 30 decisions one time, results in the same mathematical expectation. In cases where an expected value cannot be justified, criteria such as minimax may be a good approach.

Dominance In certain cases, it is possible to eliminate some alternatives from evaluation because they are *inferior to* or *dominated by* other alternatives. For example, assume that alternative a_4 is added as an investment alternative to Exhibit 5.18, where the yields are 10 in the case of solid growth, 6 in the case of stagnation, and 2 in the case of inflation. If we compare this alternative (a_4) to a_1 we get

	Solid Growth	Stagnation	Inflation
a_1	12	6	3
a_4	10	6	2

Clearly using **dominance** as a decision criterion, the decision maker should not consider a_4 at all, because no matter what state of nature occurs, the decision maker will be as well off or better with a_1. Thus, it is said that *a_1 dominates a_4*.

Payoff Variability In spite of the usefulness of expected value as a decision criterion for risky situations, it is always advisable to also consider the **variability** of the payoffs. For example, it may not be worthwhile to select an alternative that has an insignificantly higher expected return if it also entails the possibility of significant loss. Thus, the variability of returns should always be considered.

Also note that the concept of expected value is based on the long-run consequences. You lose one, you win one; what is important is the long-run average. Now, assume that there is a 30 percent chance that you lose. If you lose once, twice, or maybe even three times in a row, you may be out of business. That is, you cannot always afford to wait for the long run.

Sensitivity Analysis

Assume that a company is examining a venture in which it can make $45,000 or lose $15,000 but the probabilities of each are not known. Also assume that the company uses expected value as the criterion for decision making; if the EMV of this venture is larger than zero, the company will undertake the project, but if the EMV is zero, they will be indifferent.

Denote the unknown probability of success by p. Therefore:

$$\text{EMV} = p(45{,}000) + (1 - p)(-15{,}000)$$

If we set EMV to zero and solve for p, we will get $p = 0.25$. That is, if the probability of success is larger than 0.25, the EMV will be larger than zero. Thus, the company should undertake the project.

Now suppose that we have estimated the probability of success to be 0.60; obviously the company will undertake the project. Furthermore, the probability of success could drop all the way to 0.25 before the company would change its decision. As the probability approaches 0.25, the EMV gets smaller and smaller, shrinking to exactly zero at 0.25 and becoming *negative* for values less than 0.25. The value of $p = 0.25$ is therefore called the *critical probability*.

5.6 Decision Trees for Risk Analysis

Decision making under risk, as discussed thus far, has been limited to a single decision over one period of time. A decision was to be made at the beginning of the period, and the future consequences were then estimated. All the information was presented in the form of a decision table. There are times, however, when a decision cannot be viewed as an isolated, single occurrence, but rather as the first of a sequence of several interrelated decisions over several future periods. Therefore, the decision maker must consider the whole series of decisions simultaneously. Such a situation is called a **sequential** or *multi-period* **decision process.**

Using decision tables to analyze these decisions becomes too cumbersome. The tool that was developed instead is called a **decision tree,** which is basically a graphical exposition of decision tables in the form of a tree. A decision tree shows, at a glance, when decisions are expected to be made, in what sequence, their possible consequences, and what the resultant payoffs are expected to be. The results of the computations can be depicted directly on the tree, simplifying the analysis.

Structure of a Decision Tree

A decision tree is composed of the following elements (see Exhibit 5.21): decision points, alternatives, chance points, states of nature, and payoffs. We describe each in turn.

Decision Points At a **decision point** (also called a *decision node, act node,* or *decision fork*), usually designated by a square, the decision maker must select *one alternative course of action* from a finite number of available actions. The alternative courses of action are shown as *branches* or *arcs* emerging out of the right side of the decision point. When there is a cost or profit associated with the alternative, it is necessary to keep track of it along the path. Each alternative branch may result in a payoff, another decision point, or a chance point.

Chance Points A **chance point** (also known as an *event fork* or *chance node*), designated by a circle, indicates that a chance event is expected at this point in the process. That is, one of a finite number of *states of nature* is expected to occur. The states of nature are

EXHIBIT 5.21 The General Structure of a Decision Tree

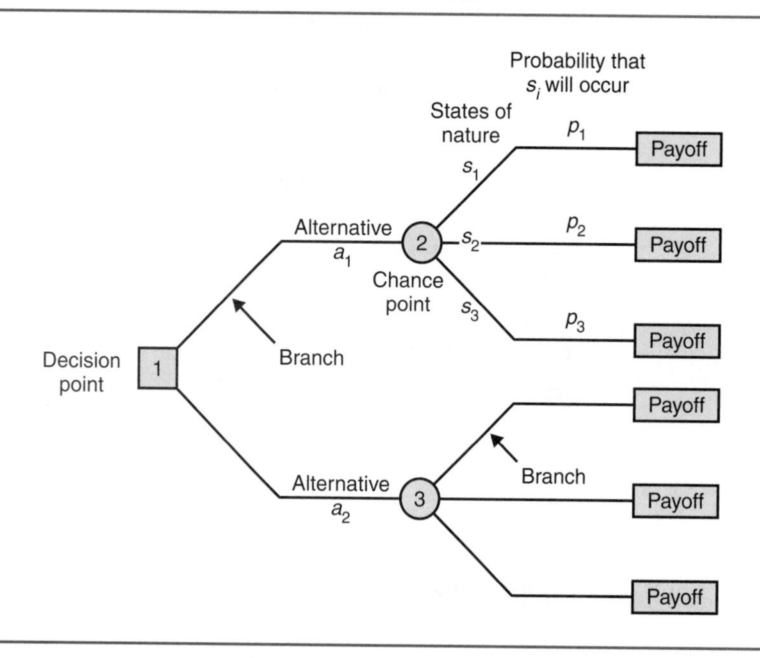

shown on the tree as branches to the right of the chance points. Because decision trees depict decision making under risk, the assumed probabilities of the states of nature are written above the branches. Each state of nature may be followed by a payoff, a decision point, or by another chance point.

Constructing a Tree A tree is started at the *left* of the page with a decision point, as in Exhibit 5.21. Once the decision point is constructed, all possible alternatives are drawn branching out to the right. Then, a chance point or other decision points are added, corresponding to events or decisions that are expected to occur after the initial decision. Each time a chance point is added, the appropriate states of nature, with their corresponding probabilities, branch out of it to the right. The tree continues to branch from left to right until the payoffs are reached. Exhibit 5.21 shows the general structure of a small tree. Larger trees involve a sequence of several decision and chance points, representing several decision periods, as shown later. The tree shown in Exhibit 5.21 represents a single decision and as such is equivalent to a decision table.

The process of constructing a tree may be divided conceptually into three steps:

1. Build a *logical tree,* which includes all decision points, chance points, and emerging arcs or branches, arranged in chronological order.
2. Introduce the probabilities of the states of nature on the branches, thus forming a *probability tree.*
3. Finally, add the conditional payoffs, thus forming the completed *decision tree.*

The Equivalence of Decision Trees and Decision Tables Reconsider the situation in Exhibit 5.18. This table is presented as a decision tree in Exhibit 5.22. Here we see the initial decision situation, with each alternative emanating as one branch from the decision node. At the right of each branch is the chance node, with the three possible state of nature outcomes, their probabilities, and the corresponding payoffs for that alternative and state of nature. Notice that the use of decision trees makes the sequence of events ex-

EXHIBIT 5.22 A Decision Tree for Exhibit 5.18

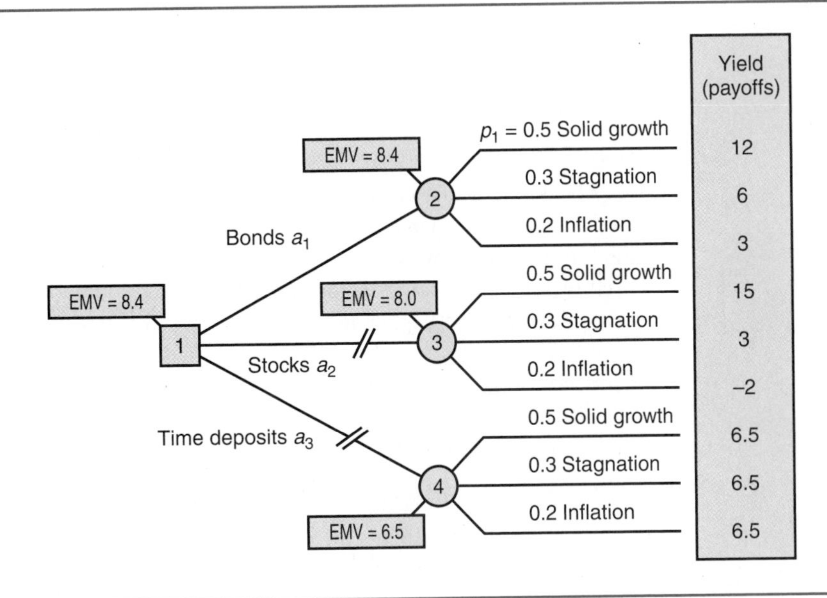

plicit. In this case a decision must be made, and then a particular state of nature occurs. We next look at the general process for evaluating such a tree.

Evaluating a Decision Tree

In order to solve a tree, it is customary to divide it into segments. Two types of segments are considered: *decision points* with all their alternatives (Exhibit 5.23*a*), and *chance points* with all their emerging states of nature (Exhibit 5.23*b*).

The solution process starts with those segments ending in the final payoffs at the *right side of the tree,* and continues to the left, segment by segment, in the *reverse* direction from which the tree was drawn.

1. **Chance Point Segments** The expected value of all the states of nature emerging from a chance point must be computed (multiply payoffs by their probabilities and sum up the results). The expected value is then written near the chance point inside a

EXHIBIT 5.23 Segments of a Tree

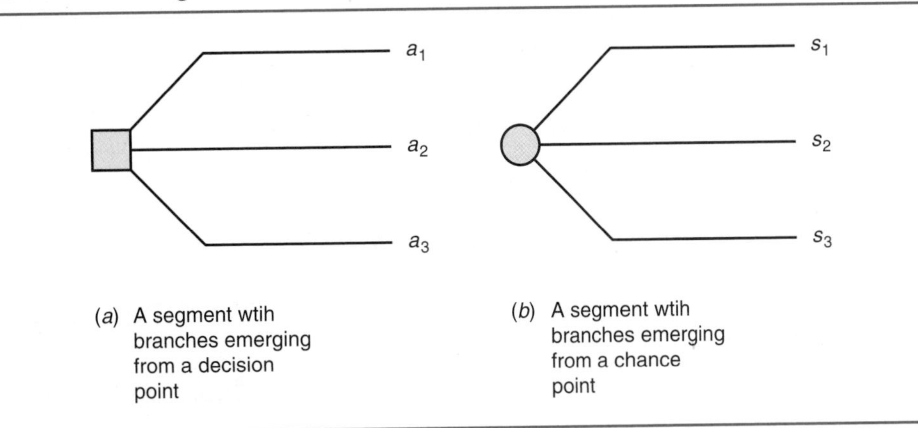

(*a*) A segment wtih branches emerging from a decision point

(*b*) A segment wtih branches emerging from a chance point

rectangle (labeled EMV in Exhibit 5.22). These expected values are considered as payoffs for the next branch to the left.

2. **Decision Point Segments** At a decision point, the payoffs given (or computed) for each alternative are compared and the best one is selected. All others are disregarded. A disregarded alternative is marked by the symbol // directly on the branch (see Exhibit 5.22).

Thus, the decision maker *must* select one alternative at each *decision* point and discard (prune) *all* other alternatives. The computation process continues from the right to the left. Pruning slowly reduces the size of the decision tree until only one alternative, the final recommendation, remains at the last decision point on the left side of the tree. As an example of the process, let us evaluate the tree of Exhibit 5.22.

Computations at a Chance Point The segments at the right are considered first. They are all chance points, and therefore *expected values* are computed. The expected values (designated in Exhibit 5.22 as EMV) are:

$$\begin{aligned} &\text{For point 2: EMV} = 12(0.5) + 6(0.3) + 3(0.2) &&= 8.4 \\ &\text{For point 3: EMV} = 15(0.5) + 3(0.3) - 2(0.2) &&= 8.0 \\ &\text{For point 4: EMV} = 6.5(0.5) + 6.5(0.3) + 6.5(0.2) &&= 6.5 \end{aligned}$$

The EMVs are entered inside a rectangle above each chance point. They are now considered as *payoffs* for the next step.

Computations at a Decision Point Exhibit 5.24 shows the situation in Exhibit 5.22 after EMVs for all chance points have been computed. At decision point 1, all alternatives are compared with the EMVs considered as payoffs. Alternative a_1, the choice with the highest payoff, is recommended.

The example just presented showed a decision tree for a single-decision period (equivalent to a decision table). However, decision trees are especially useful in sequential decisions, perhaps made over multiple periods.

The Multiperiod, Sequential Decision Case

Decision trees involving a sequence of decisions are nothing but a collection of smaller decision trees, each representing a single time period. All grow horizontally from left to right; the trunk is at the left and the branches are at the right. The tree can be extended to the limit of forecasting ability.

To illustrate the solution process in this case, consider the following detailed example. The Microflange Company is facing heavy demand for one of its flanges. The existing manufacturing facility is currently working at full capacity on a normal shift. The

EXHIBIT 5.24 Computation at a Decision Point

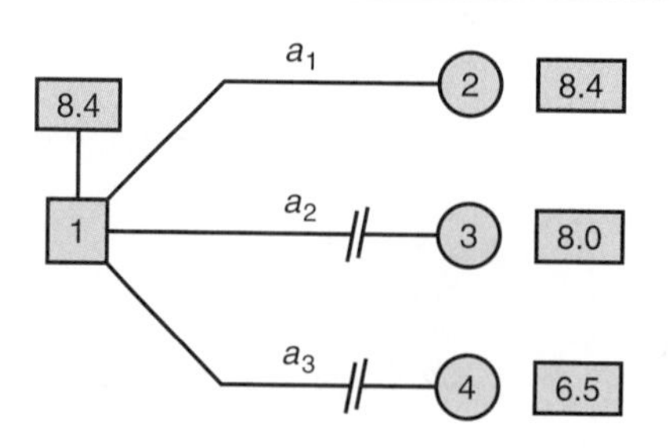

firm has two options to meet the heavy demand: either instituting overtime, an option that will cost \$2,000, or installing new machinery, at a cost of \$20,000. There is insufficient volume to justify a second shift. The choice between the options depends mainly on what happens to sales over the next 2 years. During the *first year,* management estimates that there is a 70 percent chance that sales will rise and a 30 percent chance that they will fall.

The information given so far is sufficient to start building a decision tree (Exhibit 5.25 points 1, 2, and 3). After one year of operation, management will be faced with another decision, which will depend on the action taken initially, the events during the year, and the projection of second-year sales.

The Decisions After One Year Depending on the action at time zero and the future states of nature, management could be at either point 4, 5, 6, or 7.

First Decision Point (4 in Exhibit 5.25). If a new machine had been installed at time zero and sales had risen, then management could either install a second machine or institute overtime. The decision at point 4 depends on the anticipated payoffs after two years. These depend on sales forecasts, which can be either high (20 percent chance), medium (70 percent), or low (10 percent). In the event that a second machine is installed, the anticipated payoffs are \$80,000, \$60,000, and \$50,000, respectively. We assume that these figures and the rest of the payoffs and expenses in this problem are given in present values, so they can be combined and compared. This information is entered on the right side of the trees. The expected value is then computed (\$63,000) and entered above chance point 8.

However, if overtime is instituted, the profits are estimated to be \$60,000 for high sales, \$50,000 for medium sales, and \$40,000 for low sales. This information is used to compute the expected value, which is then entered at chance point 9 in Exhibit 5.25 (\$51,000).

Second Decision Point. If a new machine had been installed initially and sales had fallen, then the decision maker would be at point 5 after one year. At that point, management would have no choice but to use the existing capacities to the fullest extent. Anticipated results are shown on the tree at point 10.

Third Decision Point. If overtime had been instituted initially and sales had risen, then management would be at decision point 6 with two alternatives open: install a new machine or install a new machine *and* use overtime. The anticipated payoffs are shown at points 11 and 12.

Fourth Decision Point. If overtime had been instituted initially and sales had fallen, then management would be at decision point 7, where only one alternative is assumed to be available: institute overtime. The anticipated results are shown at point 13.

The problem is to find the best course of action the company should take *initially* and at *intermediate* stages, knowing all the preceding information.

Evaluation Using the procedure previously outlined, the expected values at all chance points are computed (starting from the right).

Point 8.

$$\text{EMV} = (0.2 \times 80{,}000) + (0.7 \times 60{,}000) + (0.1 \times 50{,}000) = \$63{,}000$$

Point 9.

$$\text{EMV} = (0.2 \times 60{,}000) + (0.7 \times 50{,}000) + (0.1 \times 40{,}000) = \$51{,}000$$

EXHIBIT 5.25 The Microflange Company Decision Tree

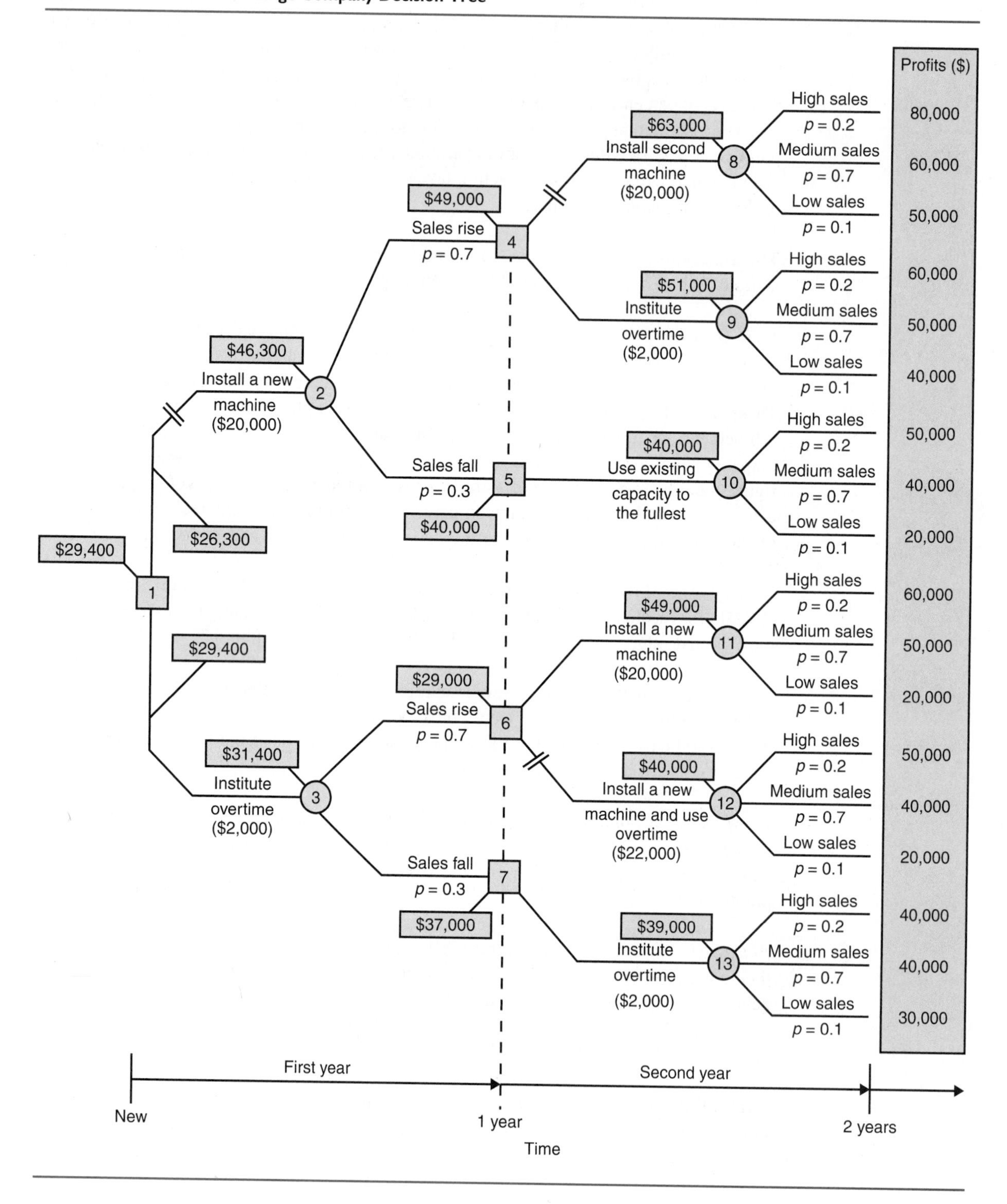

Similarly, for the other points the EMVs are listed here:

Point	10	11	12	13
EMV	$40,000	$49,000	$40,000	$39,000

Next, the computation moves leftward, thus reaching decision points 4, 5, 6, and 7.

Point 4. At this decision point, the alternative of a second machine ($63,000 – $20,000 = $43,000 profit) is compared with the alternative of overtime ($51,000 – $2,000 = $49,000 profit). Because the latter is more profitable, it is selected, and the EMV of $49,000 is entered above point 4.

Point 5. There is only one alternative. The expected value of point 10 is thus recorded at point 5.

Point 6. At this decision point, there are two alternatives: install a new machine ($49,000 – $20,000 = $29,000) or overtime plus a new machine ($40,000 – $22,000 = $18,000). The first one is better, so an EMV of $29,000 is recorded at point 6.

Point 7. There is only one alternative at this point. The EMV from point 13, $39,000, is recorded (less the $2,000 expense) at point 7.

Computation of the Left Side. At this stage, only the left side of the tree is considered. This information is presented in Exhibit 5.26.

$$\text{EMV of point 2} = (0.7 \times \$49{,}000) + (0.3 \times \$40{,}000) = \$46{,}300$$
$$\text{EMV of point 3} = (0.7 \times \$29{,}000) + (0.3 \times \$37{,}000) = \$31{,}400$$

Finally, decision point 1 is considered. The expected value of installing a new machine is

$$\text{EMV} = \$46{,}300 - \$20{,}000 = \$26{,}300$$

The expected value with overtime is

$$\text{EMV} = \$31{,}400 - \$2{,}000 = \$29{,}400$$

EXHIBIT 5.26 Left Side Exhibit 5.26

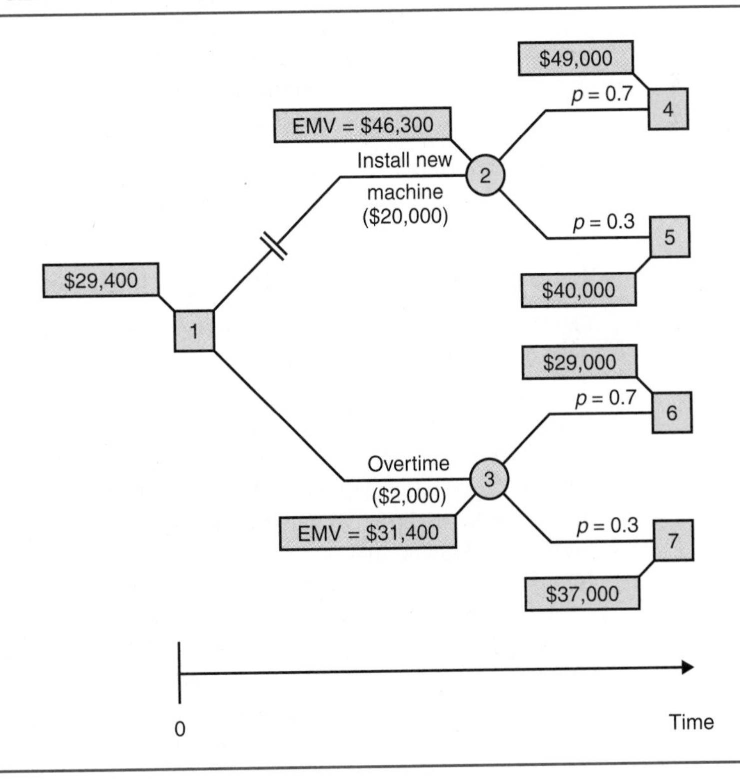

Thus, it is better to plan now on overtime, for an expected gain of $29,400. The final decision, therefore, is to use overtime this year and, if sales rise, install a new machine the second year. If sales fall, however, the overtime should continue.

5.7 The Value of Additional Information

Examining the situations of certainty, risk, and uncertainty, it is clear that there is a difference between the three situations in terms of the amount of information available for making the decision. As noted earlier, in the case of certainty it is assumed that we have complete information. That is, each time an alternative is selected, there is only one possible outcome, which is known. In the case of risk, there are several possible outcomes associated with each alternative, and their chances of occurring are assumed to be known. Finally, the case of uncertainty, there are again several possible outcomes associated with each alternative, but the decision maker does not know the likelihood of each of these occurring. That is, there is even less information available than in risk.

Of course, managers prefer to make decisions with as much information as possible. That is, they prefer to make decisions under certainty, if at all possible. Otherwise, they will accept risk. Only in extreme cases do they make decisions under *uncertainty*. For this reason, in many decision situations, a subdecision has to be made: whether to acquire additional information so as to convert the current situation to a more certain one.

Whenever a decision about the acquisition of additional information is to be made, the following questions should be considered.

1. What information is needed, and is it available?
2. Is there time to acquire the information?
3. What is the quality of the additional information? Information can have very different levels of quality. This issue will be explored later.
4. What is the value of the information?
5. What is the cost of obtaining the information?
6. Should the information be acquired?

Our analysis here will be based on the assumption that additional information is available and that there is sufficient time to acquire and use it.

Information Quality: Perfect Versus Imperfect Information

Our discussion here will be limited to information quality in terms of its accuracy, although there are several other dimensions of information quality that can also be important.

Perfect Information Both perfect and imperfect information relate to decision making *under risk* where the long-run probability of an outcome, given a repetitive decision situation, is assumed to be known. For example, suppose you are about to make a decision about whether to bring an umbrella with you. A simplified decision table is shown in Exhibit 5.27. Assume that, on the average, there is a 30 percent chance of rain each day. Based on this information, if we can quantify the outcomes in the table, we can use an expected value approach to develop a *policy* that will tell us whether to take an umbrella.

However, it may make more sense to examine the weather every morning. This requires more information. If we can find a predictor that can tell us whether it will rain that day, and if such a predictor is *always correct* (i.e., possesses *perfect* information), then we must decide whether to acquire such perfect information at its current cost. The

EXHIBIT 5.27 Simple Decision Table

Action \ Event	0.3 Rain	0.7 No Rain
Take umbrella	Dry, happy	Frustrated Why did I bring it?
Do not take umbrella	Wet, mad	Thankful

EXHIBIT 5.28 The Quality of Information in Risk Situations

(*a*) Perfect

Predicted	Actual State of Nature: *Rain*	Actual State of Nature: *No Rain*
Rain	1	0
No rain	0	1

(*b*) Imperfect

Predicted	Actual State of Nature: *Rain*	Actual State of Nature: *No Rain*
Rain	0.8	0.2
No rain	0.2	0.8

concept of perfect information as compared with imperfect information is shown in Exhibit 5.28. *Note:* The number of predictions may be the same as the number of states of nature (two each, here) or it may be larger (for example, "not sure" can be a prediction).

Imperfect Information Here, the prediction is not perfect, as in Exhibit 5.28*b*. In this situation, the predictions for the case of imperfect information are right more often than wrong (e.g., 80 percent), but the opposite can happen occasionally as well. For that matter, a *perfectly incorrect* prediction is highly valuable because you know exactly what the outcome will be—exactly the opposite of the prediction! Obviously, the situation of predictions being wrong much more often than right is rare.

The Value of Perfect Information

Suppose that additional information can be obtained; what will its value be to the decision maker? This discussion will be confined to the case where the obtained information changes the situation from one of *risk* to one of *certainty*. That is, the decision maker, just prior to the time a decision is to be made, is assumed to acquire perfect information. Consider the following example.

Let us analyze the investment decision presented in Exhibit 5.22 as reproduced in Exhibit 5.29. The problem, solved by the expected value approach, indicated that the best

EXHIBIT 5.29 The Investment Decision (in Percent Yield)

	A	B	C	D	E
1		States of Nature			Expected
2		Solid Growth (s_1)	Stagnation (s_2)	Inflation (s_3)	Value
3	Probabilities →	0.5	0.3	0.2	Percent
4	a_1 Bonds	12	6	3	8.4
5	a_2 Stocks	15	3	-2	8.0
6	a_3 Time deposits (CDs)	6.5	6.5	6.5	6.5
7					
8	Cell E4	=SUMPRODUCT($B$3:$D$3,B4:D4) Copy to cells E5:E6			

alternative was a_1, yielding an average of 8.4 percent. Let us now assume that each percent of yield equals $10,000; that is, a decision maker who uses the expected value as a criterion will make, over the long run, 8.4 × $10,000 = $84,000 per decision.

Selecting an Alternative Course of Action, Given Perfect Information Assume that we deal with a repetitive situation. Suppose that in advance of making a specific decision, a market research firm is able to predict, with certainty, the state of the economy that will prevail for that decision. Thus, the decision maker can make a choice with complete certainty. The choice depends on what the research firm predicts:

- If the research firm predicts growth, then the best choice is stocks (a_2).
- If the research firm predicts stagnation, the choice will be time deposit (a_3).
- If the prediction is for inflation, the choice will again be time deposit (a_3).

Assuming that the frequency distribution of the states of the economy does not change over the long run, then 50 percent of the time the research firm must predict *growth* as the next state. (That is, the research firm does not have the power to *alter* the outcomes or states of nature; they can only *predict* them.) The decision maker, now knowing what is going to happen for the next decision, will select a_2 and realize a 15 percent yield. Similarly, 30 percent of the time stagnation will be predicted and the decision maker will make 6.5 percent by selecting a_3, and 20 percent of the time inflation will be predicted, yielding 6.5 percent for selecting a_3. These choices are circled in Exhibit 5.29. The decision maker's new average (expected) yield, per decision, using these perfect predictions will now be:

$$(0.5 \times 15) + (0.3 \times 6.5) + (0.2 \times 6.5) = 10.75 \text{ percent}$$

Using the $10,000 per 1 percent equivalence, the average return per decision will be $107,500.

Should the perfect information be acquired? If we compare the expected yield with perfect information ($107,500) with the expected yield under regular conditions ($84,000), we see an increase due to using this information of $23,500. In general, decisions with perfect information normally yield much better results than decisions without it. The *difference* of $23,500 is called the *expected value of perfect information (EVPI)* and is used to answer the question of whether perfect information should be acquired. If it costs more than this, it should not; if less, then we will benefit by acquiring it.

The mathematical expression of the expected value of perfect information in the case of maximization is

$$\text{EVPI} = \sum_{j=1}^{n} p_j \max_i(o_{ij}) - \max_i \sum_{j=1}^{n} p_j o_{ij}$$

where

p_j = probability of state of nature j

o_{ij} = the payoff when action a_i is taken and state of nature j occurs

In this equation, the first term after the equality represents the expected yield with perfect information and the second term represents the expect yield without perfect information. In the case of minimization, the equation is represented this way:

$$\begin{aligned}\text{EVPI} &= [\text{expected cost without perfect information}] \\ &\quad - [\text{expected cost with perfect information}] \\ &= \min_i \sum_{j=1}^{n} p_j o_{ij} - \sum_{j=1}^{n} p_j \min_i(o_{ij})\end{aligned}$$

Note that the EVPI is the average (per decision) improvement in the value of the objective function. As noted, if this figure is compared against the cost of acquiring the information, management can make a decision regarding the acquisition of the information. For example, if the research firm charges $15,000 per prediction, the investor stands to gain $23,500 – $15,000 = $8,500, on average, by using the service. But if the marketing firm charges $23,500 or more for this service, then the arrangement would not be profitable.

Thus, the EVPI tells the decision maker the *upper limit* one should be willing to pay for perfect predicting information—information that is 100 percent reliable. If decision makers decide to buy the perfect information, they should, of course, wait for the result of each prediction and then make the appropriate, best choice.

The Equivalence of EVPI and EOL The expected value of perfect information (EVPI) in the investment example was $23,500, or 2.35 percent of yield. Examining Exhibit 5.19, the reader will find that the expected regret or expected opportunity loss (EOL) of the *best alternative* is also 2.35 percent yield. Is this a coincidence? The answer is no! As a matter of fact, the EVPI is *always equal to the best* (smallest) EOL. The reason is that in computing EVPI we actually compute the regret, which is the difference between the best choice with perfect information and the best choice without it.

5.8 Imperfect Information and Bayes' Theorem

The value of perfect information is computed with the assumption that a perfect predictor exists. In other words, each time that a prediction for a state of nature is made, this prediction comes true. In reality, however, this is a very rare case. The typical case involves predictions that may not come true all the time. Such a case is called prediction under imperfect (or sample) information, as discussed next.

Bayes' Theorem

Bayes' theorem is a procedure used to *revise* the probabilities of the states of nature. This helps us decide whether to acquire the additional information needed to revise the prior (initial) probabilities. Consider the following example.

American Ecology, Inc. estimates that its new product, E-3, has a 0.8 chance (80 percent) of being a winner and a 0.2 chance of being a loser. However, before the company makes a production commitment, it would like to further investigate the situation. This is because these initial estimates (*prior probabilities*) may not be accurate; hence, the desirability of calling on a market research firm to conduct a special survey.

From previous experience, it is known that such a special survey can predict either *success, failure,* or be *inconclusive.* Statistically, it is known that of all the new products that were eventually found to be successful, 70 percent of the time the previously conducted surveys had correctly predicted success (S), 10 percent of the time the surveys had incorrectly predicted failure (F), and 20 percent of the time the surveys were inconclusive (I). On the other hand, an examination of all the cases that were actual failures (losers) indicated that in 85 percent of these cases the surveys correctly predicted failure, in 10 percent they were inconclusive, and success was incorrectly predicted in the remaining 5 percent. This information is summarized in Exhibit 5.30.

The probabilities in Exhibit 5.30 are *conditional probabilities* (see Chapter 2) because they tell us the historical distribution of predictions before the actual outcomes (winners or losers). For example, according to the exhibit there is a 70 percent chance

EXHIBIT 5.30
Reliability of the Surveys

	Actual State of Products	
Results of Survey	**Winners (*W*)**	**Losers (*L*)**
Predicted success (*S*)	0.70	0.05
Inconclusive (*I*)	0.20	0.10
Predicted failure (*F*)	0.10	0.85
Total	1.00	1.00

that the survey will indicate success in cases where the product ultimately turned out to be a winner. Thus, they are an indication of the *reliability* of the surveys or the *track record* of the forecaster. These probabilities may be based on personal knowledge of the past accuracy of the forecaster, reports from other customers or users, or a simple subjective assessment. It is assumed that the forecaster's future prediction success will be identical to that in the past.

Using Revised Probabilities with Imperfect Information

The *revised probabilities* of the future states of nature depend on the results of the additional information, which can be determined only if such information is acquired. Thus, any analysis we conduct, prior to the actual acquisition of the information, will not help us prescribe the best decision alternative. However, it can help us decide what to do *if* the additional information makes certain predictions. From this, we derive a **decision policy** that recommends specific alternatives, one for every possible outcome of the prediction. The process involves the following nine steps:

Step 1 Evaluate the decision situation with the prior probabilities.
Step 2 Assuming that it is impossible to obtain perfect information, check the possibility of acquiring partial information.
Step 3 Compute revised (posterior) probabilities, one for each possible outcome of the prediction.
Step 4 Compute the probabilities of each of the research outcomes ("indicators"). (Step 4 can precede Step 3.)
Step 5 Construct a decision tree.
Step 6 Solve the tree.
Step 7 Compute the expected value of the additional information.
Step 8 Decide whether to acquire the information.
Step 9 Select an alternative in the original problem.

We illustrate the steps through the following example.

Production Unlimited, Inc.

The marketing department of Production Unlimited, Inc. is considering whether to develop a new product. All the relevant information is shown in Exhibit 5.31.

Step 1. Initial Evaluation Using the prior probabilities, the expected values can be computed. Accordingly, the product *should not* be developed (Exhibit 5.31).

Step 2. Check Track Record A consultant has been called in to predict, through a survey, what state of nature in the consultant's opinion will occur next. The consultant asks

EXHIBIT 5.31 Marketing Payoffs (Dollars)

	A	B	C	D	E
1			States of Nature		
2		*A*	*B*	*C*	Expected Value
3	Alternatives	0.2	0.5	0.3	
4	a_1 Develop	300,000	200,000	-600,000	-20,000
5	a_2 Do not develop	0	0	0	0
6					
7	Cell E4	=SUMPRODUCT($B$3:$D$3,B4:D4) Copy to cell E5			

EXHIBIT 5.32 Consultant's Reliability (Conditional Probability Matrix)

	Actual State of Nature		
Indicators	**A**	**B**	**C**
Survey predicted A (call this prediction A_p)	$P(A_p \mid A) = 0.80$	$P(A_p \mid B) = 0.10$	$P(A_p \mid C) = 0.10$
Survey predicted B (call this prediction B_p)	$P(B_p \mid A) = 0.10$	$P(B_p \mid B) = 0.90$	$P(B_p \mid C) = 0.20$
Survey predicted C (call this prediction C_p)	$P(C_p \mid A) = 0.10$	$P(C_p \mid B) = 0$	$P(C_p \mid C) = 0.70$

$50,000 for the survey. The track record of the consultant indicates that the consultant's surveys have the prediction reliabilities shown in Exhibit 5.32. This information is called the *conditional probabilities* of the source.

For example, of all the past cases in which *B* actually occurred, 90 percent of the time the survey correctly predicted that *B* would occur, 10 percent of the time the wrong prediction of *A* was made, and 0 percent of the time the wrong prediction of *C* was made.

Step 3. Compute Revised Probabilities Given (in Exhibit 5.31) the prior probabilities and the reliability of the survey (Exhibit 5.32), the revised probabilities can be calculated with the aid of Bayes' equation. Recall that the *conditional* probability $P(A_p \mid A)$ means the probability of A_p, *given that A has occurred.* (The general form of Bayes' equation is discussed further below and illustrated with another example.)

For branch A_p:

$$P(A \mid A_p) = \frac{P(A)P(A_p \mid A)}{P(A)P(A_p \mid A) + P(B)P(A_p \mid B) + P(C)P(A_p \mid C)}$$

In our case:

$$P(A \mid A_p) = \frac{0.2 \times 0.8}{(0.2 \times 0.8) + (0.5 \times 0.1) + (0.3 \times 0.1)} = \frac{0.16}{0.24} = 0.667$$

Similarly,

$$P(B \mid A_p) = \frac{0.5 \times 0.1}{(0.5 \times 0.1) + (0.2 \times 0.8) + (0.3 \times 0.1)} = \frac{0.05}{0.24} = 0.208$$

and

$$P(C \mid A_p) = \frac{0.3 \times 0.1}{(0.3 \times 0.1) + (0.2 \times 0.8) + (0.5 \times 0.1)} = \frac{0.03}{0.24} = 0.125$$

For branch B_p:

$$P(A|B_p) = \frac{0.2 \times 0.1}{(0.2 \times 0.1) + (0.5 \times 0.9) + (0.3 \times 0.2)} = \frac{0.02}{0.53} = 0.038$$

$$P(B|B_p) = \frac{0.45}{0.53} = 0.849$$

$$P(C|B_p) = \frac{0.06}{0.53} = 0.113$$

For branch C_p:

$$P(A|C_p) = \frac{0.2 \times 0.1}{(0.2 \times 0.1) + (0.5 \times 0) + (0.3 \times 0.7)} = \frac{0.02}{0.23} = 0.087$$

$$P(B|C_p) = 0$$

$$P(C|C_p) = \frac{0.21}{0.23} = 0.913$$

Step 4. Compute the Marginal Probabilities In order to determine whether to use the consultant, it is necessary to use the probabilities that the consultant's survey will actually predict events A_p, B_p, or C_p. The probabilities (derived as the *denominators* of the Bayes' equations) are:

$$P(A_p) = 0.24$$
$$P(B_p) = 0.53$$
$$P(C_p) = 0.23$$

Step 5. Construct a Decision Tree Management must now decide whether to use the consultant. Then they must decide about the new product (a_1—develop, a_2—do not develop). The situation is shown in the form of a decision tree in Exhibit 5.33.

The upper part of the tree shows the information presented in Exhibit 5.31. The lower part of the tree presents the situation of using the consultant. There exist three branches: branch A_p for when the survey predicts *A*, branch B_p for when the survey predicts *B*, and branch C_p for when the survey predicts *C*. The revised probabilities from Step 3 and the marginal probabilities from Step 4 are next entered on the tree (Exhibit 5.34).

Step 6. Solve the Decision Tree (Exhibit 5.34) The expected values at all chance points are computed. Then the best alternative at every decision point is selected. In this case, if the consultant is not used, the expected value (top branch) is zero. If the consultant is used, the expected value (rounded) using posterior probabilities is

$$0.24(\$166{,}700) + 0.53(\$113{,}400) + 0.23(\$0) = \$100{,}110$$

Step 7. Compute the Expected Value of Imperfect Information The information computed in Bayes' analysis is less than perfect, because its prediction reliability is not 100 percent. Therefore, it is interesting to determine the value of such information.

The expected value per decision of the imperfect sample or survey information (EVSI) is as follows:

EVSI = [Best expected value of the decision with the imperfect information (revised probabilities) before payment is made for the information]
– [Best expected value of the decision using the prior probabilities]

EXHIBIT 5.33 A Decision Tree for Bayes' Analysis—General Structure

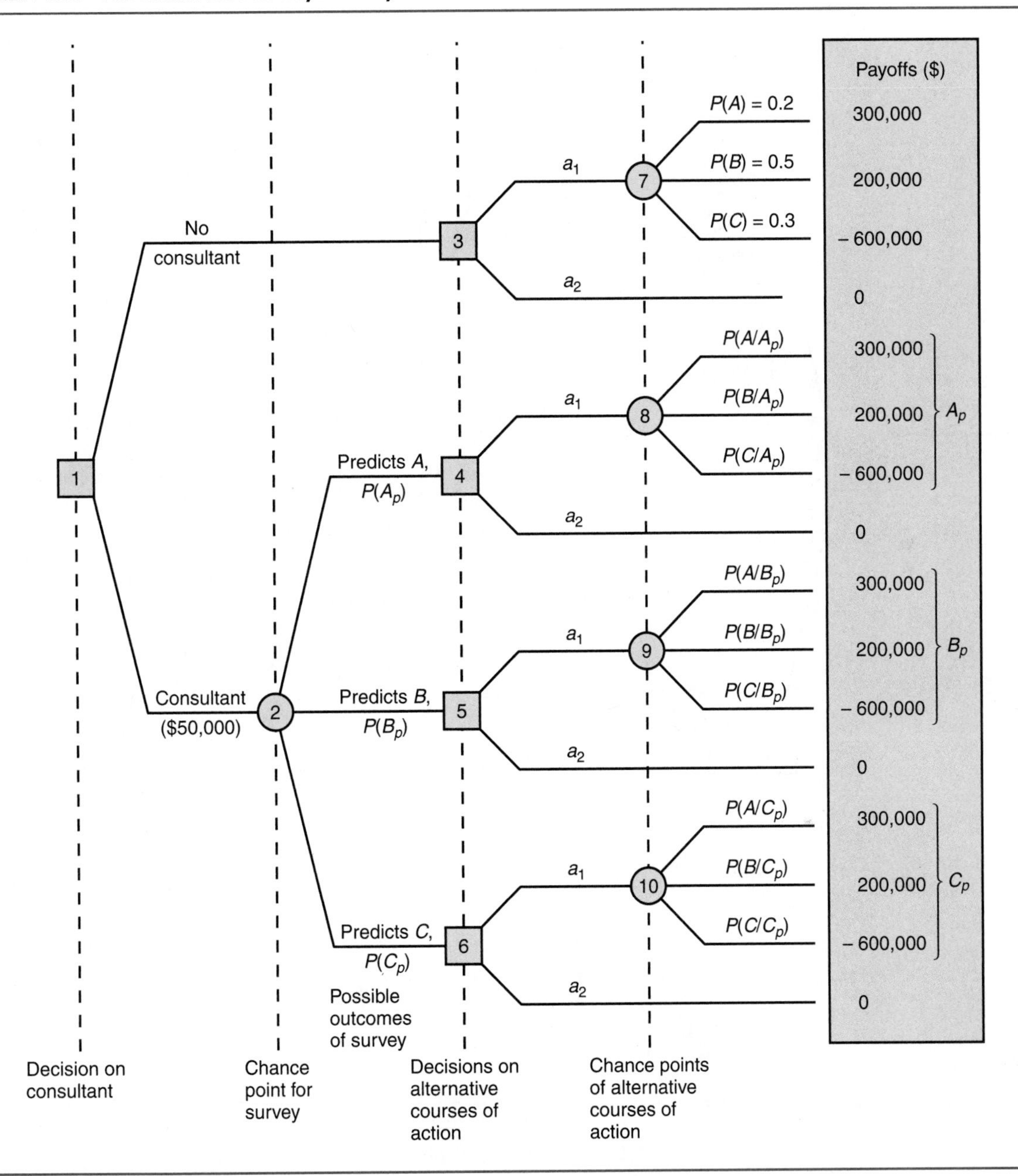

Note: In the case of minimization, the equation would contain these values:

EVSI = [Best expected cost using prior probabilities]
– [Best expected cost using revised probabilities]

In our example:

Expected value with imperfect information = $100,110
Expected value with prior probabilities = $0

EXHIBIT 5.34 Decision Analysis with Revised Probabilities

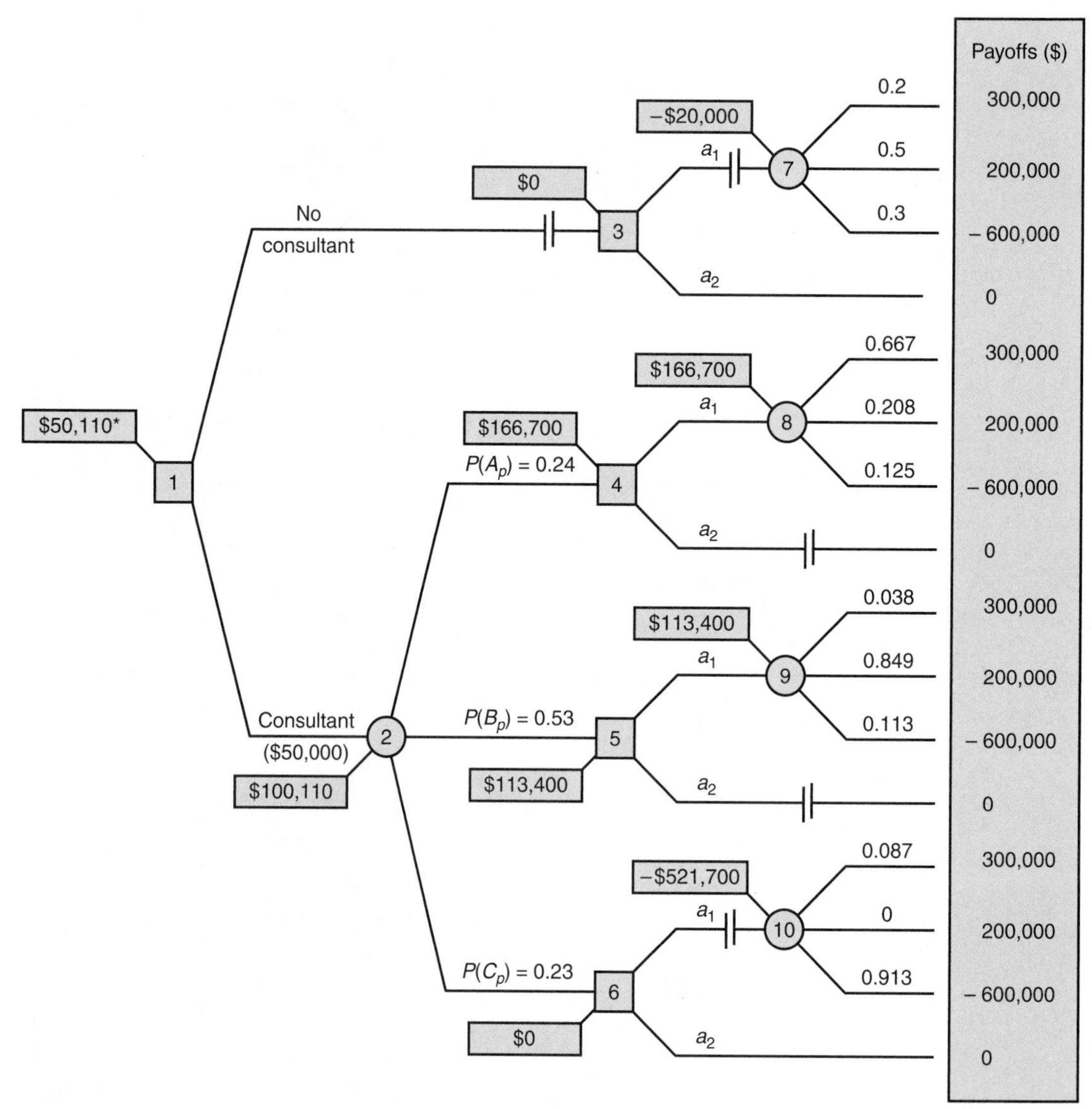

*Use of consultant is recommended.

Thus, EVSI = \$100,110 – 0 = \$100,110. As in the case of EVPI, here, too, we get an indication regarding the upper limit that management should be willing to pay for the additional information. Note that the expected value of imperfect information is *smaller than* the expected value of perfect information. The reason for this is that the imperfect information is less reliable. The value of perfect information serves as an *upper limit* for the value of imperfect information. The ratio of EVSI/EVPI is called the *efficiency* of the imperfect information.

Step 8. Decide Whether to Acquire the Information The expected value of the sample information is now compared with the cost of acquiring it.

If EVSI > Cost, acquire the information
If EVSI < Cost, do not acquire the information

In our case, EVSI = \$100,110 and the cost is \$50,000. Therefore, the information should be acquired.

Step 9. Select an Alternative If the consultant is not hired, select the alternative with the best expected value. If the consultant *is* hired, wait for the additional information and make a final decision accordingly. In our case, the decision policy is as follows:

If the survey indicates A_p, select alternative a_1
If the survey indicates B_p, select alternative a_1
If the survey indicates C_p, select alternative a_2

Note: In addition to using imperfect information for accepting or rejecting an opportunity to conduct research to gain additional information, one can use the methodology to select a research firm (or consultant) from among several. To do so, the track record of each contender is examined and the expected value of the sample information (EVSI) is compared against the cost.

Calculating Revised Probabilities

We use the American Ecology example from earlier in this section to further illustrate the calculation of revised probabilities.

Let $P(W)$ be the prior probability of the product being a winner. Given:

$P(W) = 0.80$
$P(L) = 0.20$ (the probability of the product being a loser)

The conditional probabilities of Exhibit 5.30 are given here:

1. In the event the product is actually a winner:

$P(S|W)$ = probability of the survey predicting success, given the product is actually a winner = 0.70
$P(I|W)$ = probability of the survey being inconclusive, given the product is actually a winner = 0.20
$P(F|W)$ = probability of the survey predicting failure, given the product is actually a winner = 0.10

2. In the event the product is actually a loser:

$P(S|L)$ = probability of the survey predicting success, given the product is actually a loser = 0.05
$P(I|L)$ = probability of the survey being inconclusive, given the product is actually a loser = 0.10
$P(F|L)$ = probability of the survey predicting failure, given the product is actually a loser = 0.85

The point to remember here is that all this information is known *before* the survey is actually taken.

Revision of the Probabilities of the States of Nature Suppose the survey is taken and it predicts success. The prior probability of $P(W)$ will be changed now to a posterior probability, $P(W|S)$, which is the probability of the product being a winner given that the survey predicts success. This probability can be computed by using the Bayes' formula for two variables:

$$P(W|S) = \frac{P(W)\,P(S|W)}{P(S)} = \frac{P(W)\,P(S|W)}{P(W)\,P(S|W) + P(L)\,P(S|L)}$$

This formula is adapted from the general Bayes' formula:

$$P(N_i|B) = \frac{P(N_i)\,P(B|N_i)}{\sum_{i=1}^{n} P(N_i)\,P(B|N_i)} = \frac{P(N_i)\,P(B|N_i)}{P(N_1)\,P(B|N_1) + P(N_2)\,P(B|N_2) + \cdots + P(N_n)\,P(B|N_n)}$$

where

B = outcome predicted by the research (or new information), which is, in our example, either success (S), inconclusive (I), or failure (F)

N_i = a possible state of nature (either a winner [W] or loser [L] in our example)

$i = 1, 2, 3, \ldots n$, where n = the number of states (two here: winner and loser)

This equation states that the posterior probabilities of the states of nature (N_i: winner or loser), after observing some survey evidence (B: success, inconclusive, or failure), is proportional to the product of the prior probability of N_i and the conditional probability of B given state N_i.

Computing the Revised Probabilities

In the event the survey predicts success, use the Bayes' formula for two variables to get

$$P(W|S) = \frac{0.8 \times 0.7}{(0.8 \times 0.7) + (0.2 \times 0.05)} = \frac{0.56}{0.57} = 0.9825$$

That is, the probability of a winner is increased from 80 to 98.25 percent due to the fact that the additional information predicted success. Similarly:

$$P(L|S) = \frac{P(L)\,P(S|L)}{P(L)\,P(S|L) + P(W)\,P(S|W)} = \frac{0.2 \times 0.05}{(0.2 \times 0.05) + (0.8 \times 0.7)} = 0.0175$$

It is possible to compute $P(L|S)$ in a shorter manner. Because the product can be either a winner or a loser, $P(W|S) + P(L|S)$ must sum to 1.0. Therefore:

$$P(L|S) = 1 - P(W|S) = 1 - 0.9825 = 0.0175$$

An important question that one may ask is, What is the probability of the survey predicting success? This probability is designated $P(S)$ and is computed from

$$P(S) = P(W)P(S|W) + P(L)P(S|L)$$

Note that this value is exactly the denominator in the Bayes' formula for two states. In our example:

$$P(S) = (0.8 \times 0.7) + (0.2 \times 0.05) = 0.57$$

In the event the survey is inconclusive, we use the general Bayes' formula for i different states. We get

$$P(W|I) = \frac{P(W)\,P(I|W)}{P(W)\,P(I|W) + P(L)\,P(I|L)} = \frac{0.8 \times 0.2}{(0.8 \times 0.2) + (0.2 \times 0.1)} = 0.8889$$

Similarly: $P(L|I) = 1 - 0.8889 = 0.1111$. Also, the possibility of the survey predicting "inconclusive," $P(I)$, is computed as 0.18 (the denominator of the $P(W|I)$ equation). *In the event the survey predicts failure,* use this equation:

$$P(W|F) = \frac{P(W)\,P(F|W)}{P(W)\,P(F|W) + P(L)\,P(F|L)} = \frac{0.8 \times 0.1}{(0.8 \times 0.1) + (0.2 \times 0.85)} = 0.32$$

Notice the drastic revision, from 80 percent down to 32 percent! And similarly,

$$P(L|F) = 1 - 0.32 = 0.68$$

The chance of the survey predicting "failure," $P(F)$, is 0.25 (the denominator of the $P(W|F)$ equation).

What Will the Survey Predict? We showed that it is possible to revise the initial probabilities *without actually taking the survey.* Of course, the answers that we received are *conditional,* depending on the outcome of the survey. For example, if the survey predicts success, then the probability of having a winner, $P(W|S)$, is 98.25 percent, and so on. In decision making, it is important to find out *before* the survey is taken, the chance that the survey will predict success, failure, or will be inconclusive. In deriving this solution, it was found that

$$P(S) = \text{probability that the survey will indicate success} = 0.57$$
$$P(I) = \text{probability that the survey will be inconclusive} = 0.18$$
$$P(F) = \text{probability that the survey will indicate failure} = 0.25$$

Note that because these are the only possible survey outcomes, they must sum to 1.0. Exhibits 5.35 and 5.36 summarize the process and the results obtained in this example.

EXHIBIT 5.35 Summary of the Bayes' Process

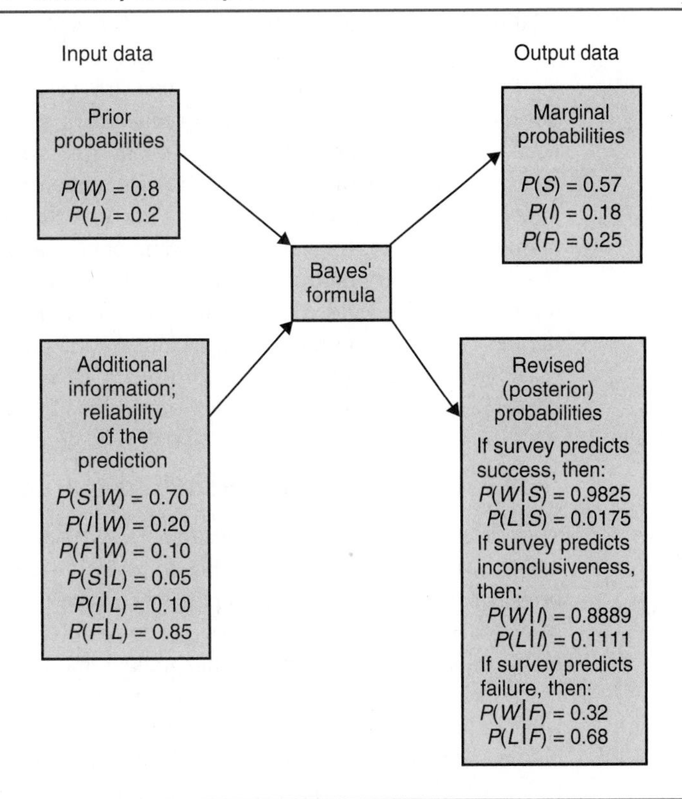

EXHIBIT 5.36 Tree Representation of the Bayes' Process

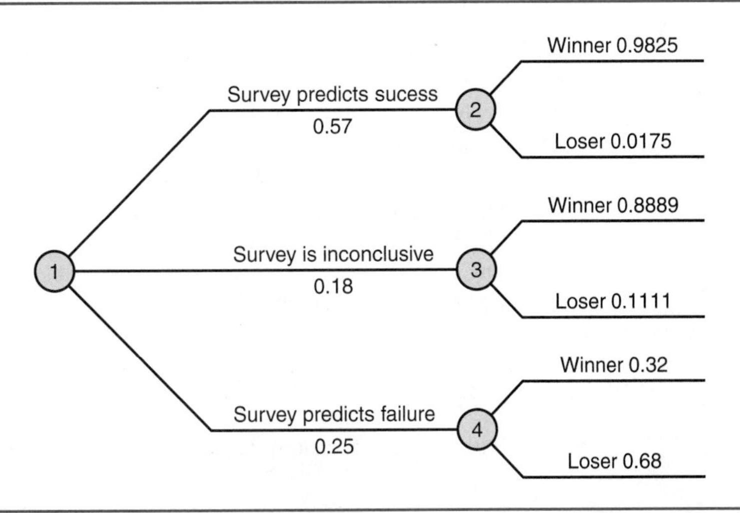

5.9 Detailed Modeling Example

To illustrate the use of decision analysis in the broader modeling context, consider M & S Securities, a regional stock brokerage firm with branches located throughout the southeast. M & S focuses on recruiting experienced financial advisors who can provide clients with superior investment advice. As a full service broker that conducts its own proprietary research, M & S's commission structure is significantly higher than discount brokers and online brokers that primarily execute trades but offer no advice. Thus, a key to its past success was the ability of its financial advisors to develop and cultivate long-term relationships with the clients.

Initially, top management at M & S dismissed the use of the Web as a potential vehicle for trading stocks. Their primary concern was a feeling that most investors simply do not have the knowledge or savvy to make their own investment decisions and therefore need the advice of experienced investment professionals. However, as a result of the increasing popularity of the Web, M & S did implement a password-protected Web site that allows its clients to view and track their investment portfolios, obtain real-time stock quotes, e-mail their financial advisor, and obtain research reports.

When industry brokerage giant Merrill Lynch began offering online trading, M & S's top management decided to reevaluate its e-commerce strategy. A further motivation was the rapid growth of online trading, with the total number of online accounts in the United States increasing from 1.5 million in 1996 to 7.5 million by early 1999. In addition, online commissions are projected to triple from 1998 to 2002.[3]

Step 1: Opportunity/Problem Recognition

As a result of the recent trends toward increased usage of the Web for stock trading, management at M & S Securities would like to assess the extent this threatens its traditional stock brokerage business and identify options to ensure its long-term viability. More specifically, M & S has identified three options: (1) do not offer online trading but continue to monitor online trading trends; (2) offer online trading through its existing business; and (3) create a new, separate subsidiary that specializes in online trading. The president and CEO of M & S believes that three scenarios or states of nature are possible: (1) there will be a major shift to online trading over the next 5 years; (2) there will be a moderate shift to online trading; and (3) there will be only limited acceptance of online trading 5 years from now.

Step 2: Model Formulation

An ad-hoc committee was set up to analyze the situation and make recommendations. The committee first determined that a less ambiguous definition was needed for the three states of nature. Eventually the team agreed to the following definitions: (1) a major shift to online trading was defined as more than 75 percent of stock trades being conducted online; (2) a moderate shift was defined as between 30 percent to 75 percent of stock trades being conducted online; and (3) limited acceptance was defined as less than 30 percent of stock trades being done online in 5 years.

Having a more precise definition of the possible states of nature, the committee decided to split into two subcommittees. The first committee was charged with determining the probabilities of the three states of nature. After conducting interviews with a variety of industry analysts and consulting with several academic sources, the committee reached the following consensus:

- A 30 percent chance that more than 75 percent of trades would be done online in 5 years

- A 60 percent chance that between 30 percent and 75 percent of trades would be done online
- A 10 percent chance that fewer than 30 percent of the trades would be done online in 5 years.

The other subcommittee developed the influence diagram shown in Exhibit 5.37 to identify the key factors affecting profitability. Then, based on these factors they conducted a detailed financial analysis for each option/state of nature combination. As the influence diagram shows, conducting this analysis required making detailed assumptions about the number of clients the company would have, how many trades each client would make, and what the commission structure would be. The committee ultimately realized that for each option/state of nature combination, there was a range of possible profits. To illustrate, for the state of nature of more than 75 percent of stock trades being conducted online, the analysis showed that M & S's profits would be vastly different if 75 percent of trades were conducted online, versus 95 percent. Therefore, the committee concluded that simply reporting the expected profit for each option/state of nature combination could be somewhat misleading and opted to also include the standard deviation to give a more complete picture of the likely outcome.

Step 3: Data Collection

Exhibit 5.38 summarizes the results of the financial analysis. For example, if more than 75 percent of trading is done online in 5 years, and M & S does not offer online trading, the expected loss is $100 million per year, with a standard deviation of $20 million. The subcommittee constructed histograms for the payoffs for each option/state of nature combination and in all cases the distributions were approximately normally distributed.

Step 4: Analysis of the Model

The expected profits for each option/state of nature combination and the probabilities for each state of nature were entered into the spreadsheet shown in Exhibit 5.39. Then formulas were entered in column E to calculate the expected value for each option.

EXHIBIT 5.37 Influence Diagram Identifying Relevant Factors that Impact M & S's Profitability

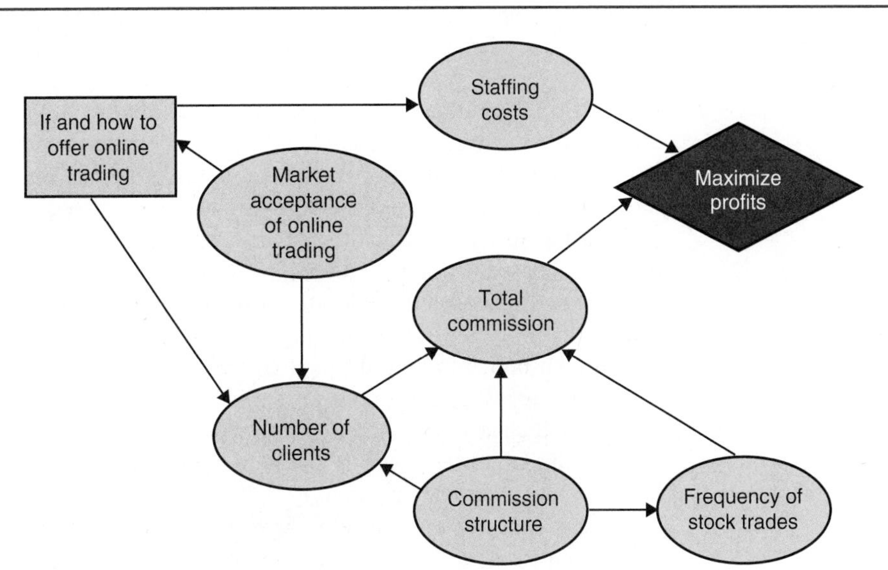

EXHIBIT 5.38 Summary of Payoffs for Each Option/State of Nature Combination ($ Millions)

	> 75% of Trades Done Online	30% to 75% of Trades Done Online	< 30% of Trades Done Online
Don't offer online trading (mean, σ)	–100, 20	–25, 7.2	135, 34.3
Offer online trading (mean, σ)	24, 4.2	18, 3.7	–3, 0.4
Create new subsidiary (mean, σ)	32, 10.2	16, 8.7	–4, 2.1

EXHIBIT 5.39 Payoff Table for M & S Investments

	A	B	C	D	E
1		States of Nature			
2		Significant	Moderate	Minimal	Expected
3	Alternatives	Shift	Shift	Shift	Value
4	Probabilities ⟶	0.3	0.6	0.1	
5	Don't offer online trading	-100	-25	135	-31.5
6	Offer online trading	24	18	-3	17.7
7	Create new subsidiary	32	16	-4	18.8
8					
9	Cell E5	=SUMPRODUCT($B$4:$D$4,B5:D5)			
10		Copy to cells E6:E7			

According to the results shown in Exhibit 5.39, creating a new subsidiary that specializes in online trading appears to be the most attractive option with an expected value of $18.8 million. Offering online trading through its existing business has a slightly lower expected value of $17.7 million. Not offering online trading at all appears to be highly undesirable with an expected loss of $31.5 million.

As the committee investigated the situation further, they realized that while the new subsidiary had the highest expected value, it also entailed more risk because there was more variability in the payoffs associated with this option. For example, referring to Exhibit 5.38, if there was a moderate shift to online trading, offering online trading through the existing business would have an expected annual profit of $18 million, with a standard deviation of only 3.7 million. Alternatively, for this same state of nature, creating a new subsidiary has a slightly lower expected profit of $16 million, but has considerably more variability with a standard deviation of $8.7 million. Stated another way, based on the information provided in Exhibit 5.38, the following can be easily shown: If there is a moderate shift to online trading, then there is a 95 percent chance that M & S would earn a profit in the range of about $11 million to $25 million if it offered online trading through its existing business. On the other hand, if M & S creates a new subsidiary and there is a moderate shift to online trading, there would be a 95 percent chance that annual profits would range from –$1 million to $33 million. In general, the larger the range of likely outcomes, the riskier the option becomes due to less certainty about the outcome.

Because the expected value of the new subsidiary option was only slightly higher than offering online trading through its current business but was significantly more risky, the committee determined that additional analysis was needed. To analyze the situation further and to account for the differences in variability across the options, the committee decided to perform a quick "simulation" of the situation (simulation will be discussed in more detail in Chapter 7). To accomplish this, the spreadsheet shown in Exhibit 5.40 was developed.

The committee initially decided to *replicate* (i.e., run, or repeat) the decision 50 times. Thus, the first column of the spreadsheet shown in Exhibit 5.40 keeps track of the replication number. The second column corresponds to the state of nature. Since the probability of a significant shift to online trading is estimated to be 30 percent, 15 (i.e., 30 percent × 50) of the replications correspond to this state of nature. Likewise, 30 (60 percent × 50) replications correspond to a moderate shift to online trading, and five replications (10 percent × 50) correspond to a minimal shift to online trading.

Column C in Exhibit 5.40 contains randomly generated outcomes for the offer online trading option. For example, the values in cells C4:C18 were generated using Excel's random number generator and specifying that the values should be generated from a normal distribution with a mean of 24 and a standard deviation of 4.2 (see Exhibit 5.38). The remaining values in columns C and D were generated in a similar fashion. For example, the values in cells C19:C48 were generated using Excel's random number generator and specifying a mean of 18 and standard deviation of 3.7.

Interestingly, when the variability of the options is considered, offering online trading through the current business becomes the preferred choice. To understand how this could have happened refer to offering online trading (column C) and the state of nature corresponding to a moderate shift to online trading. Calculating the mean of the randomly generated profits for this option/state of nature combination (cells C19:C48) gives 18.7 (which is close to the specified mean of 18). Further examination reveals that while the average of the values generated was close to the expected payoff, the range of payoffs is quite large, from a low of 8.9 to a high of 23.9.

Of course, the simulation model developed here was replicated a very small number of times and the results might have been different had it been replicated more times. The point of this exercise is to demonstrate a fundamental limitation of making decisions purely on the basis of precise mathematical and financial models. Although these models often involve making precise calculations, the calculations are made on the basis of data that are often subject to a high degree of uncertainty. Therefore, the wise manager not only develops estimates for unknown parameters but also considers the variability or the amount of uncertainty associated with the estimate in evaluating decision alternatives. Furthermore, the importance of considering variability increases as the number of variables that depend on one another increase, and as the distribution becomes more skewed. However, even in the example presented here with only one randomly generated variable and the perfectly symmetrical normal distribution, we were able to demonstrate the importance of considering the variability of our estimates.

In addition to considering the variability of the payoffs, sensitivity analyses can be performed on the payoffs and on the probabilities of the states of nature. Exhibit 5.41 shows the results of a sensitivity analysis investigating how much the probabilities would have to be altered for the alternative of offering online trading through the existing business to become the preferred choice.

Step 5: Implementation

In this situation, the decision maker is faced with a dilemma. Namely, should the option with the highest expected value and also the highest risk be chosen, or should an option with a slightly lower expected value and significantly less risk be selected? There is no right answer to this question, and it largely depends on the attitude toward risk of the decision maker. Exhibit 5.42 illustrates in the form of a memo how the results of this analysis might be communicated to top management at M & S Securities. The companion Web site for this book contains a sample PowerPoint presentation of the results of the analysis.

EXHIBIT 5.40 Simulation Analysis of M & S Securities

	A	B	C	D
1			Offer	Create
2		State of	Online	New
3	Replication	Nature	Trading	Subsidiary
4	1	Significant shift	30.9	25.5
5	2	Significant shift	34.2	33.4
6	3	Significant shift	22.1	21.5
7	4	Significant shift	23.6	33.3
8	5	Significant shift	24.9	28.9
9	6	Significant shift	21.8	24.5
10	7	Significant shift	22.3	26.3
11	8	Significant shift	28.2	18.2
12	9	Significant shift	22.4	32.2
13	10	Significant shift	25.4	36.0
14	11	Significant shift	23.3	43.4
15	12	Significant shift	21.9	30.9
16	13	Significant shift	25.0	34.3
17	14	Significant shift	22.2	23.3
18	15	Significant shift	29.6	20.5
19	16	Moderate shift	21.9	-9.9
20	17	Moderate shift	20.1	4.5
21	18	Moderate shift	17.7	20.0
22	19	Moderate shift	19.7	20.1
23	20	Moderate shift	20.8	28.9
24	21	Moderate shift	18.3	11.5
25	22	Moderate shift	22.5	17.9
26	23	Moderate shift	14.7	22.2
27	24	Moderate shift	23.4	36.0
28	25	Moderate shift	8.9	1.4
29	26	Moderate shift	20.1	6.8
30	27	Moderate shift	19.5	0.2
31	28	Moderate shift	17.6	12.0
32	29	Moderate shift	14.9	16.9
33	30	Moderate shift	22.7	10.3
34	31	Moderate shift	17.7	19.4
35	32	Moderate shift	11.2	14.1
36	33	Moderate shift	15.7	2.5
37	34	Moderate shift	23.9	14.6
38	35	Moderate shift	18.8	17.1
39	36	Moderate shift	20.0	15.6
40	37	Moderate shift	18.3	13.2
41	38	Moderate shift	23.9	18.5
42	39	Moderate shift	14.1	20.3
43	40	Moderate shift	17.9	26.7
44	41	Moderate shift	17.3	5.1
45	42	Moderate shift	21.2	9.6
46	43	Moderate shift	16.1	13.5
47	44	Moderate shift	20.0	11.3
48	45	Moderate shift	23.2	34.6
49	46	Minimal shift	-3.5	-3.7
50	47	Minimal shift	-3.2	-4.0
51	48	Minimal shift	-3.1	-3.5
52	49	Minimal shift	-2.4	-4.5
53	50	Minimal shift	-3.1	-3.1
54	**Average**		**18.49**	**16.96**

EXHIBIT 5.41 Sensitivity Analysis of the Probability Estimates

	A	B	C	D	E
1		States of Nature			
2		Significant	Moderate	Minimal	Expected
3	Alternatives	Shift	Shift	Shift	Value
4	Probabilities ⟶	0.1	0.8	0.1	
5	Don't offer online trading	-100	-25	135	-16.5
6	Offer online trading	24	18	-3	16.5
7	Create new subsidiary	32	16	-4	15.6
8					
9	Cell E5	=SUMPRODUCT($B$4:$D$4,B5:D5)			
10		Copy to cells E6:E7			

EXHIBIT 5.42 Sample Memo to M & S Top Management

MEMO

To: Jonathan Mars, managing partner
From: Ad-Hoc Online Trading Committee
CC: Senior and associate partners
Date: 3/28/02
Re: Online stock trading strategy

Introduction

We have concluded our analysis of three online stock trading strategies: (1) not offering online trading at the present time but continuing to monitor trends in online trading; (2) offering online trading through our current business; and (3) creating a new, separate subsidiary that specializes in online trading. Based on the analysis performed, our committee recommends that online trading be offered through the existing business. Although the expected profitability associated with this option is slightly less than that of creating a new subsidiary, it also entails significantly less financial risk.

Analysis

For the purpose of this analysis, the committee considered three scenarios: (1) a major shift to online trading which was defined as over 75 percent of stock trades being conducted online within five years; (2) a moderate shift to online trading defined as 30 percent to 75 percent of stock trades being conducted online within five years; and (3) limited acceptance of online trading with less than 30 percent of stock trades being conducted online within five years. Based on interviews with a variety of industry analysts and academic sources, the committee estimates that the

(continued)

EXHIBIT 5.42 Continued

probabilities of these three scenarios occurring are 30 percent, 60 percent, and 10 percent, respectively.

A detailed financial analysis was conducted for each scenario–alternative combination. Conducting this analysis required making many detailed assumptions about the number of clients the company would have, the frequency clients would trade, and what the commission structure would be for various types of clients. It was soon recognized that for each scenario–alternative combination, a range of possible profits existed. Therefore, rather than simply reporting the average profit for each scenario–alternative combination, the standard deviations are also included in Table 1 to provide a more complete picture of the likely outcome. To illustrate, if more than 75 percent of the trades are done online in five years and M & S does not offer online trading, our analysis indicates the company would incur an annual loss of $100 million with a standard deviation of 20 million.

TABLE 1
Summary of Annual Profit Projections for Each Scenario–Alternative Combination ($ Millions)

	> 75% of Trades Done Online	30% to 75% of Trades Done Online	< 30% of Trades Done Online
Don't offer online trading (mean, σ)	−100, 20	−25, 7.2	135, 34.3
Offer online trading (mean, σ)	24, 4.2	18, 3.7	−3, 0.4
Create new subsidiary (mean, σ)	32, 10.2	16, 8.7	−4, 2.1

Based on the financial analysis summarized in Table 1 and the committee's estimates of the probability of each scenario occurring, the expected profit associated with each decision alternative was calculated. Accordingly, it is estimated that the expected annual loss of not offering online trading would be $31.5 million. Likewise, if M & S were to offer online trading through its existing business, the expected profit would be $17.7 million and the expected profit of creating a new subsidiary specializing in online trading is $18.8 million.

Although it is true that creating a new subsidiary does have the highest expected profit, note that it is also entails significantly more risk than offering online trading through our current business unit. To illustrate, if there is a moderate shift to online trading, then our analysis indicates there is a 95 percent chance that M & S would earn profits of $11 million to $25 million if it opted to offer online trading through its existing business. On the other hand, under this same scenario, if a new subsidiary were created there is a 95 percent chance that annual profits would range from –$1 million to $33 million. Therefore, we deem creating a new subsidiary as more risky, given the much larger range of possible outcomes, which ultimately translates into greater uncertainty.

Recommendations

In conclusion, because the expected profitability of creating a new subsidiary is only slightly higher but entails significantly more financial risk, our committee recommends that online trading be offered through the existing business. In fact, when the variability was explicitly considered in a simulation model developed by the committee, offering online trading through the existing business became the preferred choice.

Assumptions/Limitations

Given the dynamic nature of the competitive environment it is important to recognize that many of the assumptions made in this analysis may need to be updated as new information becomes available. In particular, key assumptions were made regarding the likelihood of the alternative scenarios occurring, the size of the online trading market, the frequency of trades, and the future commission structure. Even modest deviations from our assumed values could lead to vastly different results. Therefore, it is strongly recommended that the assumptions be periodically revisited and updated as needed, and the analysis be rerun to reflect these adjustments.

QUESTIONS

1. Describe the difference between risk and uncertainty; between uncertainty and ignorance.
2. Since the techniques for decision making under uncertainty are clearer and easier to invoke, why not also use those approaches for decision making under risk?
3. Do the criteria of optimism, pessimism, or regret have a natural appeal to your instincts?
4. Discuss the logic of using an expected value criterion versus a most probable state of nature criterion for a one-shot decision situation.
5. What sensitivity analyses might be useful to a manager in the decision under risk tables?
6. Is a decision situation posed as a decision tree easier to visualize than in a table?
7. Explain how to find the value of either perfect or imperfect information. Where would a manager go to find such information?
8. Bayes' theorem is useful only in particular circumstances. Explain what these circumstances are.
9. Legend has it that when Akio Morita, the former co-founder and head of Sony, commissioned a marketing survey of potential consumers of his Walkman conception, the survey results were dismal. No one, it seems, was interested in walking around with earphones on and a cord running down their body, plugged into a radio or cassette player strapped to their pants or skirt. Nevertheless, Mr. Morita was not dissuaded by the negative results. He went ahead with marketing his innovation. Discuss the logic of paying for survey information and then selecting an alternative that has the lowest expected value according to the revised probabilities. Is there some wisdom in Mr. Morita's decision?

EXPERIENTIAL EXERCISES

1. An article that appeared in *The Wall Street Journal* in early 1999 discussed a dilemma facing Merrill Lynch & Co. (ML).[4] On the one hand, it charges substantially higher commissions than discount and online brokerage firms to support its army of 15,000 stockbrokers that earn six- and seven-figure incomes and have resisted offering its clients an online trading service. At the same time, it may be facing what is often described as a once-in-a-lifetime revolution in technology. Being the largest and most profitable firm on Wall Street also means that ML has the most to lose by not adequately responding to current trends. It has been estimated that approximately $2 billion of its $17 billion in revenues are derived from commissions paid by individual investors.

   Perhaps ML's current position is best summarized in the words of the head of its individual investor operations department: "The do-it yourself model of investing, centered on Internet trading, should be regarded as a serious threat to American financial lives."

   For the purpose of this exercise, assume that ML's position has not changed and that it does not offer online trading. You have been contacted to advise ML regarding its e-commerce strategy. As such, you have been charged with the following:

   - Develop a list of options for ML to pursue. At a minimum, this should include not offering online trading, offering online trading through its current operations but not changing its current commission fee structure, offering online trading through its current operations and changing its fee structure to be more competitive with other discount and online stock brokerage firms, and purchasing an online brokerage firm. For the last option, identify one or more online discount brokerage firms and estimate what the acquisition cost would be. Also note that many stock brokerage firms charge different fees, depending on how an order is placed (e.g., phone versus Web).
   - Identify possible future states of nature and estimate the probability of each. Be prepared to support your probability estimates with appropriate sources and analysis.
   - Evaluate the impact that the occurrence of each state of nature would have on each option you identified. Clearly define any assumptions made, including increases/decreases in the number of clients, changes in the frequency of trades, changes to the commission structure, and so on.
   - Calculate the expected payoff for each option.
   - Write a memo to ML management overviewing your analysis and offering specific recommendations. Also include key assumptions and limitations of your analysis.

2. With reference to the Amazon example at the beginning of the chapter, complete the following tasks:

- Select a technology company such as a dot-com or biotechnology firm that you are familiar with for analysis.
- Research the company using analyst reports, articles in the business press, and/or interviews with knowledgeable individuals. Based on your research, develop an influence diagram similar to Exhibit 5.1 but specific to your firm, identifying key factors that will drive its market value.
- Identify various likely scenarios and calculate your firm's market value for each scenario.
- Investigate the impact of assigning alternative probabilities to the various scenarios.
- Based on your analysis, write a memo to the manager of a technology mutual fund regarding your recommendation about adding this company to the fund.

MODELING EXERCISES

1. An $800,000 property has a 0.1 percent chance of catching fire that will cause damages of $100,000; and a 0.05 percent chance of catching fire that will completely destroy the property. Management decided to insure the property, and it is reviewing two possible insurance policies. The first is a policy with $50,000 deductible; that is, the insurance company covers all damages except the initial $50,000. The annual premium for such a policy is $750. The second option is a no-deductible (fully paid) policy with an annual premium of $1,000.
   a. If the company's objective is cost minimization, which policy should it purchase? Build both an opportunity loss table and a regular payoff decision table for the situation and solve both of them.
   b. Suppose the company decides *not* to insure, a practice that is called *self-insurance.* What will then be the expected cost?
   c. Why is the expected cost of self-insuring *lower* than that of insuring?
   d. Why do companies insure rather than self-insure even though the expected cost of self-insuring is much lower?
2. The yearly demand for a seasonal, profitable item follows the distribution below:

Demand (Units)	Probability
1,000	0.20
2,000	0.30
3,000	0.40
4,000	0.10

   The manufacturer of the item can produce it by one of three methods:
   a. Use existing tools at a cost of $6 per unit.
   b. Buy special equipment for $1,000. The value of the equipment at the end of the year (salvage value) is zero. The cost is $3 per unit.
   c. Buy special equipment for $10,000 that can be depreciated over 4 years (one fourth of the value each year). The cost of using this equipment is $2 per unit.

   Which method of production should the manufacturer follow in order to maximize profit? *Hint:* Compare total annual costs. Assume production must meet all demand; each unit demanded and sold means more profit.
3. Find the best alternative in the following decision tables: a and b are *profits,* and c is *cost* data. Use both an expected value and an expected opportunity loss (EOL) approach. Find the expected value of perfect information in each of the three tables. Compare the results to the EOL.

a.

Alternatives \ States of Nature	0.3 s_1	0.5 s_2	0.2 s_3
a_1	5	8	3
a_2	6	5	7

b.

Alternatives \ States of Nature	0.6 s_1	0.1 s_2	0.2 s_3	0.1 s_4
a_1	3	5	8	−1
a_2	6	5	2	0
a_3	0	5	6	4

c.

Alternatives \ States of Nature	0.1 s_1	0.6 s_2	0.3 s_3
a_1	5	2	1
a_2	4	3	3
a_3	2	6	1

4. A firm needs temporary business space. It can lease the desired space for $5,000 for 1 year or $9,000 for 2 years (all rent is paid in advance). Alternatively, it can rent the space for 1 year for

$5,000 and then if it wishes to re-rent, pay $5,500 on the first day of the second year (rent for the second year). The firm estimates that there is a 25 percent chance that it will have to depart from the city after a year. In that case, if they have rented the space for 2 years, they can sublet it for the second year. The chance of subletting is 60 percent, and the firm would receive $5,500, paid on the first day of the second year. If it is unable to sublet the space, it will remain empty for the entire second year. The interest rate that the company uses for evaluating such decisions is 10 percent. Should the company rent the space for 1 year or 2? Why?

5. Greenwood Groceries buys fresh fruit daily for $10 a crate. Crates sold the same day bring $15 profit contribution, each. Crates that are not sold in a day are sold later as animal food for $2.50 each. The demand for fruit fluctuates, according to the following distribution (data collected over the last 300 days):

Demand (Number of Crates)	Number of Days
10	120
11	90
12	75
13	15
	Total 300

a. How many crates should the store order if Greenwood wants to maximize profit from selling fruit? Use an expected value approach.
b. What will be the average daily profit?
c. Assume the store can buy information that will enable it to predict the daily demand with certainty. How much should the store be willing to pay for such information?

6. A consultant plans to work in Bolivia, where the exchange rate is 100 pesos for $1. He plans to be there for several months, and his expenses for the period are estimated at 250,000 pesos. Skyrocketing inflation is temporarily in check while the government attempts to get a loan, the effect of which will be to lower the exchange rate by 10 percent (i.e., $1 = 90 pesos). If the loan is refused, the exchange rate will increase by 20 percent ($1 = 120 pesos). Suppose it is known that the probability of the government receiving the loan is 0.80. The consultant considers the following alternatives:

a_1 Immediately convert enough dollars into pesos to meet expenses for the entire period.

a_2 Wait until the loan is either granted or refused; in the meantime, hold dollars.

a_3 Hedge by converting part of the dollars to 125,000 pesos now and holding enough in dollars until the loan is either granted or refused, and then buy 125,000 additional pesos.

Assume that the decision on the loan is to be made prior to the consultant's arrival in the foreign country. Assume, also, that after the change in the exchange rate he will still need 250,000 pesos (regardless of the exchange rate). If the consultant wants to minimize his dollar expense, which course of action should he take? Find the value of perfect information and comment on it.

7.
a. Write the following decision table in tree form:

Alternatives \ Futures	Fire	No Fire
Insure	$ 100	$100
Do not insure	8,000	0

These figures are cost data. There is a 0.01 chance of a fire.

b. What action should an individual take to minimize cost?

8. Given the decision tree below, find the best alternative and its expected value. The outcomes shown are *costs* and the investment expenses are in parentheses.

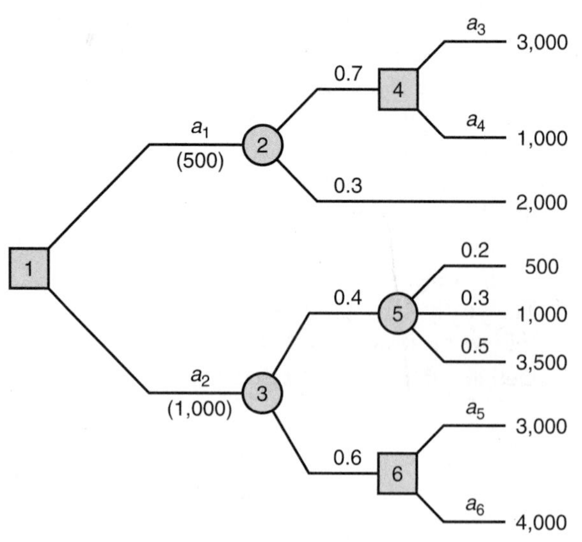

9. A machine shop received an order for 2,000 units to be made on one of its automated machines. This is a multistation operation and once it is started, it runs without interruption. The machine shop makes $2 profit on each part of acceptable quality. Each unit that is classified as "defective" needs rework at an additional cost of $3.50 before it is considered "acceptable."

Historical data indicate that if the machine is used without any special preparation, it produces 1 percent defectives (this happens 50 percent of the time), 2 percent defectives (this happens 30 percent of the time), 3 percent defectives (this happens 12 percent of the time), or 5 percent defectives in the remaining cases. With a minor adjustment that costs $42, the defective rate *above* 2 percent is reduced to 2 percent. (That is, there is *no* chance of a 3 percent or 5 percent defect rate.) With a major adjustment (cost $100), the defective rate is 1 percent. What adjustment policy should the machine shop adopt in order to maximize profit?

10. An oil explorer, commonly called a wildcatter, must decide whether to drill a well or sell his rights to a particular exploration site. The desirability of drilling depends on whether there is oil beneath the surface. Before drilling, the wildcatter has the option of taking seismographic readings that will give him further geological and geophysical information. This information will enable him to deduce whether subsurface structures usually associated with oil fields exist in this particular location. However, some uncertainty about the presence of oil will still exist after seismic testing because oil is sometimes found where no subsurface structure is detected, and vice versa.

    The wildcatter estimates that the cost of drilling a well would be $250,000 (in net present value terms, after making allowance for all taxes). The yield that would be expected from a typical oil well is estimated to be $1.2 million (in net present value terms, net of all taxes and operating costs, but excluding drilling costs). Seismic tests would cost $50,000 per test.

    The wildcatter could sell his rights for $230,000 before either drilling or testing. However, if he should decide to carry out seismographic readings and no subsurface structure is indicated, the site will be considered almost worthless by other wildcatters, in which case he will barely be able to sell the rights for $10,000. If substructure is indicated by the test, he can sell his rights for $300,000. If no oil is found, the value of the exploration site is considered to be zero.

    The probability of getting oil from the site without any test is 41 percent. If he carries out the seismic test, he feels that the test will indicate subsurface structure with a 40 percent probability. In case of structure, he can drill with a 65 percent chance of finding oil. In the case of no structure, he can drill with a 75 percent chance of finding the well dry.

    What should the wildcatter do? Draw a decision tree to solve the problem.

11. Todo Corp. operates two complex production lines on an 8-hour-per-day basis. These lines fail frequently due to the extensive workload. The daily probability of line no. 1 failing is 0.10 and that of line no. 2 is 0.15. When any line fails, the cost is 10,000 yen each hour; when both lines are down at the same time, the cost is 14,000 yen per hour. The company is trying to decide whether to hire one or two repairpersons. The time for one person to repair a line is 5 hours. If two are hired, they can work either as a team or each on a separate line. If they work as a team, the repair time is 3 hours per line. If a repairperson costs 8,000 yen per day, what would you recommend Todo do? *Note:* Repairpersons are salaried and not paid overtime, even if their workload some days requires it.

12. Given two decision tables below, find the best alternative in each by the following criteria:
    a. Equal probabilities (Laplace)
    b. Pessimism
    c. Optimism
    d. Coefficient of optimism (Hurwicz) with $\alpha = 0.4$
    e. Regret (Savage)

TABLE 1

	s_1	s_2	s_3	s_4
a_1	5	8	3	1
a_2	7	4	5	2
a_3	3	6	6	4

TABLE 2

	s_1	s_2	s_3
a_1	7	2	−1
a_2	3	6	2
a_3	0	3	8

13. The manager of an advertising agency has to decide between three available programs (a_1, a_2, a_3). There are three possible futures that can be expected: s_1 = market rises, s_2 = market falls, s_3 = no change in the market. The manager can estimate the yields in each case (given in the following table, in percent of return) but cannot estimate the probabilities of the various futures occurring.

Programs \ Futures	s_1	s_2	s_3
a_1	3	6	−1
a_2	8	5	4
a_3	−4	7	12

Which program will the manager select if she uses the following decision approaches:

a. Equal probabilities (Laplace)
b. Pessimistic approach
c. Optimistic approach
d. Hurwicz criterion with $\alpha = 0.55$
e. Minimax regret (Savage)

14. Italian Investor's Bank is evaluating two investment proposals involving 3 billion lire. The first is to buy Italian class A bonds with a 7.3 percent return. The second is to buy some land in Easton, Pennsylvania. The land is intended for development into an industrial park, in which case a 17 percent return is expected. However, the land is close to a planned new highway, and the government may purchase the land to build a rest area. In this case, the government will pay the bank 4.5 percent above the purchase price.

    Assuming a one-year decision horizon, what would you advise the bank to do if the probability of the government action is unknown? What assumption must be made in order to solve this problem? Although the probability of the government action is unknown, there exists a theoretical probability that will make the two alternatives *equal.* Find that probability.

15. Given a payoff matrix (profits in thousands of dollars):

	s_1	s_2
a_1	5	8
a_2	9	3
a_3	−1	10

The probability of s_1 is 0.35 and that of s_2 is 0.65. A research company's track record showed that it had three predictions (p_1, p_2, and p_3) with respect to the states of nature as shown below.

	s_1	s_2
p_1	0.8	0
p_2	0.1	0.1
p_3	0.1	0.9

a. Draw a decision tree for the situation, including the decision regarding the employment of the research company.
b. Show all revised (posterior) probabilities.
c. What is the probability of predicting p_3?
d. The research company charges $2,700 per prediction; would you advise using it?

16. A patient calls her doctor complaining that she is sick. Based on the described symptoms and the knowledge of the diseases currently epidemic in the city, the physician suspects that there is a 40 percent chance that the patient has disease D_1 and a 60 percent chance that she has disease D_2.

    On arrival at the physician's office, the patient is subjected to a test that has either a *negative* or a *positive* result. The physician knows that there is a 0.6 likelihood that the results discovered in the test are associated with D_1. That is, in all past cases of D_1, the test was positive 60 percent of the time. Also, the physician knows that there is only a 0.2 likelihood that the results are associated with D_2.

    a. Find the revised probabilities of D_1 and D_2. Assume that the patient has either D_1 or D_2.
    b. What is the probability that a test will yield negative results?

17. The following is the payoff profit matrix for two alternate plans:

Alternatives \ Futures	$p = 0.75$ Market Receptive	$p = 0.25$ Market Unfavorable
Plan a	$20,000	$6,000
Plan b	25,000	3,000

a. Which plan do you recommend adopting, using the expected value criterion?
b. What is the expected value of perfect information?
c. It is known from past experience that of all the cases when the market was receptive, a research company predicted it in 90 percent of the cases. In the other 10 percent, they predicted an unfavorable market. Also, of all the cases when the market proved to be unfavorable, the research company predicted it correctly in 85 percent of the cases. (In the other 15 percent of the cases, they predicted it incorrectly.) Find the posterior probabilities of all states of nature.
d. Using the posterior probabilities, which plan would you recommend now?
e. How much should one be willing to pay (maximum) for the research survey?

18. Management is considering replacing an energy-saving device with a new one. The new device has a probability of 60 percent of being superior to the old one. A testing service is called on to test it. From experience, it is known that when a new device was actually superior, the testing service predicted this superiority in 80 percent of the cases. However, when the new device was really inferior, the testing service predicted it to be superior 50 percent of the time.

a. Suppose a test is undertaken and superiority for the new device is predicted. What will management's revised probabilities be of the device being superior?
b. Suppose the test indicates that the device is inferior. How would the probabilities be changed now?
c. What is the probability that the test will indicate a superior device?

19. e-Toy Corp. specializes in Web-related products. A new product, ET3M, sells for $19.95 and is planned for the forthcoming holiday season. Projected sale figures are 10,000 units if the economy is strong (45 percent chance), 7,000 in a moderate economy (35 percent), and 4,000 in a weak economy. Because of the seasonal nature of the product, the company will have to discount all *unsold* items by 55 percent at the end of the season in order to clear the stock. The company can use one of two production methods. If the items are produced with the existing equipment, the cost is $8 per unit, plus $10,000 fixed cost. Alternatively, special equipment can be leased at $25,000. In this case, the cost per unit is $5. Marketing and overhead costs in either case are $4 per unit.
   a. What production method would you recommend, and why?
   b. What quantity should be produced?
   c. What is the maximum the company should be willing to pay for a perfect prediction of the demand? What assumptions must be made in such a case?
   d. A marketing research company charges $10,000 per prediction. Their track record is good, but not perfect. In the past, of the 15 times sales were strong, the company predicted it correctly 14 times (one time they predicted moderate sales). Of the 10 times that sales were moderate, they predicted it correctly 8 times (one time they overestimated and one time they underestimated). Of the six times that sales were a bust, they predicted it correctly four times (twice they overestimated). Would you advise using the marketing research service?
   e. Should the marketing research service be used, what quantity should the company produce, and what production method should be used?

20. A manufacturer must decide whether to manufacture and market a new seasonal novelty that has just been developed to sell at $1.50 per unit. If he decides to manufacture it, he will have to purchase special machinery that will be scrapped after the season is over. If a machine costing $1,000 is bought, the variable cost of manufacturing will be $1 per unit; if a machine costing $5,000 is bought, the variable manufacturing cost will be $0.50 per unit. In either case, it will be possible to manufacture in small batches as sales actually occur and there will be no danger of having unsold merchandise left over at the end of the season. The manufacturer's probability distribution for sales volumes is shown in the following table:

Sales Volume	Probability
1,000	0.5
5,000	0.25
10,000	0.25
	1.00

   a. Draw up a payoff table, remembering that there are *three* possible acts.
   b. Find the "best" act.
   c. Set up the opportunity loss (regret) table.
   d. The manufacturer has received an offer from The Great Transcendental Swami, a soothsayer. Swami claims to be able to make a perfect sales forecast. Assuming this claim is valid, determine how much the manufacturer should be willing to pay (at most) for Swami's forecast. (Use the opportunity loss [regret] table.)

21. It was a hot summer day in Los Angeles, and Dave Greenhouse was trying to make a decision before 5 P.M. Dave was in the business of buying repossessed condominium apartments from lending institutions such as savings and loan associations and banks. The lending institutions' objectives were to get rid of the property as soon as possible. Dave would buy the apartments and then sell them, hopefully with a nice profit.

   This time, the Aztec Savings and Loan Association offered him three units (he must take all of them or nothing) at a nonnegotiable price of $720,000. It was the last day that the offer was valid and Dave knew that he must make a fast decision. He had already had the assets appraised. The estimated selling price that he could get for the units is shown below:

Unit 1 $269,000
Unit 2 25 percent chance of $250,000; 50 percent chance of $260,000; and 25 percent of $270,000
Unit 3 30 percent chance of $250,000; 40 percent chance of $260,000; and 30 percent chance of $270,000

There was also a selling cost of $10,000 per unit (advertising, legal, financial, and so forth).

Dave hoped to sell the units within 60 days. This was the time limit the savings and loan association gave him to pay the $720,000. Dave estimated that there was a 70 percent chance that he could do it. Any unit that was not sold within 60 days would be sold, for certainty, within the next 30 days. However, in that case, there would be a financial charge for late payment of $4,400 per apartment. Present the situation in a decision table and advise Dave on what to do.

22. Referring to the "Sensitivity Analysis" discussion at the end of Section 5.5, develop a one-input data table to identify the probability of success in which you would be indifferent with regard to undertaking the project.
23. Tara Allison is planning to purchase a house in the next couple of months. She would be very happy if she could find a suitable house for $200,000 but is willing to spend $250,000. Currently, the annual interest rate on a 30 year fixed mortgage is 7.75 percent. Some analysts are projecting that these rates may drop by 0.50 percent in the near term, while others project that they may increase as much 0.75 percent. Develop a two-input data table for Tara showing the impact on her monthly mortgage payment to changes in both her house purchase price and interest rates. Assume that in all cases, Tara makes a 20 percent down payment.

CASES

Maintaining the Water Valves

Because of the pressure to hold down costs, local governments across the country have been very cost conscious. Sharon Brown, Evergreen's city manager, was under continuous pressure to contain costs.

The city of Evergreen owns and operates a water system. A major expense item is water valve repair. These valves are currently repaired by the city's maintenance department. In preparing next year's budget, the water system manager, Bill White, was faced with the following situation.

The average number of valves repaired in one year is 4,120. The average time needed to repair a valve is 42 minutes. It is estimated that the labor cost next year will be $14 an hour. Parts and supplies are estimated at $3 per valve. Overhead shop expenses are computed at 40 percent of the total labor and parts cost. Some of the valves being repaired also need reworking, because of poor material or mistakes made by employees. Historical data indicate that the percentage of repaired valves that need reworking varies according to the following distribution:

Reworking Needed (%)	Probability
3.0	0.50
4.0	0.40
4.6	0.10

A reworked valve is 100 percent reliable because such valves go through a special quality control check. The reworking cost is estimated to be $20 per valve.

Last month, the city manager called in all her assistants and advised them about some cost containment programs that were soon to be implemented. Specifically, she requested they each check on the possibility of eliminating inefficient in-house services that could be contracted out. Accordingly, Bill White advertised the valve repair job on a contractual basis. Western Maintenance Corp. (WMC), a reputable company, came up with the lowest bid: a $38,000 flat yearly fee plus $5 a valve, with all reworking done at no additional charge.

City manager Ms. Brown requested that Mr. White evaluate WMC's proposal in terms of possible dollar savings. Mr. White's response is given in the following memo:

MEMO

To: Sharon Brown, city manager
From: William White, manager, Water System
Subject: WMC's proposal on valve repair jobs

After careful evaluation of the proposal, I recommend that the contract be awarded to WMC. However, due to the budget squeeze, I will not be able to reassign the employees involved in any other jobs. Therefore, I recommend that the contract be awarded to WMC if and only if they will hire all displaced employees.

Upon receipt of this memo, Sharon Brown called the president of WMC. His response was as follows: "WMC is currently fully staffed, so hiring the city employees, some of whom are not highly skilled, will increase our cost. Therefore, WMC will be able to hire them only if the terms of the contract are changed to a flat fee of $40,000 plus $5 per valve, or else a $38,000 flat fee and $5.50 per valve."

Use a decision tree approach to advise Sharon Brown about what to do.

1. Construct a decision tree and include the alternative of WMC not hiring the city employees.
2. Solve the tree; discuss the results.
3. Suppose the city agrees to the $40,000 flat fee and $5 per valve proposal, provided that the $40,000 is paid in four quarterly payments. The first payment is made with the signing of the contract. If Evergreen's cost of capital is 10 percent compounded quarterly, how will this affect the decision?

The Air Force Contract

Nova Aviation, Inc., is considering its bidding strategy for a U.S. Air Force contract. Because of the specialization of this job, there is only one other contractor currently approved to bid on it—Sun Aircraft. The Air Force determines bid eligibility after lengthy investigation, and is, therefore, unlikely to approve of any additional contractors in the near future.

Due to the repetitive nature of such bids, Nova usually has special meetings to consider bidding strategy. They are considering two options: bidding "high" or bidding "low." Marketing Director Judy Scher explained the logic behind the options as follows:

"Our company's policy is to bid every time, whereas Sun's policy is to bid only 60 percent of the time. And we have only two choices: a high bid or a low one. If we bid high and Sun doesn't bid, we get the job and make $1.1 million profit. But if Sun bids, then we might get the contract or we might not, depending on their bidding level. All in all, our profit, when we place a high bid and Sun bids as well, averages out to zero. If we place a low bid, our per-bid profit averages out to $500,000 *regardless* of Sun's action."

Steve Green, Nova's director of finance, thought for a few moments. "If what you say is true, it seems that we should bid 'high' whenever our competitor doesn't bid and 'low' whenever they do."

"Exactly," replied Ms. Scher. "But," said Steve, "how can we find out what Sun might do? We could try to buy this information from someone at Sun, but not only is that unethical, it is probably illegal as well."

"Wait a minute," said Ms. Scher. "Suppose it were legal; how much should we offer for such information, assuming it were 100 percent reliable?"

"That's a good question," replied Mr. Hunt, the company's president. "It seems to me, however, that we're getting off the track. What do you propose to do, Judy?"

"We could use an expected value approach to solve this problem," said Ms. Scher, "but that approach would assume that our competitor will behave in the future as they have behaved in the past, and how can we make such an assumption? How can we know what they'll do?"

"Well," said the president, "we've tried before to predict their moves. Aviation Research Co. has provided such information before, at $100,000 a shot. Their track record isn't bad: of 30 times that Sun bid, Aviation Research predicted it correctly 24 times. On the other hand, of the 20 times that Sun didn't bid, Aviation called it right 17 times."

"What you're saying, Mr. Hunt," said Steve, "is that this research company isn't 100 percent reliable."

"You may infer that, but Aviation Research is still considered the best research company available in this market. It's getting late, and we have to make a decision."

You are Steve Green. Prepare a short report advising your boss how to handle this decision. Specifically, use the concepts of the value of perfect and imperfect information and decision policy. Work out the actual numerical values.

ENDNOTES

1. Market capitalization is calculated as the product of the total number of shares outstanding times the stock's current selling price.
2. D. Desmet, T. Francis, A. Hu, T. M. Koller, and F. A. Riedel, "Valuing Dot-Coms," *The McKinsey Quarterly,* 2000, Number 1.
3. Charles Gasparino and Randal Smith, "Internet Trades Put Merrill Bull on Horns of a Dilemma," *Wall Street Journal* (February 12, 1999), C1, C10.
4. Ibid.

BIBLIOGRAPHY

Bertsimas, D., and R. M. Freund. *Data, Models, and Decision: The Fundamentals of Management Science.* Cincinnati, OH: South-Western, 2000.

Camm, J. D., and J. R. Evans. *Management Science and Decision Technology.* Cincinnati, OH: South-Western, 2000.

Clemen, R. *Making Hard Decisions with Decision Tools Suite.* Pacific Grove, CA: Brooks/Cole, 2000.

Hertz, D. B. *Risk Analysis and Its Applications.* New York: Wiley, 1983.

Keeney, R. *Value-Focused Thinking,* reprint ed. Cambridge, MA: Harvard University Press, 1996.

Lootsma. F. A. *Multi-Criterion Decision Analysis via Ratio and Difference Judgement.* Amsterdam: Kluwer, 1999.

Marshall, K. T., and R. M. Oliver. *Decision Making and Forecasting with Emphasis on Model Building and Policy Analysis.* New York: McGraw-Hill, 1995.

Ragsdale, C. T. *Spreadsheet Modeling and Decision Analysis,* 3rd ed. Cincinnati, OH: South-Western, 2001.

Sharpe, P., and T. Keelin. How SmithKline Beecham makes Better Resource Allocation Decisions. *Harvard Business Review* (March–April 1998), 45–57.

Winston, W. L., and S. C. Albright. *Practical Management Science,* 2nd ed. Pacific Grove, CA: Duxbury, 2001.

Queuing Theory

6

© PHOTODISC, INC.

INTRODUCTION

As the newest partner of your consulting firm, you have become aware of the increasing amount of time the consultants in the office are waiting to get their printouts from the shared network custom printer. Specially designed for your firm some years back, this custom printer smoothly executes a number of special functions impossible for generic printers to duplicate. Since it is more trouble to return to their offices and come back later—or forget to come back—the consultants simply wait at the printer until their special job prints out. You think that a brief study is needed to determine whether the printer should be replaced with a newer model that prints at twice the current speed.

First, you sketch out an influence diagram (Exhibit 6.1) to identify the various factors that affect the total costs related to this situation. According to the exhibit, total costs are driven by the cost of the printer and the waiting costs of the consultants. The diagram further indicates that waiting costs are influenced by the consultants' salaries and the amount of time they must wait for their printouts. Waiting time is, in turn, dependent on the arrival rate of new print jobs and the speed of the printer. Finally, we see that our decision related to the selection of a printer directly impacts the speed and cost of the printer. According to Exhibit 6.1, further analysis of this situation will require data on printer speed, the arrival rate of new print jobs, and the consultants' salaries.

EXHIBIT 6.1 Influence Diagram for Printer Wait Problem

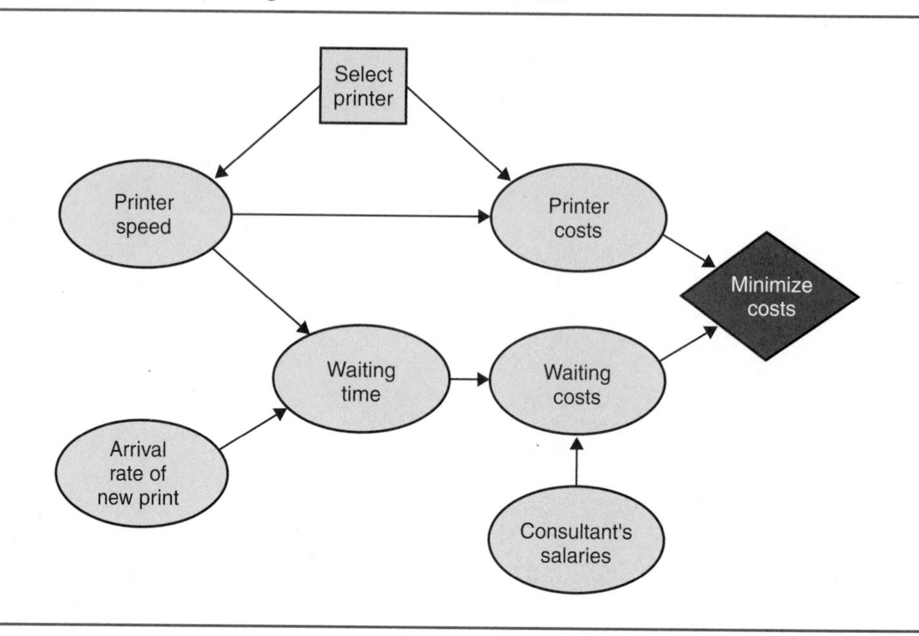

Next, you decide to talk with the network administrator to find out what data are currently being collected about the printer. According to the network administrator, the current network operating system does collect and store data related to when print jobs enter the print queue, when they start printing, and when the job is complete. She agrees to provide you with a report containing this information for all the jobs sent to the printer over the last four weeks.

After analyzing the data, you determine that there is a fair amount of daily "seasonality" in the data. For example, from 9:00 A.M. and 11:00 A.M. the 25 consultants working out of the office submit an average of 162.5 print jobs while on average only 47 jobs are submitted between 11:00 A.M and 1:00 P.M. Further analysis also indicates that an average print job takes 40 seconds to print.

First, you inspect the distributions for the pattern of submissions to the printer and the printer processing times to verify that they are consistent with the assumptions of standard *queuing* (or *waiting line*) theory. Verifying these, you refer back to the queuing theory chapter of your old quantitative modeling textbook that you kept for future reference. Using the equations you find in Section 6.4, "Analysis of the Basic Queue," you develop the spreadsheet shown in Exhibit 6.2. To begin the analysis, estimates for the rate at which print jobs arrive at the printer and the rate at which the printer is able to process jobs are needed in the same time units. For the purpose of this analysis, you decide to express these rates on an hourly basis. Thus, the 162.5 jobs that are submitted during the two-hour period between 9:00 A.M. and 11:00 A.M. works out to an average arrival rate of 81.25 (162.5 ÷ 2) print jobs per hour. Likewise, the printer's ability to process a job in 40 seconds is equivalent to processing 90 jobs per hour.

Having expressed the arrival rate and processing rate in the same units, we can then use the standard queuing theory formulas to gain important insights into this situation. According to the analysis shown in Exhibit 6.2, the custom printer is utilized or busy 90 percent (cell B8) of the time during the peak usage period. On average, there are 8.38 jobs waiting to be printed (cell B9) and a total of 9.29 jobs (cell B10) either waiting to be printed or being printed at any given time. Finally, a newly submitted job can expect to wait 0.10 hour or 6 minutes (cell B11) for the printer to become free and will require, on average, 0.11 hour from the time it is submitted until it is completed (cell B12).

EXHIBIT 6.2 Spreadsheet to Analyze Laser Printer Usage

	A	B
1	**Custom Printer Study (Peak Usage Time)**	
2		
3	**Model Inputs**	
4	Mean arrival rate per hour	81.25
5	Mean service rate per hour	90
6		
7	**Model Outputs**	
8	Printer utilization	0.90
9	Average number of print jobs waiting in the queue	8.38
10	Average number of jobs in the system	9.29
11	Expected time waiting for the printer	0.10
12	Expected total time to complete the job	0.11
13	Probability printer is idle	0.10
14	Probability printer is busy	0.90

Based on this analysis, can we conclude whether purchasing a new custom printer is justified? Unfortunately, the answer is no. Unlike the prescriptive optimization models discussed in Chapter 4 that dictate the specific actions to be taken, queuing theory models are descriptive and therefore only portray the situation being analyzed. To answer the question of whether it is justified to replace the custom printer with a new one, we need to take the insights obtained based on standard queuing theory a step further.

More specifically, one way we can determine whether a new custom printer is justified is to compare the costs of obtaining the new printer to the costs of *not* purchasing a new printer. As was noted earlier, if the current custom printer is replaced with a new one, the service rate will be doubled to 180 print jobs per hour. The spreadsheet originally developed to describe the current situation has been enhanced to characterize the operation of a new printer and is shown in Exhibit 6.3. According to these new results,

EXHIBIT 6.3 Potential Savings If a New Custom Printer Is Purchased

	A	B	C	D
1	**Custom Printer Study (Peak Usage Time)**			
2			**New**	
3	**Model Inputs**		**Printer**	**Savings**
4	Mean arrival rate per hour	81.25	81.25	
5	Mean service rate per hour	90	180	
6				
7	**Model Outputs**			
8	Printer utilization	0.90	0.45	
9	Average number of print jobs waiting in the queue	8.38	0.37	8.01
10	Average number of jobs in the system	9.29	0.82	8.46
11	Expected time waiting for the printer	0.10	0.00	0.10
12	Expected total time to complete the job	0.11	0.01	0.10
13	Probability printer is idle	0.10	0.55	
14	Probability printer is busy	0.90	0.45	

replacing the current printer with a new one will reduce the average number of jobs waiting for the printer by 8.01 to just 0.37 jobs. Also, the amount of time the consultants would have to wait for their print jobs to be completed would be reduced from 6.6 minutes to 0.6 minute, a reduction of 6 minutes, or 91 percent!

Are these savings sufficient to justify the purchase of a new printer? In order to make a more direct comparison it is necessary to translate these time savings into dollars. Mathematically, the cost savings can be computed as time savings/job × number of jobs × wage rate of the consultants. In your new, young firm, the consultants earn an average of $50/hour. Thus, the savings to the consulting firm during the peak two-hour period would be $812.50 (0.10 hour time savings per job × 162.5 jobs × $50/hour). If the cost of the new custom printer is estimated to be $35,000, then the custom printer would pay for itself in 35,000/812.5 = 43.08 worker-days, or about two months. Of course, the custom printer will pay for itself in even less time than this because we have only included the savings from the *peak* usage period between 9:00 and 11:00 A.M., not the entire day. Obviously, purchasing a new custom printer is well justified in this case!

6.1 The Modeling Process for Queuing Studies

As discussed in the preceding example, the queuing methodology is basically a *descriptive* tool of analysis rather then *prescriptive,* like linear programming. As such, the major objective of waiting line theory is *predicting* the behavior of a system as reflected in its *operating characteristics* or *measures of performance,* such as average waiting time per customer, average number of customers in the waiting line, or utilization of the service facility. This information is needed by management to determine the most appropriate service level for the system. Thus, the use of queuing theory to address opportunities or problems is more a "satisficing" approach than an optimizing one. Although queuing theory is basically descriptive, it can also be used at times to determine the optimal number of service facilities or the optimal speed of a facility, particularly if there is a limited number of feasible alternatives to consider.

Step 1: Opportunity/Problem Recognition

Recognizing a waiting line, or a queuing situation without an obvious "line," is critical to applying this type of quantitative model. For example, fast food restaurants form obvious lines for service whereas fancy restaurants take names and table preferences, then ask you to wait in the bar where you'll be out of the sight of new customers (and spending money on their more profitable drinks, they hope). Sometimes it isn't people who are waiting, it is machines (e.g., waiting for a repair crew) or jobs (waiting for someone to work on them) or calls (waiting for service). Sometimes the queue is obvious, like a line of parts, but other times it is abstract, with the queue being in computer memory, or simply implied (automobiles that run fine but need their oil changed soon). Although the chapter opener posed a situation where the need to wait in queue was a problem, the astute modeler will realize there are similar situations where no one is complaining but a valuable opportunity is at hand for reducing costs, improving service, or increasing profits.

Step 2: Model Formulation

We will cover the basic queuing models in this chapter, which can be applied fruitfully in a great variety of situations if the assumptions behind the models are met. And although there are a few other queuing situations that can also be modeled with equations, the equations quickly become extremely complex. In these more complex situations, simulation (Chapter 7) is commonly used to give insight into or resolve the situation.

All of these queuing models assume the situation being studied is in **steady state**—the arrival distribution stays the same over the period of interest, the service distribution stays the same, and the overall situation stays the same. However, many queuing situations are rarely in this type of steady state for long. For example, steady state at a fast food restaurant may last only a half hour, such as when the pre-lunch crowd hits, followed by the lunch crowd (which may stay in steady state for perhaps an hour), followed by the post-lunch crowd, followed by an empty period, and so on.

The queuing models we will describe, however, are highly dependent on their assumptions being met. Violation of some of the regular assumptions, such as those concerning customer behavior (e.g., leaving the queue is prohibited) or the system being in *steady state* can result in the inapplicability of most analytical queuing models. In nonsteady state or *transient* conditions, the regular distributions of arrivals and services do not usually hold; indeed, there may be no pattern to them at all. Validation of the chosen model may be more important in queuing analysis than in other quantitative models because of the high likelihood that some of the numerous queuing assumptions will be violated. (See Chapter 7, "Simulation," for solution methods to more complex queuing situations, as well as the references in the bibliography.)

Step 3: Data Collection

Given that the brief nature of most steady-state queuing situations is understood beforehand, necessitating multiple data collection sessions, collecting queuing data is largely straightforward. The situation, or "type" of queue and service process, must be clearly perceived beforehand, of course, or the wrong data will be collected. However, the only difficulty usually encountered is trying to measure the occasional qualitative benefits, such as satisfaction with the service process, or the value of a reduction in waiting time. Thus, in many situations the qualitative benefits are often left unquantified and the managers make the trade-offs between cost and qualitative benefits within the decision situation.

Step 4: Analysis of the Model

In queuing analysis, there are usually only a small number of alternatives to be evaluated, which facilitates the task of data collection as well as analysis. For example, in a decision about the number of elevators to be constructed in a new building, 10 possibilities would be a realistic consideration, but not 5,000. The number of feasible alternatives in a service system is usually small because of human, technical, financial, or legal constraints. Alternatives may differ in the size of the facility, the number of servers, the speed of service, the priorities given to customers, or the operating procedures. For each alternative, the measures of effectiveness must be computed.

The alternative solutions are then compared on the basis of their overall effectiveness. One approach is the use of a total cost curve, as shown in Exhibit 6.4. As noted earlier, however, a major data collection problem in this step may be determining the cost, particularly the cost of waiting. In some cases, therefore, a qualitative comparison of the alternatives is performed based on measures such as waiting time and no attempt is made to perform a cost analysis.

Step 5: Implementation

The common situation being studied in queuing analyses typically involves determining how much capacity to devote to servicing a queue. To some degree, the implementation is relatively straightforward because the situation is usually well defined: Should we add

EXHIBIT 6.4 The Queuing System

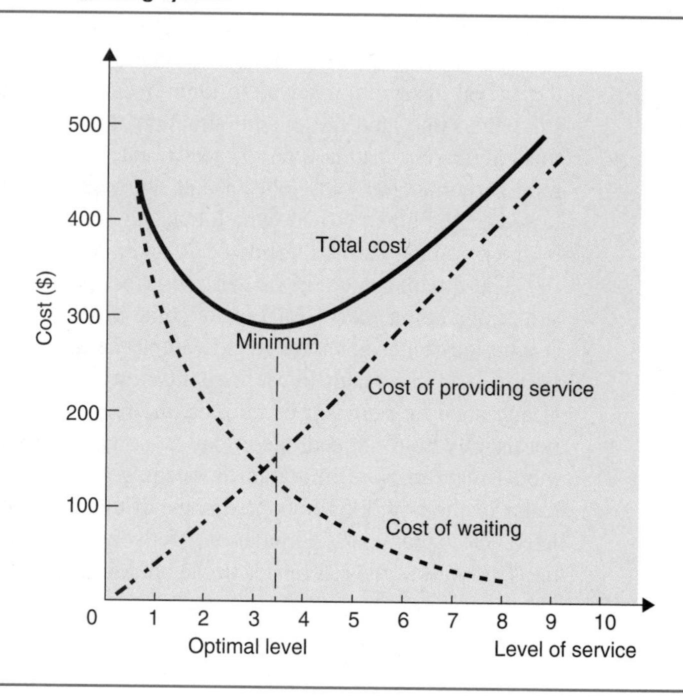

another station during rush hours? Should we lease another machine to speed up service? Do we need a priority or quick-check lane? However, sometimes the problem is more complicated because the costs and benefits reside in different departments, and although the organization as a whole would benefit from a particular action, the department making the decision would not benefit, so the action isn't taken.

6.2 The Queuing Situation

The more society becomes interdependent psychologically, economically, and technically, the more we encounter waiting lines, or queues, in our daily lives. People queue at doctors' offices, supermarkets, gasoline stations, connecting to their Internet service provider, and tollbooths. Waiting lines may also involve nonhumans as customers: networked printers, airplanes circling airports, and machines waiting for repairs. The problem of managing waiting lines is complex, because the cost of providing services of all kinds is rapidly increasing. The objective is to determine the appropriate level of service. The method of analyzing waiting line problems illustrated in this chapter is called **queuing theory.**

Characteristics of Waiting Line Situations

The printer incident at the consulting firm in the chapter introduction is typical of a situation that arises in the delivery of services. On the one hand, the demand for services is unstable and there are some foreseeable fluctuations during certain time periods. In addition, there may also be unforeseeable changes in the pattern of demand. On the other hand, the length of service may vary due to particular requirements of those requesting the service. The result is difficulty in meeting demand immediately on request, especially during heavy demand or rush periods.

The only way that demand can be immediately supplied, all the time, is to build a large service capacity that can always meet peak demand. Such a situation confronts the electric utility industry. But in some industries, it is very expensive to build, operate, and maintain a service facility that can meet all demands, all the time, on request. It may also be expensive to constantly *change* the capacity in an existing service facility to meet the demand, especially when people provide the service. Instead, service facilities are usually designed so that their capacity is less than the maximum demand. As a result, whenever demand exceeds capacity, a waiting line, or queue, is formed; that is, the customers do not get service immediately on request, but must wait. Yet, on other occasions such as nonrush hours, the service facility will be idle. Thus, at a high level of service capacity, the people served will have only a limited wait, but the service facility will frequently be idle (at considerable expense). At a lower, less expensive level of service capacity, there will be less idle time and cost, but people will have to wait longer.

The management of services is indeed complicated. Although management would like to satisfy the customer ("the customer is always right"), it is very expensive, sometimes even impossible, to satisfy everyone immediately, all the time. Therefore, management is interested in finding the *appropriate* level of service. The theory applied to this problem, *queuing theory,* was pioneered by A. K. Erlang, a Danish engineer in the telephone industry in the early 1920s. Queuing theory was extended in application, especially after World War II, to a large number of situations:

- Determining access to telecommunications networks
- Determining the capacity of an emergency room in a hospital
- Determining the number of runways at an airport
- Determining the number of elevators in a building
- Determining the number of traffic lights and their frequency of operation
- Determining the number of flights between two cities
- Determining the number of first-class seats in an airplane
- Determining the size of a restaurant
- Determining the number of employees in a storeroom, in a typing pool, or in a nursing team
- Determining the number of cashiers in a supermarket
- Determining the number of drive-through windows at banks and fast food restaurants

The Structure of a Queuing System

A queuing system (Exhibit 6.5) is composed of the following parts, each of which is described in more detail in the following section on modeling the queuing process.

EXHIBIT 6.5 The Major Components of a Queuing System

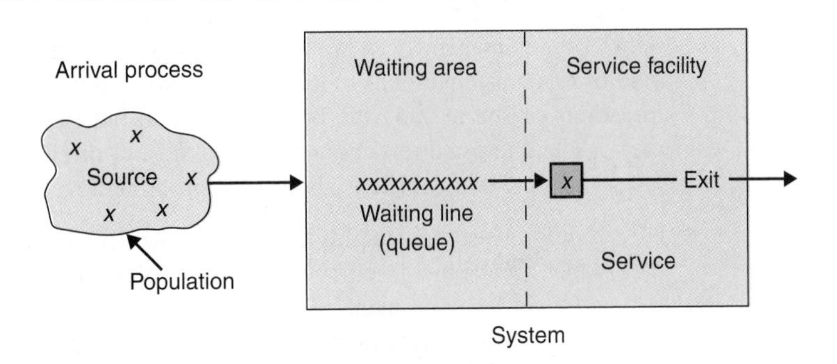

The Customers and Their Source Customers are defined as those in need of service. Customers can be people, airplanes, machines, jobs, or raw materials. The customers are generated from a **population** or a source, which can be considered either infinite (that is, so large that our model is unaffected by the size) or finite. As an example of the former, a hospital's "population" would be the people in the surrounding community and the "customers" would be those people in the population requiring hospitalization for sickness or accidents. An example of a finite population might be a set of machines, or elevators, that are serviced by a repair crew.

The Arrival Process The manner in which customers show up at the service facility—individually or in batches, scheduled or unscheduled—is called the **arrival process.**

The Service Facility and the Service Process The service is provided by one or more **service facilities.** This may be a person (bank teller, barber), a machine (elevator, gasoline pump, printer), or a space (airport runway, parking lot, hospital bed), to mention just a few. A service facility may include one person or several people, operating either as a team or in parallel (multiple service channels).

The Queue Whenever an arriving customer finds that the service facility is busy, a **queue,** or waiting line, is formed. There may be single, multiple, or even priority lines, and the order of service may be random, by appointment, first-come–first-served, or some other order. The queue can also be considered to be either infinite or finite. An example of the former is the taking of orders for products to be sent out. An example of the latter is when there is only a limited number of waiting spaces available, such as parking spaces for cars. Here we list several examples:

System	Queue	Service
Computer	Jobs	Printer
Bank	People	Tellers
Telephone	Callers	Switchboards
Library	Books to be shelved	Librarian's assistants
Freeway	Automobiles	Number of lanes
Airplane	People	Seats, flights
Airport	Circling planes	Runways

The Managerial Problem

Reconsidering the printer problem, we can now better articulate the managerial problem of waiting lines. As a responsible manager, you realize that waiting for a printer is costing the firm valuable consultant time. On the other hand, there is a cost attached to the provision of faster service, in this case buying another printer. In general, the basic problem of the management of waiting lines is: What is an "appropriate" level of service? In addition, management has to make several related decisions regarding such factors as the priorities of service and the operating hours, but these decisions will not be discussed here. (The interested reader is referred to the end-of-chapter bibliography.)

There are two decisions about an appropriate level of service:

1. If only one service facility exists, then the decision involves the *speed* of service, which can be increased by adding more personnel and/or equipment or using faster personnel or equipment.
2. If additional service facilities can be added, then the decision must be made as to how many more should be added, and at what speed they should operate.

In making such decisions, the objective would be to minimize total cost. Management must consider both the cost of providing the service and the cost of customers' waiting (lost time, loss of goodwill, lost sales, and so on). Unfortunately, these costs are in direct opposition to each other, as shown in Exhibit 6.4. That is, the cost of providing the service increases with the service level, while the waiting time (and its cost) declines with the service level. Furthermore, the cost of waiting, in many cases, cannot be expressed in terms of dollars. Thus, management may hold either or both of the following objectives in mind when making decisions about an appropriate service level.

Cost Minimization In cases where it is possible to ascribe a cost to the waiting time (usually when a company is serving its own employees or equipment), management will provide a service level such that the total cost of waiting and service is minimized, as described in the chapter opener. In that case, the cost of service was that of purchasing a new printer that had much greater capacity.

Achieving a Specified Performance Level (Service Goal) Instead of (sometimes in addition to) minimizing costs, management will strive to achieve a certain level of service. Determining the desired level of service is a matter of organizational policy and is commonly influenced by many external factors such as competition and consumer pressures. There are many examples of specified service levels:

- Telephone companies desire to repair 99 percent of all inoperative telephones within 24 hours.
- Fast-food restaurants advertise that you will not have to wait more than 3 minutes for breakfast.
- Banks avoid having more than six cars in any lane of their drive-through windows at a time.
- Service facilities should be in use at least 60 percent of the time.
- Call-in service centers strive to not keep customers waiting on the phone longer than 90 seconds.

The Costs Involved in a Queuing Situation

The two primary costs in queuing are the cost of providing the service and the cost of waiting. We address each in turn.

The Facility Cost The cost of providing a service is typically much easier to calculate than the cost of waiting. It includes

1. Cost of construction (capital investment), as expressed by interest and amortization rates
2. Cost of operation: labor, energy, and materials required for operations
3. Cost of maintenance and repair
4. Other costs: insurance, taxes, rental of space, and other fixed costs

The cost of providing the service is typically composed of both fixed and variable costs. The annual fixed cost (amortization, insurance, taxes) and the variable (hourly) cost must both be converted into the same time units so that the cost components can be added together. The cost of service can be given in any of the following ways:

- Per hour (or other time unit) of service (e.g., $20 per hour).
- Per customer served (e.g., $5 per customer). If four customers are served during the hour, then the hourly cost is $4 \times \$5 = \$20$.
- Per unit capacity of service (e.g., $2 for each customer that can be served). If the facility can serve 10 customers per hour, the hourly cost is $\$2 \times 10 = \$20$.

The Cost of Waiting Customers In situations such as retail sales, external customer "ill will" may be involved. Then the cost of waiting is very difficult to assess and may involve several factors. For example, a waiting customer may get impatient and leave, thus resulting in a loss of revenue and possible loss of repeat business due to his or her dissatisfaction. There may also be an ill-will cost incurred; that is, when talk is spread about the poor service given at a facility, other customers might not come. A more extreme situation would be that of a patient waiting for surgery. If the patient waits too long, she might die. In addition to the loss of revenue to the hospital and the cost of ill will, there is the additional "cost" to the customer of her own life.

The difficulties in expressing the cost of waiting are especially severe in cases where the customers are *external* to the organization providing the service, and especially when the provider of the service is a *nonprofit* organization. In such a case, one may raise such questions as, "Cost to whom (e.g., to the patient, to the doctor, to the community, or to society)?" Or, "Is the cost directly proportional to the waiting time?" The answers to such questions are not simple, because they involve personal values, social priorities, and other qualitative factors. For this reason, decisions concerning queuing systems are often made from the perspective of achieving a specified performance level.

In other cases, such as when the customers belong to the same organization providing the service, it is easier to assess the waiting cost. For example, when employees are waiting in line to use a copying machine or get tools, the cost, at minimum, is their wasted wages. In addition, there is their lost contribution to profit. If, like in the chapter opener, this is the case, then queuing theory may be applied. Different service systems can then be compared on the basis of their total cost (TC), which is composed of two components: the facility cost (C_F) and the total cost of waiting customers (C_W).

$$TC = C_F + C_W$$

Costs are computed on one of two bases: either as "cost per unit of time (hourly, daily)," or as "cost per customer served." In this text, the cost-per-unit-of-time basis is used.

Two different waiting cost components can be distinguished. First, the cost of the *total wait* (W), including the time of service, may be of relevance (e.g., in industrial settings, where employees are being paid for the time they are away from their work while waiting for tools or other services). In the second case, typical of retail situations, only the time *while waiting in the queue* (W_q) is relevant, because consumers harbor virtually no ill will while being served in a normal fashion. Here we will assume that the waiting cost (regardless if it is based on W or W_q) is *proportional* to the waiting time; that is, if the cost of waiting for an hour is \$12, then the cost of waiting two hours is \$24.

6.3 Modeling Queues

To describe the process for modeling queues, we first discuss the wide variety of queuing situations and the nomenclature developed to represent them. In the process, we also consider the deterministic case, a very special type of queue. We then consider in detail the three major components of queuing models—the arrival process, the service process, and the waiting line. Last, we describe the queuing methodology for the probabilistic case.

Queuing Model Notation

Due to the large number of possible queuing systems, a notation set was developed D. G. Kendall in 1953. This notation system is commonly used today to label queuing models.

EXHIBIT 6.6
Queuing System Symbols

Item	Value	Notation Used
Arrival process	Poisson	*M*
	Erlang, shape parameter *k*	*Ek*
	Constant	*D*
	Normal	*N*
	Uniform	*U*
	Only mean and variance known	*G*
Service process	Exponential	*M*
	Erlang, shape parameter *k*	*Ek*
	Constant	*D*
	Normal	*N*
	Uniform	*U*
	Only mean and variance known	*G*
Number of servers	One or more	*K* (actual number)
Queue discipline	First-come, first-served	*FCFS*
	Nonpreemptive priority	*PRI*
Maximum queue length	No limit	∞
	Finite limit	*n* (actual number)
Calling population size	Infinite	∞
	Finite	*n* (actual size)

Six items of information are necessary to completely define a queuing system. The *arrival process, service process,* and *number of servers* in the system are first defined. The way in which arriving units are accepted for service (the *queue discipline*), the *maximum size* permitted for the queue, and the *number of people (or units) in the population* served by the queuing system are then specified. The symbols are shown in Exhibit 6.6. These are also used in many computer packages.

When a system is described, its symbols are summarized using a simple two-part system based on Kendall's notation. The items are shown in two sets of three items, separated by slashes. For example, a system with a Poisson arrival process; exponential service time distribution; two servers; first-come, first-served discipline; no queue size limit; and serving a large (presumed infinite) population would be labeled *M/M/2 FCFS/∞/∞*. The most common queuing systems are listed in Exhibit 6.7.

There are many variations of queuing situations, for example, queues with priorities, cyclic queues, and others that will not be treated here. See the bibliography for more discussion.

Deterministic Queuing Systems

The simplest of all waiting line situations involves constant arrival rates at predetermined (such as by appointment) times and constant service times. Three cases can be distinguished:

1. **Arrival Rate Equals Service Rate** Assume that people arrive every 10 minutes, to a single server, where the service takes exactly 10 minutes. Then the server will be utilized continuously (100 percent utilization) and there will be no waiting line.

EXHIBIT 6.7
Notation for Common Queuing Systems

Descriptive Label		Comments
M/M/1	*FCFS/∞/∞*	Standard single-server model
M/M/K	*FCFS/∞/∞*	Standard multiserver model
M/Ek/1	*FCFS/∞/∞*	Single Erlang service model
M/G/1	*FCFS/∞/∞*	Service time distribution unknown
M/M/1	*PRI/∞/∞*	Priority service
M/M/K	*PRI/∞/∞*	Multiserver priority service
M/M/1	*FCFS/n/∞*	Finite queue, single server
M/M/K	*FCFS/n/∞*	Finite queue, multiserver
M/M/1	*FCFS/∞/n*	Limited source, single server
M/M/K	*FCFS/∞/n*	Limited source, multiserver

2. **Arrival Rate Larger than Service Rate** Assume that there are six arrivals per hour (one every 10 minutes) and the service rate is only five per hour (12 minutes each). Therefore, one remaining arrival cannot be served each hour, and a waiting line will continuously build up at the rate of one per hour. Such a waiting line will grow and grow as time passes, and is termed an **explosive queue.**
3. **Arrival Rate Smaller than Service Rate** Assume that there are six arrivals per hour and the service capacity is eight per hour. In this case, the facility will be utilized only 6/8 = 75 percent of the time. There will never be a waiting line (*if* the arrivals come at equally scheduled intervals).

Note that in the first case, there is no waiting line and no idle facility. In the second case, the facility is fully utilized but a waiting line is formed. In the third case, there is no waiting line but the facility is not fully utilized. *In all these deterministic cases, there cannot be a waiting line and underutilization in the same situation.* However, when non-scheduled arrivals are involved, it is common to have *both* idle facilities at times and waiting lines at other times in the same service system, as will be shown later.

The Arrival Process

Description of Arrivals Arriving customers are classified according to the following criteria.

Finite Versus Infinite Source. Two cases are of primary interest: when the source (population) is basically *infinite* (or unlimited), such as the number of people visiting Niagara Falls, or when it is *finite,* as when a business school's information systems group repairs computers in faculty offices and computer labs. Unless otherwise specified, queuing theory assumes an infinite population. (*Note:* No population is really infinite. What is meant is that the population is *large enough* that the probability of one arrival is not significantly changed by the number of previous arrivals.)

Batch Versus Individual Arrivals. Customers may arrive in *batches* (such as the arrival of a family at a restaurant) or *individually* (such as the arrival of an airplane at an airport). In this text, individual arrivals are assumed in all cases.

Scheduled Versus Unscheduled Arrivals. Customers arrive at a service facility either on a scheduled basis (by appointment) or without prior notification. If they come without

EXHIBIT 6.8 Arrivals at the Toolroom

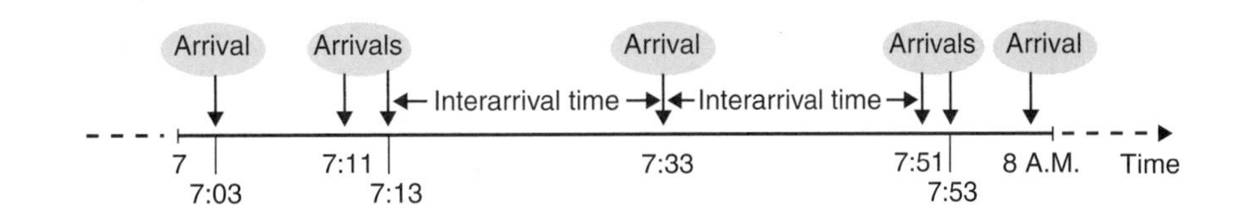

prior notification, their arrival times are not exactly known, but historical data enable us to describe arrivals by some **frequency distribution.**

In scheduled arrivals, the arrival rate is relatively predetermined, whereas in unscheduled arrivals, the times are random variables and we must therefore talk about *averages* and *frequency distributions* of the times. In both cases, the arrival process can be described by either the *average arrival rate* (the average number of arrivals per unit of time) or by the *average interarrival time* (the reciprocal of the average arrival rate).

Measures for Unscheduled Arrivals The number of unscheduled arrivals to a service facility in some fixed period of time can be considered to be a random variable having a particular statistical distribution such as the Poisson. To help model this process we commonly use either the **average arrival rate** or the **average interarrival time.** As an example, consider the situation at a toolroom (an area for storage and dispersal of tools) between 7 and 8 A.M. Exhibit 6.8 shows that seven employees arrived during the hour. Therefore, the *arrival rate* is seven per hour. The times between two consecutive arrivals vary. For example, there are eight minutes between the first arrival and the second, but there are two minutes between the second and third arrivals. These times between arrivals are called the *interarrival times.* The *average* (or *mean*) interarrival time during the first hour is 60/7 = 8.6 minutes. Let us examine these times through examples.

Average Arrival Rate, λ. Assume that the arrival times of employees to the toolroom were recorded over a period of 100 hours. Of these 100 hours, there were 5 hours within which there were 0 arrivals (5 percent of all cases), 6 hours where there was only 1 arrival, and so on. Such results can be described in the form of a frequency distribution, or *histogram,* as shown in Exhibit 6.9. The shape of Exhibit 6.9 is similar to the theoretical

EXHIBIT 6.9 The Frequency Distribution of Arrivals

Number of Arrivals per Hour	Frequency (probability)
0	5.000%
1	6.000%
2	10.000%
3	8.000%
4	7.000%
5	4.000%
6	3.000%
7	2.750%
8	2.500%
9	2.250%
10	2.000%
11	1.750%
12	1.500%
13	1.250%

Frequency (probability): 10.000%, 8.000%, 6.000%, 4.000%, 2.000%, 0.000%
Number of Arrivals per Hour: 0 1 2 3 4 5 6 7 8 9 10 11 12 13

Poisson distribution (Chapter 2), a distribution very common in queuing systems. As was discussed in Chapter 2, creating frequency distributions for raw data can be easily accomplished using Excel's histogram function (Tools/Data Analysis. . ./Histogram).

When customers in a given period of time arrive at random, the distribution of arrivals often follows this Poisson distribution. Random arrival means that even if the mean number of arrivals in a time period is known, the exact moment of arrival cannot be predicted. Thus, each moment in the time span has the same chance of having an arrival. Such behavior is observed when arrivals are independent of each other; namely, when the arrival time is unaffected by preceding or future arrivals. Examples of such arrivals are customers to gasoline stations and failures of machines. The *average* arrival rate is usually designated by the Greek letter "lambda," λ.

Average Interarrival Time, $1/\lambda$. As an alternative to the arrival rate, the interarrival time can also be used. After many observations of a certain service system, it may be possible to say that in 15 percent of the cases the time between two consecutive arrivals was 5 minutes, in 12 percent of the cases it was 7 minutes, in 10 percent of the cases it was 17 minutes, and so on. This information is recorded by the dots in Exhibit 6.10.

If the arrival rate of Exhibit 6.9 follows the Poisson distribution, then the interarrival times of Exhibit 6.10 are distributed according to the negative exponential distribution (see Chapter 2). The average interarrival time is designated by $1/\lambda$; for example, if $\lambda = 5$ per hour then the average interarrival time will be 1/5 of an hour, or 12 minutes.

The Service Process

There are several possible ways of providing service:

- A single facility, such as a dentist's chair
- Multiple, parallel, identical facilities (a *multifacility*), either with a single queue (e.g., seats on a Ferris wheel) or with multiple queues (e.g., a gasoline station)
- Multiple, parallel, but not identical facilities, such as express and regular checkout counters in a supermarket
- Service facilities arranged in a series (*serial* arrangement)

EXHIBIT 6.10 The Interarrival Times

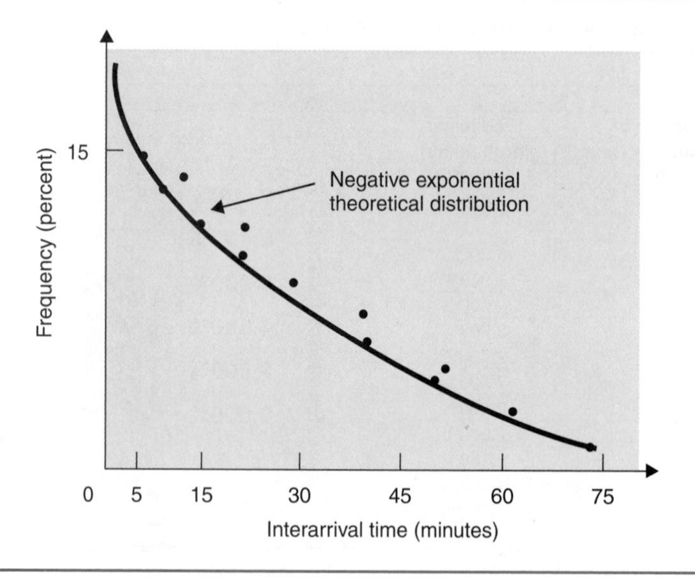

In a series arrangement, the customer enters the first facility and gets a portion of the service, then moves on to the second facility, and so on, as though he is in an assembly line. An example of this arrangement is a cafeteria style restaurant, where you first select a main course, select side dishes at the next station, drinks at a third station, and pay at the last station. There are also facilities that offer a combination of these service options.

Description of Service Facilities and the Service Exhibit 6.11 illustrates some of these possible service arrangements. The arrows into the boxes represent arriving and waiting customers, the boxes represent the service facilities, and the arrows out represent served customers. The service is rendered by a **server** which can be a person, group, machine, or a person–machine combination. There is a server in each facility. The service facility is also called a **channel.** Thus, the multifacility system such as *b* in Figure 6.11 is also called a **multichannel** system.

Measures for the Service

The service given in a facility consumes time. The length of time of the service can once again be considered a random variable characterized by some statistical distribution. It may, for example, be *constant* (e.g., exactly 10 minutes for each service), or it

EXHIBIT 6.11 Different Arrangements of Service Facilities

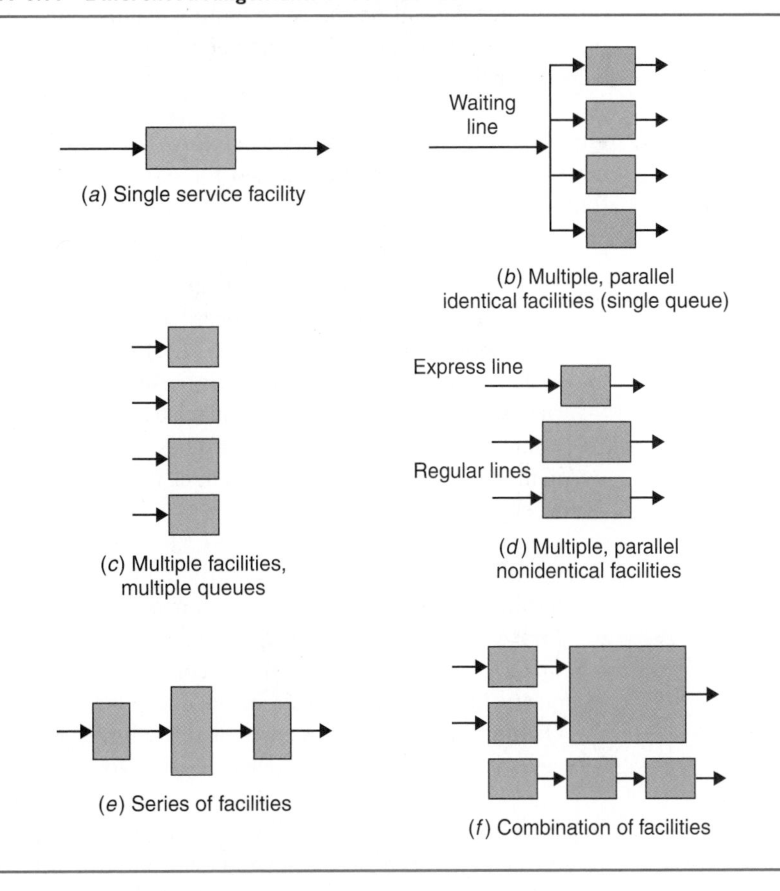

(*a*) Single service facility

(*b*) Multiple, parallel identical facilities (single queue)

(*c*) Multiple facilities, multiple queues

(*d*) Multiple, parallel nonidentical facilities

(*e*) Series of facilities

(*f*) Combination of facilities

may *fluctuate.* A fluctuating service time may be described by a frequency distribution. There are two ways of describing fluctuating service times:

1. **The Average Length of Service (Service Time), $1/\mu$** A fluctuating service time may follow one of several statistical distributions. Most common is the *negative exponential* (as in Exhibit 6.10). For example, the length of telephone calls is distributed in this fashion. A less common distribution is the *normal distribution* (Chapter 3), such as might be used in describing the time required for repairing a car. The average service time is usually designated by $1/\mu$. (*Note:* Service times may, in actuality, be affected by the length of the waiting line. If the line is long, the servers may work faster because of psychological and social pressures. In this text, the length of the line is assumed *not* to affect the service rate.)
2. **The Average Service Rate, μ** The service rate measures the service capacity of the facility in terms of customers per unit of time. The mean service rate, μ, is the *inverse* of the mean service time. For example, if the average service time is half an hour, then the mean service rate is $1/0.5 = 2$ customers per hour. Note that the service rate *inherently* assumes the facility is busy; that is, it is *not* the average number of customers served over some period of time. If the service time is exponentially distributed, then the service rate can be shown to be Poisson distributed. Other service time distributions can also exist, such as constant, and *arbitrary* (which means that the mean service time and its standard deviation are known).

The Waiting Line

A queue is formed whenever customers arrive and the facility is busy. The characteristics of the queue depend on the rules and regulations governing it (termed the *queue discipline*), the allowable length of the queue, the organization of the queue, and customers' behavior in the queue.

Queue Discipline The **queue discipline** describes the policies that determine the manner in which customers are selected for service. Examples of some common disciplines follow:

- **A Priority System** Priority is given to selected customers. For example, those with 10 items or less in a supermarket can go to the express lane. The handicapped and passengers with small children board airplanes first.
- **Emergency (Preemptive Priority) System** An emergency (*preemptive* priority) system is one in which an important customer not only has priority in entrance, but can even interrupt a less important customer in the middle of her service. For example, in an emergency case in a hospital, the doctor may leave the regular patient in the middle of the treatment. That is, the regular patient is preempted by the emergency patient.
- **Last-Come, First-Served (*LCFS*)** In *LCFS,* last arrivals are served first. This system is commonly used with parts and materials in a warehouse, because it reduces handling and transportation, the last ones being easier to reach or closer.
- **First-Come, First-Served (*FCFS*)** In *FCFS,* customers that arrive first are served first, a "fairness" discipline that is commonly used when humans constitute the queue. In this text, the *FCFS* queue discipline is assumed.
- **Queue Length** In some cases there are limits on the length of the queue; in these cases we talk about a *finite* queue length. Normally we assume no such limitations.

Organization of the Queue Queues may be organized in various ways. For example, customers may be screened at a main gate and then referred to one of several lines (such as in some theaters or banks), depending on the service and the queue discipline. In other cases, there is one line for several parallel service facilities. This is the organization generally assumed here.

Behavior in a Queue There are some interesting human behaviors observed in queues:

- **Balking** Customers refuse to join the queue, usually because of its length.
- **Reneging** Customers tire and leave the queue before they are served.
- **Jockeying** Customers switch between waiting lines (a common scene in supermarkets).
- **Combining or Dividing** Customers combine or divide queues at certain queue lengths (e.g., when a counter is closed or opened in a supermarket).
- **Cycling** Customers return to the queue immediately after obtaining service (e.g., children taking turns using the slide at a playground, or ore cars cycling at a mine).

In this text, we assume that a customer enters the system, stays in the line (if necessary to receive service), receives the service, and leaves permanently. If a customer behaves otherwise (according to any of the above behaviors), the queuing system becomes very complex, requiring simulation for its analysis.

6.4 Analysis of the Basic Queue (*M/M/1 FCFS/∞/∞*)

The classical and probably best known of all waiting line models is the common **Poisson-exponential single-server system.**

Poisson-Exponential Model Characteristics

Arrival Rate The arrival rate is assumed to be random and is thus described by the Poisson distribution. The average arrival rate is designated by the Greek letter lambda, λ.

Service Time The service time is assumed to follow the *negative exponential distribution.* The average service rate is designated by the Greek letter mu, μ, and the average service time by $1/\mu$. The major assumptions for the operation of such a single-server system are:

- There is an infinite source of population.
- Treatment is first-come, first-served.
- The ratio λ/μ is smaller than 1. This ratio is designated by the Greek letter rho, ρ. The ratio is a measure of the utilization of the system. If ρ is greater than or even equal to 1.0, the waiting line will increase without bound (explosive), a situation that is unacceptable to management. This is the case even when $\rho = 1$ because of the variability of the arrivals. Note that this is *not* the deterministic case where people arrive on schedule. Here they are *not* precisely on time, and λ is only the *average* arrival rate. If they arrive early they will probably have to wait, and if late, the server will be idle, but since the service rate is precisely equal to the arrival rate, idle service time accumulates and the line grows without bound.
- Steady state (equilibrium) exists. The system is in a *transient* state when its measures of performance are still dependent on the initial conditions. Our interest is in the "long-run" behavior of the system, commonly known as steady state, which is reached when the measures become independent of time.
- There is unlimited space and no limit on the queue length.

Measures of Performance (Operating Characteristics)

A queuing system is usually evaluated by one or more of the following measures of performance (given with their respective formulas). These measures depend on only two given variables, λ and μ, which must be stated in the same time dimensions.

Average Waiting Time, W The average time a customer spends in the system, waiting for service and being served, is

$$W = \frac{1}{\mu - \lambda}$$

Average Waiting Time in the Queue, W_q This is the average time a customer will wait, in the queue, before the service starts.

$$W_q = \frac{\lambda}{\mu(\mu - \lambda)}$$

Average Number of Customers in the System, L The average number of customers in both the queue and being served is:

$$L = \frac{\lambda}{\mu - \lambda}$$

Average Number of Customers in the Queue, L_q The length of the waiting line, not including the customer being served, is

$$L_q = \frac{\lambda^2}{\mu(\mu - \lambda)}$$

Probability of an Empty (Idle) Facility, $P(0)$ The probability that there are no customers in the system, either waiting or being served, is:

$$P(0) = 1 - \frac{\lambda}{\mu}$$

Probability of the System Being Busy, P_w The probability the system is busy is the same as *not* finding an empty system:

$$P_w = 1 - P(0) = \frac{\lambda}{\mu}$$

Probability of Being in the System (Waiting and Being Served) Longer than Time t The probability of taking more than time t to go through the system to completion is

$$P[T > t] = e^{-(\mu - \lambda)t}$$

where $e = 2.718$ (the base of the natural logarithms)
t = the specified time
T = time in the system

Note: $P[T \leq t] = 1 - P[T > t]$

Probability of Waiting for Service Longer than Time t_q The probability of taking more than time t_q to start receiving service is:

$$P[T_q > t_q] = \rho e^{-(\mu - \lambda)t_q}$$

Probability of Finding Exactly N Customers in the System, $P(N)$

$$P(N) = \rho^N (1 - \rho)$$

Probability that the Number of Customers in the System, N, Exceeds a Given Value, n

$$P[N > n] = \rho^{n+1}$$

This latter measure is used to determine the **balking rate.** The assumption here is that if you can accommodate, say, only six people in the line and in service, then when customers arrive and see that there is no more room to wait (in systems with a limited waiting area), they will leave. Thus, the probability of finding more than six people in the system will be the balking rate (probability of not staying). When people balk, the characteristics of the system change and must be recomputed based on more complex equations, or simulation. Note that $P[N < n] = 1 - P[N > (n - 1)]$. For example, $P[N < 3] = 1 - P[N > 2]$, since $P[N < 3]$ implies $N = 0$, 1, or 2, and $P[N > 2]$ implies $N = 3, 4, \ldots$

The following relationships, developed by the scholar J. D. C. Little, are extremely important because they hold for any queuing system and enable us to find L, L_q, W, and W_q as soon as any one is computed.

$$L = \lambda W$$
$$L_q = \lambda W_q$$
$$W = W_q + \frac{1}{\mu}$$

As an example of the use of these equations, consider the following. The toolroom at All-American Aviation Co. is staffed by one clerk who can serve 12 production employees, on average, each hour. The production employees arrive at the toolroom every 6 minutes, on average. To find the measures of performance of this queuing situation, we must first change the time dimensions of λ and μ to a common denominator, hours. The problem states that $\mu = 12$ per hour. For λ, the arrival rate of one employee every 6 minutes means $\lambda = 10$ per hour. Hence, the *toolroom utilization,* $\rho = 10/12 = 0.833$.

- *Average waiting time at the toolroom,* $W = \dfrac{1}{12 - 10} = 0.5$ hour, per employee
- *Average waiting time in the line,* $W_q = \dfrac{10}{12(12 - 10)} = 0.417$ hour, per employee
- *Average number of production employees at the toolroom,* $L = \dfrac{0.833}{1 - 0.833} = 5$
- *Average number of production employees in the line,* $L_q = \dfrac{0.833^2}{1 - 0.833} = 4.16$
- *Probability the toolroom clerk will be idle,* $P(0) = 1 - 0.833 = 0.167$
- *Probability the system is busy,* $P_w = 0.833$
- *Probability of waiting longer than 0.5 hour in the system,*
  $P[T > 0.5] = e^{-(12-10)0.5} = 1/e = 0.368$
- *Probability of exactly four production employees in the system,*
  $P(4) = 0.833^4(1 - 0.833) = 0.0804$
- *Probablility of more than three production employees in the system,*
  $P[N > 3] = (0.833)^4 = 0.481$

EXHIBIT 6.12 Goal Seek Dialog Box

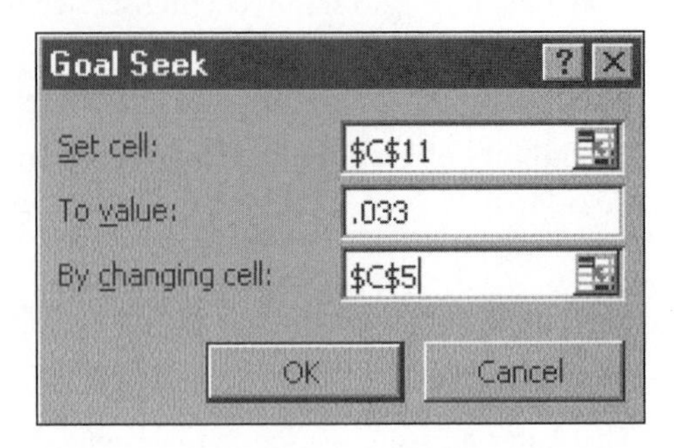

Managerial Use of the Measures of Performance

These measures of performance can be used in a cost analysis, or just to aid in determining service level policies. For example:

- A fast-food restaurant wants to design its service facility so that a customer will not wait, on average, more than 2 minutes (i.e., $W_q \leq 2$ minutes) before being served.
- A telephone company desires that the probability of any customer being without telephone service more than 2 days be 3 percent (i.e., $P[T > 2 \text{ days}] = 0.03$).
- A bank's policy is that the number of customers at its drive-in facility will exceed 10 only 5 percent of the time (i.e., $P[N > 10] = 0.05$).
- A city information service should be busy at least 60 percent of the day (i.e., $P_w \geq 0.6$).

Excel can also aid in determining the proper service level through its Goal-Seeking function, as illustrated next.

Using Excel's Goal Seek Function

Excel's Goal-Seeking feature varies the value in a specified cell until a formula that depends on that cell returns the desired result. Referring to the example at the beginning of the chapter, let us assume that you are interested in purchasing a printer such that print jobs would not have to wait more than 2 minutes (or 0.033 hour) on average. What output rate would the new printer need to have to achieve this goal?

We can answer this question using Excel's Goal-Seeking function. Referring to Exhibit 6.3, we desire to find the mean service rate entered in cell C5 such that the expected time waiting for the printer calculated in cell C11 is equal to 0.033. To do this, select Tools from the main menu at the top of the screen and then Goal Seek. After Goal Seek is selected, the Goal Seek dialog box is displayed as shown in Exhibit 6.12. Then the values for the Set cell, To value, and By changing cell were entered, as shown in Exhibit 6.12 and the OK button was clicked. According to the results shown in Exhibit 6.13, the new printer would be required to print 104.8 (cell C5) jobs per hour in order to reduce the wait time for the printer to an average of 2 minutes.

6.5 More Complex Queuing Situations

In this section, we will illustrate the solution process for two other common but more complex queuing systems: multifacility (also called multichannel) queues and serial queues.

EXHIBIT 6.13 Solution found by Excel's Goal Seek

	A	B	C	D	E	F	G	H
1	**Custom Printer Study (Peak Usage Time)**							
2			**New**					
3	**Model Inputs**		**Printer**	**Savings**				
4	Mean arrival rate per hour	81.25	81.25					
5	Mean service rate per hour	90	104.848					
6								
7	**Model Outputs**				***Key Formulas***			
8	Printer utilization	0.90	0.77		Cell B8	=B4/B5		
9	Average number of print jobs waiting in the queue	8.38	2.67	5.71	Cell B9	=(B4^2)/(B5*(B5-B4))		
10	Average number of jobs in the system	9.29	3.44	5.84	Cell B10	=B9+B8		
11	Expected time waiting for the printer	0.10	0.03	0.07	Cell B11	=B9/B4		
12	Expected total time to complete the job	0.11	0.04	0.07	Cell B12	=B11+(1/B5)		
13	Probability printer is idle	0.10	0.23		Cell B13	=1-B8		
14	Probability printer is busy	0.90	0.77		Cell B14	=1-B13		
15					Cells C8:C14	Copy cells B8:B14 to C8:C14		
16								
17								
18								

Want to find service rate so that jobs don't wait more than 2 minutes for printer

Goal

Multifacility Queuing Systems ($M/M/K\ FCFS/\infty/\infty$)

The **multifacility queues** considered here are composed of several identical and parallel service facilities. Such a situation is depicted in Exhibit 6.14. Note that there is only one waiting line that feeds the multiple service facilities. Whenever a server becomes free, the next customer in the queue goes to that service facility. Examples of such situations are an IRS (Internal Revenue Service) office or a restaurant, where arrivals get a number as they arrive and then the numbers are called sequentially as the examiners or tables become free.

There are several managerial problems in a multifacility system. For example, management is concerned with determining the proper number of servers, whether to use identical or nonidentical servers (e.g., whether to open an express lane in the supermarket), and the organization of the waiting line (one line for all servers, one line per server). Such decisions are based on the computed measures of performance.

EXHIBIT 6.14 The Multifacility Waiting Line System

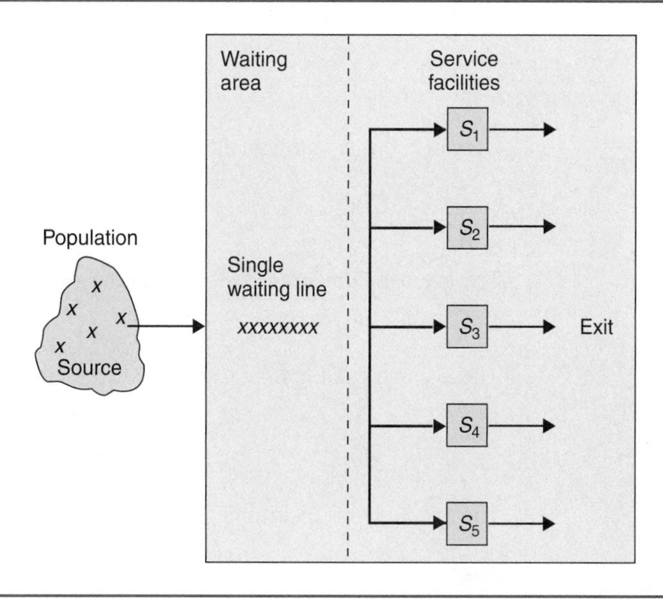

In this section, the simplest multichannel system is analyzed. In such a system, the following assumptions are made:

1. It is a Poisson-exponential system.
2. The service facilities (channels) are identical.
3. Only one waiting line exists.
4. The arrival rate λ is smaller than the combined service rate ($K\mu$) of all service facilities.

Formulas for Computing the Measures of Performance

Let:

λ = mean arrival rate
μ = mean service rate of *each* facility
K = number of servers (or service facilities)
ρ = utilization factor λ / μ of each facility (just as in the single facility system)
$\bar{\rho}$ = utilization factor of the entire system where $\bar{\rho} = \frac{\rho}{K} = \frac{\lambda}{K\mu}$

Assuming that $\lambda < K\mu$ (a necessary condition to avoid an explosive queue), then some of the most common measures of performance are:

1. The probability of finding no customer in the system (an idle system):

$$P(0) = \frac{1}{\frac{\rho^K}{K!(1-\bar{\rho})} + \sum_{i=0}^{K-1} \frac{\rho^i}{i!}}$$

where i = index of summation.

2. The probability of finding exactly N customers in the system:

$$P(N) = \begin{cases} P(0)\frac{\rho^N}{N!} \text{ when } N \le K \\ \frac{P(0)\bar{\rho}^N K^K}{K!} \text{ when } N > K \end{cases}$$

3. The average number of customers in the waiting line:

$$L_q = \frac{P(0)\rho^K \bar{\rho}}{K!(1-\bar{\rho})^2}$$

Given L_q, this equation yields the following for $P(0)$:

$$P(0) = \frac{L_q K!(1-\bar{\rho})^2}{\rho^K \bar{\rho}}$$

4. The average number of customers in the system:

$$L = L_q + \rho$$

5. The average waiting time in the queue per customer, before service:

$$W_q = \frac{L_q}{\lambda}$$

6. The average time a customer spends in the system for both waiting and service:

$$W = \frac{L}{\lambda} = W_q + \frac{1}{\mu}$$

Use of a Table to Solve Multifacility Problems In order to save computational time, a table for $P(0)$ can be used. The process, illustrated in Exhibit 6.15, is simple: $P(0)$ is read from the table, Exhibit 6.16, for various values of λ, μ, and K. Then L, L_q, W, or W_q can be computed.

E X A M P L E

Multichannel Queue

Given $\lambda = 36$ per hour, $\mu = 10$ per hour, and $K = 5$, find $P(0)$, L, L_q, W, and W_q.

Solution. Find $\lambda/K\mu = \bar{\rho} = 36/50 = 0.72$. From Exhibit 6.16, the value of $P(0)$ that corresponds to $K = 5$ is $P(0) = 0.0228$. From $P(0)$ compute the other variables:

$$\rho = 36/10 = 3.6$$

$$L_q = \frac{P(0)\rho^K\bar{\rho}}{K!(1-\bar{\rho})^2} = \frac{0.0228(3.6)^5 0.72}{5!(1-0.72)^2} = 1.055$$

$$L = 1.055 + 3.6 = 4.655$$

$$W_q = 1.055/36 = 0.029$$

$$W = 4.655/36 = 0.129$$

E X A M P L E

Multichannel Queue at Macro-Market

Macro-Market is considering how many of its eight identical check-out stations to staff during the night. Past experience indicates that the average number of random arrivals to the check-out area (there are always a number of customers coming into and wandering about the store) during the 11 P.M. to 6 A.M. period are 16 per hour. Each customer brings the market a revenue of $15. The service time for evening customers is usually short, three minutes on average, and follows a negative exponential distribution. Long lines create ill will and also discourage customers from entering the market at night. Therefore, management of the market has estimated that each customer-hour of waiting time in the market effectively costs $30. The operating cost of staffing each check-out station is $15 per hour. How many stations should be staffed so that total profit is maximized?

Solution. Given this more complex situation, we construct an influence diagram as shown in Exhibit 6.17. Note that at this stage in our analysis we assume that the number of stations staffed does not impact the time to service a particular customer or the rate of arrivals to the check-out stations. Of course the model could be expanded to consider these relationships. For example, if past experience suggests that customers tend to leave the store immediately when they observe long lines upon entry (i.e., they balk), we could modify the arrival rate to the check-out counters to be a function of the queue length.

EXHIBIT 6.15 The Multifacility Solution Process

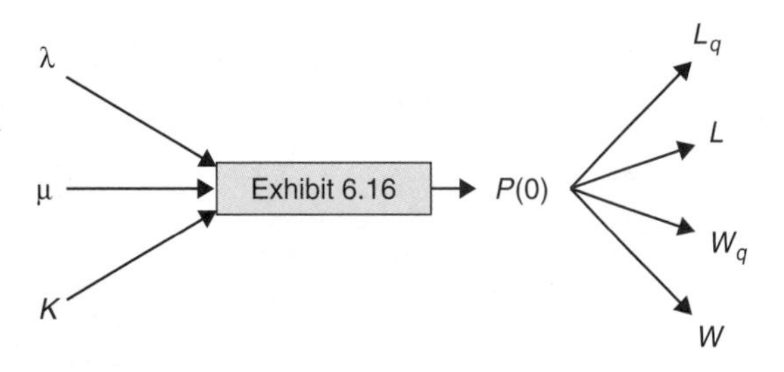

EXHIBIT 6.16
***P*(0) for the Multichannel Queue**

$\frac{\lambda}{K\mu}$	Number of Channels, *K*								
	2	**3**	**4**	**5**	**6**	**7**	**8**	**10**	**15**
0.02	0.9608	0.9418	0.9231	0.9048	0.8869	0.8694	0.85214	0.81873	0.74082
0.04	0.9231	0.8869	0.8521	0.8187	0.7866	0.7558	0.72615	0.67032	0.54881
0.06	0.8868	0.8353	0.7866	0.7408	0.6977	0.6570	0.61878	0.54881	0.40657
0.08	0.8519	0.7866	0.7261	0.6703	0.6188	0.5712	0.52729	0.44933	0.30119
0.10	0.8182	0.7407	0.6703	0.6065	0.5488	0.4966	0.44933	0.36788	0.22313
0.12	0.7857	0.6975	0.6188	0.5488	0.4868	0.4317	0.38289	0.30119	0.16530
0.14	0.7544	0.6568	0.5712	0.4966	0.4317	0.3753	0.32628	0.24660	0.12246
0.16	0.7241	0.6184	0.5272	0.4493	0.3829	0.3263	0.27804	0.20190	0.09072
0.18	0.6949	0.5821	0.4866	0.4065	0.3396	0.2837	0.23693	0.16530	0.06721
0.20	0.6667	0.5479	0.4491	0.3678	0.3012	0.2466	0.20189	0.13534	0.04979
0.22	0.6393	0.5157	0.4145	0.3328	0.2671	0.2144	0.17204	0.11080	0.03688
0.24	0.6129	0.4852	0.3824	0.3011	0.2369	0.1864	0.14660	0.09072	0.02732
0.26	0.5873	0.4564	0.3528	0.2723	0.2101	0.1620	0.12492	0.07247	0.02024
0.28	0.5625	0.4292	0.3255	0.2463	0.1863	0.1408	0.10645	0.06081	0.01500
0.30	0.5385	0.4035	0.3002	0.2228	0.1652	0.1224	0.09070	0.04978	0.01111
0.32	0.5152	0.3791	0.2768	0.2014	0.1464	0.1064	0.07728	0.04076	0.00823
0.34	0.4925	0.3561	0.2551	0.1812	0.1298	0.0925	0.06584	0.03337	0.00610
0.36	0.4706	0.3343	0.2351	0.1646	0.1151	0.0804	0.05609	0.02732	0.00452
0.38	0.4493	0.3137	0.2165	0.1487	0.1020	0.0698	0.04778	0.02236	0.00335
0.40	0.4286	0.2941	0.1993	0.1343	0.0903	0.0606	0.04069	0.01830	0.00248
0.42	0.4085	0.2756	0.1834	0.1213	0.0800	0.0527	0.03465	0.01498	0.00184
0.44	0.3889	0.2580	0.1686	0.1094	0.0708	0.0457	0.02950	0.01226	0.00136
0.46	0.3699	0.2414	0.1549	0.0987	0.0626	0.0397	0.02511	0.01003	0.00101
0.48	0.3514	0.2255	0.1422	0.0889	0.0554	0.0344	0.02136	0.00820	0.00075
0.50	0.3333	0.2105	0.1304	0.0801	0.0496	0.0298	0.01816	0.00671	0.00055
0.52	0.3158	0.1963	0.1195	0.0721	0.0432	0.0259	0.01544	0.00548	0.00041
0.54	0.2987	0.1827	0.1094	0.0648	0.0382	0.0224	0.01311	0.00448	0.00030
0.56	0.2821	0.1699	0.0999	0.0581	0.0336	0.0194	0.01113	0.00366	0.00022
0.58	0.2658	0.1576	0.0912	0.0521	0.0296	0.0167	0.00943	0.00298	0.00017
0.60	0.2500	0.1460	0.0831	0.0466	0.0260	0.0144	0.00799	0.00243	0.00012
0.62	0.2346	0.1349	0.0755	0.0417	0.0228	0.0124	0.00675	0.00198	0.00009
0.64	0.2195	0.1244	0.0685	0.0372	0.0200	0.0107	0.00570	0.00161	0.00007
0.66	0.2048	0.1143	0.0619	0.0330	0.0175	0.0092	0.00480	0.00131	0.00005
0.68	0.1905	0.1048	0.0559	0.0293	0.0152	0.0079	0.00404	0.00106	0.00004
0.70	0.1765	0.0957	0.0502	0.0259	0.0132	0.0067	0.00338	0.00085	0.00003
0.72	0.1628	0.0870	0.0450	0.0228	0.0114	0.0057	0.00283	0.00069	0.00002
0.74	0.1494	0.0788	0.0401	0.0200	0.0099	0.0048	0.00235	0.00055	0.00001
0.76	0.1364	0.0709	0.0355	0.0174	0.0085	0.0041	0.00195	0.00044	
0.78	0.1236	0.0634	0.0313	0.0151	0.0072	0.0034	0.00160	0.00035	
0.80	0.1111	0.0562	0.0273	0.0130	0.0061	0.0028	0.00131	0.00028	
0.82	0.0989	0.0493	0.0236	0.0111	0.0051	0.0023	0.00106	0.00022	
0.84	0.0870	0.0428	0.0202	0.0093	0.0042	0.0019	0.00085	0.00017	
0.86	0.0753	0.0366	0.0170	0.0077	0.0035	0.0015	0.00067	0.00013	
0.88	0.0638	0.0306	0.0140	0.0063	0.0028	0.0012	0.00052	0.00010	
0.90	0.0526	0.0249	0.0113	0.0050	0.0021	0.0009	0.00039	0.00007	
0.92	0.0417	0.0195	0.0087	0.0038	0.0016	0.0007	0.00028	0.00005	
0.94	0.0309	0.0143	0.0063	0.0027	0.0011	0.0005	0.00019	0.00003	
0.96	0.0204	0.0093	0.0040	0.0017	0.0007	0.0003	0.00012	0.00002	
0.98	0.0101	0.0045	0.0019	0.0008	0.0003	0.0001	0.00005	0.00001	

EXHIBIT 6.17 Influence Diagram for Macro-Market's Staffing Decision

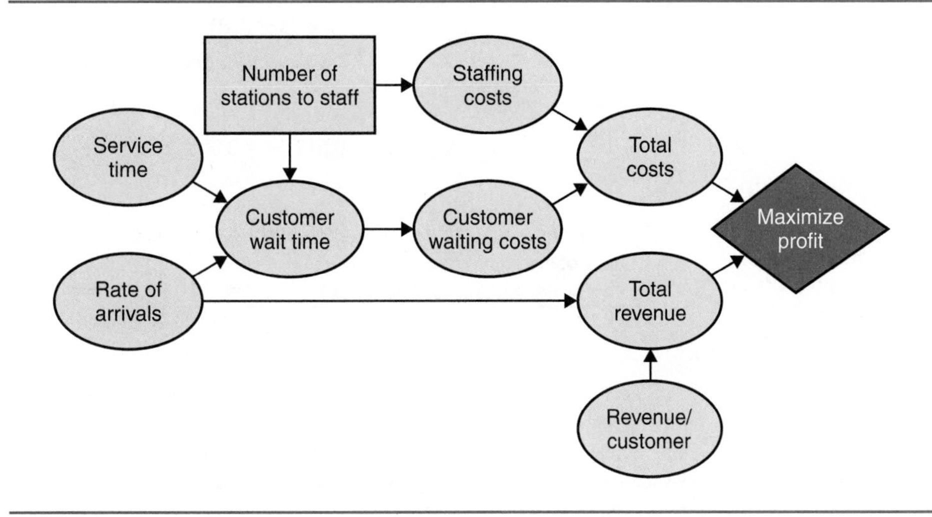

Our approach for determining the number of stations to staff will be to evaluate the profitability of staffing one more check-out station at a time, until we get to the point that the cost of staffing another check-out station is far greater than the costs saved by reducing the waiting time further. Staffing more check-out stations after this will obviously reduce the total profit, so we can stop once we reach this point, which could range anywhere from one to eight.

In this system, $\lambda = 16$, $\mu = 60/3 = 20$. The waiting time of interest is W_q because it is the line itself we are interested in rather than the total wait plus service time. We first calculate the profit for having one station open:

For one station:

$$W_q = \frac{\lambda}{\mu(\mu - \lambda)} = \frac{16}{20(20-16)} = 0.2 \text{ hour per customer}$$

Since 16 customers come to the check-out counters each hour, the cost of waiting will be 16(0.2)($30) = $96 per hour. The total profit per hour is thus:

Gross income: 16 customers ($15 each)	= $240.00
Less operating expense	= 15.00
Less waiting cost	= 96.00
Profit	= $129.00 per hour

Two stations open: To compute W_q for two stations, Exhibit 6.16 is used with $\rho = 0.8$, $K = 2$, and $\bar{\rho} = 0.4$; the result is $P(0) = 0.4286$. Thus,

$$L_q = \frac{0.4286(0.8)^2 0.4}{2(0.6)^2} = 0.152$$

and the waiting time

$$W_q = L_q/\lambda = 0.152/16 = 0.0095 \text{ hour per customer}$$

Thus,

Gross income: 16 customers ($15 each)	= $240.00
Less operating expenses: 2 (15)	= 30.00
Less waiting cost: 16(0.0095)(30)	= 4.60
Profit	= $205.40 per hour, which is considerably better than the $129 previously from one station

Three stations open: From Exhibit 6.16 with $\rho = 0.8$, $K = 3$, and $\bar{\rho} = 0.26$, the result is $P(0) = 0.4564$.

Thus,

$$L_q = \frac{0.4564(0.8)^3 0.26}{3(2)(0.74)^2} = 0.018$$

and the waiting time

$$W_q = L_q/\lambda = 0.018/16 = 0.0011 \text{ hours per customer.}$$

Thus,

Gross income: 16 customers ($15 each)	=	$240.00
Less operating expenses: 3 (15)	=	45.00
Less waiting cost: 16(0.0011)(30)	=	0.53
Profit	=	$194.47 per hour, which is less than with two stations

Four or more stations open: There is no need to check further because the cost of waiting is now trivial and staffing more stations will increase the cost $15 for each additional station. (This was actually the case with three stations also since we spent $15 to save less than $4.60.) Thus we can conclude that having two stations open is the most profitable, at the rate of $205.40 per hour. [*Note:* It is of interest to note that staffing the second check-out station was *much more* effective than just reducing the wait in half, as intuition commonly tells us. The wait was reduced from 0.2 hour per customer (12 minutes) to 0.01 hour (36 seconds)!]

Serial (Multiphase) Queues

In certain service situations, a customer receives service at a number of stations in sequence. The customer (or product) moves from station to station and possibly from queue to queue. This is known as a **serial** or *multiphase* **queue.** Under certain assumptions, such a process may be analyzed rather easily. Multiple (parallel) servers may even be included in the process.

The first two necessary assumptions are that the source is infinite and the queues at each station are not limited in length. Second, in the case of multiple servers within each station, all servers must have the same exponential service time distribution. Third, the customers at the first station arrive randomly (Poisson). Finally, $\lambda < K\mu$ at every station, where K = number of servers, so that an explosive queue is not formed anywhere in the system. Under these assumptions, the output from each station will also be Poisson, with the average rate λ. Because each station has Poisson arrivals, it may be treated independently of the others, and Exhibit 6.16 can be used for computing the measures of performance throughout the entire process.

EXAMPLE

Serial Queue: Three-Station Process

Consider a three-station process where the arrival rate λ = 5 per hour. The number of servers and the service rates are

$$K_1 = 1;\ \mu_1 = 6 \text{ (for station 1)}$$
$$K_2 = 3;\ \mu_2 = 2 \text{ (for station 2)}$$
$$K_3 = 2;\ \mu_3 = 4 \text{ (for station 3)}$$

EXHIBIT 6.18 A Serial Queue Situation (Combined with Parallel Servers)

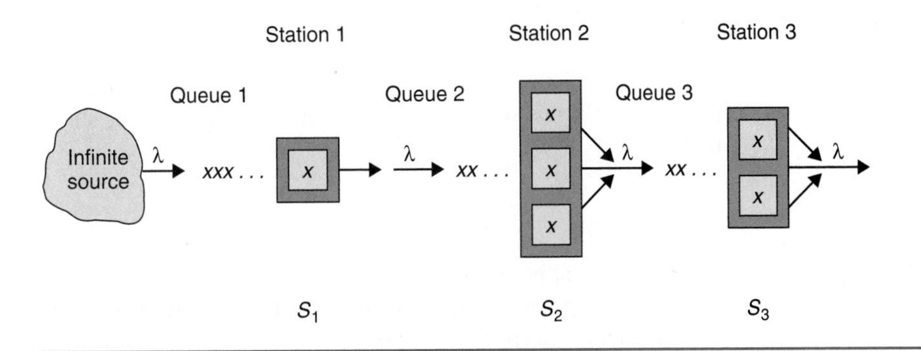

The problem is shown schematically in Exhibit 6.18. Find the waiting times within the process.

Solution (all times in hours)

Station 1 (single server, use formulas):

$$\lambda / \mu = 0.833$$

$$L_{q1} = \frac{\lambda^2}{\mu_1(\mu_1 - \lambda)} = \frac{5(5)}{6(6-5)} = 4.167$$

$$W_{q1} = \frac{\lambda}{\mu_1(\mu_1 - \lambda)} = \frac{5}{6(6-5)} = 0.833$$

Station 2 (multiple servers, use Exhibit 6.16): Given $\lambda = 5$, $\mu_2 = 2$, $K_2 = 3$

$$\rho_2 = \lambda / \mu_2 = 5/2 = 2.5$$

$$\bar{\rho}_2 = \frac{\rho_2}{K_2} = \frac{2.5}{3} = 0.833 \text{ (use 0.84 in Exhibit 6.16)}$$

From Exhibit 6.16, $P(0) = 0.0428$, and thus:

$$L_{q2} = \frac{P(0)\rho_2^{K_2}\bar{\rho}_2}{K_2!(1-\bar{\rho}_2)^2} = \frac{0.0428(2.5)^3 0.833}{(3 \times 2)(1 - 0.833)^2} = 3.333$$

$$W_{q2} = L_{q2} / \lambda = 3.333/5 = 0.667$$

Station 3 (multiple servers, use Exhibit 6.16): Given $\lambda = 5$, $\mu_3 = 4$, $K_3 = 2$

$$\rho_3 = \lambda / \mu_3 = 5/4 = 1.25$$

$$\bar{\rho}_3 = \frac{\rho_3}{K_3} = \frac{1.25}{2} = 0.625 \text{ (use 0.62 in Exhibit 6.16)}$$

From Exhibit 6.16, $P(0) = 0.2346$ and thus:

$$L_{q3} = \frac{P(0)\rho_3^{K_3}\bar{\rho}_3}{K_3!(1-\bar{\rho}_3)^2} = \frac{0.2346(1.25)^2 0.625}{(2)(1 - 0.625)^2} = 0.815$$

$$W_{q3} = L_{q3} / \lambda = 0.815/5 = 0.163$$

Thus, the total waiting time for service is:

$$W_{qSystem} = W_{q1} + W_{q2} + W_{q3} = 0.833 + 0.667 + 0.163 = 1.663 \text{ (hours)}$$

6.6 Detailed Modeling Example

The focus of this chapter thus far has been on the model formulation stage of the modeling process. In this section, we illustrate the use of queuing theory in the broader modeling context.

Step 1: Opportunity/Problem Recognition

A large discount hardware retailer receives its inventory by truck. Recently, drivers have begun to complain that they have to wait to unload their trucks more often than not. Headquarters is interested in determining if the drivers' complaints are justified, and how much it could save if the unloading operation could be speeded up. One option is to lease a piece of equipment that would permit unloading 10 trucks per hour. The cost of leasing this equipment would be $200 per day.

Step 2: Model Formulation

An influence diagram for this situation is shown in Exhibit 6.19. According to the exhibit, the analysis of this situation requires four types of data: truck arrival rates to the loading dock, the unloading service rate, the equipment cost, and the operating costs of the trucks.

Step 3: Data Collection

The hardware store receives shipments between 8:00 A.M. and 5:00 P.M. A two-week analysis was undertaken to determine the average arrival rate and truck unloading times. Initial analysis of the collected data suggested that an average of two trucks arrives each hour and that truck unloading times average 20 minutes per truck. Histograms of the data indicated that the pattern of truck arrivals closely match the Poisson distribution and that the pattern of truck unloading times resembles a negative exponential distribution. It was further verified using the chi-square goodness of fit test (see Chapter 2) that a Poisson distribution with an arrival rate of two trucks per hour is a good fit to the pattern of observed truck arrivals. Likewise, a negative exponential distribution with an average ser-

EXHIBIT 6.19 Influence Diagram for Loading Dock of Hardware Retailer

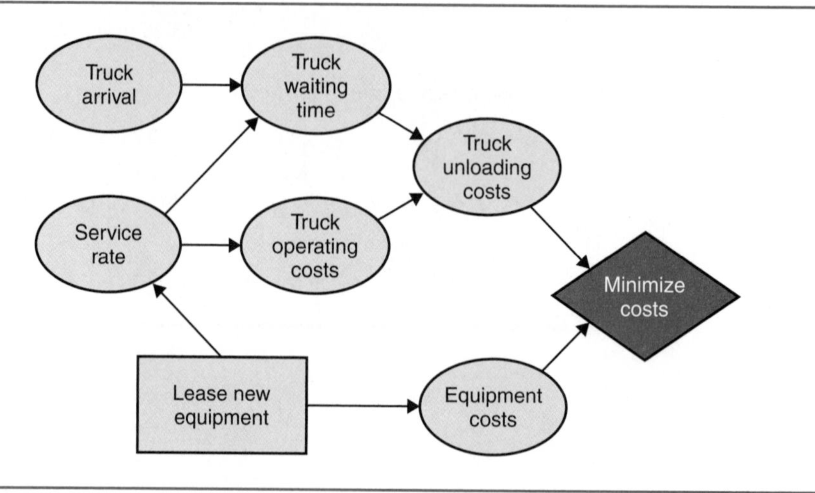

vice time of 20 minutes per truck provides a good fit to the pattern of observed truck-unloading times. Finally, corporate records indicate that operating a truck costs $30/hour, including both labor and depreciation.

A spreadsheet with the standard queuing theory equations was created and the data for the retailer entered, as shown in Exhibit 6.20. Note in the exhibit that the average service (unloading) time of 20 minutes was converted to the service rate of three trucks per hour.

Step 4: Analysis of the Model

According to the results shown in Exhibit 6.20, the likelihood of an arriving truck finding the dock occupied (probability system is busy) is 0.67, so the truck drivers are correct. At any given time, there are an average of 1.33 trucks waiting to be unloaded (two trucks in the total system being either unloaded or waiting to be unloaded). On average, a truck waits 0.67 hour prior to being unloaded and another 0.33 hour actually being unloaded, for a total waiting time of 1.0 hour.

Exhibit 6.21 describes the anticipated operation after installing the new equipment. According to Exhibit 6.21, the probability that trucks would find the loading dock occupied drops from 67 percent to 20 percent. The average number of trucks waiting to be unloaded would be reduced from 1.33 trucks to 0.05 trucks, and the total number of trucks at the retailer would decrease from 2.00 trucks to 0.25 truck. Likewise, the average time a truck spends waiting in the queue would be reduced by 0.64 hour (0.67 – 0.03), and the total time a truck spends at the retailer would be reduced by 0.87 hour (1.00 – 0.13).

Because there are 2 trucks/hour × 9 hours of operation = 18 loads a day, each of which currently requires an hour, there is a total waiting and unloading time of 18 hours each day with the current operation. Based on the truck costs of $30/hour, the total cost of waiting and unloading is thus $540/day. If the new equipment were leased, the total time would decrease to 2.34 hours per day (18 loads a day × 0.13 hour). Based on the truck costs of $30/hour, the total cost would then be reduced from $540/day to $70.20/day. This represents a savings of $469.80 per day.

EXHIBIT 6.20 Hardware Retailer's Current Loading Dock Operations

	A	B	C
1	**Hardware Retailer**		
2			
3	**Model Inputs**		
4	Mean arrival rate per hour	2	
5	Mean service rate per hour	3	
6			
7			
8	**Model Outputs**		
9	Utilization	0.67	=B4/B5
10	Average number of trucks in the queue (L_q)	1.33	=(B4^2)/(B5*(B5-B4))
11	Average number of trucks in the system (L)	2.00	=B10+B9
12	Expected waiting time in the queue (W_q)	0.67	=B10/B4
13	Expected total time in the system (W)	1.00	=B12+(1/B5)
14	Probability system is idle (P_0)	0.33	=1-B9
15	Probability system is busy (P_{busy})	0.67	=1-B14

EXHIBIT 6.21 Hardware Retailer's Operations after Installing New Equipment

	A	B	C
1	**Hardware Retailer**		
2			
3	**Model Inputs**		
4	Mean arrival rate per hour	2	
5	Mean service rate per hour	10	
6			
7			
8	**Model Outputs**		
9	Utilization	0.20	=B4/B5
10	Average number of trucks in the queue (L_q)	0.05	=(B4^2)/(B5*(B5-B4))
11	Average number of trucks in the system (L)	0.25	=B10+B9
12	Expected waiting time in the queue (W_q)	0.03	=B10/B4
13	Expected total time in the system (W)	0.13	=B12+(1/B5)
14	Probability system is idle (P_0)	0.80	=1-B9
15	Probability system is busy (P_{busy})	0.20	=1-B14

Step 5: Implementation

Since the $469 savings resulting from shorter truck waiting times and faster unloading times exceeds the $200 cost of the new equipment, purchasing the new equipment is clearly justified. Exhibit 6.22 illustrates in the form of a memo how the results of this analysis might be communicated to management at the retailer.

EXHIBIT 6.22 Sample Memo to Management of Hardware Retailer

MEMO

To: Nikki Scott, VP supply chain management
From: George Samuelson, senior operations analyst
CC: Jennifer Allison, store manager
Date: 3/28/02
Re: Loading dock operations

Introduction

I have concluded my analysis of the loading dock operations of store #132 regarding the complaints the truck drivers have raised about wait times to unload their trucks. In summary, I recommend that new equipment be leased for the loading dock. While the cost of this equipment is estimated to be $200 per day, my analysis indicates that this equipment will generate savings of $469.80 per day, for a net savings of $269.80 per day.

(continued)

EXHIBIT 6.22 Continued

Analysis

After analyzing the loading dock operations over a two-week period, I found that an arriving truck is likely to find the dock occupied 67 percent of the time. At any given time, there was an average of 1.33 trucks waiting to be unloaded and 2 trucks at the store. Furthermore, trucks waited an average 0.67 hour prior to being unloaded and spent one full hour at the store. Thus, my conclusion is that the drivers have a valid concern.

If the new equipment under consideration were leased, the likelihood of the loading dock being occupied would decrease from 67 percent to 20 percent. Also, the average number of trucks waiting to be unloaded would be reduced from 1.33 trucks to 0.05 truck, and the total number of trucks at the store at any given time would decrease from 2.00 trucks to 0.25 truck. Likewise, the average time a truck would spend waiting to be unloaded would be reduced by 0.64 hour and the total time a truck spends at the retailer would be reduced by 0.87 hour.

Assumptions/Limitations

Several limitations of this analysis should be pointed out. First, the truck arrival patterns and unloading times were based on a two-week study. Given the seasonality inherent in our business, the results discussed here are not generalizable to other patterns of demand. Additional data would need to be collected and analyzed to gain similar insights into the loading dock operations for other truck arrival patterns and other service rates. Second, this analysis was based on specific assumptions about the distributions of arrivals and service times. Although these assumptions were validated, if it was later determined that these assumptions are no longer valid, the analysis would need to be repeated, perhaps using a different methodology. Finally, the results presented here are simply averages. The performance of the loading dock will vary within a given day and from day to day.

QUESTIONS

1. What alternatives might services that must always meet all demand, such as electric power utilities, consider in order to maintain the necessary capacity to meet peak demand at a reasonable cost?
2. How might you go about trying to determine the cost of ill will?
3. Your manager is frustrated by the number of complaints she is receiving concerning the long waiting lines to get show tickets. Yet her studies show that the ticket office is, on average, idle 70 percent of the time. How would you explain this apparent contradiction to your manager? Do people instinctively assume a probabilistic system will act like a deterministic one? Would this comparison be helpful in your explanation?
4. Consider the chores and activities you do (or observe) and the queue disciplines you (or others) employ. For what activities do you use a LCFS discipline? A priority discipline? What other disciplines do you use?
5. Reconsider the five types of common human behavior in queues described in the chapter. Do any of these explain why we would have trouble conducting a queuing study at a multichannel supermarket checkout? Would your answer change if we studied tollbooths on a freeway?

6. Our queuing models in this chapter all assume the system is in steady state. How valid is this assumption, in general, do you think? Would it be valid for a bank? For an airport?
7. In a fast-food restaurant, does the owner have to be more concerned with service at the drive-through, the indoor lines, or both equally? If not equally, which requires more attention, and why?
8. Common sense tells us that $W = \mu L$. That is, if you join a queue of length L and the service time averages μ, wouldn't your expected wait be just μL? What is wrong with this logic, given the equations relating W and L in the chapter?
9. If managers can ascribe accurate costs to a queuing situation, why would they ever employ other measures such as line length or waiting time?
10. In the multichannel queuing model, why must we assume a single waiting line?

EXPERIENTIAL EXERCISE

Select a fast-food restaurant with a single drive-through window. Select either the lunch hour (11:30–12:30) or dinner hour (5:30–6:30) to study. Visit your chosen restaurant on at least two different occasions during your selected time period and record (with a helper) the following information:

- The time between arrivals to the drive-through (*Hint:* It is easier to simply record the time of each arrival and then later calculate the time between arrivals).
- The time it takes for the order to be taken (*Note:* This time begins when a restaurant employee first requests the order and ends when the order has been completed and confirmed).
- The time it takes to deliver the food to the customer (*Note:* This time begins when a restaurant employee first opens the drive-through window and establishes contact with the customer and ends when the customer has completely received the contents of his order.)

After collecting the data, complete the following tasks:

- Perform a chi-square test to verify that the assumption of Poisson arrivals is valid.
- Perform a chi-square test to verify that the assumption of exponential service times is valid for both the order-taking times and drive-through window processing times.
- Calculate the average waiting time, the average waiting time in queue, the average number of customers in the system, the average number of customers in the queue, the probability of an idle facility, and the probability of the system being busy.
- Calculate the probability of being in the system longer than 1 minute, 3 minutes, 5 minutes, 7 minutes, and 10 minutes.
- Calculate the probability that there are exactly 3 customers in the system. What is the probability that there are more than 3 customers in the system?
- Write a memo to the management of the restaurant you studied. In the memo include a discussion of how the data were collected, how the collected data were analyzed, what key insights were obtained, what recommendation you would make, and key limitations associated with this study.

MODELING EXERCISES

For these exercises, unless otherwise stated, assume random (Poisson) arrivals and a negative exponential service time.

1. A physician schedules check-up patients at the rate of one every 15 minutes. Assume that the patients arrive exactly on schedule. Assume a constant check-up rate of four patients per hour.
   a. Calculate the waiting line that is likely to be generated after 4 hours.
   b. Assume that the physician can see five patients in an hour. What will be the waiting time after 4 hours, and what will be the physician's rate of utilization?
   c. What will happen if the physician can see only three patients an hour? How long will the line be after 4 hours?
2. Vic's Vending Corporation operates vending machines in one town. The machines break down at an average rate of two per hour. An hour of downtime of

a machine is considered as a loss of $13. Currently, the machines are serviced by one of the company's engineers on an as-needed basis, who is capable of repairing each machine in 24 minutes, on average. The hourly cost of the engineer's time is $20. A maintenance contractor offered to take over the maintenance work. The contractor can repair three machines each hour, with an hourly charge of $40. Should management accept the contractor's offer?

3. A service system has an average interarrival time of two minutes and an average service rate of 60 per hour.
   Find:
   a. The probability that a customer will have to wait
   b. The probability that four persons are in the system
   c. The probability of finding more than three in the system
   d. The probability of waiting more than 3 minutes for service
   e. The probability that fewer than four are in the system
   f. The probability of being through the system within 10 minutes
4. A toolroom clerk at British Light Industries, Ltd. is serving a maintenance department with a large number of employees who earn £8 per hour. The clerk earns £5 per hour. The workers arrive at the toolroom at an average rate of 6.2 per hour. The average service time is eight minutes.
   a. What is the probability of finding no workers at the toolroom (either waiting or being served)?
   b. What is the average waiting time (before being served) per worker?
   c. What is the average number of workers waiting in line (excluding the one being served)?
   d. Would you recommend installing an incentive plan that will reduce the average service time to 6.4 minutes and will cost the company £2 per hour?
   e. Another clerk can be hired at £5 per hour. The two clerks will operate as a single team serving one line, with an average service time of 4 minutes. Would you recommend hiring the additional clerk? (Assume no incentives.)
   f. If you had the alternative of installing the incentive plan with the existing system or hiring a second clerk, which one would you recommend?
   g. Two more clerks can be hired (at £5 each per hour) to help the single clerk, reducing the average service time to two minutes. Would you recommend this over hiring only one more clerk? Why or why not?
5. Given an arrival rate $\lambda = 3$, find values of L_q and W_q for the following values of μ: 3.1, 3.5, 4, and 6. Above what value does speed of service (i.e., μ) become important?
6. A waiting line system has the following characteristics:

- Average interarrival times of 6 minutes
- Space necessary for accommodating a waiting customer = 5 square feet
- Cost per hour of waiting time per customer = $2

   It was also observed that the facility is idle 20 percent of the time.
   Find:
   a. The area necessary to accommodate the average waiting line
   b. If it is profitable to invest $5 an hour in the facility if the service rate can be doubled (twice as fast).
7. Sunny Engineering Corporation is designing a special machine for processing chickens. The chickens arrive from the farms on trucks, in cages, at a rate of 10 trucks per hour. If the chickens are kept in the cage in the waiting area more than 3 hours, they will start to dehydrate, lowering their quality and causing damage to the processor. Draw an influence diagram and determine the minimum average processing rate (in truckloads per hour) that must be designed for the machine in order to ensure that the cages will be processed, on the average, in 3 hours or less. That is, waiting and processing time is 3 hours or less.
8. Lima Airport currently operates with one runway for landings. The average landing time is 3 minutes. Airplanes arrive at the airport at the rate of 17 per hour. The estimated average fuel consumption for an airplane waiting for a landing is 10 liters per minute. A liter of fuel costs 2,000 Peruvian sol.
   Find:
   a. The average number of airplanes circling the airport in a holding pattern, that is, waiting for permission to land. Do not include the landing plane.
   b. The average cost of fuel "burned" by an airplane waiting to land.
   c. The chance of finding less than three airplanes in the airport vicinity (in the waiting line and landing).
   d. The utilization of the runway.
9. Five cars arrive at an emissions testing station each hour. The average service time is 6 minutes. The station can accommodate only three cars (waiting

and being served). Any car that cannot be accommodated in the station is parked in a No-Parking area on the street, where there is a 40 percent chance of being fined $10. The owner of the station pays the fines. The station is in operation 48 hours per week. Cars completed are picked up immediately by the customers. Draw an influence diagram and find the weekly fines paid (in $).

10. The Jamaican Pelican is a one-man yogurt shop where people arrive at the average rate of 20 per hour. Big Joe, the proprietor, serves a customer, on average, in 2 minutes. During the noontime rush, the arrival rate increases to one arriving customer every 2 minutes.
Find:
   a. How fast must Big Joe work to ensure that a noontime customer will not wait for service, on the average, more than 10 minutes.
   b. The probability that six or more people are in the shop during the nonrush period.
   c. The average waiting time during the nonrush hours.
11. Customers arrive at a service facility one every 12 minutes, on average. The average service time is 10 minutes. The operation of the existing system costs $5 an hour. The facility is in operation 8 hours a day.
Find:
   a. If the waiting (prior to service) area, which can accommodate three customers, is sufficient 70 percent of the time.
   b. It is proposed to speed up the service so 10 customers can be served in an hour, at an additional cost of $240 per day. If an hour waiting time (prior to service), per customer, is worth $10, is the investment justified?
   c. An alternative solution is to use three parallel, identical facilities serving one line. Each facility is capable of serving three customers per hour at a cost of $2 per hour per facility. Analyze this alternative. Would you recommend it?
12. Customers arrive at a one-person barbershop with an average interarrival time of 20 minutes. The average time for a haircut is 12 minutes.
   a. The owner wishes to have enough seats in the waiting area so that no more than 5 percent of the arriving customers will have to stand. How many seats should be provided?
   b. Suppose there is only sufficient space in the waiting area for five seats. What is the probability that an arriving customer will not find a seat?
13. The industrial engineering department of First National Bank conducted a study to determine the effectiveness of its two drive-in stations. These stations operate independently of each other, and each has its own waiting line. The study involved random observations of the number of cars in a line (including the one being served). The results of the first day of the study are given as follows:

Time	Cars in Station 1	Cars in Station 2
9:12	3	2
9:37	5	3
10:04	2	2
11:30	0	4
11:58	4	2
12:20	3	3
12:39	2	2
1:23	5	1
1:37	4	5
2:06	2	3
2:19	1	0
2:46	3	4

In addition, it was noted that line 1 served 84 customers and was open 6 hours, whereas line 2 served 79 customers and was open 6 hours and 15 minutes. Find the utilization of each line and which is better utilized.

14. The following are observations taken at South London Bank Drive-In regarding the number of cars in the waiting line (prior to service):

Time	Cars in Line
10:30	6
10:37	7
10:53	2
11:06 (0 in service)	0
12:12	1
12:44	4
1:20	3
1:40	6
1:50 (1 car in service)	0
2:06	5
2:50	2
3:00	8

The Drive-In opened at 10:30 A.M. and at 3:00 P.M. no more cars were allowed to join the line. The last car left at 3:20 P.M.
Find:
   a. The average number of cars in the system
   b. The average number of cars in the waiting line

15. One branch of the post office in Calcutta would like to know how many windows to staff so the average number of customers waiting for service does not exceed eight. The average service time is 3 minutes, and the post office uses a single queue system, as illustrated below. Draw an influence diagram and determine how many windows the post office should staff on Monday mornings, when the average arrival rate is 60 customers per hour.

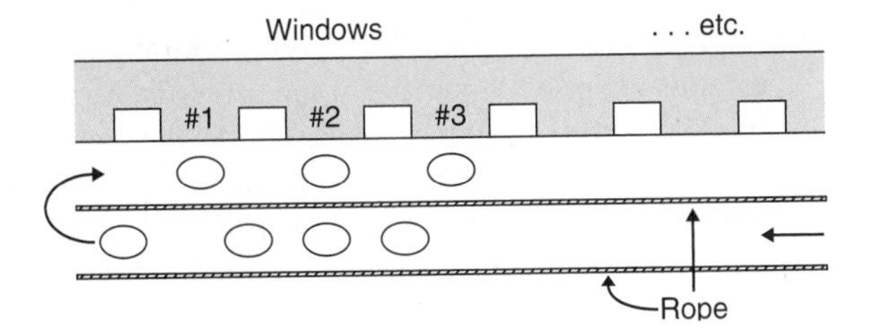

16. Mrs. Grim and Mr. Moss each have a private secretary who can type letters at the average rate of four per hour. The letters are generated by each manager at the average rate of three per hour. The managers have been wondering if they would benefit by pooling the two secretaries. What would you suggest? (Show calculations.)
17. The supervisor of the maintenance department of Everglade City is faced with a decision regarding maintenance of the city's heavy equipment. He is considering three alternatives:
    a. Hire a first-class mechanic, which will cost the city $14.20 per hour (including fringe benefits). The mechanic can repair five units per 8-hour day.
    b. Hire two second-class mechanics, each of whom will cost the city $11.50 per hour and would work separately. When the two serve one waiting line, they can each fix four units a day.
    c. Subcontract the maintenance to ABC Engineering at a cost of $50 per unit repaired. The average repair time is one hour.

    The quality of repairs in all three alternatives is considered the same. Currently, there is an average of four units of heavy equipment requiring daily repair (assume random arrivals). A unit of heavy equipment not in operation costs the city $25 per hour, because the city must lease alternative equipment. Find the total daily cost of the three alternatives.
18. Trucks are loaded by forklift at a rate of four per hour. An hour of forklift time (working or idle) costs $10. The trucks arrive at an average rate of one every 20 minutes. An idle hour (waiting for loading) for a truck is estimated to cost $16. Draw an influence diagram and find how many forklifts should be used. *Assumption:* Only one forklift can be used per truck.
19. The keen competition in the successful fast-food industry has forced the management of Burger Corporation to study the operation of its various restaurants. At its Northwestern restaurant, where it receives the most complaints, an analysis revealed the following. During rush hours, customers arrive at an average rate of one every minute. The attendants can serve, on the average, 66 customers per hour. Burger Corporation's president cannot understand why there are so many complaints about the Northwestern restaurant. As the president states: "Why do they complain? We can serve even 10 percent more without any problems."
    a. Explain to the president why you think there are so many complaints.
    b. What will happen if the average number of customers grows by 10 percent?
    c. It was proposed that two teams, each with an hourly serving capacity of 33 customers, replace the existing team, which can serve 66 customers per hour. If there is no additional cost involved in such a change, would you recommend it? Show why or why not. (Assume that the two teams serve a single waiting line.)

    Hint: Assume that the number of complaints is a function of waiting time per customer.
20. Hong Kong Machine Company has a toolroom with two clerks. Both clerks issue spare parts and tools to maintenance workers. Maintenance workers arrive at the rate of 20 per hour, each wanting either a part (40 percent) or a tool (60 percent), but not both. The average issuing time is 5 minutes per order (service by one clerk). Each clerk currently issues both parts and tools. A maintenance worker not at his bench costs the company $5 per hour. It was proposed that the two clerks be specialized, namely: One will issue spare parts only, and the other will issue tools only. Draw an influence diagram and answer the following:
    a. Would you advise specialization of the clerks if the service time does not change?
    b. Would you advise specialization if the service time was reduced to four minutes per order?

    Hint: Treat the current situation as a multiple station system with two servers. Treat the specialization as two single stations.

21. Toulouse Bakeries serves 480 customers during 8 hours of operation. The average service time per customer is 4 minutes. Currently, there are five servers at the bakery serving one line.
Find:
    a. The utilization of the bakery
    b. The average number of customers at the bakery
    c. The average waiting time (prior to service) in the bakery (per customer), in minutes
    d. The average number of customers being served
    e. The probability of finding no customers in the bakery
    f. The probability of finding exactly three customers in the bakery
22. A five-stage manufacturing process receives raw materials (in units) randomly and must process an average of 12 units each day (8 hours). The data of the system are as follows:

Stage	**Number of Parallel Servers in Stage (*K*)**	**Service Capabilities (per 8-Hour Day, per Server)**
1	2	7.0
2	1	15.0
3	5	3.0
4	3	4.5
5	1	30.0

    a. Find the *expected time* that a unit will spend in the entire processing system (from arrival until finished in stage 5).
    b. Find the average waiting time prior to entering the processing system.
    c. The cost of a server in the system is $6 per hour. Unit waiting time costs $1 per hour. Would it be profitable to add more servers to the system? If yes, in what stage(s) should they be placed?
    d. As an alternative to adding more servers, would a transfer of servers among stages be profitable?

Assume:
(1) The arrival rate cannot be changed.
(2) Servers are paid on an hourly basis; if they work a portion of an hour, they are paid proportionally.
23. Team Oil is a gas station with four pumps selling unleaded gasoline. The arrival rate during rush hours is 60 per hour. The service time is 3 minutes. What is the chance that an arriving customer will find no cars in the station? (Assume one waiting line.)
24. Access to the freeway system in Los Angeles is controlled by a signaling system. The system permits one car to enter the freeway every few seconds. The time is constant (e.g., 3 seconds). However, it is changed during rush hours to a larger value. Explain in queuing terminology the reason for such an arrangement. Draw an influence diagram and find how long the average patient spends in the clinic.
25. The City of Cincinnati is designing a new customer service building. The average number of customers expected to arrive at the center is 30 per hour. To serve a customer takes, on the average, 6 minutes. Management is interested in knowing how many service windows to plan for (assuming that there will be only one waiting line). The city would like to meet the following goals:
    a. There will be no more than five customers in the waiting area.
    b. The average time in the facility, including service, should be 8 or fewer minutes.

    Construct an influence diagram and determine the best answer for management.
26. Customers arrive at the First Western drive-in window at a rate of 10 per hour during *regular* periods. The average interarrival time during *peak hours* is 4 minutes. The average service time is 4.1 minutes.
Determine:
    a. The probability a customer is in the system 12 or fewer minutes during a period of regular demand,
    b. The probability of finding fewer than six customers in the system during a period of regular demand.
    c. First Western wants to implement a policy that any customer who arrives during *peak time* will *not* have to wait (prior to service), on the average, more than 10 minutes. Is this goal achievable today? If not, determine what service rate is necessary to attain this goal.
    d. The operation of the drive-in window costs $20 an hour. The estimated loss due to decreased goodwill, because of the time spent in line, is figured at $1 per hour per customer. Management can expedite the service time to 3 minutes per customer, at a cost of $15 per hour. Would you recommend it for the regular hours? For the peak hours? Why or why not?

CASES

City of Help

Patients arrive with minor injuries at City of Help's walk-in urgency clinic at random. The average arrival rate is 12 per hour. The urgency room is organized as follows:

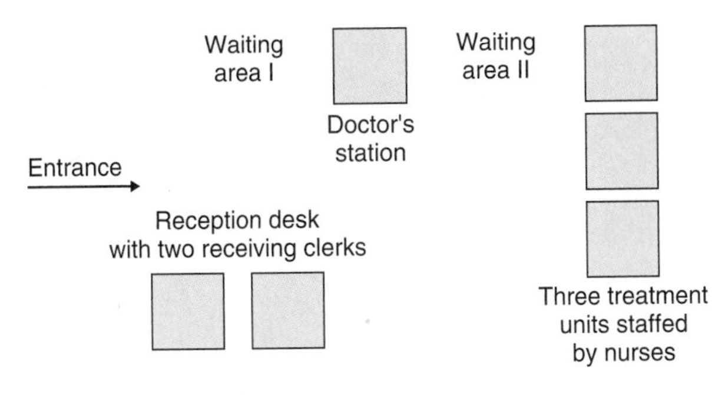

Patients enter the urgency room and go to the reception desk, where two clerks fill out the necessary paperwork (independently). The average time required to do the paperwork, per patient, is 7.5 minutes. The patient is then given a number and seated in the waiting area. When the number is called, the patient sees the doctor (the average examination time is 4 minutes). The patient then waits for the first available treatment unit; here, a nurse treats the patient (following the doctor's orders).

Questions

1. Draw an influence diagram for this problem/opportunity. What is the goal? What could be the decision variables?
2. Describe the type of queuing situation this represents.
3. If the nurse treatment time in the treatment unit averages 8 minutes, what it the total average time spent in the clinic per patient? How many chairs are needed in waiting area II, on average?
4. During the 30-minute lunch break, the nurses take turns away from the treatment units so that two of the three units are always staffed. What effect does this have on total patient time in the clinic?
5. Bearing in mind that the transient lunch break only runs for 90 minutes, how many chairs will now be needed in waiting area II?
6. How long will it take to work off the extra patients in waiting area II following lunch?

Newtown Maintenance Division

The city of Newtown, like many other cities today, is caught in a severe financial squeeze. Up to now, previous city managers have taken a short-term approach to Newtown's maintenance and repair services in the expectation that future tax receipts would improve. However, Newtown's voters have just rejected the fourth proposed tax increase on the ballot for city services in as many years, and the expected relief of future tax revenues now looks hopeless.

The new city manager has decided that a long-term policy must finally be established that recognizes the reality of continued low funding for city services. One portion of this problem is the manner of making daily repairs to the city's streets. Calls for repairs arrive randomly at the average rate of two per day. It is the mayor's declared policy that the city will respond to all calls for street repairs within one week (five 8-hour working days) of the call.

The city manager has two alternatives for servicing street repairs:

1. He can use any number of standard city crews, which cost \$79 per hour and can each repair a street, on the average, in 10 hours.
2. He can lease special heavy-duty street repairing equipment and use smaller crews, resulting in an hourly cost of \$90 per crew. These special crews can repair the average street in only 7 hours.

Make a recommendation to the city manager concerning these two alternatives. How soon can the city respond to calls under the cheaper of the above two policies? What would happen if the mayor insisted that the response time be reduced to *two* working days?

BIBLIOGRAPHY

Gelenbe, E., and G. Pujolle. *Introduction to Queuing Networks.* New York: Wiley, 1998.

Gibson, R. Merchants Mull the Long and the Short of Lines, *Wall Street Journal* (Sept. 3, 1998), B1.

Gross, D., and C. M. Harris. *Fundamentals of Queuing Theory.* 3rd ed. New York: Wiley, 1997.

Kleinrock, L., and R. Gail. *Queuing Systems: Problems and Solutions.* New York: Wiley, 1996.

Ng, C., and N. Hook. *Queuing Modeling Fundamentals.* New York: Wiley, 1997

Prabhu, N. U. *Foundations of Queueing Theory.* Amsterdam: Kluwer, 1997.

Ragsdale, C. T. *Spreadsheet Modeling and Decision Analysis.* 3rd ed., Cincinnati, OH: South-Western, 2001.

Simulation

7

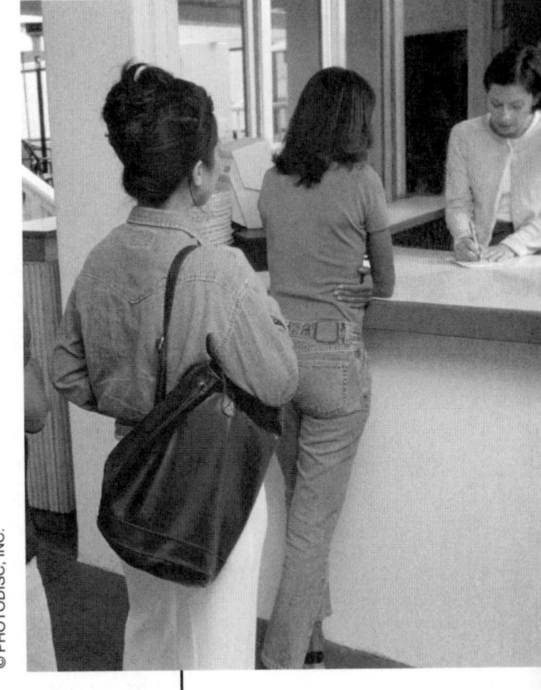
© PHOTODISC, INC.

INTRODUCTION

Banyon State University (BSU) operates a walk-in medical clinic to meet the acute medical needs of its 13,000 students, 1,200 faculty and staff members, and covered relatives. The clinic is staffed by one doctor and one nurse and operates 8 hours a day, 5 days a week. The doctor and nurse do not take a lunch break, but rather, use gaps between patient arrivals to eat lunch and take other short breaks. Because patients often do not arrive right when the clinic opens and because they must visit with a nurse before seeing the doctor, the doctor's official start time is 45 minutes after the clinic opens. Patients arriving at the clinic are served on a first-come, first-served basis.

As part of a new total quality management (TQM) initiative, BSU conducted an in-depth 4-month study of its current operations. A key component of the study was a survey, distributed to all students, faculty, and staff. The purpose of the study was to identify and prioritize areas most in need of improvement. An impressive 44 percent of the surveys were returned and deemed usable. Follow-up analysis indicated that the respondents to the survey were representative of the population served by the clinic. After the results were tabulated, it was determined that the walk-in medical clinic was located at the bottom of the rankings, indicating a great deal of dissatisfaction with the clinic.

Preliminary analysis of the respondents' comments indicated that people were reasonably satisfied with the treatment they received at the clinic but were very dissatisfied with the amount of time they had to wait to see a care giver. To gain additional insight into the problem, a team of students was asked to study the problem as part of a course project. In addition to determining the general issues, they were asked to determine the desirability of a new, computerized patient record system (CPRS) to aid in reducing waiting times. The student team initially collected data on the pattern of arrivals at the clinic and the various service times (discussed in more detail later). The team determined that on a typical day, interarrival times were uniformly distributed between 6 and 20 minutes. After arriving at the clinic, patients complete a form that requests background information and the reason for the visit. The staff collect these forms and retrieve the patients' records from the basement. The team determined that the time to retrieve patient records follows a normal distribution with a mean of 4 minutes and a standard deviation of 0.75 minute. Retrieved patient records are placed in a pile for the clinic's nurse in the order that the patients arrived at the clinic.

When the nurse finishes with the current patient, the file of the next patient is selected and the patient is directed to the nurse's station. Here the nurse further documents the problem and takes some standard measurements such as temperature and blood pressure. Then the nurse places the patient's file at the bottom of a stack of files for the doctor. When the doctor finishes with a patient, the file on the top of the stack is selected and the next patient is called to the examining room. The team determined that the processing times of the nurse closely approximate a normal distribution with an average of 10 minutes and standard deviation of 2.3 minutes. Likewise, it was determined that the time required for the doctor to examine and treat the patients also closely approximates a normal distribution with a mean of 17 minutes and a standard deviation of 3.4 minutes. The team's influence diagram for the clinic is shown in Exhibit 7.1.

It may have already occurred to you that the queuing models discussed in Chapter 6 are not appropriate for this situation. For example, the pattern of arrivals to the clinic appear to follow a uniform distribution and not a Poisson distribution, as is assumed in the queuing theory models discussed in Chapter 6. Similarly, the processing times in the clinic follow a normal distribution, not an exponential distribution. Therefore, a more flexible tool is needed: *simulation.*

EXHIBIT 7.1 Influence Diagram for BSU's Walk-in Medical Clinic

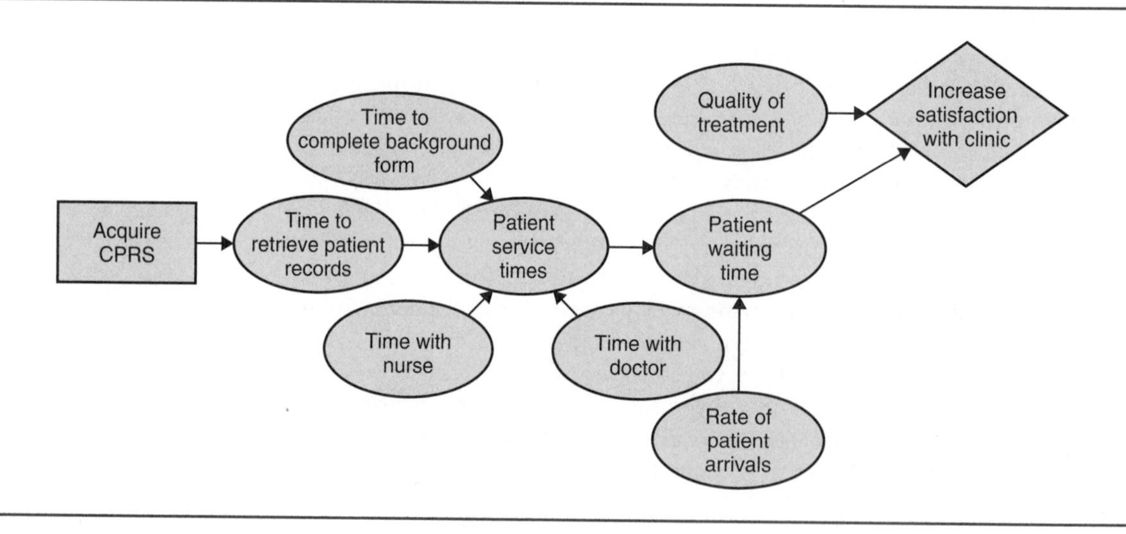

As we demonstrate in this chapter, simulation is a versatile, yet powerful, modeling technique for decision-making situations. It is particularly appropriate in situations where one or more of the assumptions associated with an analytical model is violated. This is the case in the BSU example. It is also particularly appropriate in situations characterized by risk. For example, in our discussion of optimization in Chapter 4, it was assumed that all the coefficients were known with certainty. If management is not comfortable with this assumption, techniques that are able to incorporate this uncertainty into the model should be employed, such as simulation. We will return to the analysis of BSU's walk-in clinic at the end of the chapter after we have had a chance to better acquaint you with the simulation methodology.

7.1 General Overview of Simulation

As was noted previously, applications of the quantitative modeling tools described in earlier chapters are frequently limited to relatively straightforward managerial problems. When managerial problems become complex, they often do not fit the standard problem classifications that can be solved with previously described tools. Development of special optimization models to handle such problems may be too costly, take too long, or even be impossible. For these cases, simulation models are useful.

Simulation has many meanings, depending on where it is being used. To simulate, according to the dictionary, means *to assume the appearance or characteristics of reality.* In quantitative modeling, it generally refers to *a technique for conducting experiments with a computer on a model of a management system over an extended period of simulated time.*

Simulation does more than just *represent* reality through a model, it *imitates* it. In practical terms, this means that there are generally fewer simplifications of reality in a simulation than in other models. Also, simulation is a technique for conducting experiments. Therefore, simulation involves the testing of specific values of the decision variables and observing the impact on the dependent variables.

Simulation is a descriptive rather than a prescriptive or normative tool; there is usually no automatic search for an optimal solution, although optimization models or processes may be a part of a simulation.[1] In general, simulation describes or predicts the characteristics of a given system under varying circumstances. Once these characteristics are known, the best policy can be selected. However, the true optimal policy, if such even exists, may not be considered at all in the simulation. The simulation process often consists of the repetition of an experiment many, many times to obtain an estimate of the overall effect of certain actions. It can be executed manually in some cases, but a computer is usually needed for realistic situations.

In the main body of this chapter we employ Excel to illustrate the simulation process. However, for larger, more realistic situations there are many specialized software packages available, one of which—Crystal Ball 2000—is illustrated in the appendix to this chapter. Like Crystal Ball, @Risk 4.0 is an Excel add-in, and both are well suited for performing a variety of financial analyses. There are also a number of packages that are particularly well suited for modeling manufacturing operations, including APS Virtual Planning, Arena, Awesim, ProModel, Simscript II.5, and Taylor Enterprise Dynamics. In addition to being useful for modeling manufacturing operations, many of these packages are well suited for modeling supply chains, business processes, and service operations. However there are also specialized packages for modeling service operations (e.g., amusement parks, call centers, telecom and networking, and airlines), such as Service Model. Many of the packages just listed include sophisticated statistical and animation capabilities. The interested reader is referred to the Web site of the Institute for Operation Research and the Management Sciences (www.informs.org), where computer simulation packages are regularly surveyed and compared on a variety of dimensions (www.lionhrtpub.com/software-surveys.shtml).

Finally, simulation may be called for when the problem under investigation is too complex to be treated by analytical models or by numerical optimization techniques. Complexity here means that the problem cannot be formulated mathematically (e.g., because the assumptions do not hold), there are too many interacting random events to predict, or the formulation is too involved for a practical or economic solution.

Types of Simulation

There are several types of simulation. We cover the major ones in this chapter:

- Deterministic and probabilistic simulation
- Time dependent and time independent simulation
- Visual interactive simulation
- Business games
- Corporate and financial simulation
- System dynamics

Deterministic and Probabilistic Simulation **Deterministic simulation** is used when a process is very complex or consists of multiple stages with complicated (but known) procedural interactions between them. Formulating a mathematical model that finds the measures of performance of such a system would be extremely detailed and time consuming. Formulating the process as a simulation with fixed procedures and interactions (an *algorithmic* model) allows the determination of the outcome and measures of performance in a much more straightforward way. Note that in such a simulation, there is no doubt about when something will happen, or in what amounts or degrees. That is, there are no probabilistic elements in the model.

In **probabilistic simulation,** one or more of the independent variables (e.g., the arrival rate in a queuing situation) is probabilistic; that is, it follows a certain probability distribution. Two subcategories exist: discrete distributions and continuous distributions (see Chapter 2).

1. **Discrete Distributions** These involve a situation with a limited number of events or variables that can only take a finite number of values.
2. **Continuous Distributions** These refer to a situation involving variables with an unlimited number of possible values that follow familiar density functions such as the normal distribution.

Probabilistic simulation is conducted with the aid of a technique called *Monte Carlo,* described in detail in Section 7.2.

Time Dependent and Time Independent Simulation **Time independent simulation** refers to a situation where it is not important to know exactly when the event occurred. For example, we may know that the demand is three units per day, but we do not care *when* during the day the item was demanded. On the other hand, in some situations, such as waiting line problems, it is important to know the precise time of arrival (to know if the customer will have to wait). Then we are dealing with a **time dependent situation.**

Visual Interactive Simulation This is one of the more interesting and successful recent developments in computer graphics and quantitative modeling. **Visual interactive**

simulation (VIS) uses computer graphics displays to present the impact of various managerial decisions. The decisions are implemented interactively while the simulation is running. These simulations can show dynamic systems that evolve over time in terms of animation. The user watches the progress of the simulation in an animated form on a graphics terminal and can alter the simulation as it progresses. For an interesting example, see Lembersky and Chi (1984).

Business Games **Business games** are simulation models involving several participants who are engaged in playing a role in a game that simulates a realistic competitive situation. Individuals or teams compete to achieve their goal, such as profit maximization, in competition or cooperation with the other individuals or teams. Games exist for a variety of specific situations, such as manufacturing, hospitals, banks, nonprofit organizations, and so on. For example, a team running a hospital must make decisions concerning staffing, room rates, expansion, fund drives, and so forth. A popular business game utilized in numerous MBA programs is called *The Beer Game* (Hammond 1994). This business game simulates the process of managing a supply chain. Both computerized and manual game board versions exist.

The two primary purposes of these games are for training and for research. The advantages for training are that the participant learns much faster and the knowledge and experience gained are more memorable than passive instruction. In addition, complexities, interfunctional dependencies, unexpected events, and other such factors can be introduced into the game by the game administrator to evoke special circumstances. And the time compression—allowing many years of experience in only minutes or hours—lets the participants try out actions that they would not be willing to risk in an actual situation and see the result in the future.

In the research role, the games provide insight into the behavior of organizations, the decision making process, and the interactions within a team. Observing the dynamics of team decision making sheds light on important issues, such as the roles assumed by individuals on the teams, the effect of personality types and managerial styles, the emergence of team conflict and cooperation, and so on. For an example of this use, see *The Executive Game* (Henshaw and Jackson 1990).

Corporate and Financial Simulations One of the more important applications of simulation is in corporate planning, especially the financial aspects. Corporate planning involves both long- and short-range plans. The models integrate production, finance, marketing, and possibly other functions into one model, either deterministic or, when risk analysis is desired, probabilistic. Many large corporations (e.g., Sears, General Motors, and United Airlines) have developed such models.

System Dynamics One of the most interesting types of simulation, **system dynamics,** is represented by the software package Dynamo, developed in the 1960s by J.W. Forrester (1971). Regular simulation models are most commonly meant to be evaluated in steady state conditions but the real world is rarely in steady state for long. Thus, there is a need for continuous simulation models that allow dynamic behavior. System dynamics is an engineering-oriented method of simulation based on the concept that complex systems are usually composed of chains of causes and effects known as *feedback loops.* And in contrast to other simulation models that usually deal with decision-making situations, system dynamics deals more with macroeconomic policies. Sets of equations capture these policies and describe how the various elements and loops of the system interact. A decision or action in one area produces an effect in another area, which, in turn, creates another effect or produces the need for another decision or action. System dynamics has been used to study social, political, corporate, governmental, and even world systems.

Uses of Simulation

Because of its flexibility, simulation has been used to study a wide variety of situations, including helping a bakery minimize its transportation costs (Martin 1998), evaluating intervention strategies for preventing the heterosexual spread of HIV in an African city (Bernstein et al. 1998), assisting the Department of Energy compare alternative hazardous waste remediation alternatives (Toland et al. 1998), and the design of manufacturing operations (Mollaghasemi 1998). There are other familiar situations that can be addressed with simulation as well:

- Urban transportation systems, including their costs as well as their travel times, congestion, and pollution. This information can help in designing optimal throughways, such as one-way streets, lane conversions, and traffic signal settings.
- Plant and warehouse location studies that simulate both incoming materials as well as shipments of finished goods and replacement parts.
- Determination of the proper size of repair crews for expensive equipment that breaks down and the costs incurred by each.
- Airport runway takeoffs and landings in order to improve productivity (throughput) as well as minimize costs and maximize profits.

You can see that simulation is one of the most flexible techniques in the tool kit of quantitative business modelers. It can be applied to many different types of situations and yields a great deal of information concerning the effectiveness of different operating policies under various conditions and assumptions.

Advantages and Disadvantages of Simulation

The increased acceptance of simulation at higher managerial levels is probably due to a number of factors:

1. Simulation theory is relatively straightforward.
2. Simulation is descriptive rather than normative, allowing managers to ask broad, "what-if" questions and to test wide-ranging policies.
3. An accurate simulation model requires an intimate knowledge of the situation, thus forcing the modeler to constantly interface with the manager.
4. The simulation model is built for one particular situation and, typically, will not address any other situation. Thus, no generalized understanding is required of the manager—every component in the model corresponds one to one with a part of the real-life system.
5. Simulation can handle an extremely wide variation in problem types (e.g., inventory and staffing), as well as higher managerial level functions like long-range planning. Thus, it is "always there" when the manager needs it.
6. The manager can experiment with different factors to determine which are important and experiment with different policies and alternatives to determine which are the best. The experimentation is done with a model rather than by interfering with the system.
7. Simulation, in general, allows for inclusion of the real-life complexities of problems; simplifications are not necessary. For example, simulation can utilize real-life probability distributions rather than approximate theoretical distributions.
8. Due to the nature of simulation, a great amount of time compression can be attained, giving the manager some feel as to the long-term effects of various policies, in a matter of minutes.
9. The great amount of time compression enables experimentation with a very large sample. Therefore, as much accuracy can be achieved as desired at a relatively low cost.

The primary disadvantages of simulation are these:

1. An optimal solution cannot be guaranteed.
2. Constructing a simulation model is frequently a slow and costly process.
3. Solutions and inferences from a simulation study are usually not transferable to other problems. This is due to the incorporation in the model of the unique factors of the situation.
4. Simulation is sometimes so easy to use and explain to managers that analytical solutions that can yield optimal results are often overlooked.
5. Validating a simulation model relative to the situation it is supposed to represent can be difficult.
6. Analysis of a simulation output can sometimes be extremely difficult and time consuming.

7.2 The Modeling Process for Monte Carlo Simulation

Monte Carlo simulation is named for its random nature, similar to the famous gambling spot. The modeling process follows much the same generic process as with other models except that the analysis step involves conducting repetitive experiments on the model. We elaborate the various steps as follows.

Step 1: Opportunity/Problem Recognition

Recognizing a real-world opportunity or problem is identical to that described with other models except that simulation is usually called into play when the assumptions required for the other models are not satisfied, or there is no appropriate model developed for the situation. For example, a queuing situation may be of interest but the arrival and/or service processes do not meet the random assumptions required to use queuing theory. This was the case for the BSU walk-in medical clinic. Or an optimization situation may not involve linear relationships. Or the situation may not fit into one of the standard models of quantitative business modeling, and a special model will be required to model the situation.

Step 2: Model Formulation

This task involves developing the procedural steps in the model of the process. For simulation studies, the influence diagram is particularly useful because developing a simulation model requires an intimate understanding of the relationships among the elements of the system being studied. A good example is Exhibit 7.1 for the medical clinic. The influence diagram may be redrawn in a variety of different ways and simulation models for each of the diagrams may be formulated until one seems better, or more appropriate, than the others. Even after one has been chosen, it may be modified many times before a final formulation is acceptable.

Another issue in simulation is deciding whether a transient or steady state model is desired. Many situations do not fit our regular models because they represent transient phenomena. These can be modeled with simulation, but the concern is how to determine good managerial policies when the situation keeps changing. In these cases, it is important to be able to identify the range of transient behavior and test the managerial policies against the full range of situations that may occur. More typically, we try to develop good managerial policies for situations that reach a steady state. Although we discuss a bit later how to determine whether the simulation has reached a steady state, it may well be the case that the real-world situation never reaches a steady state. For example, the noontime rush at a fast food restaurant may never reach a steady state condition, either building up in arrivals and service personnel, or decreasing in one or the other.

Step 3: Data Collection

The data collection process for Monte Carlo simulation is similar to that with other types of quantitative models. However, the modeler must be careful to collect sufficient data to fully describe the situation because it might not fit the modeling assumptions of the other models. For example, in a queuing situation we assume that the arrivals follow the Poisson process and thus collect enough data to verify that this assumption is correct. If it is not, however, then more data may be needed to ascertain the actual arrival distribution, or range of distributions if no one distribution fits the data, as was done for BSU's medical clinic.

This burden of data collection is even greater for situations that are less well defined; that is, not a clear queuing situation, or optimization situation, or regression or decision analysis or other standard type of modeling situation. Then a great deal of data must be collected to even describe the situation, and more may be needed as the model being constructed is analyzed, requiring a return trip(s) to gather the additional data.

Step 4: Analysis of the Model

In this step, we divide the analysis into four segments that are particularly important for simulation studies:

1. Test and validate the model.
2. Design the experiment.
3. Conduct the experiment.
4. Evaluate the results.

Let's look at each segment in more detail.

Testing and Validating the Model Obviously, the simulation model must properly reflect the system under study. This requires validating the model by comparing it to the actual system. A valid simulation model should behave similarly to the underlying phenomenon. This is a necessary validation condition, but by itself may not be sufficient to allow us to rely on its predictive abilities. Theoretical insights into the underlying phenomena that govern the behavior of the business, economic, and social system that is being simulated are critical to the construction of a valid model.

Validation may be viewed as a two-step process. The first step is to determine whether the model is internally correct in a logical and programming sense (**internal validity**). The second is to determine whether it represents the phenomenon it is supposed to represent (called **external validity**). The first step thus involves checking the equations and procedures in the model for accuracy, both in terms of mistakes or errors as well as in terms of properly representing the phenomenon of interest. The task of verifying a model's internal validity can often be simplified if the model is developed in modules and each module is tested as it is developed. Focusing on a particular module rather than trying to evaluate the logic of the entire model all at once facilitates identifying the source of errors and correcting them.

Once the internal validity has been established, the model is then tested by inputting historical values into the model and seeing if it replicates what happens in reality. If the model passes this test, *extreme* values of the input variables are entered and the model is checked by management for the reasonableness of its output. When a model is intended to simulate a new or proposed system for which no actual data are available, there is no way to verify that the model, in fact, represents the system based on historical data, so managers must rely on their own or expert opinions. And, of course, it is always necessary to test the model thoroughly for logical or programming errors (especially at extreme values of the data) and be alert for any discrepancies or unusual characteristics in the results obtained from the model.

Some simulation models, as noted with visual interactive simulation, display a visual representation of the results that gives managers a better feel for what is happening in the model. Then the manager can suggest changes in the assumptions or input data and see the effect on the outputs. This improves the validity testing process and thereby bolsters the chances for the model's eventual implementation.

Designing of the Experiment Once the model has been proven valid, the next task is to design the simulation experiment. **Experimental design** refers to controlling the conditions of the study, such as the variables to include. This is in contrast to situations where observations are taken but the conditions of the study are not controlled. With designed experiments, interpreting the results of the study is often more straightforward because the impact of extraneous factors and variables has been controlled.

More specifically, this step involves determining what factors should be considered fixed in the model and what factors will be allowed to vary, what levels of the factors to use, what the resulting dependent measures are going to be, how many times the model will be replicated, the length of time of each replication, and other such considerations. For example, in a simple queuing simulation we may decide to fix the arrival and service rates but vary the number of servers and then evaluate the customer waiting times, the dependent variable. Clearly, many of these issues have important implications regarding the actual development of the simulation model. Therefore, decisions made at this stage may require making modifications to the computer model. Alternatively, more experienced modelers often address experimental design issues prior to or concurrent with model development.

However, some simulation experiments may be much more complex because the number of factors that must be investigated is very large. For example, consider investment problems with 10 possible investment alternatives (e.g., stocks, bonds), each of which may assume only five values. All together, there are 5^{10} different possibilities (close to 10 million). To simulate 10 million runs is time consuming and costly. Thus, we would instead pick some critical combinations of variables to investigate that, we hope, would give us an intuitive feel for what was happening. If we were then able to gain some intuition about the situation, we would use that to further investigate particular variable combinations of interest. All in all, the design of simulation experiments is similar to the usual design of experiments. Issues such as the *structure, sample size, cost, quality,* and the use of *statistical tools* to analyze the results are frequently involved.

Also included in the experimental design task is determining how long to run the simulation (when to stop the experiment). Sometimes, the length of a run is based on the length of an actual phenomenon; for example, the length of the boating season in Chicago is 10 weeks. This is called a **terminating simulation** and is illustrated in Section 7.6. In a similar fashion to the way increasing the sample size reduces the standard error and therefore increases the accuracy of a survey, longer simulation runs and more replications of the simulation model increase the accuracy of the results. Furthermore, *stopping rules* can be developed using statistical theory to determine the appropriate number of replications to achieve a specified confidence level for the results. For example, statistical theory can be used to determine the number of days to simulate arrivals to a particular ATM in order to be 95 percent confident that the true average customer waiting time is within ±20 seconds.

Several techniques are available for decreasing the variance of the distribution of the measures of performance (*variance reduction*). Reducing the variation of the distribution of the performance measures helps to increase the precision of the simulation results. Perhaps the most common variance reduction technique is the use of **common random numbers.**

For example, assume a company was interested in comparing alternative investment strategies over an extended period of time. To compare these strategies, a number of economic variables need to be randomly generated, such as the rate of inflation, interest rates, and so on. If the company generated separate random numbers for each investment strategy, then the fact that one investment strategy performed better than the others could simply be the result of the random numbers generated. Had the same set of random numbers been used to compare all the investment strategies, the company would have more confidence that observed differences in the results were due to differences in the strategies themselves and not the result of the random numbers generated. While the use of common random numbers is an effective way to reduce the variation across the various designs in an experiment, it is important to consider how the results will be analyzed prior to using this approach. Many statistical analysis techniques such as analysis of variance (ANOVA) are based on the assumption that the observations are independent of one another. The use of common random numbers clearly violates this assumption.

Another consideration in the experimental design task is whether to consider *all* the data or to ignore the "transient" start-up data. It is usually necessary to wait until the model stabilizes before conducting the simulation, whereupon the start-up data are discarded. For example, in simulating the operation of a factory, at the beginning of the simulation there is no work-in-process. As simulated time elapses, the work in process in the factory will gradually build up and approach its steady state level. If at the end of the simulation run the average work-in-process is calculated starting at time zero, this average will be smaller than the actual amount of average inventory once the shop reaches steady state because the average will include data from the transient period when the factory was approaching steady state. To avoid this problem, data during the start-up period is typically discarded and not included in the calculation of the performance measures. The start-up period can be determined statistically or sometimes even visually (e.g., see Welch's method in Law and Kelton 2000). In the preceding example, a plot of the work-in-process level at fixed intervals could be used to determine the length of time required to *warm-up* the system.

We illustrate many of these issues in the example in Section 7.4.

Conducting the Experiment Conducting an experiment involves running the model for the length determined in the previous step and inspecting the output measures. This step also involves deciding whether to run independent replications of the model or to run it once for a long time and break this long run into several runs or "batches" (called the **batch means approach**). The advantage of the batch means approach is that there is only one warm-up period required, thereby reducing the amount of runs needed. The major drawback of this approach is that the replications created by the batch means approach are not truly independent, as they are when independent replications are used, which can complicate the statistical analysis of the output.

Evaluating the Results The final task of the analysis step, prior to implementation, is evaluating the results. Here, we deal with issues such as: "What constitutes a significant difference?" "What do the results mean?" "Do more runs need to be made?" "Should we change the model and repeat the experiment?" To help answer such questions, we often rely heavily on statistical tools such as *t* tests, ANOVA, and regression (see Chapters 2 and 3).

We may also conduct a sensitivity analysis (in the form of "what-if" questions). Sensitivity analysis is performed in simulation in two ways. First, using a trial-and-error approach, one can change the input values of the simulation (especially the uncontrollable parameters) to find how sensitive the proposed solutions are to changes in the input data. This is usually done by rerunning the simulation, either by using a what-if feature or simply using the computer's editing capabilities to change data values. Second, there is the issue of the value of additional information. One should explore the issue of whether and where effort should be directed to obtain better estimates of parameter values. The latter may be done either quantitatively (if possible) or qualitatively.

Step 5: Implementation

Implementing the simulation results involves the same issues as any other implementation. However, the chances of implementation are often better with simulation because the manager is usually more involved in the simulation process and the models are closer to the manager's reality. And as noted earlier, many simulation packages allow the actual simulation to be visually displayed in two and even three dimensions on a computer screen, giving the manager more confidence in recommending the implementation of the simulation results.

7.3 The Monte Carlo Methodology

Managerial systems of decisions under risk include chance elements in their behavior. As such, they can be simulated with the aid of a technique called Monte Carlo simulation. The technique involves random sampling from the probability distributions that represent the real-life processes. Let us give an example.

The Tourist Information Center

Alisa Goldman was delighted with her new job as director of the Tourist Information Center for the city of Miami Beach. She had completed her graduate work in the Hotel and Entertainment Services program of a highly rated college in New York and had accepted an offer for this new position from her former internship employer, the city of Miami Beach.

The city manager, Sean Bushnell, had been impressed with Alisa's analytical skills during her summers as an intern working at the Senior Citizens Center. There, Alisa had been instrumental in instituting programs that raised the quality of the center's services while simultaneously cutting their costs. Sean had been straightforward in his expectations when offering Alisa the permanent position of director for this new center: The center was severely underfunded, yet the city council had high expectations for the center. If the first year were successful, the center would be much better funded the second year. If not, the city council might well cancel the entire project.

Alisa saw her first task as determining the needs for service at the center. This required statistics concerning the tourists' arrival rates, their waiting times, and the service times involved in meeting their needs. Following this, Alisa would look into more details concerning the variety of services the tourists required. Special brochures and posters might handle a significant portion of their information requirements, for example. Or perhaps some form of "express line" for frequently asked questions (FAQs) was desirable.

Alisa's first approach to the data collection problem was to log tourist arrivals and services in the facility. Her results for the first 10 tourists are shown in Exhibit 7.2. Based on this quick preliminary sample, Alisa concluded that the average tourist waited 0.7 minute and the employee was busy during 0.82 minute, or 82 percent of the time. Several questions came to Alisa's mind:

- How long should she clock the operation of the information clerk?
- How do the employees feel about being clocked?

EXHIBIT 7.2
Tourist Information Center Data

Tourist Number	Arrival Time	Start of Service	End of Service	Tourist Waiting From–To	Employee Idle From–To
1	9:00	9:00	9:08	—	9:08–9:10
2	9:10	9:10	9:14	—	—
3	9:12	9:14	9:17	9:12–9:14	—
4	9:13	9:17	9:20	9:13–9:17	—
5	9:20	9:20	9:23	—	—
6	9:22	9:23	9:28	9:22–9:23	9:28–9:31
7	9:31	9:31	9:34	—	9:34–9:35
8	9:35	9:35	9:40	—	—
9	9:40	9:40	9:45	—	9:45–9:48
10	9:48	9:48	9:50	—	—
Total				7 minutes	9 minutes

- How do the tourists feel about being clocked?
- What other kinds of measurements should she take?

Unfortunately, as we will show, this situation cannot be solved by the standard queuing theory formulas because the arrival rate does not follow the Poisson distribution, nor is the service time exponential. What Alisa did not know was that she could conduct all her experiments on a model of the Tourist Center and get answers to each of her questions by using the technique of simulation. (We return to Alisa's situation a bit later.)

In this example, the arrivals and lengths of service are probabilistic. What the Monte Carlo method does is to generate *simulated* arrival times and service times from a given distribution by the use of *random sampling*. Thus, the Monte Carlo mechanism is not a simulation model per se, although it has become almost synonymous with probabilistic simulation. It is a mechanism used in the process of a probabilistic simulation. Before we show how this is done, let us define some basic terms.

Simulation Terminology

Uniform Distribution A **uniform distribution** is one where each value of the variable has exactly the same chance of occurring (see Exhibit 7.3). This distribution is shown graphically in Exhibit 7.4.

Random Numbers and Random Variates A *random number* (RN) is a number picked, at random, from a population of uniformly distributed numbers and used to generate a *random variate* for the probability distribution of interest. That is, each random

EXHIBIT 7.3
An Example of the Uniform Distribution

Demand	Probability
5	0.25
6	0.25
7	0.25
8	0.25

EXHIBIT 7.4 A Uniform Distribution Shown Graphically

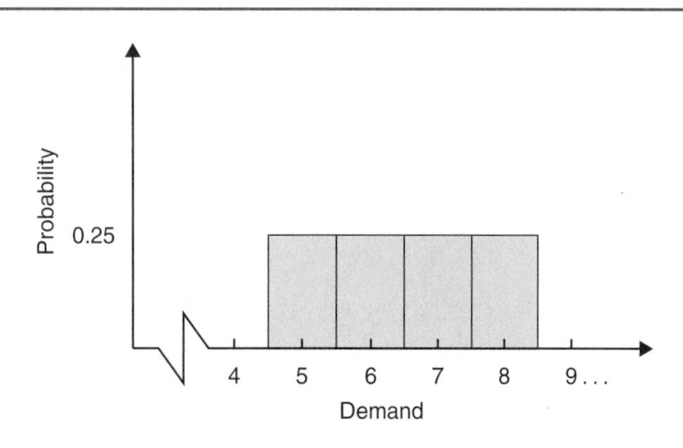

number in the population has an equal probability of being selected. Before spreadsheets such as Excel came into widespread use, conducting Monte Carlo simulations often depended on having available tables of random numbers. A key advantage of using spreadsheets to generate random variates is that they can generate these numbers from a variety of probability distributions. Also, they can generate as many random variates as are needed. In Excel, the generation of random variates from random numbers is automatic, and the entire process is referred to as **random number generation.**

To illustrate, assume that we would like to generate 10 random variates from a uniform distribution between 0 and 60. To display the Random Number Generation dialog box in Excel, select the menu items Tools, Data Analysis, and then Random Number Generation. Next, to generate the 10 random numbers and place them in cells A1:A10, enter the information in the Random Number Generation dialog box as shown in Exhibit 7.5 and click on the OK button. In the Random Number Generation dialog box, the Number of Variables field tells Excel how many columns to fill with random variates. For example, if a 2 had been entered in this field along with the other information shown in Exhibit 7.5,

EXHIBIT 7.5 Using Excel to Generate Ten Random Variates from a Uniform Distribution

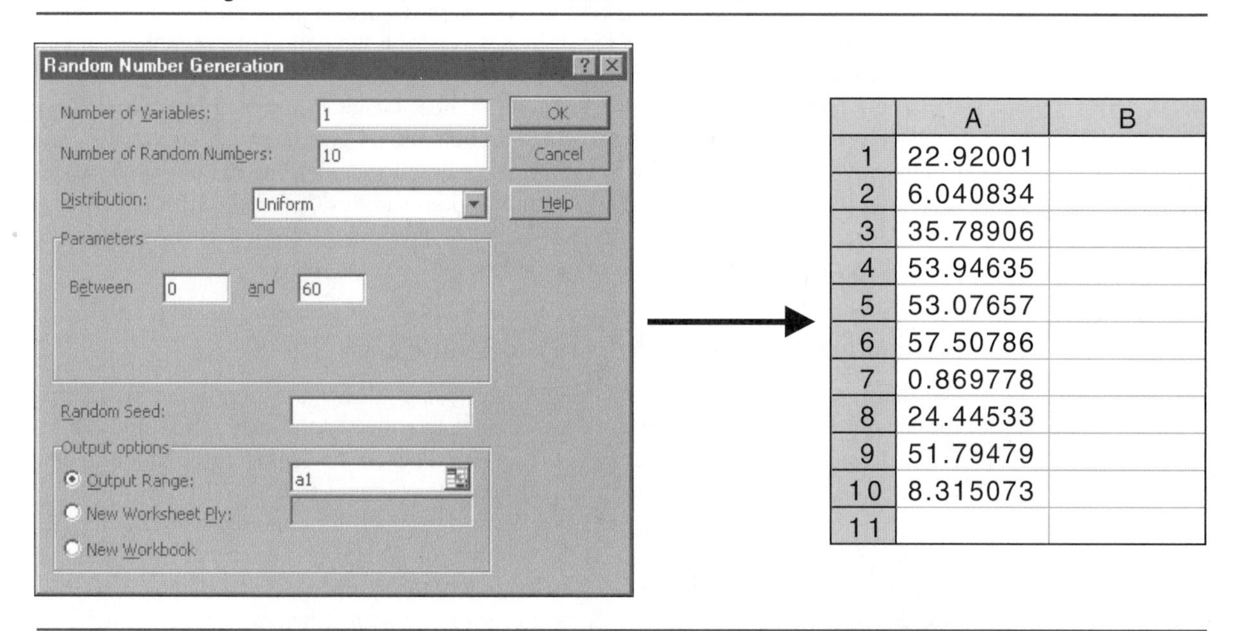

	A	B
1	22.92001	
2	6.040834	
3	35.78906	
4	53.94635	
5	53.07657	
6	57.50786	
7	0.869778	
8	24.44533	
9	51.79479	
10	8.315073	
11		

EXHIBIT 7.6 Probability Distributions Available in Excel

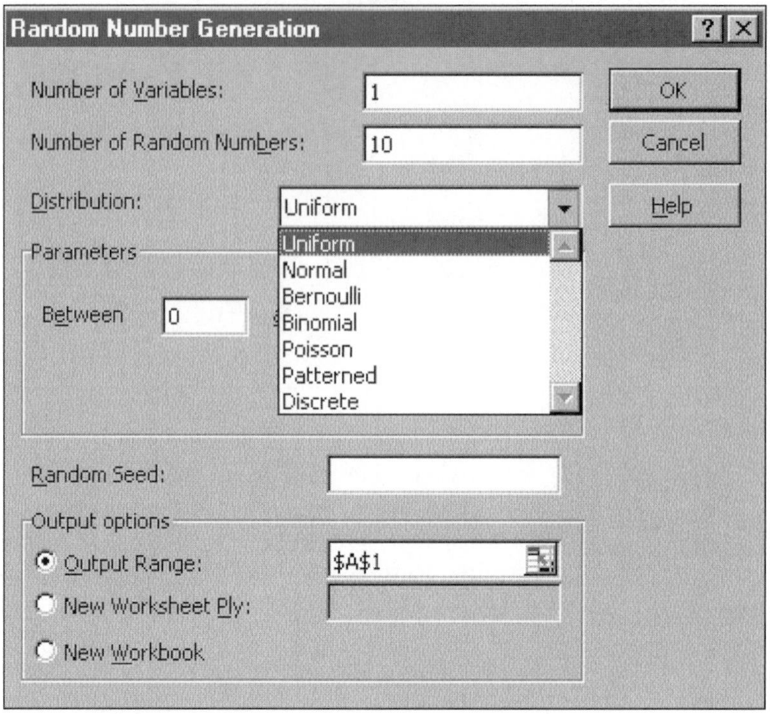

20 random variates from a uniform distribution would have been generated and placed in cells A1:B10. The Number of Random Numbers field tells Excel how many random variates to generate in each column.

The third field in the Random Number Generation dialog box provides you with choices on which probability distribution you want to use to generate the random variates. As Exhibit 7.6 illustrates, Excel provides you with the option of generating random variates from seven common probability distributions.

The Parameters section in the Random Number Generation dialog box will change depending on the distribution from which you choose to generate random variates. For example, when generating random variates from a uniform distribution you will be asked to specify the lower and upper range for the random variates (Exhibit 7.5). Alternatively, if you specify that the random variates should be generated from a normal distribution, you will be asked to specify the mean and the standard deviation.

Generating Random Variates in the Monte Carlo Process

We illustrate the generation of random variates using Alisa's situation in the Tourist Information Center. Assume that more historical data collected through a time study or by estimation make it possible to express the service times in the Tourist Information Center by the probability distribution shown in the second column of Exhibit 7.7.

The spreadsheet shown in Exhibit 7.8 was developed to randomly generate the service times for 15 visitors to the Tourist Information Center based on the distribution of services times shown in Exhibit 7.7 (the top part of the spreadsheet in Exhibit 7.8). As we will demonstrate shortly, Excel will use this information to generate the service times in

EXHIBIT 7.7 Service Times at the Tourist Information Center

Service Time (Minutes)	Probability
3	0.156
4	0.287
5	0.362
6	0.195

EXHIBIT 7.8 Randomly Generating Service Times for 15 Visitors to the Tourist Information Center

	A	B
1	Service	
2	Time	Probability
3	3	0.156
4	4	0.287
5	5	0.362
6	6	0.195
7		
8	Arrival	Service
9	Number	Time
10	1	4
11	2	3
12	3	6
13	4	3
14	5	5
15	6	4
16	7	3
17	8	5
18	9	5
19	10	5
20	11	4
21	12	4
22	13	3
23	14	5
24	15	6

proportion to the probabilities listed. The bottom of the spreadsheet contains the random service times generated by Excel.

To generate the 15 service times shown in Exhibit 7.8, the menu items Tools/Data Analysis. . ./Random Number Generation were selected to display the Random Number Generation dialog box shown in Exhibit 7.9. Since we want to place the random variates in one column, a one was entered in the Number of Variables field. Next, since we want to generate 15 service times, 15 was entered in the Number of Random Numbers field. Since we want to base the randomly generated service times on the discrete probability distribution

EXHIBIT 7.9 Generating 15 Service Times for the Tourist Information Center

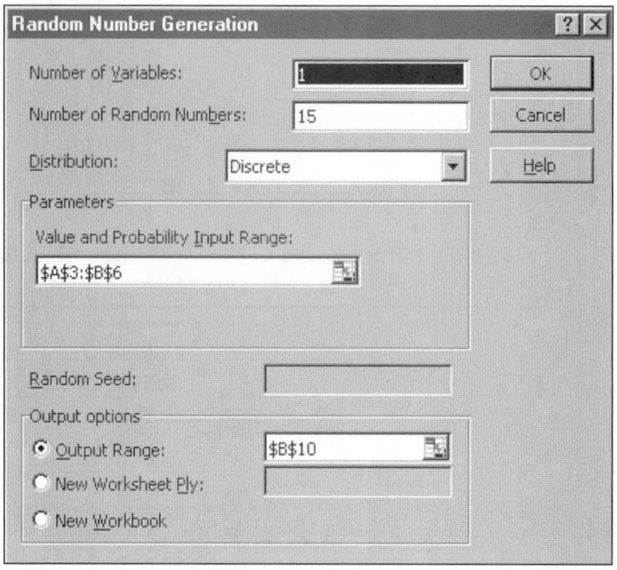

Random Number Generation

Number of Variables: 1

Number of Random Numbers: 15

Distribution: Discrete

Parameters

Value and Probability Input Range:

$A$3:$B$6

Random Seed:

Output options

Output Range: $B$10

New Worksheet Ply:

New Workbook

OK

Cancel

Help

shown in Exhibit 7.7, we specify that a discrete distribution will be used in the Distribution field. After indicating that a discrete distribution will be used, the Parameters section of the Random Number Generation dialog box requests the range for the Value and Probability Input Range. This range always consists of two columns. The first column contains the possible outcomes that can occur and the second column contains the probability of each outcome actually occurring. Excel uses this information to randomly generate the outcomes according to the probabilities specified. For example, referring to the spreadsheet in Exhibit 7.8, a service time of 5 minutes has a 0.362 chance of being generated by Excel based on specifying the cells A3:B6 as the Value and Probability Input Range. Finally, cell B10 was entered for the Output Range. It is sufficient to enter only the cell in the upper left-hand corner of the range when specifying an output range. Alternatively, the entire range B10:B24 could have been specified in the Output Range field.

On closer examination of the random service times shown in Exhibit 7.8, it can be observed that a 4-minute service time was generated four times, representing approximately 27 percent (4/15) of the generated service times. This is relatively close to our target of having 28.7 percent of the service times be 4 minutes. On the other hand, a 3-minute service time was also generated four times, but this is relatively far from our target of 15.6 percent of the service times being 3 minutes. We comment that this is the result of such a small sample size and, in general, as our sample size is increased, it will more closely conform to the specified distribution. This is one reason why it is important to run the simulation model for a sufficiently long period.

7.4 Time Independent, Discrete Simulation

Following is a list of specific steps detailing some of the major simulation tasks described earlier, but for the time independent discrete simulation process. Following the list, we offer an example to illustrate the steps. In Section 7.5, we address the time dependent simulation process.

1. Describe the system and obtain the probability distributions of the relevant elements of the system. This is a crucial step requiring intimate familiarity with the system. Frequently, incorrect assumptions are made at this point that invalidate the rest of the simulation.
2. Define the appropriate measure(s) of system performance. If necessary, write it in the form of an equation(s).
3. Set up the initial simulation conditions (e.g., insert the values needed to start the simulation).
4. For each probabilistic element, generate a random value and determine the system's performance.
5. Derive the measures of performance and their variances.
6. If steady-state results are desired, repeat steps 4 and 5 until the measures of system performance "stabilize," as described in the following example.
7. Repeat steps 4–6 for various managerial policies. Given the values of the performance measures and their confidence intervals, decide on the appropriate managerial policy.

This procedure will be demonstrated with an inventory control example.

EXAMPLE

Marvin's Service Station

Marvin's Service Station sells gasoline to boat owners. The demand for gasoline depends on weather conditions and fluctuates according to the following distribution:

Weekly Demand (Gallons)	Probability
2,000	0.12
3,000	0.23
4,000	0.48
5,000	0.17

Shipments arrive once a week. Because Marvin's Service Station is in a remote location, it must order and accept gasoline once a week. Joe, the owner, faces the following problem: If he orders too small a quantity, he will lose, in terms of lost profits and goodwill, 12 cents per gallon demanded and not provided. If he orders too large a quantity, he will have to pay 5 cents per gallon shipped back due to lack of storage space for what he ordered but had to return. For each gallon sold, he makes 10 cents profit. Joe now receives 3,500 gallons at the beginning of each week before he opens for business. He thinks that he should receive more, maybe 4,000 or even 4,500 gallons. The tank's storage capacity is 5,500 gallons. The problem is to find the best order quantity.

Joe *could* solve his problem by trial and error. That is, he could order different weekly quantities for periods of, say, 10 weeks, and then see which worked best by comparing the results. However, simulation can give an answer in a few minutes and a simulated loss is only a loss on paper. This section will explain how to solve Joe's dilemma.

Solution by Simulation

To find the appropriate ordering quantity, it is necessary to compute the expected profit (loss) for the existing order quantity (3,500 gallons) and for other possible order quantities such as 4,000 and 4,500 (as suggested by Joe), or any other desired figure. Assume that today is the first day of the week, there were 300 gallons remaining after business last week, and a shipment has just arrived, resulting in an inventory of 3,800 gallons. (*Note:* All quantities in this example are in gallons.)

To clarify our thinking about this situation, and to specify the exact relationships in the simulation, we first construct a diagram. In this case, we will include the equations in the diagram, which is then known as a **flow diagram** of the relationships. A flow diagram is a schematic presentation of all computational activities used in the simulation. Exhibit 7.10 shows the flow diagram for this inventory situation. We discuss the equa-

EXHIBIT 7.10 Flow Diagram for the Inventory Example

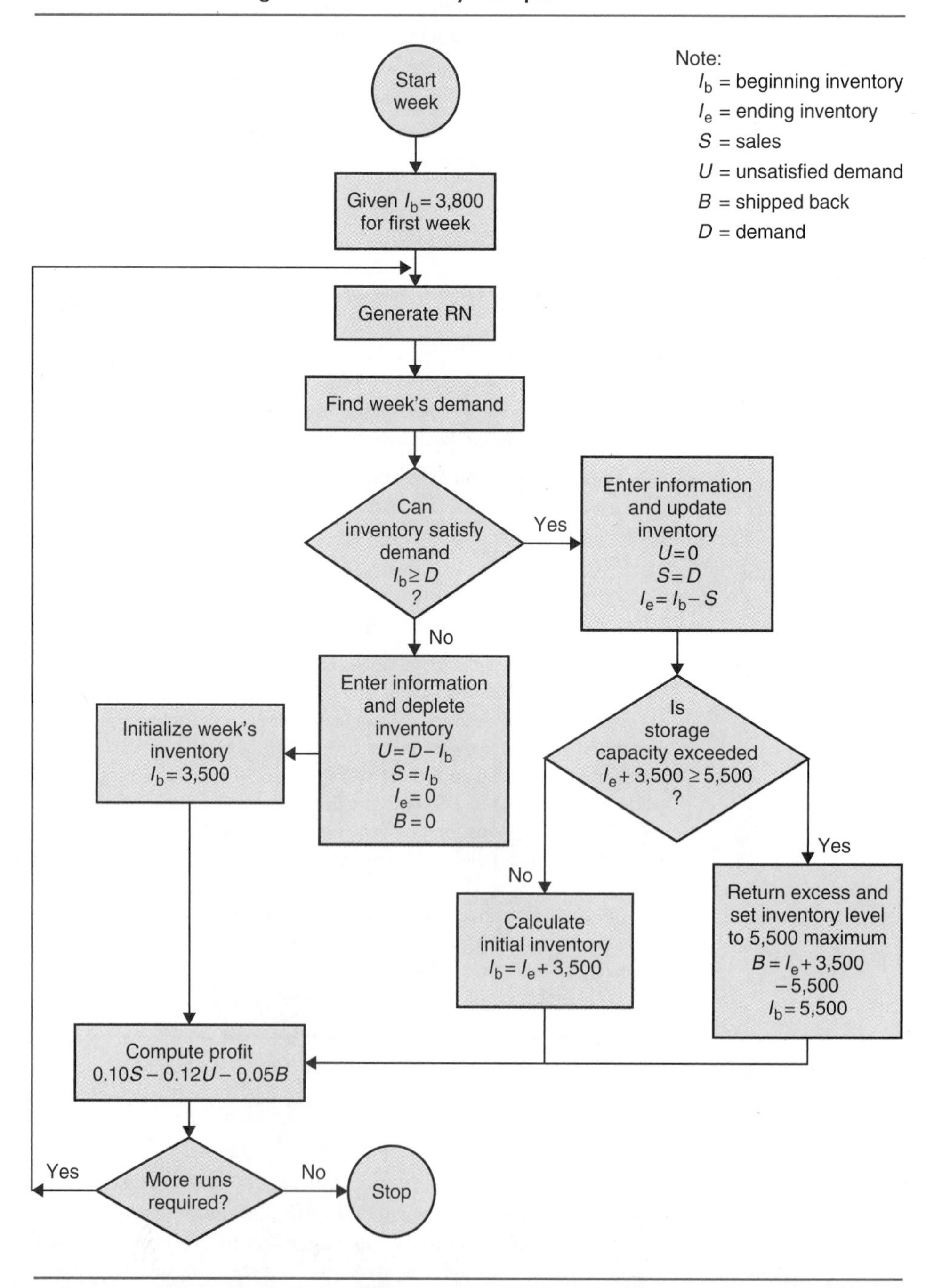

tions and variables later, but the logic flow for the simulation process is clear. Therefore, let us begin the simulation in a step-by-step fashion to follow the flow diagram.

Step 1: Describe the System and Determine the Probability Distributions There is only one probability distribution in this case; it describes the demand. In more complicated Monte Carlo simulations, there may be several distributions involved.

Step 2: Decide on the Measure of Performance The primary measure of performance is the average weekly profit, which is computed as:

$$\text{Average weekly profit} = \$0.10(\text{sales}) - \$0.12(\text{unsatisfied demand}) - \$0.05(\text{quantity shipped back})$$

Several less important measures such as the average shortage are discussed at the end of this example.

Step 3: Set Initial Conditions The simulation starts with a set of initial conditions. For example, the inventory at the beginning of the week, $I_b = 3{,}800$ in our example (that is, the 3,500 shipment added to the existing 300 gallons from the previous week). In some cases, arbitrary numbers may be used. The stabilization (step 6) will take care of such situations.

Step 4: Generate Random Numbers and Compute the System's Performance The first inventory system that will be considered is the current ordering policy of 3,500 gallons per week. For purposes of demonstration, step 4 is repeated here only 10 times to simulate 10 weeks. In reality, it should continue until the measure of performance (average weekly profit) achieves *stability,* as will be explained in step 6. A simulation model was created in Excel and is shown in Exhibit 7.11.

The top of Exhibit 7.11 contains the Value and Probability Input Range for Excel to use in generating weekly demand. Right below this is a section for entering key parameters, including the beginning inventory level at the start of the simulation, the size of the weekly shipment, and the storage capacity.

The main part of the spreadsheet shown in Exhibit 7.11 is a table for simulating the operation of the service station for 10 weeks. Column A of this table keeps track of the week number. The inventory available at the beginning of the week is calculated in column B. While the formula entered for the first week is slightly different from the other weeks, the logic is the same. Specifically, the inventory available at the beginning of the week is equal to the minimum of the inventory at the end of the previous week plus the weekly shipment, or 5,500. The inventory at the end of the previous week is in column E.

The weekly demand shown in Column C was generated using Excel's Random Number Generation tool, with discrete distribution specified. As noted earlier, the Value and Probability Input Range for generating weekly demand is contained in cells A3:B6.

The amount sold is calculated in column D. Two cases may occur.

1. The demand is equal to or smaller than the inventory on hand. In this case, sales equal demand as in weeks 1, 2, 3, and 4.
2. Demand is *larger* than the inventory on hand. In this case, sales are limited to the inventory on hand. To capture this logic the IF function was used. For example, in cell D15 the formula

=IF(B15>=C15,C15,B15)

was entered. In English, this formula says that if the inventory available in week 1 (B15) is greater than or equal to the demand in week 1 (C15), then the amount sold in week 1 is equal to the demand in week 1. On the other hand, if the inventory available in week 1 is less than the demand in week 1, then sales are limited to what is available in week 1. This formula is copied into D16:D24.

EXHIBIT 7.11 The Simulation for 10 Weeks

	A	B	C	D	E	F	G	H	I	
1	**Weekly**									
2	**Demand**	**Probability**								
3	2,000	0.12								
4	3,000	0.23								
5	4,000	0.48								
6	5,000	0.17								Input
7										
8	Beg. inv.	300								
9	Shipment	3,500								
10	Capacity	5,500								
11										
12		*Inventory*			*Inventory*				*Average*	
13	*Week*	*at Beginning*	*Simulated*		*at End*	*Unsatisfied*	*Shipped*	*Weekly*	*Weekly*	
14	*Number*	*of Week*	*Demand*	*Sold*	*of Week*	*Demand*	*Back*	*Profit*	*Profit*	
15	1	3,800	3,000	3,000	800	0	0	$300.00	$300.00	
16	2	4,300	2,000	2,000	2,300	0	0	$200.00	$250.00	
17	3	5,500	4,000	4,000	1,500	0	300	$385.00	$295.00	
18	4	5,000	5,000	5,000	0	0	0	$500.00	$346.25	
19	5	3,500	4,000	3,500	0	500	0	$290.00	$335.00	
20	6	3,500	4,000	3,500	0	500	0	$290.00	$327.50	Output
21	7	3,500	3,000	3,000	500	0	0	$300.00	$323.57	
22	8	4,000	4,000	4,000	0	0	0	$400.00	$333.13	
23	9	3,500	4,000	3,500	0	500	0	$290.00	$328.33	
24	10	3,500	3,000	3,000	500	0	0	$300.00	$325.50	
25	Total	40,100	36,000	34,500	5,600	1,500	0	$3,255.00		
26	Weekly									
27	Average	4,010	3,600	3,450	560	150	30	$325.50		
28	*Key Formulas*									
29	B15	=B8+B9								
30	B16	=MIN(E15+$B$9,5500) {copy to cells B17:B24}								
31	D15	=IF(B15>=C15,C15,B15) {copy to cells D16:D24}								
32	E15	=IF(B15>=C15,B15-C15,0) {copy to cells E16:E24}								
33	F15	=IF(C15>=B15,C15-B15,0) {copy to cells F16:F24}								
34	G15	=IF(B8+$B$9>$B$10,B8+$B$9-$B$10,0)								
35	G16	=IF(E15+$B$9>$B$10,E15+$B$9-$B$10,0) {copy to cells G17:G24}								
36	H15	=(0.1*D15)-(0.12*F15)-(0.05*G15) {copy to cells H16:H24}								
37	I15	=AVERAGE(H$15:H15) {copy to cells I16:I24}								

In column E the inventory at the end of the week is calculated. Since it is not possible to have a negative amount of inventory, the IF function is used to calculate the inventory at the end of week 1 in cell E15 as follows:

=IF(B15>=C15,B15–C15,0)

According to this formula, if the inventory available at the beginning of week 1 (B15) is greater than or equal to the demand in week 1 (C15), then inventory at the end of week 1 is equal to the inventory available at the beginning of week 1 minus the demand during week 1. Alternatively, if the inventory available at the beginning of week 1 is less than the demand during week 1, then the inventory at the end of week 1 would be zero. Since unsatisfied demand (column F) is the opposite of ending inventory, it is calculated in a similar fashion.

The amount shipped back is calculated in column G. Such a situation occurs when the "end-of-the-week inventory" plus the shipment (3,500 gallons in the system under study) exceed the 5,500-gallon tank capacity. In this case, the excess supply is shipped back and the beginning inventory is 5,500. For example, in week 3, the shipment of 3,500, added to the weekend inventory of week 2 of 2,300, gives a total of 5,800 gallons. Therefore, 5,800 – 5,500 = 300 gallons are shipped back.

EXHIBIT 7.12 Profit Results

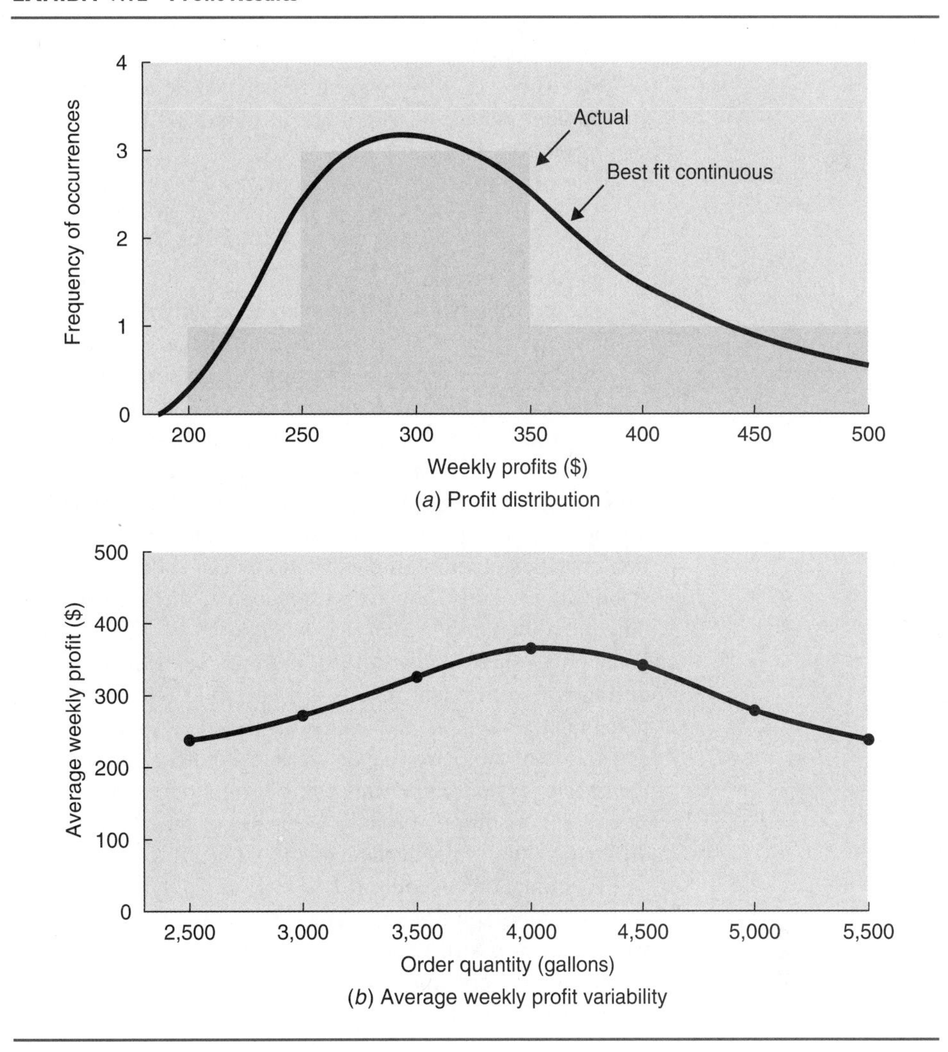

(*a*) Profit distribution

(*b*) Average weekly profit variability

Column H shows the measure of performance in this problem—profit. The profit is calculated, every week, according to the formula:

$$\$\text{ Profit} = 0.10(\text{sales}) - 0.12(\text{unsatisfied demand}) - 0.05(\text{shipped back})$$

The resulting values are plotted in Exhibit 7.12*a* and compared to a "theoretical" continuous distribution that may, in reality, be the underlying distribution of weekly profit.

Column I represents the *average* weekly profit at any week, computed by totaling the weekly profits up to that week (cumulative profit) and dividing it by the number of weeks. Notice how the absolute cell reference was used in the first term of the formula entered in cell I15 so that it could be copied to the other cells in column I.

Step 5: Compute the Measures of Performance Each simulation run is composed of repeat, multiple *trials*. The question of how many trials to have in one run (i.e., finding the *length* of the run) involves statistical analysis. The longer the run, the more accurate the results, but the higher the simulation time and cost. This issue concerns what are called *stopping rules,* discussed further in Step 6.

The simulation performed thus far indicated an average weekly profit of $325.50. In addition to total profit, some other *measures of performance* can be computed:

a. *The probability of running short and the average size of the shortage.* In 3 out of the 10 weeks, there was an unsatisfied demand. Therefore, there is a 3/10 = 30 percent chance of running out of stock. The average shortage, per week, is 1,500/10 = 150 gallons.
b. *The probability of shipping back and the average quantity shipped back.* In 1 out of the 10 weeks, some gasoline was shipped back. On the average there is a 1/10 = 10 percent chance of shipping back; the average amount is 300/10 = 30 gallons per week.
c. *The average demand.* The average weekly demand is computed as 3,600, which is close to the expected value of the demand (data in Exhibit 7.8) of 3,700. (In a stabilized process, these two numbers will be very close.)
d. *The average beginning inventory* is computed as 4,010 gallons.
e. *The average weekly sales* are computed as 3,450 gallons.
f. *The average ending inventory* is computed as 560 gallons.

Step 6: Continue the Simulation Process Until It Stabilizes In all the examples in this text and the homework problems, we use only the early data generated by the simulation to achieve brevity in presentation, even if the measures of performance have not stabilized. In reality, however, we recognize that the simulation begins to represent reality only after *stabilization* has been achieved. Stabilization is equivalent to what is commonly called *steady state.* Therefore, we distinguish a **start-up transient period** during which the data results are not valid for steady state conclusions. Usually, though not always, the modeler is interested in finding the long-run, steady-state average values of the performance measures rather than the short-run, transient values. If the transient values are of interest, it is important to begin the simulation with the *correct* initial conditions because every set of initial conditions will generally result in different values of the dependent, output measures.

Examination of column I in Exhibit 7.11 indicates that the process, although close to stabilizing, has not yet stabilized (see Exhibit 7.13). That is, the average weekly profit is still fluctuating. Notice, however, that after 6 weeks the differences are becoming very small and we are approaching the end of the transient period. As noted earlier, this transient period can be assessed statistically. Once this period is

EXHIBIT 7.13 Stabilization of the Simulation Process

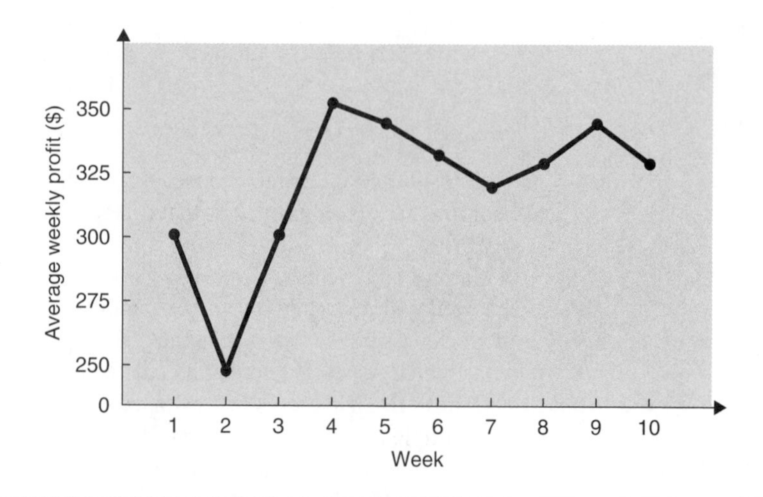

determined, *then* simulation runs are made to extend past this point into the stabilization period, and the measures of performance and their variances are recorded to determine their average values and confidence intervals.

If there exist several measures of performance, then the stabilization analysis must be performed for *each* measure. Only after stabilization is achieved in *all* measures of performance (or at least in all *important* measures) should the simulation be stopped.

Step 7: Find the Best Ordering Policy Steps 4, 5, and 6 are now repeated for other ordering policies in order to find the best. In the example just presented, the ordered quantity Q was 3,500; other values of Q (e.g., 3,300, 3,700, 4,000) should be considered next. Each value of Q constitutes an independent system for which the various measures of effectiveness such as average profit, average sales, and unsatisfied demand are computed. Normally, the same set of random numbers is used for all trials in order to increase their comparability (refer to discussion of design of experiments in Section 7.2). Each such experiment is called a **simulation run.** The results for average weekly profit are shown in Exhibit 7.12*b;* the best results seem to occur at about 4,100 gallons.

In this case, the most important measure of performance has been assumed to be the average profit, and therefore the policy with the highest average profit will be selected. In other systems, two or more measures of performance may have to be compared such as the probability of a stockout.

7.5 Time Dependent Simulation

As you may recall from earlier in the chapter, Alisa needed some tools to support her job as director of the Tourist Information Center. To use simulation required extensive historical data concerning both the demand for services as well as the service capabilities. She has now collected the following information: The Tourist Information Center is staffed by one employee and is open from 9 A.M. to 5 P.M. The length of service required by tourists varies according to the probability distribution in Exhibit 7.14*a,* and they arrive at the center according to the distribution in Exhibit 7.14*b.*

Alisa now wishes to find the following:

- The average waiting time per tourist, in minutes
- The percentage of time that the employee is busy (utilization)

EXHIBIT 7.14
Service and Arrival Distributions

(*a*) Service		(*b*) Arrivals	
Length of Service (Minutes)	**Probability (%)**	**Time Between Two Consecutive Arrivals (Interarrival Time, Minutes)**	**Probability (%)**
3	15.6	3	20.2
4	28.7	4	23.6
5	36.2	5	31.2
6	19.5	6	18.4
		7	6.6

EXHIBIT 7.15 Influence Diagram for Tourist Information Center

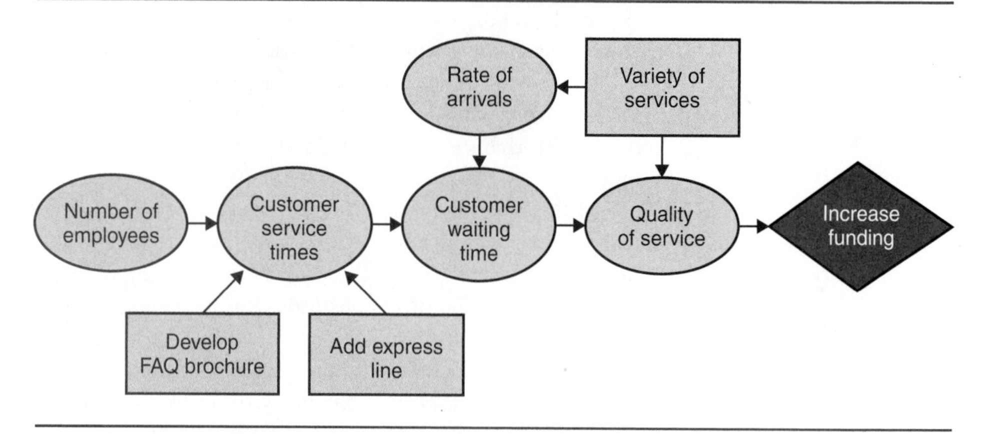

- The average number of tourists in the center
- The probability of finding two tourists in the center

Simulation Analysis with Discrete Distributions

Before starting the simulation, we create an influence diagram for Alisa's situation, shown in Exhibit 7.15. Based on this diagram, we develop the spreadsheet shown in Exhibit 7.16 to simulate the processing of arriving tourists, using random numbers generated by Excel. (For illustration purposes, we limit the simulation to only 10 tourists.) To simulate this situation two sets of random numbers are needed: the time required to service each tourist and the time between tourist arrivals. The time of the first arrival is given as 9:00 A.M. (For the purpose of this example, from now on we will ignore the hour and consider only the number of minutes that have elapsed since 9:00 A.M.) The interarrival times for the other 9 tourists (column B) and the service times (column D) for all 10 tourists were generated from the discrete distributions entered at the top of the spreadsheet shown in Exhibit 7.16 using Excel's Random Number Generation tool.

In column C the predicted arrival time of each tourist is calculated. As noted earlier, the first tourist is assumed to enter at 9:00 A.M., or zero minutes after the information center opens. The interarrival time for the second tourist was generated to be 6 minutes after the arrival of the first tourist. To calculate the time of the second tourist's arrival, the formula =C12+B13 was entered in cell C13 and then copied to cells C14:C21.

For service to begin for a given tourist, two events must occur: (1) the employee must finish with the preceding tourist and (2) the tourist must arrive at the center. In other words, for the service employee to begin helping the second tourist, the service for the first tourist must be finished and the second tourist must be physically present at the information center. To capture this logic, the formula =MAX(F12, C13) was entered in cell E13 and then copied to cells E14:E21.

The time that service ends is calculated simply as the time service begins (column E) for a particular tourist plus the randomly generated service time (column D) for the tourist. Two measures of system performance are calculated in columns G and H.

Average Waiting Time Since the first arriving tourist will not have to wait, formulas were entered to calculate only the waiting time of tourists two through nine. In the spreadsheet shown in Exhibit 7.16, waiting time is calculated by subtracting the time the tourist arrived (column C) from the time service begins (column E). For the arriving

EXHIBIT 7.16 Simulation of Tourist Information Center

	A	B	C	D	E	F	G	H
1	**Service**			**Interarrival**				
2	**Time**	**Probability**		**Time**	**Probability**			
3	3	0.156		3	0.202			
4	4	0.287		4	0.236			
5	5	0.362		5	0.312			
6	6	0.195		6	0.184			
7				7	0.066			
8								
9		**Predicted**						
10	**Tourist**	**Interarrival**	**Time**	**Service**	**Service**	**Service**	**Wait**	**Idle**
11	**Number**	**Time**	**Arriving**	**Time**	**Start**	**End**	**Time**	**Time**
12	1	0	0	4	0	4		
13	2	6	6	4	6	10	0	2
14	3	3	9	6	10	16	1	0
15	4	5	14	4	16	20	2	0
16	5	5	19	5	20	25	1	0
17	6	6	25	3	25	28	0	0
18	7	4	29	5	29	34	0	1
19	8	5	34	4	34	38	0	0
20	9	3	37	3	38	41	1	0
21	10	7	44	4	44	48	0	3
22	**Average**						**0.56**	**0.13**
23								
24	*Key Formulas*							
25	C13	=C12+B13 {copy to cells C14:C21}						
26	E12	=C12						
27	E13	=MAX(F12,C13) {copy to cells E14:E21}						
28	F12	=E12+D12 {copy to cells F13:F21}						
29	G13	=E13-C13 {copy to cells G14:G21}						
30	H13	=E13-F12 {copy to cells H14:H21}						
31	G22	=AVERAGE(G13:G21)						
32	H22	=SUM(H13:H21)/F21						

tourists, the average waiting time per tourist was 0.56 minute (excluding the first tourist who will never have to wait, a *startup transient* phenomenon).

Utilization of the Service Facility Idle time is calculated as the difference between the start of service of a particular tourist and the end of service of the preceding tourist. In Exhibit 7.16, the center was simulated for a total of 48 minutes. During this period, there were 6 minutes of idle time. Thus, the employee was idle 13 percent of the time (6/48) or alternatively, busy 87 percent of the time (42/48).

Average Number of Tourists in the Center During 6 minutes, there were no tourists in the center, but during 5 minutes, there were two (during times of waiting). During the remaining 37 minutes (48 – 6 – 5 = 37), there was one tourist. On the average, the number of tourists either waiting or being served were

$$L = [0(6) + 1(37) + 2(5)]/48 = 0.98$$

This average corresponds to the symbol L in Chapter 6.

Probability of Finding Two Tourists in the Center This situation occurred in 5 of the 48 minutes, or 10.4 percent of the time. In a similar manner, other measures of performance for this service system could be calculated.

Simulation with Continuous Probability Distributions

In the preceding case, both the interarrival times and the service times followed discrete distributions. If one or both of these follow a continuous distribution, such as the normal distribution or the uniform distribution, we can use Excel's Random Number Generation tool to generate random numbers in a similar way. Beyond specifying a different distribution and the parameters that are unique to the distribution, the procedure is identical to the procedure described. In Section 7.7 we simulate a situation that requires generating random variables from several continuous distributions.

7.6 Risk Analysis

In Chapter 5, we presented simple examples of risk analysis in the form of a decision tree or a decision table. Simulation can deal with much more complicated risk analysis problems. Such problems may involve many possible combinations and probabilities, and even some constraints. Thus, the standard decision analysis approach is insufficient. The example we use here is fairly simple but illustrates the applicability of simulation in risk analysis.

Let us assume that we want to predict the profit from product M-6 where the profit is given by the following formula:

$$\text{Profit} = [(\text{unit price} - \text{unit cost}) \times \text{volume sold}] - \text{advertising cost}$$

Now let us assume that the unit selling price can take three levels: either $5, $5.50, or $6, depending on market conditions. We also assume that the probabilities of these market conditions are known. Similarly, the unit cost may assume several levels (depending on the commodity markets). The volume is a function of the economic conditions, and the advertising cost depends on competitors' actions. All this information is summarized in Exhibit 7.17.

Using Excel to generate the random numbers, we can simulate the four random variables and compute the profit or any other measures of performance such as the probability of having a loss, the probability of making $10,000 or more, and so on. The first 10 trials are then shown in Exhibit 7.17. For example, in trial 1, the profit = (5.00 – 3.50)18,000 – 30,000 = – 3,000.

This information is then summarized in a *risk profile* probability distribution and a cumulative probability distribution, such as in Exhibit 7.18. Such functions are extremely important in risk analysis. What these figures show is that the range of profit varies between a loss (–$3,000) and $50,000 profit. The mean is about $23,000 (based on 100 trials; the true mean based on expected values is $23,140). If we compute the *most likely* profit (based on the most likely values of the variables) we would get

$$(5.50 - 3.00)18{,}000 - 20{,}000 = \$25{,}000$$

a $2,000 difference compared to the long-run mean. The cumulative probability curve also shows us that there is a 2 percent chance of *losing* money on this product and a 14 percent chance of making less than $10,000. On the other hand, there is a 15 percent chance of making more than $30,000 and a 5 percent chance of making more than $40,000.

EXHIBIT 7.17 Ten Simulated Trials of M-6 Product

	A	B	C	D	E	F	G	H
1	**Selling**		**Unit**				**Advertising**	
2	**Price**	**Probability**	**Cost**	**Probability**	**Volume**	**Probability**	**Cost**	**Probability**
3	$5.00	0.20	$2.50	0.35	15,000	0.30	$20,000	0.50
4	$5.50	0.50	$3.00	0.50	18,000	0.45	$25,000	0.30
5	$6.00	0.30	$3.50	0.15	20,000	0.25	$30,000	0.20
6								
7							*Cumumlative*	
8							*Average*	
9	*Trial*	*Price*	*Cost*	*Volume*	*Advertising*	*Profit*	*Profit*	
10	1	5.00	3.50	18,000	30,000	-3,000	-3,000	
11	2	5.00	3.50	18,000	20,000	7,000	2,000	
12	3	5.50	2.50	18,000	25,000	29,000	11,000	
13	4	5.50	3.50	18,000	20,000	16,000	12,250	
14	5	5.50	3.50	20,000	25,000	15,000	12,800	
15	6	5.50	3.00	18,000	25,000	20,000	14,000	
16	7	5.00	3.00	18,000	20,000	16,000	14,286	
17	8	5.50	2.50	18,000	20,000	34,000	16,750	
18	9	5.50	3.00	18,000	20,000	25,000	17,667	
19	10	5.50	3.00	20,000	25,000	25,000	18,400	
20								
21	*Key Formulas*							
22	F10	=((B10-C10)*D10)-E10 {copy to cells F11:F19}						
23	G10	=AVERAGE(F$10:F10) {copy to cells G11:G19}						

EXHIBIT 7.18 Risk Profile

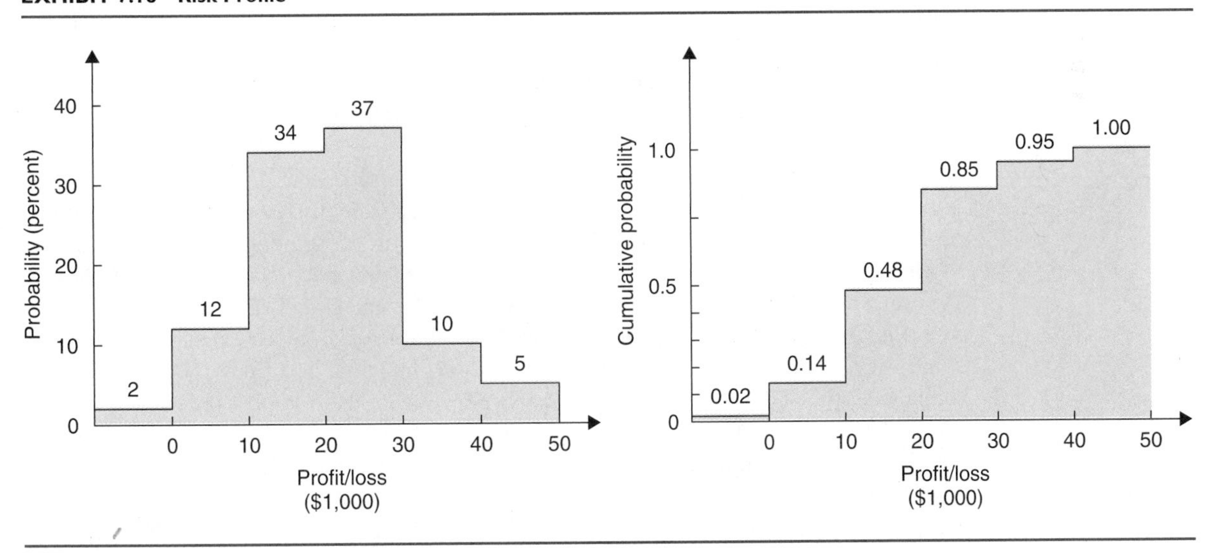

7.7 Detailed Modeling Example

In this section we return to the Banyon State University (BSU) example to illustrate simulation in the broader modeling context. We overviewed this case in the chapter opener; now we will look at the situation in more detail.

Step 1: Opportunity/Problem Recognition

Respondents to a survey indicate that the university's walk-in clinic is one of the areas with which they are most dissatisfied. Analysis of the respondent's comments suggests that although the university community is reasonably satisfied with the quality of care they receive at the clinic, they are quite dissatisfied with the amount of time they spend at the clinic. A student team was asked to study the impact that a computerized patient record system would have on the clinic's operations. With this system, all patient records would be stored electronically and could be immediately accessed by the doctor or nurse via computers at their respective work areas.

Steps 2 and 3: Model Formulation and Data Collection

A team of students was selected to study the problem as part of a course project and developed the preliminary influence diagram of the situation, as illustrated earlier in Exhibit 7.1. In order to study the clinic in more depth, the team collected data on the pattern of arrivals and patient processing times to the clinic over a 2-week period. To collect this data, a student from the team recorded the time each patient arrived on a time card designed by the team. Analysis of the arrival data indicated that patient interarrival times were uniformly distributed between the times of 6 and 20 minutes.

Next, the time when the patients' file was retrieved was recorded on the card. The difference between when the patient arrived and when the patient's record was retrieved was used to estimate the time required to retrieve the records. An analysis of the data collected over the 2-week period indicated that the time to retrieve patient records followed a normal distribution, with a mean of 4 minutes and standard deviation of 0.75 minute.

Once a patient's record was retrieved, the card for recording times was attached to it with a paper clip. When the nurse got to a particular patient, she would record the time the file was picked up. Likewise, the nurse would record the time when she was finished with this patient. Analysis of this data suggested that the nurse processing times followed a normal distribution with a mean of 10 minutes and a standard deviation of 2.3 minutes.

Finally, a similar process was followed by the doctor. The doctor recorded both the time each medical examination began and was completed. Then the doctor placed the completed time card in a designated box for the student team to pick up. An analysis of the doctor treatment times indicated that they also closely followed a normal distribution, with a mean of 17 minutes and a standard deviation of 3.4 minutes.

Rather than jumping right in and trying to develop a simulation model of the clinic's operations, the student team first developed the flow diagram shown in Exhibit 7.19 to gain a better understanding of its operations. Based on this flow diagram, the student team decided to develop a model in Excel to simulate 8 hours (480 minutes) of operation of the clinic, both as it currently operates and how it would likely operate if the computerized patient record system were implemented. Thus, this is a terminating rather than steady state simulation due to the nature of the situation being simulated. The spreadsheet is shown in Exhibit 7.20.

EXHIBIT 7.19 Flow Diagram of Walk-In Clinic's Operations

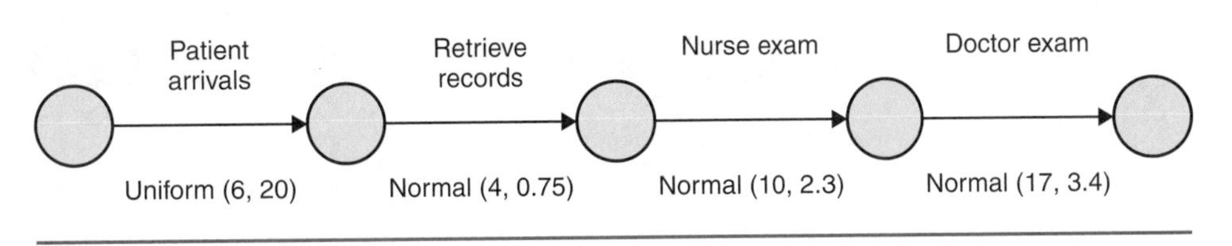

EXHIBIT 7.20 Simulation Model Results of Walk-In Clinic

	A	B	C	D	E	F	G	H	I	J	K	L	M	N	O	P
1						Present System						New Computerized Patient Record System				
2		Time	Time of	Time to	Nurse	Time for	Doctor	Time for	Time	Time	Total	Nurse	Doctor	Time	Time	Total
3	Patient	Between	Patient	Retrieve	Exam	Nurse	Exam	Doctor	Patient	In	Waiting	Exam	Exam	Patient	In	Waiting
4	Number	Arrivals	Arrival	Record	Starts	Examination	Begins	Examination	Finished	System	Time	Starts	Begins	Finished	System	Time
5	1	10.5	10.5	2.9	13.4	8.7	45.0	16.1	61.1	50.6	25.8	10.5	45.0	61.1	50.6	25.8
6	2	13.4	23.9	5.6	29.5	10.9	61.1	16.9	78.0	54.1	26.3	23.9	61.1	78.0	54.1	26.3
7	3	13.9	37.8	3.0	40.8	8.3	78.0	13.5	91.5	53.7	31.9	37.8	78.0	91.5	53.7	31.9
8	4	18.7	56.5	4.1	60.6	10.8	91.5	22.9	114.4	57.9	24.2	56.5	91.5	114.4	57.9	24.2
9	5	15.4	71.9	3.4	75.3	14.1	114.4	20.9	135.3	63.4	28.4	71.9	114.4	135.3	63.4	28.4
10	6	7.1	79.0	3.3	89.4	14.7	135.3	17.2	152.5	73.5	41.6	86.0	135.3	152.5	73.5	41.6
11	7	6.8	85.8	4.7	104.1	9.9	152.5	17.4	169.9	84.1	56.8	100.7	152.5	169.9	84.1	56.8
12	8	6.2	92.0	2.8	114.0	13.8	169.9	17.1	187.0	95.0	64.1	110.6	169.9	187.0	95.0	64.1
13	9	12.0	104.0	4.1	127.8	6.5	187.0	18.5	205.5	101.5	76.5	124.4	187.0	205.5	101.5	76.5
14	10	18.8	122.8	3.0	134.3	8.8	205.5	21.1	226.6	103.8	73.9	130.9	205.5	226.6	103.8	73.9
15	11	8.1	130.9	4.6	143.1	8.0	226.6	13.8	240.4	109.5	87.7	139.7	226.6	240.4	109.5	87.7
16	12	9.9	140.8	4.6	151.1	8.3	240.4	19.0	259.4	118.6	91.3	147.7	240.4	259.4	118.6	91.3
17	13	6.7	147.5	2.8	159.4	12.2	259.4	17.1	276.5	129.0	99.7	156.0	259.4	276.5	129.0	99.7
18	14	15.8	163.3	3.2	171.6	12.1	276.5	15.7	292.2	128.9	101.1	168.2	276.5	292.2	128.9	101.1
19	15	7.9	171.2	2.5	183.7	9.5	292.2	14.8	307.0	135.8	111.5	180.3	292.2	307.0	135.8	111.5
20	16	11.2	182.4	3.7	193.2	11.7	307.0	18.7	325.7	143.3	112.9	189.8	307.0	325.7	143.3	112.9
21	17	10.3	192.7	6.0	204.9	9.4	325.7	13.7	339.4	146.7	123.6	201.5	325.7	339.4	146.7	123.6
22	18	9.0	201.7	3.1	214.3	9.4	339.4	14.3	353.7	152.0	128.3	210.9	339.4	353.7	152.0	128.3
23	19	18.6	220.3	3.4	223.7	7.5	353.7	18.7	372.4	152.1	125.9	220.3	353.7	372.4	152.1	125.9
24	20	13.9	234.2	4.1	238.3	7.8	372.4	20.3	392.7	158.5	130.4	234.2	372.4	392.7	158.5	130.4
25	21	8.2	242.4	2.9	246.1	10.5	392.7	21.1	413.8	171.4	139.8	242.4	392.7	413.8	171.4	139.8
26	22	18.4	260.8	4.3	265.1	9.3	413.8	18.5	432.3	171.5	143.7	260.8	413.8	432.3	171.5	143.7
27	23	17.7	278.5	3.5	282.0	10.1	432.3	12.6	444.9	166.4	143.7	278.5	432.3	444.9	166.4	143.7
28	24	6.9	285.4	4.0	292.1	13.5	444.9	13.7	458.6	173.2	146.0	288.6	444.9	458.6	173.2	146.0
29	25	9.7	295.1	3.5	305.6	10.2	458.6	16.2	474.8	179.7	153.3	302.1	458.6	474.8	179.7	153.3
30	26	9.5	304.6	3.7	315.8	11.2	474.8	16.9	491.7	187.1	159.0	312.3	474.8	491.7	187.1	159.0
31	27	8.6	313.2	4.4	327.0	6.2	491.7	23.3	515.0	201.8	172.3	323.5	491.7	515.0	201.8	172.3
32	28	7.8	321.0	3.5	333.2	9.6	515.0	16.4	531.4	210.4	184.4	329.7	515.0	531.4	210.4	184.4
33	29	10.9	331.9	3.9	342.8	10.9	531.4	17.1	548.5	216.6	188.6	339.3	531.4	548.5	216.6	188.6
34	30	7.8	339.7	3.4	353.7	8.6	548.5	21.9	570.4	230.7	200.2	350.2	548.5	570.4	230.7	200.2
35	31	7.1	346.8	3.4	362.3	10.6	570.4	22.2	592.6	245.8	213.0	358.8	570.4	592.6	245.8	213.0
36	32	7.5	354.3	5.0	372.9	9.8	592.6	22.4	615.0	260.7	228.5	369.4	592.6	615.0	260.7	228.5
37	33	16.6	370.9	3.9	382.7	9.3	615.0	28.0	643.0	272.1	234.8	379.2	615.0	643.0	272.1	234.8
38	34	19.9	390.8	5.1	395.9	11.0	643.0	17.3	660.3	269.5	241.2	390.8	643.0	660.3	269.5	241.2
39	35	6.8	397.6	3.3	406.9	11.3	660.3	17.3	677.6	280.0	251.4	401.8	660.3	677.6	280.0	251.4
40	36	19.3	416.9	3.2	420.1	9.5	677.6	23.7	701.3	284.4	251.2	416.9	677.6	701.3	284.4	251.2
41	37	10.8	427.7	3.6	431.3	9.4	701.3	22.4	723.7	296.0	264.2	427.7	701.3	723.7	296.0	264.2
42	38	12.2	439.9	4.0	443.9	11.5	723.7	20.4	744.1	304.2	272.3	439.9	723.7	744.1	304.2	272.3
43	39	10.6	450.5	3.4	455.4	10.7	744.1	15.6	759.7	309.2	282.9	451.4	744.1	759.7	309.2	282.9
44	40	14.2	464.7	3.8	468.5	9.6	759.7	22.8	782.5	317.8	285.4	464.7	759.7	782.5	317.8	285.4

Column A in Exhibit 7.20 simply keeps track of the order in which patients arrive at the clinic on a particular day. In column B the interarrival times for the patients were randomly generated from a Uniform distribution over the range of 6 to 20 minutes, using Excel's Random Number Generation tool (see Exhibit 7.21). According to Exhibit 7.20, the first patient arrives 10.5 minutes after the clinic opens, the second patient arrives 13.4 minutes after the first patient, and so on.

In column C, the time the patient actually arrives is calculated. For the purpose of simulating the clinic's operations, the clinic is assumed to open each day at time zero. Also, since trying to keep track of both hours and minutes can become tedious in spreadsheets, all times are expressed in the number of minutes since the clinic opened. The interarrival time for the first patient represents the time between when the clinic opens and when the patient arrives, so the formula =B5 was entered in cell C5. Referring to Exhibit 7.20, we observe

EXHIBIT 7.21 Generating Random Patient Interarrival Times with Excel

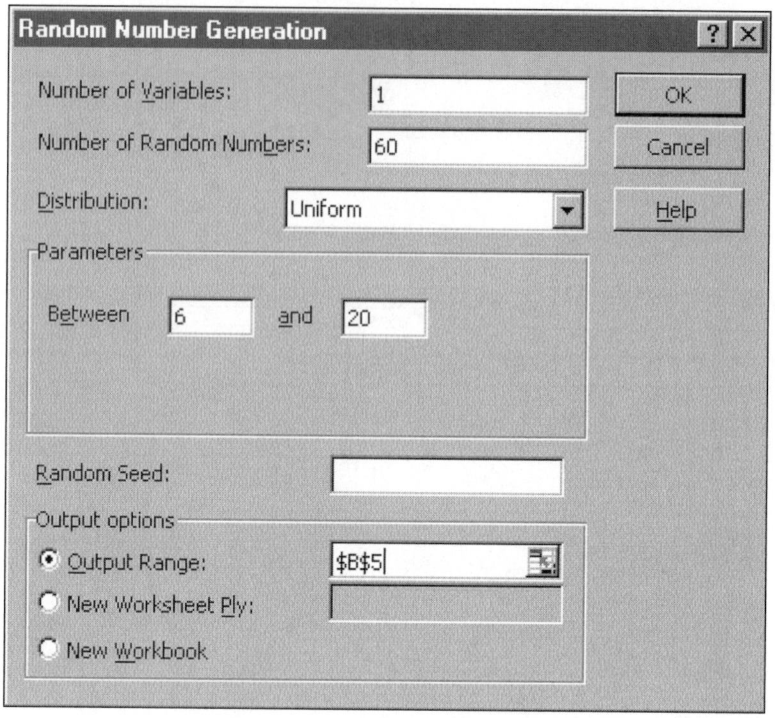

Random Number Generation

Number of Variables: 1

Number of Random Numbers: 60

Distribution: Uniform

Parameters

Between 6 and 20

Random Seed:

Output options

Output Range: $B$5

New Worksheet Ply:

New Workbook

OK

Cancel

Help

that the first patient arrived 10.5 minutes after the clinic opened. The interarrival time of 13.4 generated for the second patient indicates that the second patient of the day arrived 13.4 minutes after the arrival of the first patient. Therefore, to calculate the actual arrival time of the second patient, the formula =C5+B6 was entered in cell C6. Since this logic repeats for the other arriving patients during the day, the formula entered in cell C6 was copied down to the other cells in column C.

In column D, Excel's Random Number Generator tool was used to generate the time to retrieve each patient's medical record from a normal distribution with a mean of 4 minutes and standard deviation of 0.75. The process of doing this is similar to generating the interarrival times for the patients, the major difference being that the random numbers were generated from a normal distribution as opposed to being generated from a uniform distribution.

The time the nurse starts examining the patient is calculated in column E. Since the first patient does not have to wait for the nurse to finish with other patients, the time the nurse begins examining the first patient is equal to the amount of time required to retrieve the patient's file after the patient arrives. Therefore, the formula =C5+D5 was entered in cell E5. For all the patients that arrive after the first patient, two events have to occur before the nurse can begin examining a particular patient. First, the patient must arrive at the clinic and have his or her medical record retrieved. Second, the nurse must finish with the preceding patient. The time required to retrieve the second patient's medical record can be calculated as C6+D6. The time the nurse finishes with the first patient is calculated as E5+F5 (the time the exam begins + the time required to complete the exam). Since both of these events must be completed before the exam can be started for the second patient, we enter the formula =MAX(C6+D6,E5+F5) in cell E6. This formula can then be copied to the remaining cells in column E.

Excel's Random Number Generator tool was used to generate the time for the nurse to complete her exam of a particular patient from a normal distribution with a mean of 10 minutes and standard deviation of 2.3 in column F.

The time the doctor begins examining a given patient is calculated in column G. Since the doctor does not arrive until 45 minutes after the clinic opens, the first patient will not be seen until both the clinic has been open for 45 minutes and the nurse exam of the first patient has been completed. Therefore, the time the first patient is seen by the doctor is calculated as =MAX(C5+D5+F5,45) in cell G5. For the remaining patients that visit the clinic on a particular day, the doctor cannot begin the exam until two events are completed. First, the doctor must finish with the preceding patient. Second, the nurse must finish with the current patient. Referring to the second patient that arrives at the clinic, the time the doctor finishes with the first patient can be calculated as G5+H5. Likewise, the time the nurse finishes with the second patient is calculated as E6+F6. Hence, the time that the doctor begins examining the second patient is calculated as =MAX(G5+H5,E6+F6) in cell G6. The formula entered in cell G6 can be copied to the remaining cells in column G.

Excel's Random Number Generator tool was used to generate the time required for the doctor to perform an exam from a normal distribution with a mean of 17 minutes and standard deviation of 3.4. The time the patient treatment was completed is calculated as =G5+H5 in cell I5 for the first patient. This formula can be copied to the remaining cells in column I.

Two performance measures are calculated in columns J and K, respectively. In column J the time the patient spends at the clinic is calculated by subtracting the time the patient arrived from the time the doctor finished with the patient. For the first patient, the total time spent in the clinic is calculated as =I5–C5 in cell J5. The total amount of time a patient spends waiting is calculated as the total time the patient spends in the clinic less the time the patients spends with the doctor and nurse. For the first patient, the amount of waiting time is calculated as =J5–F5–H5 in cell K5. The formulas entered in cells J5 and K5 can be copied to the remaining cells in their respective columns.

Finally, columns L through M correspond to a model that simulates the operation of the clinic if the new computerized patient record system were implemented. Note that an advantage to this simulation model is that the same patient arrivals, nurse processing times, and doctor processing times are used to compare the performance of the manual record retrieval system and the computerized record retrieval system. That is, *common random numbers* are used to compare both options. Had different sets of random numbers been used, the decision maker would not be sure if observed differences in performance were due to one option actually being superior to the other or due to the fact that different sets of random numbers were used.

The formulas entered to simulate the operation of the clinic with the computerized patient record system are similar to the formulas discussed for the model developed to simulate the clinic's current operations. The only difference between the two systems is that there is no delay for retrieving patient records if the computerized system is implemented.

Step 4: Analysis of the Model

The simulation model shown in Exhibit 7.20 was manually replicated 10 times. Each replication required generating new random numbers for patient interarrival times (column B), record retrieval times (column D), nurse examination times (column F), and doctor examination times (column H). The results of the 10 replications are summarized in Exhibit 7.22.

Upon further examination of the results summarized in Exhibit 7.22, it can be observed that there is no apparent difference in the performance of the clinic after the computerized

EXHIBIT 7.22 Summary of Walk-In Clinic Simulation Results

Replication	Number of Patients	Time in System (Current System)	Time in System (Computerized Records)	Waiting Time (Current System)	Waiting Time (Computerized Records)
1	41	175.32	175.32	146.77	146.77
2	35	109.99	109.99	82.89	82.89
3	37	124.83	124.83	98.46	98.46
4	34	94.56	94.56	67.58	67.58
5	38	122.10	122.10	95.08	95.08
6	36	122.24	122.24	88.25	88.25
7	37	114.00	114.00	86.69	86.69
8	36	110.91	110.91	85.53	85.53
9	39	155.43	155.43	127.85	127.85
10	35	93.13	93.13	66.63	66.63

patient record system has been implemented. You are probably wondering how this can be! After all, implementing the computerized patient record system completely eliminates the delay associated with patient record retrieval.

To explain how this is possible, note that the pace of the slowest operation governs the rate of output of sequential production and service processes such as the walk-in clinic. In the case of the walk-in clinic, the examination with the doctor is the slowest task. The patients' average interarrival time of 13 minutes, or (6 + 20)/2, is equivalent to an average of 4.8 patients arriving at the clinic each hour, but the doctor can only process an average of 3.5 patients per hour. Since the patients are arriving faster than the doctor can process them, a queue will build up and the last patients arriving at the clinic will have to wait the longest, as seen in Exhibit 7.20. Clearly, we are simulating a transient state situation in this terminating simulation because the queue grows without bound and the system never reaches steady state. In addition, since all patients must see the doctor, the rate that the clinic can process patients is limited entirely by the rate that the doctor can treat patients. Since computerizing patient records had no impact on the rate at which the doctor was able to treat patients, implementing this system had no impact on the clinic's performance. The net result was that instead of patients dividing their total waiting time between waiting for their records to be retrieved, waiting for the nurse, and waiting for the doctor, in the system with the computerized record system the patient's total waiting time was now spent waiting for the doctor and nurse. Therefore, implementing the computerized record system only changes how patient waiting time is allocated. It would not, however, do anything to *reduce* this waiting time.

Although the simulation study does not support the implementation of the computerized patient record system, the results do provide some interesting insights into the operation of the clinic. According to the results, anywhere from 34 to 41 patients are expected to visit the clinic on any given day. Referring to Exhibit 7.22, in the best case patients can be expected to spend more than 93 minutes at the clinic, and 67 minutes of this waiting. In the worst case, patients can be expected to spend almost 3 hours at the clinic with more than 2 hours of this being spent waiting to be seen by the doctor or nurse.

Step 5: Implementation

Implementing the computerized patient record system cannot be justified on the basis of improved patient service times. Of course there may be other reasons why such a system would be justified, including more legible and accurate records, lower labor costs associated with handling the records, and so on. Other options that could be investigated to improve patient service times include hiring another doctor, staffing the clinic with more highly trained doctors that can treat patients faster, and perhaps seeing patients on an appointment-only basis. As can be seen in Exhibit 7.19, with two doctors, the average doctor exam time would be reduced in half (8.5 minutes) and would then be more in line with the duration of the other activities (better balanced). Exhibit 7.23 illustrates in the form of a memo how the results of this analysis might be communicated to the president of BSU. The companion Web site for this book contains a sample PowerPoint presentation of the results of this analysis.

EXHIBIT 7.23 Sample Memo to President of BSU

MEMO

To: President Thomas Kourpias
From: Walk-In Clinic Student Task Force
CC: Dr. Marsha Williams, VP University Health Services
Date: 4/4/03
Re: Computerized Patient Record Storage System

Introduction

We have concluded our investigation of BSU's walk-in clinic. In summary, we do not recommend that a computerized patient record system be implemented at this time. While such a system could prove beneficial in terms of improving the legibility and accuracy of patient records, and could reduce the labor costs associated with handling the records, it would not provide any benefit in terms of lowering patient service times. Other options that could be investigated to improve patient service times include hiring another doctor, staffing the clinic with more highly trained doctors who can treat patients faster, and perhaps moving to seeing patients on an appointment-only basis.

Analysis

Our detailed analysis of the clinic indicates that the slowest operation at the clinic is the examination of the patient by the doctor. More specifically, while an average of 4.8 patients arrive at the clinic each hour, the doctor is only able to examine and treat approximately 3.5 patients an hour. Therefore, it is the examination with the doctor that is creating the large waiting times, not the time to retrieve patient records. On the surface it may appear that patients are actually delayed while their records are being retrieved. In actuality, they are really waiting for the doctor. Our analysis clearly indicates that even if record retrieval times were eliminated, total patient service times would not be affected. Basically, in the current system, some of the time the patient spends waiting for the doctor is currently being allocated to the record retrieval process.

(continued)

EXHIBIT 7.23 Continued

Computer simulation models were developed to analyze the clinic's current operations and its operations if the computerized record system were implemented. The results of the simulation study do support the complaints expressed in the survey. In the best case, patients can be expected to spend more than 93 minutes at the clinic, and 67 minutes of this waiting. In the worst case, patients can be expected to spend almost 3 hours at the clinic with more than 2 hours of this being spent waiting to be seen by the doctor or nurse.

Recommendations

Since the computerized record system will not help reduce current patient waiting times, we recommend that this option not be implemented at this time. The computerized record system may provide other benefits such as increased record accuracy, lower handling costs, improved legibility, and higher consistency, and it is in the context of these benefits that this system should be evaluated. In the short-term, the best course of action would be to increase the number of doctors or perhaps shift some of the doctor's responsibilities to a nurse or nurse practitioner.

Limitations

There are several limitations of this analysis that should be pointed out. First, the study was based on data collected over a single 2-week period. Based on the data collected, assumptions were made about the distribution of patient arrivals. Similar assumptions were also made about the record retrieval times, nurse examination times, and doctor examination times. It is likely that there is significant seasonality in the need for medical care and the type of care needed. Therefore, the results of the study may not be generalizable to conditions that differ substantially from those that were studied. Second, the study only considered the impact on patient service times. Clearly, the performance of the walk-in clinic should consider a number of other factors, including quality of care, cost of care, and so on. Finally, the results discussed here are based on averages. The performance of the clinic will vary within a given day and from day to day.

Appendix: Crystal Ball 2000

This appendix illustrates how Crystal Ball 2000, an Add-In to Excel created by Decisioneering Inc., can be used to facilitate the development of simulation models and their analysis. To illustrate this package, consider the experience of a fictitious Internet Service Provider (ISP) that needs to upgrade its server computers. Company management has identified the following two options: (1) shift to a Windows-based platform from its current Unix-based platform, or (2) stick with a Unix-based platform. The company estimates that if it migrates to Windows, the new server hardware could cost as little as $125,000 or as much as $210,000. The technical group's best estimate is that the hardware costs will be $140,000 if the Windows option is pursued. Likewise, based on historical data of similar software conversion projects, the company estimates the software

EXHIBIT 7.24 Comparison of Web-Server Options

	A	B	C
1		**Windows**	**Unix**
2		**Platform**	**Platform**
3	Hardware cost	$140,000	$120,000
4	Software conversion cost	$275,000	$275,000
5	Employee training	$11,000	$11,000
6	Total project cost	$426,000	$406,000

conversion costs to migrate to Windows follows a normal distribution with a mean of $275,000 and standard deviation of $25,000. Finally, employee training costs associated with converting to Windows are estimated to range between $8,000 to $14,000, with all costs within this range expected to be equally likely.

If the company sticks with Unix, the new server hardware will most likely cost $120,000, but could cost as little as $75,000 or as much as $215,000. Software conversion and upgrade costs for Unix projects tend to be normally distributed with a mean of $275,000 and standard deviation of $70,000. Employee training costs should fall between $6,000 to $16,000 with all values in this range being equally likely.

Returning to the task at hand, the spreadsheet shown in Exhibit 7.24 was created. The most likely cost estimates or the mean values were entered in cells B3:C5, while the costs of pursuing each project are calculated in cells B6 and C6 by summing up the hardware, software, and training costs.

Using Crystal Ball to run Monte Carlo simulation requires defining two types of cells in the spreadsheet. The cells that are associated with variables or parameters are defined as *assumption cells.* The cells that correspond to outcomes of the simulation model are defined as *forecast cells.* Forecast cells typically contain formulas that depend on one or more assumption cells. Simulation models can have multiple assumption and forecast cells, but they must have at least one of each. Referring to Exhibit 7.24, the variables that we will make assumptions about are contained in cells B3:C5. Likewise, the outcomes, or in this case project costs, are calculated in cells B6 and C6.

To illustrate the process of defining an assumption cell, refer to cell B3 corresponding to the hardware costs for the Windows-based platform project. In this example, management has developed three estimates for the hardware cost: an optimistic cost, a most likely cost, and a pessimistic cost. Both the beta distribution and the triangular distribution are well suited for modeling variables with these three parameters. However, because the beta distribution is quite complex and not particularly intuitive, we assume that a triangular distribution provides a reasonably good approximation for the hardware costs. Recall that the most likely cost for hardware if the firm migrates to a Windows-based platform is estimated to be $140,000, with a range of $125,000 to $210,000.

The process of defining an assumption involves the following six steps:

1. Click on cell B3 to identify it as the cell the assumption applies to.
2. Select the menu option Cell at the top of the screen.[2]
3. From the dropdown menu that appears, select Define Assumption . . .
4. Crystal Ball's Distribution Gallery dialog box is now displayed as shown in Exhibit 7.25. As you can see, Crystal Ball provides a wide variety of probability distributions to choose from. Double click on the Triangular box to select it.
5. Crystal Ball's Triangular Distribution dialog box is now displayed as shown in Exhibit 7.26. In the Assumption Name textbox at the top of the dialog box, enter a

EXHIBIT 7.25 Crystal Ball 2000's Distribution Gallery

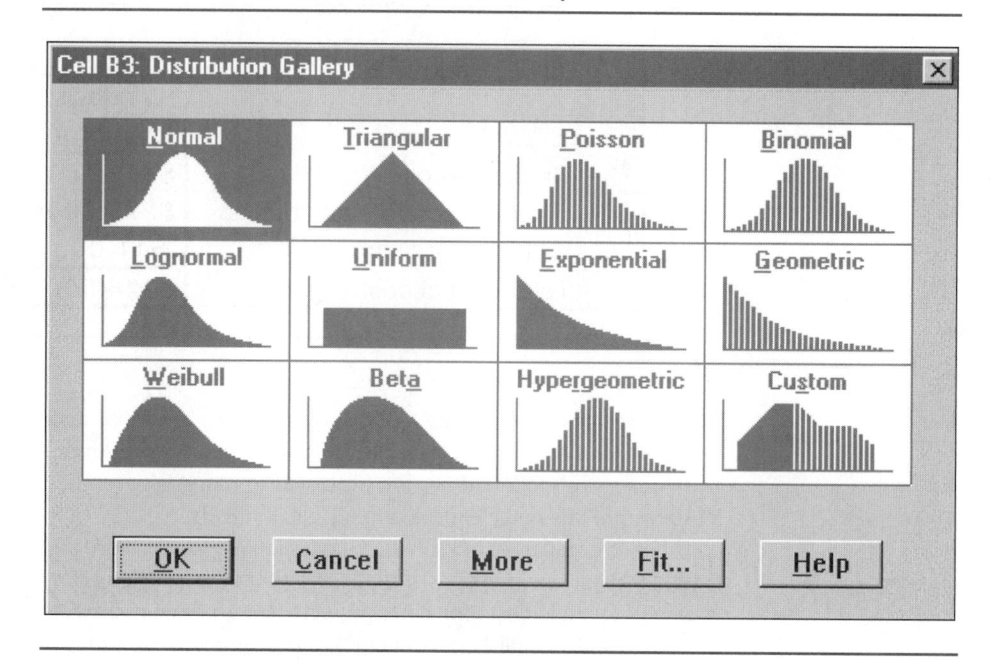

EXHIBIT 7.26 Triangular Distribution Dialog Box

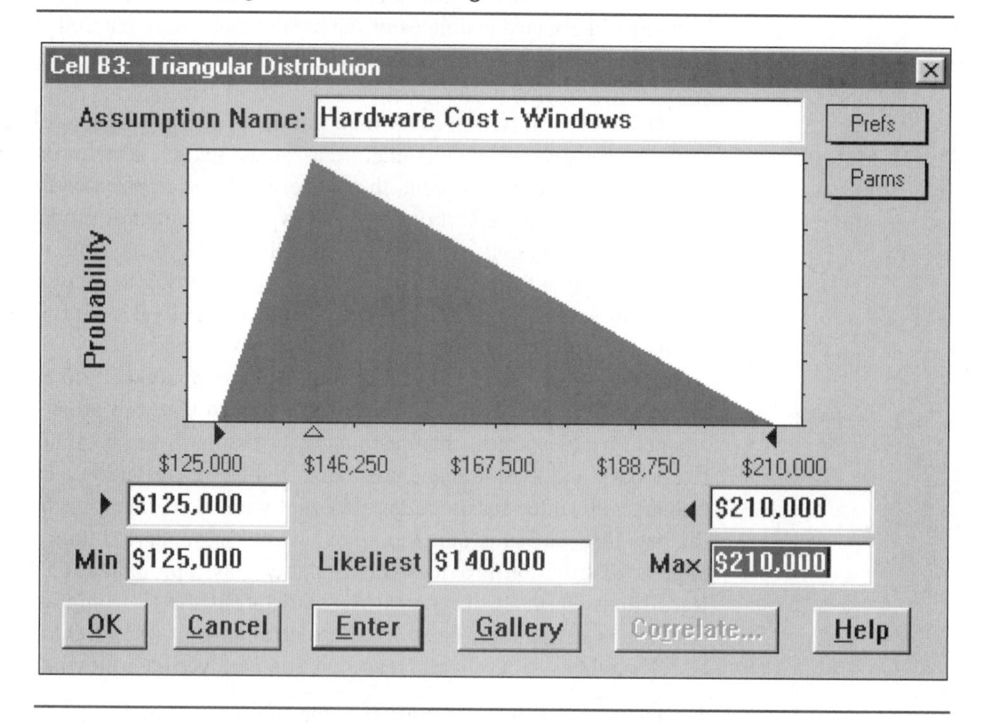

descriptive name such as *Hardware Cost—Windows* to label the assumption. Then enter the optimistic, most likely, and pessimistic costs of $125,000, $140,000, and $210,000 in the Min, Likeliest, and Max textboxes, respectively.

6. Click on the OK button.

Repeat steps 1–6 for the hardware cost of the Unix project (cell C3) but changing the information entered in step 5.

EXHIBIT 7.27 Normal Distribution Dialog Box

Cell B4: Normal Distribution

Assumption Name: Software Conversion Cost - Windows

Prefs

Parms

Probability

$200,000 $237,500 $275,000 $312,500 $350,000

-Infinity +Infinity

Mean $275,000 Std Dev $25,000

OK Cancel Enter Gallery Correlate... Help

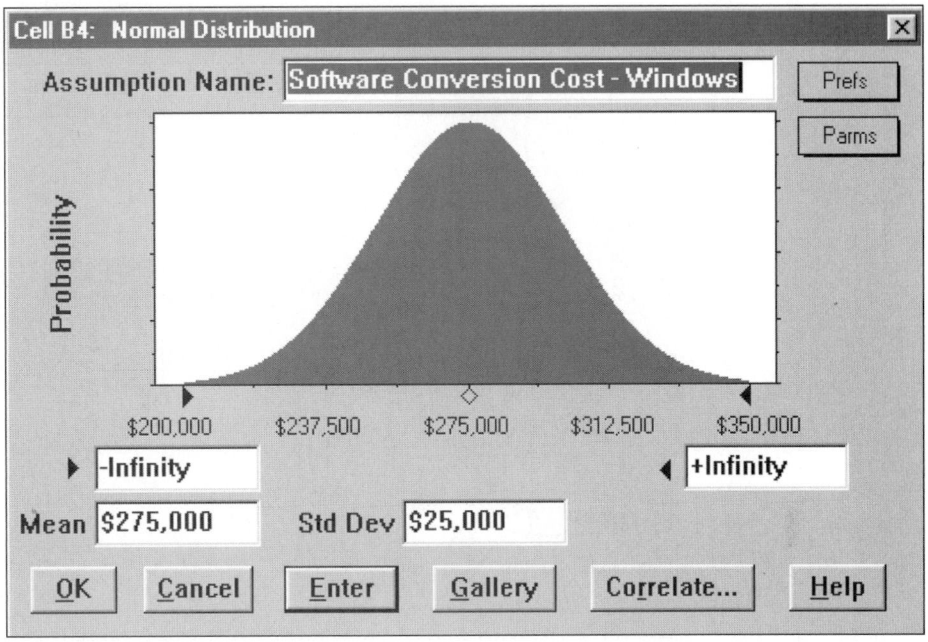

We next turn our attention to defining the assumption cells for the software conversion costs. The procedure is the same as outlined earlier, except now we select the Normal distribution from the Distribution Gallery in step 4, as opposed to the Triangular distribution that was used to model hardware costs. After making this selection, the Normal Distribution dialog box is displayed as shown in Exhibit 7.27. In the Assumption Name textbox at the top of the dialog box, the descriptive name *Software Conversion Cost—Windows* was entered to label the assumption. Then the mean and standard deviation of $275,000 and $25,000 were entered in the Mean and Std Dev textboxes, respectively. Finally, click on the OK button to define the assumption cell.

Defining the assumption cells for employee training costs (cells B5 and C5) are done in a similar fashion, except that these costs are expected to be uniformly distributed. Thus, in defining Assumption Cells B5 and C5, the Uniform distribution is selected from the Distribution Gallery. In Exhibit 7.28, the information for training costs if the Windows platform is chosen (cell B5) has been entered into the Uniform Distribution dialog box.

Having defined the assumption cells, we now turn our attention to defining the forecast or outcome cells. In our example, we are interested in comparing the costs of the two projects. The process of defining a forecast cell involves the following five steps:

1. Click on cell B6 to identify it as containing an outcome we are interested in.
2. Select the menu option Cell at the top of the screen.
3. From the dropdown menu that appears, select Define Forecast . . .
4. Crystal Ball's Define Forecast dialog box is now displayed as shown in Exhibit 7.29. In the Forecast Name textbox, enter a descriptive name such as *Total Project Cost—Windows* to label the result. Then enter a descriptive label such as *Dollars* in the Units: textbox.
5. Click on OK.

Repeat steps 1–5 for cell C6.

One iteration of the simulation model involves randomly generating values for each of the assumption cells based on the specified probability distributions and then calculating

EXHIBIT 7.28 Uniform Distribution Dialog Box

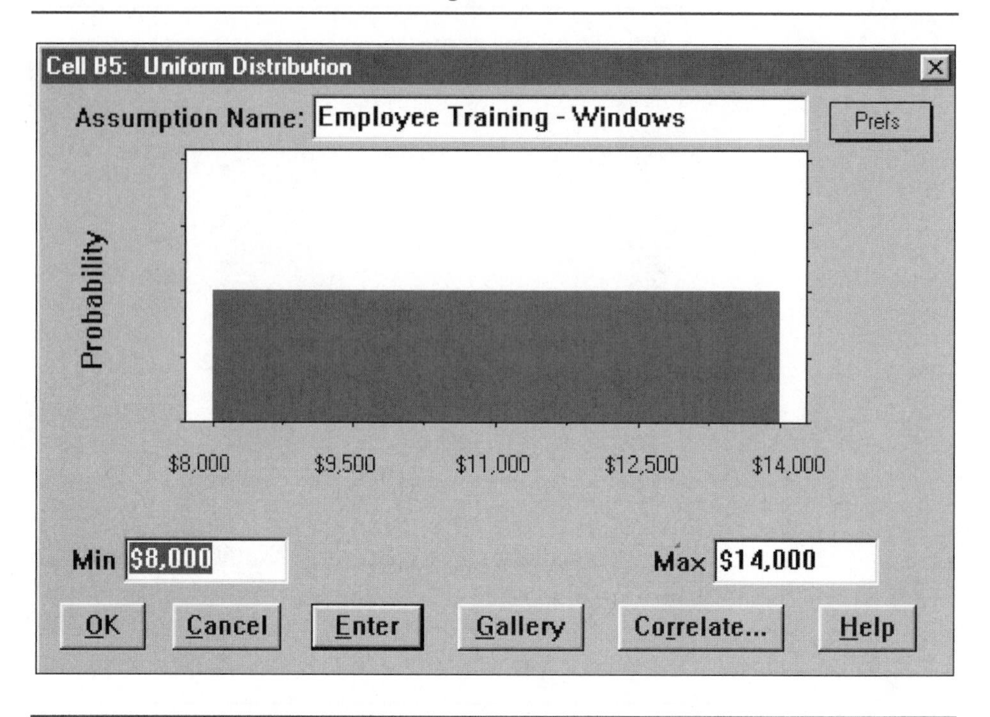

EXHIBIT 7.29 Define Forecast Dialog Box

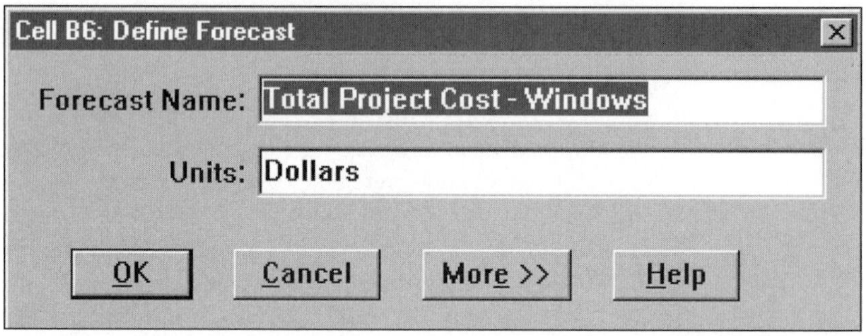

Cell B6: Define Forecast

Forecast Name: Total Project Cost - Windows

Units: Dollars

OK Cancel More >> Help

the total project costs. By repeating this process hundreds or perhaps thousands of times we can get a sense of the distribution of possible outcomes.

To simulate the completion of these projects 1,000 times select the Run menu item at the top of the screen. Next, in the dropdown dialog box that appears, select Run Preferences . . . In the Run Preferences dialog box that appears, enter 1,000 in the Maximum Number of Trials textbox and then click OK. Next, to actually replicate the simulation model 1,000 times select the Run menu item again and then Run from the drop-down menu. Crystal Ball summarizes the results of the simulation models in the form of frequency charts as the model is being executed, as shown in Exhibits 7.30 and 7.31. Crystal Ball provides other summary information about the forecast cells in addition to the frequency chart including percentile information, descriptive statistics, the cumulative chart, and a reverse cumulative chart. For example to see the descriptive statistics for a forecast cell, select the View menu option in the Forecast window and

EXHIBIT 7.30 Frequency Chart for Total Project Costs (Windows-Based Platform)

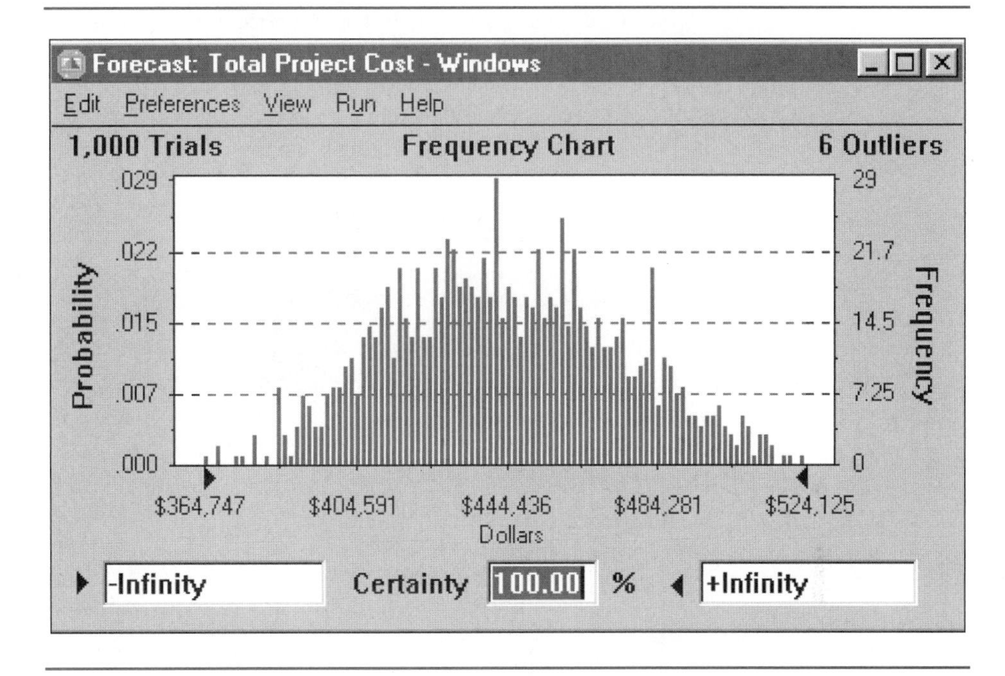

EXHIBIT 7.31 Frequency Chart for Total Project Costs (Unix-Based Platform)

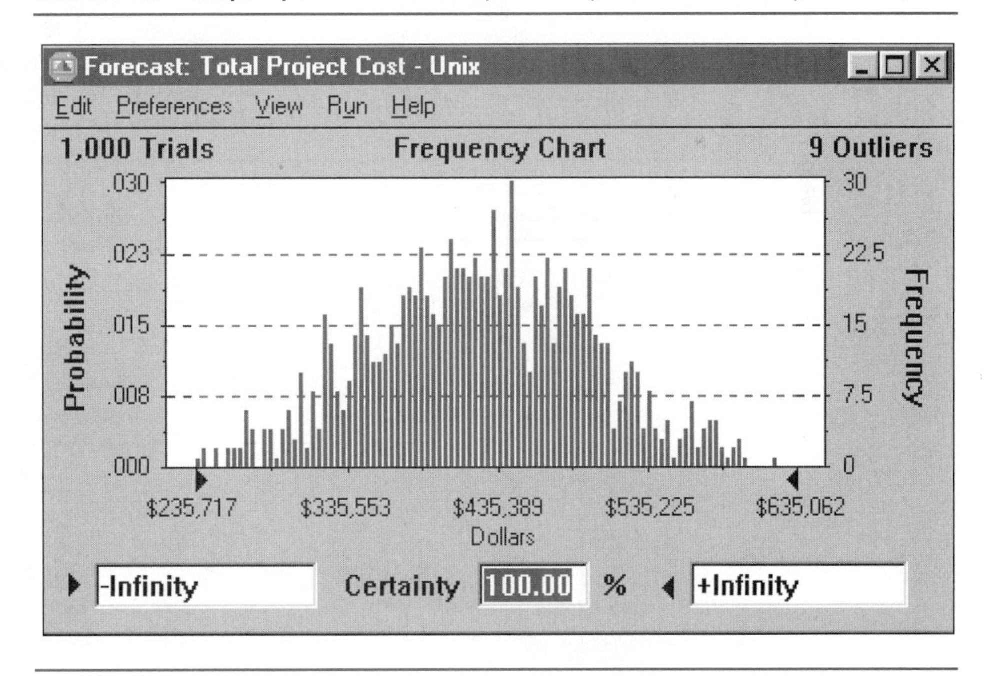

then select Statistics from the drop-down menu that appears. The Statistics view for both the Windows-based platform and Unix-based platform projects are shown in Exhibits 7.32 and 7.33, respectively.

Reviewing Exhibits 7.32 and 7.33, we observe that the average (mean) total cost of the Windows project across the 1,000 replications of the simulation model is $445,009, while the average cost of the Unix project is $423,837. On this basis alone, the Unix project appears to be the lower cost project. However, upon examining Exhibits 7.32 and 7.33 further, we can observe that the total cost of the Windows project is expected to range between $341,545 and $542,613. In contrast, the Unix project might cost as little as $202,635 or as much as $660,888. The wider range of possible costs associated with the Unix project is an indication of greater uncertainty and therefore greater risks. What should management do in this case? Should it select the Unix project because on average it is expected to cost $21,172 less than

EXHIBIT 7.32 Summary Statistics for Project If Windows Platform Adopted

Forecast: Total Project Cost - Windows

Edit Preferences View Run Help

Cell B6 Statistics

Statistic	Value
Trials	1,000
Mean	$445,009
Median	$442,931
Mode	---
Standard Deviation	$30,853
Variance	$951,877,484
Skewness	0.16
Kurtosis	2.72
Coeff. of Variability	0.07
Range Minimum	$341,545
Range Maximum	$542,613
Range Width	$201,068
Mean Std. Error	$975.64

EXHIBIT 7.33 Summary Statistics for Project If Continue with Unix Platform

Forecast: Total Project Cost - Unix

Edit Preferences View Run Help

Cell C6 Statistics

Statistic	Value
Trials	1,000
Mean	$423,837
Median	$425,709
Mode	---
Standard Deviation	$73,981
Variance	$5,473,177,953
Skewness	-0.09
Kurtosis	2.86
Coeff. of Variability	0.17
Range Minimum	$202,635
Range Maximum	$660,888
Range Width	$458,253
Mean Std. Error	$2,339.48

the Windows project but could potentially end up costing more than $660,000? Or should it select the Windows project knowing that it is very unlikely the project costs will exceed $542,000? Clearly, these types of questions only become apparent when the distributions of possible outcomes are considered.

Crystal Ball provides a feature that helps answer these types of questions. For example, assume that the firm cannot afford to spend any more than $500,000. One dimension on which to compare the two projects, then, is the probability that they will exceed the $500,000 budget limitation. We can use the Forecast window provided by Crystal Ball to answer these types of questions. More specifically, in this case we are interested in finding the probability that each project's total cost will exceed $500,000 or that its total costs will be between $500,000 and infinity. To calculate this probability for each project, the information shown in Exhibits 7.34 and 7.35 was entered in the textboxes at the bottom of each Forecast window. According to the calculations, the Windows project has a 4.1 percent chance of exceeding the maximum available funds, while the Unix project has a much higher 14.6 percent chance.

One final comment is in order. Referring to Exhibit 7.27, we see that according to the parameters entered, the hardware costs could range from $200,000 to $350,000. Because the normal distribution is symmetrical, we also assume that it is just as likely for hardware cost to exceed $275,000 as it is for it to cost less than $275,000. If, based on past experience, one or both of these conditions do not seem reasonable, then one or both of the distribution parameters could be altered to try to obtain a better approximation, or perhaps a new distribution should be used to model the variable. Also, values can be entered in the textboxes to truncate the upper and lower values returned. For example, if a survey of vendor hardware prices suggest that the new hardware will cost at least $225,000, then 225,000 could be entered for the lower bound, as shown in Exhibit 7.36.

In this appendix we briefly overviewed how Crystal Ball can facilitate the creation of simulation models and their analysis. While our intention was not to make you expert users of this Excel add-in, we believe that having an awareness of some of its capabilities will serve you well as your modeling skills progress.

EXHIBIT 7.34 Calculating the Probability That the Windows Project Total Cost Exceeds $500,000

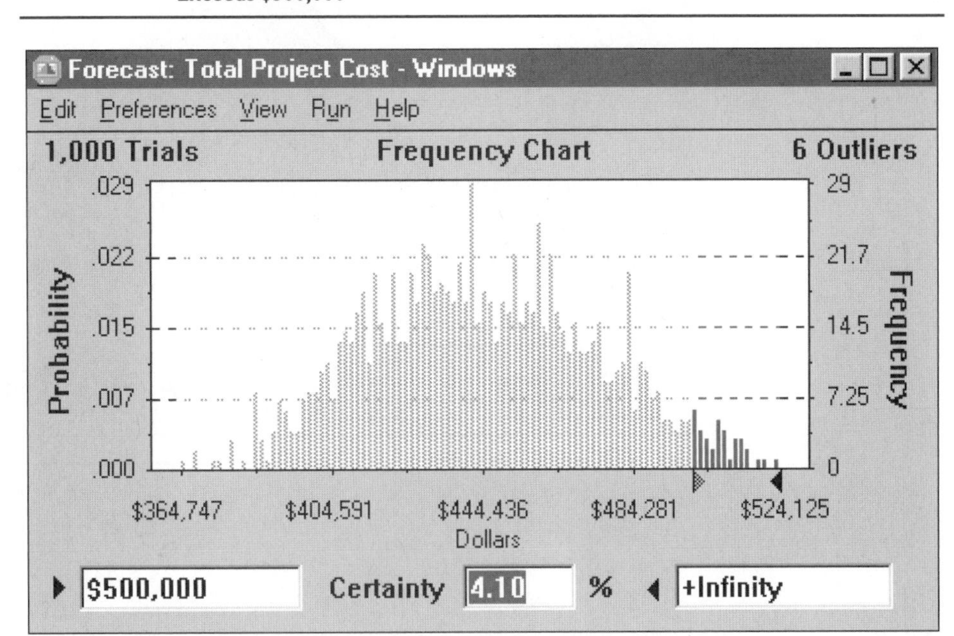

EXHIBIT 7.35 Calculating the Probability That the Unix Project Total Cost Exceeds $500,000

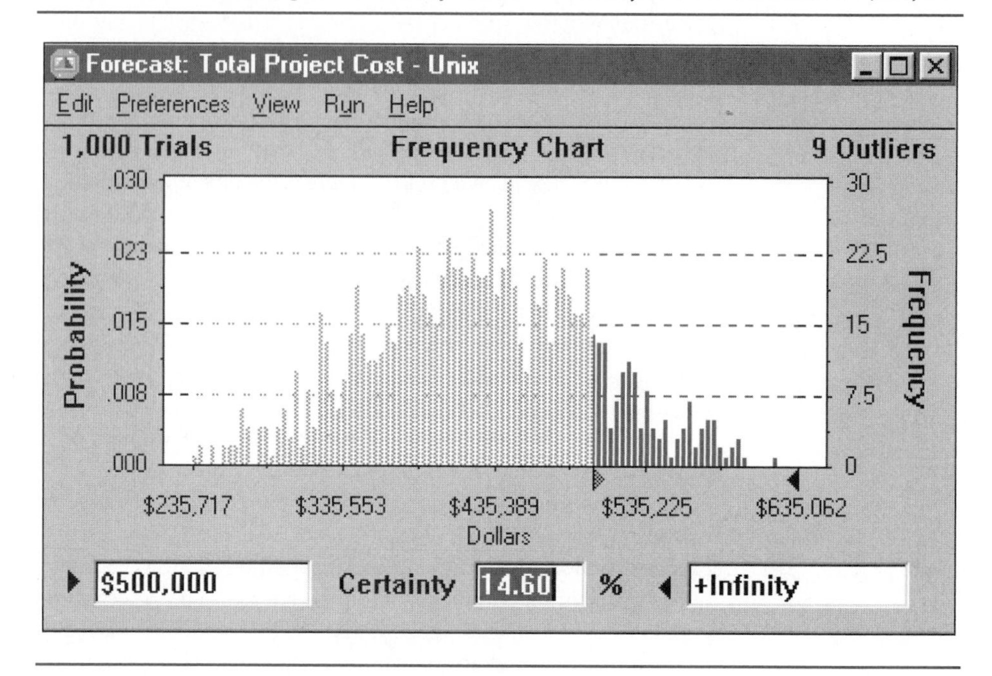

EXHIBIT 7.36 Truncating the Values Returned by a Distribution

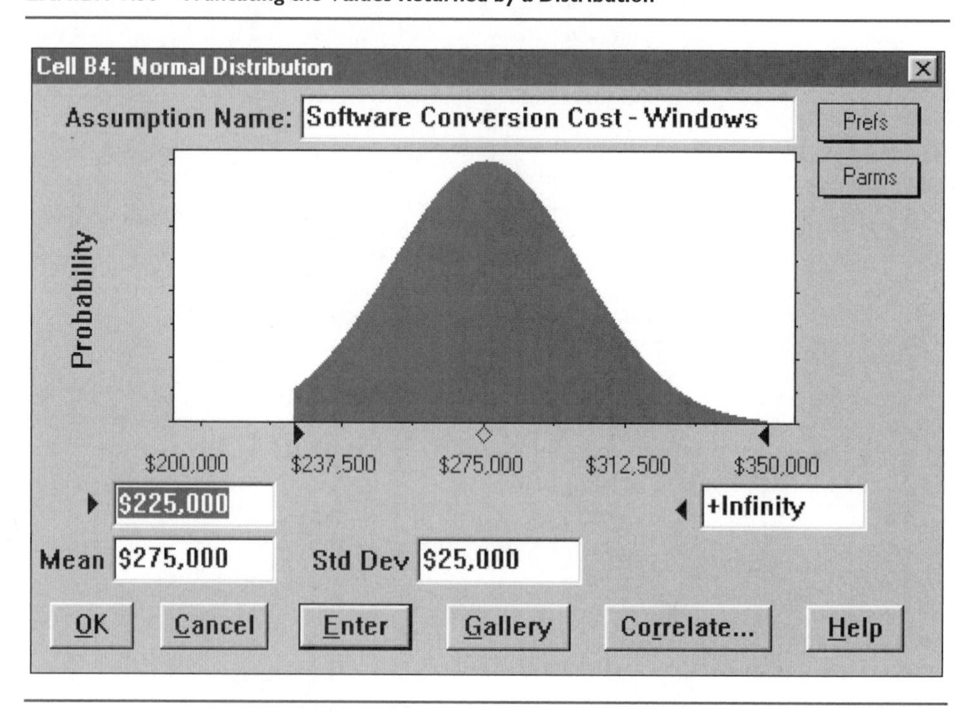

QUESTIONS

1. Simulation is an easy, flexible, powerful tool for helping managers analyze opportunities and problems facing them. Why use other tools if simulation can do the same job, and has better acceptance by management?
2. The point was made in the chapter that simulation is not, per se, a model because models represent reality whereas simulation imitates it. Explain. Does this mean that models are not useful in simulation?
3. Distinguish between internal and external validity. Which one is more important?
4. Sometimes modelers derive functions that imitate the results of complex, expensive, time-consuming simulations. In essence, they simulate the simulation. How might this be done, and what would its value be?
5. How would you design a simulation model to analyze a transient period of a process or system? How would you know if your results were valid?
6. How might a simulation sensitivity analysis differ from, say, a linear programming sensitivity analysis? What about the reasons for doing a sensitivity analysis in the first place?
7. Compare the Tourist Center results that Alisa found in Exhibit 7.2 with those from the simulation later in the chapter. What can you conclude from this?
8. Why conduct deterministic simulations (that is, without any probabilistic variables in the model) if the variables are known with certainty? When might such a model be useful?
9. It is common in risk analysis to consider various managerial policies that might affect the shape of the risk profiles in Exhibit 7.18, such as shifting them to the right or left, or squeezing them into a narrower range or spreading them out. Would a policy change that shifted the cumulative probability curve to the right be more or less desirable? What about a change that squeezed the range closer toward the median? Closer toward the left limit? Toward the right?
10. Reconsider the Banyon State clinic example in Exhibit 7.20. Note that, although the first patient is ready for the doctor 22 minutes after the clinic opens, the doctor doesn't begin work until 45 minutes after the clinic opens. Since this is an overloaded system to begin with (i.e., the arrival rate exceeds the service rate), this 23 minute initial delay will be added to *every* patient's waiting time, plus any additional delay due to normal queuing. Thus, one very simple managerial policy change might be to have the doctor arrive earlier, or if this is impossible, shift the opening and closing time of the clinic to half an hour later. Speculate on how this would change the waiting times for all the patients coming to the clinic.
11. In the Banyon State clinic example of Section 7.7, suppose that the clinic had initially only accepted appointments for every 17 minutes. Would the output rate of the clinic change? What would have been the waiting time of a student in that process? If they then computerized the record retrieval process, what would the waiting time be?

EXPERIENTIAL EXERCISE

If you completed the Experiential Exercise in Chapter 6, skip to Part B.

Part A Select a fast-food restaurant with a single drive-through window. Select either the lunch hour (11:30–12:30) or dinner hour (5:30–6:30) to study. Visit your chosen restaurant on at least two different occasions during your selected time period and record the following information:

- The time between arrivals to the drive-through (*Hint:* It is easier to simply record the time of each arrival and then later calculate the time between arrivals).
- The time it takes for the order to be taken (*Note:* This time begins when a restaurant employee first requests the order and ends when the order has been completed and confirmed).
- The time it takes to deliver the food to the customer (*Note:* This time begins when a restaurant employee first opens the drive-through window and establishes contact with the customer and ends when the customer has completely received the contents of his or her order).

After collecting the data complete the following tasks:

- Perform a chi-square goodness of fit test to verify that the assumptions of exponential interarrival times is valid.

- Perform a chi-square goodness of fit test to verify that the assumptions of exponential service times are valid for both the order-taking times and drive-through window processing times.
- Using the formulas discussed in Chapter 6, calculate the average waiting time, the average waiting time in queue, the average number of customers in the system, the average number of customers in the queue, the probability of an idle facility, and the probability of the system being busy.

Part B Develop a model in Excel to simulate the arrival of 100 customers to the drive-through. Plot your customer interarrival data and service time data and determine an appropriate distribution to use to approximate each time. Validate your chosen distribution using the chi-square goodness of fit test. *Note:* If your chi-square goodness of fit tests indicate that an exponential distribution is a good approximation for customer interarrival times and/or service times you can use the following formula to calculate random numbers from an exponential distribution since Excel's Random Number Generation tool does not include an option for calculating random numbers from an exponential distribution:

–m*ln(rand())

where *m* is the mean of the exponential distribution, ln is an Excel function for calculating the natural logarithm, and rand() is an Excel function that generates a random number between zero and one. Note that the rand() function returns a new random number each time the spreadsheet is recalculated (e.g., copying the formula to a new range of cells and pressing the F9 key). Therefore, one advantage of using the rand() function to generate random numbers over the Random Number Generation tool is that replicating the model is much simpler. As we saw before, replicating a simulation model when the Random Number Generation tool is used requires using the tool to regenerate a new column of random numbers. This can become tedious when there are several columns of random numbers. In the detailed example discussed in Section 7.7, replicating the model 10 times required entering information in the Random Number Generation dialog box 40 times (10 replications × 4 random variables). Pressing the F9 key is quite a bit easier than entering information into a dialog box 40 times!

Using the simulation model complete the following:

- Calculate the average time a customer spends in the system and the average time a customer spends waiting.
- Plot the average time in the system values. How many customers must arrive before the model reaches steady state? Calculate the average time a customer spends in the system after the model reaches steady state. How does this average compare to the average when all 100 customers are included in the average? How do these averages compare to the average you calculated using the queuing theory formula? What are the implications of this?
- Identify a possible improvement that management at the restaurant could make to the drive-through operation. Estimate the impact this change would have (e.g., reduce order-taking time by 15 percent). Replicate (10 times) the simulation model of the current operation and how you would imagine the drive-through would operate if your change were implemented.
- Write a memo to the management of the restaurant you studied. In the memo include a discussion of how the data were collected, how the collected data were analyzed, what key insights were obtained, what recommendation you would make, and key limitations associated with this study.

MODELING EXERCISES

1. The U.S. Department of Agriculture estimates that the yearly yield of limes per acre is distributed as follows:

Yield in Bushels per Acre	Probability
350	0.10
400	0.18
450	0.50
500	0.22

The estimated average price per bushel is $7.20.

   a. Find the expected annual per acre lime crop yield generated over the next 10 years. Use simulation and compare to the theoretical expected value results.
   b. Find the average yearly revenue.

2. The following information is known to you: A rainy day in Paris has a 40 percent chance of being followed by a rainy day. A nonrainy day has an 80 percent chance of being followed by a nonrainy day.

a. Use simulation to predict what the weather is going to be over the next 20 days. (Today is a nonrainy day.)
b. Based on the information collected in part a, estimate the number of rainy days in Paris in one year (365 days).

3. A company has two cars. Car 1 is in use 40 percent of the time and car 2, 30 percent of the time. The president wishes to go somewhere; what is the chance that there will be a car available?
   a. Draw a simulation flowchart.
   b. Manually simulate for 20 periods.
   c. Find the theoretical answer (use probability theory) and compare with the simulation results.
   d. Show graphically the stabilization process for 20 runs. What are your conclusions?
   e. If a company car is unavailable, a cab is used, with an average cost of $15 per ride. Find the annual cost if the president makes 100 trips per year.
4. The B & T car dealership sells 13 cars, on the average, each week. The sales statistics show that of all cars sold, 20 percent are small, 45 percent are of medium size, and the remainder are large. The profit from the sale of the cars and the commission paid are shown in the following table:

Type of Car	Price per Car	Per Car Profit to Dealership	Per Car Commission Paid
Small	$5,200	$310	$62
Medium	6,050	425	70
Large	7,800	500	80

   Draw an influence diagram and find with simulation (simulate for one week):
   a. The size of the last car sold during the week
   b. The commissions paid in an average week
   c. The total dollar sales volume generated in one month (four weeks)
   d. The chance of selling two large cars in a row
5. A newscarrier sells newspapers and tries to maximize profit. The exact number of papers purchased daily by customers can't be predicted. An elaborate time study was performed on the demand each day, and the following table was developed (for 125 days):

Demand per Day	Number of Times
15	10
16	20
17	42
18	31
19	12
20	10
Total	125

   The following ordering policy was used by the newscarrier: The amount ordered each day is equal to the quantity demanded the preceding day. Assume that demand the previous day was 18. A paper costs the carrier 15 cents; the carrier sells it for 30 cents. Unsold papers are returned and the carrier is credited 8 cents per paper out of the 15 cents she paid. An unsatisfied customer is estimated to cost 7 cents in goodwill. Determine the average daily profit if the newscarrier follows the ordering policy. Also determine the average loss of goodwill. Draw an influence diagram and simulate for 15 days.
6. Kojo Corporation stocks motors for their textile machines. The weekly demand for the motors follows the distribution below:

Demand	5	6	7	8
Probability	0.2	0.3	0.4	0.1

   Motors arrive at the end of each week (after the plant is closed for the weekend) either 6 in a package (60 percent chance) or 10 in a package (40 percent chance). Draw an influence diagram to help simulate 15 weeks of operation to find:
   a. The average inventory on hand (at the beginning of the week).
   b. The probability of stockout (in terms of number of times).
   c. The inventory at the end of 15 weeks (before the last shipment arrives).
   d. If the process has stabilized (check for the inventory level situation).
   e. Comment on the existing inventory policy.

   Assume (1) The current inventory, at the beginning of week 1, is 5; and (2) Unsatisfied demand is provided from stock whenever a supply arrives.
7. Customers arrive at a service facility according to a normal distribution with a mean of 5 arrivals per hour and a standard deviation of 1. The service time is always exactly 11 minutes. The facility opens for service at 8 A.M. At the opening time, the first customer is there waiting. Simulate for 15 customers (round to the nearest integer) and find:
   a. Average waiting time in the queue, W_q.
   b. Average time in the system, W.
   c. Average number of customers in the system.
   d. The utilization ratio.
8. Southern Airline has 15 daily flights from Miami to New York. The average profit per flight is $6,000. Each flight requires one pilot. Flights that do not have a pilot are canceled (passengers are transferred to other airlines). Because pilots get sick from time to time, the airline has a policy of keeping three

reserve pilots on standby to replace sick pilots. The probability distribution of sick pilots on any given day is uniformly distributed between 0 and 5. The reserve pilots are assumed to always be available. Draw an influence diagram and use Monte Carlo simulation to simulate 10 days and then find:

a. The average daily utilization of the reserve pilots (in percent).
b. The average daily lost revenue due to canceled flights caused by lack of pilots (in dollars).
c. The chance that one or more flights will be canceled in a day.
d. The utilization of the aircraft (in percent).
e. A standby pilot is paid $2,000 per day. Find the optimal number of standby pilots.
f. Assume now that the standby pilots may get sick too, with the same probability distribution. How should the simulation be conducted? Explain and execute.

9. A special medical diagnosis and treatment machine contains three identical radiation devices that cost $2,000 each. If any of the devices fail, the machine is shut down for one hour and the failed device is replaced. Management considers each hour of downtime as having an opportunity loss of $1,000. The life expectancy of the radiation devices, in hours of operation, is listed below.

Life Expectancy (Hours)	Probability
500	0.05
550	0.08
600	0.12
650	0.15
700	0.21
750	0.14
800	0.10
850	0.07
900	0.05
950	0.03

Management is considering three replacement policies:

a. Replace each device when it fails.
b. Replace *all* the devices whenever *one* fails.
c. Replace a failed device and at that time check the life of the other two. If a device has been in use 850 hours or more, replace it, too.

Replacing one device costs $30 labor and 1 hour downtime. Replacing two devices at a time: $40 labor, 1 hour and 20 minutes downtime. Replacing all three devices at one time: $45 labor, and 1 hour and 30 minutes downtime. Replaced devices with some remaining life are sold to a South American hospital for $15 each, regardless of age.

a. Write a flow chart for this problem.
b. Run each policy for approximately 10,000 hours of operation.
c. What are the results?

10. Conduct a risk analysis for Fiji Corporation's proposed portable word processor. The price can be set at $300 or $400; it is believed that the lower price will increase sales by 30 percent. The monthly fixed cost is $1 million and the variable costs are a function of the number of shifts used. If one shift is used, the variable cost is expected to be either $200 or $250, with a 50 percent chance of each. With two shifts, the variable cost will be either $250 or $270, again with a 50 percent chance of each. With one shift, the monthly maximum volume is 40,000 units; with two shifts, the maximum volume is 60,000 units. Assuming a $400 price, there is an equal chance of achieving either 20,000, 30,000, or 40,000 units of sales per month. Draw an influence diagram and simulate to determine what Fiji's decision should be concerning price and shifts.

11. People arrive at an elevator in groups once every three minutes (there is a pedestrian walkway in front of the building with a traffic light that switches every three minutes). The arrivals have the following frequencies:

Number of People in Group	Probability
3	0.10
4	0.15
5	0.35
6	0.25
7	0.15

The elevator can accommodate at most five persons. It takes the elevator an average of 2 minutes to go up and return to the ground floor if it doesn't go beyond the eighth floor. If it goes beyond the eighth floor, it takes an average of 4 minutes to return. In 75 percent of the trips, the elevator does not go beyond the eighth floor. If a person waits for an elevator and the elevator returns, that person takes it *without* waiting for the next group to arrive.

Draw an influence diagram and simulate for 10 arriving groups until all people are accommodated. The elevator is at the ground floor at the beginning of the simulation and should return to the ground floor at the end of the simulation. Find:

a. The average number of people arriving each time.
b. The average number in an elevator ride.
c. The average time a person has to wait for the elevator.
d. The probability that a person will have to wait for the elevator.
e. The utilization of the elevator.
f. How many trips the elevator will make in one hour.[3]

12. Of all the customers entering Swiss National Bank, 20 percent go to the receptionist, 15 percent to the loan department, 55 percent to the cashier windows, and 10 percent to the credit card department. Of all those who see the receptionist, 50 percent then go to the loan department, 25 percent go to the credit department, and 25 percent leave the bank. Of all those who go to the cashier windows, 80 percent leave the bank and 20 percent go to the loan department. Of all those who go to the credit card department, 50 percent will leave the bank. and 50 percent will go to the cashier windows. Of all those who go to the loan department, 50 percent will go to the cashier windows, 15 percent will go to the credit department, and 35 percent will leave the bank. Simulate the paths of five customers going through the bank. Assume that a customer may return to a service area more than once. Find:
    a. In how many service areas each of the customers will stop.
    b. How many times the first customer will show up at the cashier windows.
    c. The average time at the receptionist's desk is 2 minutes, at the cashier windows 3 minutes, and at the loan and credit card departments 10 minutes. Estimate how the *workload,* in terms of minutes of service, is divided among the four service areas.
    d. How many visits were paid, in total, to the cashier windows?
    e. The floor area of the bank is 5,000 square feet. How would you allocate this area to the various service areas if the area needed is considered proportional to the number of visits?

13. Pump-It-Yourself, Inc., is an independently owned and operated service station. As such, it does not receive regular deliveries from the local wholesale gasoline distributorship, which is controlled by a major oil company. Instead, the station has the opportunity of purchasing any or all of the excess gasoline remaining on the distributor's truck after all regular deliveries have been made each day. There is no delivery charge if all the gasoline remaining in the truck is taken, but if not, there is a back-shipping charge. The delivery truck stops each day *after* the station has closed for business. The amount of gasoline remaining on the truck is a random variable with the following distribution:

Gallons Remaining	Probability
0	0.20
500	0.25
1,500	0.30
2,500	0.15
3,500	0.10

    The station has a 3,000-gallon storage capacity, and the demand for gasoline is normally distributed with a mean of 1,800 gallons and a standard deviation of 500 gallons. Assume that unsatisfied demand is lost (customer goes to a competitor). Simulate the activities involving the supply and demand of gasoline for 12 days. Start with an inventory on hand of 2,100 gallons. Given:

    Profit from a gallon sold = 20 cents
    Cost of ill will = 5 cents per gallon
    Cost of shipping back = 2 cents per gallon

    Draw an influence diagram and simulate to determine:
    a. The average daily demand for gasoline (compare to the "expected value").
    b. The percentage of time that storage capacity will not be sufficient for taking all gasoline from the truck.
    c. The percentage of time that demand cannot be met.
    d. The average daily beginning inventory.
    e. The average daily ending inventory.
    f. The average daily loss of unmet demand.
    g. The percentage of times that sales equaled demand.
    h. The daily net profit.

14. Dr. Z has an appointment schedule where patients are scheduled to come every 20 minutes. The office is open from 9 A.M. to noon, four days a week. The last patient is scheduled at 11:40 A.M. each day. Assume that patients arrive exactly on time. The time required for treatment or examination is distributed as follows:

Minutes	Percent
10	14
15	25
20	41
25	20

The doctor will see all patients that are scheduled. Simulate for 2 days to find:

a. The average waiting time (prior to treatment) per patient (in minutes).
b. The average utilization of the physician (percentage that working time is of the total time in her office). *Note:* If the physician completes her examinations before noon, she will stay in the office until noon.
c. The average overtime *(in hours)* worked by the physician each week. Overtime is considered any time beyond noon.
d. The length of the last treatment on the second day (in minutes).
e. The average number of breaks the physician will have in a day between seeing patients.
f. The exact time the doctor will finish the treatment of the last patient on the second day.

15. Mexivalve, Inc. produces valves on a weekly schedule. Shipments are made to two customers, A and B. The customers enter orders by phone every Friday, after their weekly maintenance inspection is completed, and they want immediate delivery. Past experience indicated the following demand pattern: customer A orders either 15 valves (35 percent of the time) or 20 valves (65 percent of the time). Customer B orders 25 valves in 20 percent of the cases, 30 valves in 40 percent of the cases, and 35 valves in the remaining cases.

    The company would like to be able to meet all demand at least 95 percent of the time. The production manager thinks that he can do it with a weekly production schedule of 48 valves. Valves not demanded on Friday are used as safety stock, which currently stands at five units. Demand that cannot be met from production is met from the safety stock. Demand that cannot be met at all is considered a lost opportunity. Draw an influence diagram to simulate for 12 weeks and find:

    a. Is the production level of 48 sufficient to meet the company's service policy? Why?
    b. What is the average weekly profit if one valve brings 275,000 pesos?
    c. What will be the safety stock at the end of the 12th week?
    d. The average weekly number of valves demanded that are considered a lost opportunity.

16. Demand for a perishable liquid product is known to be normally distributed, with an average daily demand of 23 gallons and a standard deviation of 4 gallons.

    a. Generate demand for 10 days.
    b. Round the average daily demand found in part a to one decimal point (e.g., 25.2). Assume that an inventory of 23.6 gallons is being kept daily. If demand in a given day is more than the inventory on hand, the company incurs a loss of \$100 for each gallon short (proportion of \$100 for fraction of a gallon). If there is some left over, the company's demurrage is \$120 per gallon (proportion for a fraction). Find the average daily profit (loss) if each gallon sold contributes \$50 to profit.
    c. Based on the result of part b, estimate the chance of not meeting the demand on any specific day.

17. A survey of 100 arriving customers at a drive-in window of Northeastern Bank resulted in the following information:

Time Between Arrivals (Minutes)	**Frequency**
0.5	6
1.0	10
1.5	15
2.0	18
2.5	20
3.0	15
3.5	12
4.0	4
	100

The service time (from the time the car enters the service position until the car leaves) was distributed as follows:

Service Time (Minutes)	**Frequency**
1.0	12
1.5	18
2.0	30
2.5	25
3.0	15

Generate 12 arrivals and find:

a. The average waiting time per customer.
b. The utilization (in percent) of the drive-in window.

18. Moscow University has an information center staffed by one employee. Historical data indicate that people arrive at the center according to the following distribution:

Interrival Time (Minutes)	Frequency
4	0.223
5	0.481
6	0.275
7	0.021

The time of service is normally distributed with an average of five minutes and a standard deviation of one minute. The center opens at 8:30 sharp. Simulate for 12 customers (including the first one). Note:

a. Round the minutes to the nearest whole minute.
b. Round all other numbers to the nearest possible figure.
c. End the simulation after the last customer is served.

Draw an influence diagram and find:

a. The utilization of the center.
b. The average number of customers in the center (waiting and/or being served).
c. The probability that a customer will have to wait.
d. The probability that a customer will be in the center more than five minutes in total.
e. The average waiting time, W_q, in the queue.

19. The demand for Orange Microcomputer II for the next 10 weeks is known (from existing orders) as 520, 314, 618, 240, 590, 806, 430, 180, 300, 250. It takes 2 weeks to receive micros from the factory. The current inventory on hand is 600; an additional 400 will arrive next week. Ordering costs are $300 plus $5 per unit. Carrying cost is $10 per unit per week. Shortage cost is $30 per unit short (special rush shipment). The existing ordering policy is to order 500 units whenever the inventory is 100 units or less.

a. Draw an influence diagram and simulate for a period of 10 weeks.
b. Calculate the average inventory and shortage cost per week.
c. Design a better ordering rule and find how much money you can save.

20. In this chapter random numbers were generated primarily through the use of Excel's Random Number Generation tool. There are a number of other ways random numbers can be generated. For example, to generate random numbers from a normal distribution with a mean of μ and standard deviation of σ, the following formula could be used:

=NORMINV(RAND(),μ,σ)

Referring to Exercise 16, generating normally distributed demands with a mean of 23 and standard deviation of 4 could be accomplished as follows:

=NORMINV(RAND(),23,4)

This formula could then be copied to other cells as needed, and each cell would contain a unique random number drawn from a normal distribution with a mean of 23 and standard deviation of 4. One key advantage of generating random numbers in this fashion is that replicating the simulation model requires only a single keystroke. Namely, pressing the F9 key automatically recalculates all formulas, thereby generating a new set of random numbers. Referring to the situation in Exercise 16, develop a simulation model that uses equations to generate the random daily demands. Then use the F9 key to replicate the model 15 times.

21. Referring to Exercise 8, develop 95 percent confidence intervals for the daily utilization of reserve pilots and the average daily lost revenue due to canceled flights caused by lack of pilots. Explain the value of using these confidence intervals beyond simply reporting the point estimates.

22. Referring to Exercise 13, calculate the 95 and 99 percent confidence intervals for the average daily beginning inventory, the average daily ending inventory, the average daily loss of unmet demand, and the daily net profit. Why is one set of confidence intervals narrower than the other set? Which set of confidence intervals would you recommend management use?

23. Referring to Exercise 14, calculate the 95 percent confidence intervals for the average waiting time (prior to treatment) per patient (in minutes), the average overtime (*in hours*) worked by the physician each week, and the length of the last treatment on the second day (in minutes). Explain in language that a typical physician could understand the interpretation of these confidence intervals.

24. Referring to Exercise 8, simulate the operation of the airline for 100 days to investigate two reserve pilot policies: (1) the airline utilizes 2 reserve pilots, and (2) the airline utilizes 4 reserve pilots. Based on an analysis of variance, is one of these policies statistically better than the other?
25. Referring to Exercise 9, use analysis of variance to compare the three replacement policies.

CASES

Medford Delivery Service

Medford Delivery Service (MDS) operates delivery vans in major airport hubs to deliver airfreight from the airport to local businesses on an express basis. In September, Sean Miller, operations manager for MDS received an urgent request from Erin Carmel, traffic manager at the Chicago hub to add additional dispatchers to accommodate the growing demand they were experiencing. With only the two current dispatchers, delivery van drivers were waiting too long to receive their next delivery assignments. It not only wasted their time but also delayed deliveries to valuable customers.

The delivery operation at MDS was standardized at every hub: drivers would return from a delivery, report in on a terminal that recorded the time and disposition of the delivery, and then report to the dispatch office for the next delivery assignment. The dispatcher would assign the driver a delivery and report it on the terminal, which also recorded the time of the dispatch.

Erin had collected the following data to justify her request: The Chicago hub maintained 30 delivery vans for covering the greater Chicago region and the average duration of a delivery was two hours from dispatch to return. The drivers' hourly arrival rate back to the dispatcher's office was normally distributed with a mean of 15 and a standard deviation of 4; however, the arrivals within any hour were completely random. The dispatcher service time was also random and normally distributed with a mean of 8 minutes and a standard deviation of 3 minutes, but never less than one full minute. The wage rate of the dispatchers was $36 an hour and $30 an hour for drivers. However, the revenue generation capacity of a van with a driver was estimated to be $90 per hour.

Find the best number of dispatchers for Chicago. Draw an influence diagram and simulate the process for 2 hours. Use only the waiting times for the second hour when the system has (we will assume) reached steady state. Round all times to whole minutes. First determine the average number of arrivals in each of the 2 hours. Next determine the exact minute of arrival by multiplying the random number by 60/100 and then order the arrivals according to their arrival times.

Warren Lynch's Retirement

Since graduating from college, Warren Lynch has been teaching history for 18 years at Graham High School. As a small rural high school, Graham High does not offer its faculty a retirement plan; however, it does provide decent medical benefits. Warren's current annual salary is $32,300. His school district has prided itself on being able to consistently increase faculty and staff salaries at the same rate as inflation and thereby maintain their purchasing power. Warren and his wife, Samantha, have no children and do not plan on having any in the future. Samantha does not work outside of the home, aside from the considerable amount of volunteer work she performs. Warren loves his job and has turned down offers to move into administration, despite the opportunity to substantially increase his salary.

Warren does not have any savings set aside for his retirement. However, having recently turned 40, Warren has determined that it is now important to take a more active and formal approach to his retirement planning. His goal is to retire in 25 years at the age of 65.

Warren would like to develop a spreadsheet that will allow him to evaluate various alternative savings strategies. Based on some initial research, he has decided that he will only invest in an S&P 500 stock index fund up until the time he retires. Upon retirement, Warren has identified two possible investment strategies that he would likely pursue: maintain 80 percent of his portfolio in the S&P 500 stock index fund and invest the other 20 percent in a long-term government bond index fund, or maintain 20 percent in the S&P 500 stock fund and invest the other 80 percent in a long-term government bond fund. Exhibit A contains the total annual returns for the S&P 500 and long-term government bonds for the period 1926 to 1997.

In addition to deciding how his money will be invested, Warren also realized that he needs to determine how much money he will save each year prior to his retirement and then once retired, how much money he will withdrawal annually. In terms of savings, Warren decided that he would like to investigate the impact of saving 8 percent and 10 percent of his gross

EXHIBIT A
Total Annual Returns— S&P 500 and Long-Term Government Bonds

Year	Total Annual Return S&P 500	Total Annual Return Long-Term Government Bonds	Year	Total Annual Return S&P 500	Total Annual Return Long-Term Government Bonds
1926	11.62%	7.77%	1962	−8.73%	6.89%
1927	37.49	8.93	1963	22.80	1.21
1928	43.61	0.10	1964	16.48	3.51
1929	−8.42	3.42	1965	12.45	0.71
1930	−24.90	4.66	1966	−10.06	3.65
1931	−43.34	−5.31	1967	23.98	−9.19
1932	−8.19	16.84	1968	11.06	−0.26
1933	53.99	−0.08	1969	−8.50	−5.08
1934	−1.44	10.02	1970	4.01	12.10
1935	47.67	4.98	1971	14.31	13.23
1936	33.92	7.51	1972	18.98	5.68
1937	−35.03	0.23	1973	−14.66	−1.11
1938	31.12	5.53	1974	−26.47	4.35
1939	−0.41	5.94	1975	37.20	9.19
1940	−9.78	6.09	1976	23.84	16.75
1941	−11.59	0.93	1977	−7.18	−0.67
1942	20.34	3.22	1978	6.56	−1.16
1943	25.90	2.08	1979	18.44	−1.22
1944	19.75	2.81	1980	32.42	−3.95
1945	36.44	10.73	1981	−4.91	1.85
1946	−8.07	−0.10	1982	21.41	40.35
1947	5.71	−2.63	1983	22.51	0.68
1948	5.50	3.40	1984	6.27	15.43
1949	18.79	6.45	1985	32.16	30.97
1950	31.71	0.06	1986	18.47	24.44
1951	24.02	−3.94	1987	5.23	−2.69
1952	18.37	1.16	1988	16.81	9.67
1953	−0.99	3.63	1989	31.49	18.11
1954	52.62	7.19	1990	−3.17	6.18
1955	31.56	−1.30	1991	30.55	19.30
1956	6.56	−5.59	1992	7.67	8.05
1957	−10.78	7.45	1993	9.99	18.24
1958	43.36	−6.10	1994	1.31	−7.77
1959	11.96	−2.26	1995	37.43	31.67
1960	0.47	13.78	1996	23.07	−0.93
1961	26.89	0.97	1997	33.36	15.85

Source: "Stocks, Bonds, Bills and Inflation," Ibbotson and Associates, 1999.

salary annually. In terms of planning his withdrawals, Warren discovered after visiting a number of Web sites that retirees often withdraw from 4 percent to 8 percent of their savings in their first year of retirement and increase this amount each year based on the rate of inflation. Therefore, Warren decided to investigate the scenarios that he either withdraws 4 percent or 6 percent of his savings in his first year of retirement. In both cases, withdrawals in subsequent years would be updated based on the assumed or calculated rate of inflation. Exhibit B contains the percentage change in the consumer price index from year to year over the period of 1914 to 1998.

Finally, in developing his spreadsheet model, Warren realized that he needed to make some assumptions about how long his money would need to last. After

EXHIBIT B
Percentage Change in the Consumer Price Index

Year	CPI (% Change)	Year	CPI (% Change)	Year	CPI (% Change)
1914	1.0	1943	6.1	1971	4.4
1915	1.0	1944	1.7	1972	3.2
1916	7.9	1945	2.3	1973	6.2
1917	17.4	1946	8.3	1974	11.0
1918	18.0	1947	14.4	1975	9.1
1919	14.6	1948	8.1	1976	5.8
1920	15.6	1949	−1.2	1977	6.5
1921	−10.5	1950	1.3	1978	7.6
1922	−6.1	1951	7.9	1979	11.3
1923	1.8	1952	1.9	1980	13.5
1924	0.0	1953	0.8	1981	10.3
1925	2.3	1954	0.7	1982	6.2
1926	1.1	1955	−0.4	1983	3.2
1927	−1.7	1956	1.5	1984	4.3
1928	−1.7	1957	3.3	1985	3.6
1929	0.0	1958	2.8	1986	1.9
1930	−2.3	1959	0.7	1987	3.6
1931	−9.0	1960	1.7	1988	4.1
1932	−9.9	1961	1.0	1989	4.8
1933	−5.1	1962	1.0	1990	5.4
1934	3.1	1963	1.3	1991	4.2
1935	2.2	1964	1.3	1992	3.0
1936	1.5	1965	1.6	1993	3.0
1937	3.6	1966	2.9	1994	2.6
1938	−2.1	1967	3.1	1995	2.8
1939	−1.4	1968	4.2	1996	3.0
1940	0.7	1969	5.5	1997	2.3
1941	5.0	1970	5.7	1998	1.6
1942	10.9				

Source: U.S. Department of Labor, ftp://ftp.bls.gov/pub/special.requests/cpi/cpiai.txt, November 12, 1999.

pondering this issue for a while, he decided to simply investigate three scenarios: that he and/or Samantha would live (1) 20 years past his retirement, (2) 25 years past his retirement, or (3) 30 years past his retirement.

Questions

1. Develop an influence diagram that will help show Warren the relationship among the relevant factors.
2. For each possible scenario, determine the probability that Warren and his wife will have sufficient money to retire on. Explicitly state any assumptions you make related to the timing of cash flows.
3. Why is it particularly appropriate in this situation to use simulation, as opposed to simply using the average rates of return and average rate of inflation?
4. What are the key limitations of your analysis? What if anything can be done to overcome these limitations?

Cartron, Inc.

Cartron, Inc. has several plants located throughout the world and produces car radios for many of the major automobile manufacturers. Its North Carolina plant produces a particular radio model that is used in one of the best-selling minivans in North America. The plant currently delivers these radios on a just-in-time basis directly to the minivan final assembly plant, which also happens to be Cartron's largest customer. Currently, Cartron's plant is running at capacity and is having difficulty keeping up with orders. This has resulted in a fair amount of tension between Cartron and its largest customer. In essence, the minivan producer does not like to risk having to shut down its entire assembly line because of a shortage of car radios.

Cartron is investigating a number of options to quickly increase the capacity of its operations in North Carolina. Although expanding the plant is certainly an option, this will do little to alleviate the problem in the short run. A detailed study of the car radio assembly line revealed that the bottleneck operation was final inspection. The final inspection operation is a combination of machine-paced and human-paced activities. The machine-paced activities are computer controlled and therefore exhibit minimum variability. In addition to these machine-paced activities, the final inspection workers perform a number of supplementary functional and cosmetic inspections according to a documented ISO 9000 inspection procedure. In total, the inspection procedure requires 130 distinct evaluation criteria and operational steps.

A large amount of data were collected over a several month period related to task times at the inspection station. More specifically, data were collected on 148 workers who inspected more than 175,000 radios. Time series models were then fit to the data for each worker to try to identify the factors that impacted the workers' productivity. One interesting result that emerged was that the workers' productivity levels often decreased dramatically whenever the workers returned to the final assembly station after being assigned to another activity. Apparently, the workers were forgetting how to perform the inspection tasks after being assigned to another activity.

To reduce the effect of worker forgetting, plant management identified a new digital video-based training and information delivery system. According to the vendor of this system, this technology would reduce worker forgetting by delivering video-based task-specific content on demand to computer terminals located at each inspection station. Plant management speculated that this system could help increase the rate at which workers learned the inspection task and could help reduce the negative impact associated with worker forgetting.

To investigate the potential of this rather expensive technology, a simulation model of the final inspection operation was developed. Thirty replications of the model were run under two scenarios for an entire year. The first scenario corresponded to observed patterns of learning and forgetting exhibited by the workers during the study period. In the second scenario, adjustments were made to the observed patterns of learning and forgetting based on assumptions regarding the implementation and use of the video-based technology. The primary performance measure used in the simulation study was the number of radios completed during the year. The results of the simulation study are summarized in Exhibit C.

Questions

1. What would you recommend Cartron do regarding the video-based delivery system? Why?
2. What assumptions was your analysis based on?
3. What are the implications for future simulation studies given the required assumptions?

EXHIBIT C
Simulation Results

Replication	Finished Radios—No Change in Technology	Finished Radios—New Video System
1	466,876	470,128
2	458,423	463,355
3	472,425	474,524
4	450,058	457,016
5	426,755	440,278
6	403,832	415,261
7	496,066	499,746
8	510,435	515,963
9	497,084	499,230
10	447,974	449,064
11	463,227	464,708
12	397,828	398,799
13	436,129	436,740
14	472,404	473,479
15	495,743	496,887
16	439,863	446,047
17	468,680	476,488
18	485,304	498,968
19	509,937	510,326
20	474,365	475,095
21	485,437	488,014
22	477,346	479,088
23	439,245	442,961
24	419,772	421,354
25	458,968	461,281
26	471,704	476,120
27	476,226	477,494
28	486,806	490,294
29	459,840	466,263
30	485,918	489,511

ENDNOTES

1. Recently, however, there is a trend toward incorporating optimization techniques within simulation software programs. For example, Decisioneering has developed OptQuest as a compliment to its simulation program Crystal Ball 2000 (see chapter appendix). Briefly, OptQuest uses search techniques to find the best values for sets of variables included in a simulation model.
2 If Crystal Ball has been installed on your computer but is not running, select Tools and then Add-Ins . . . from Excel's menu. Next, click on the Crystal Ball checkbox and select OK. If the Crystal Ball Add-In has not been installed on your computer, consult your Excel manual and the CD-ROM that accompanies this book to install it.
3. Exercise 11 was developed by Professor Dieter Klein, Worcester Polytechnic University, Worcester, Mass.

BIBLIOGRAPHY

Banks, J., J. S. Carson, B. C. Nelson, and D. M. Nicol. *Discrete Event System Simulation.* Upper Saddle River, NJ: Prentice Hall, 2000.

Banks, J. (ed.) *Handbook of Simulation: Principles, Methodology, Advances, Applications, and Practice.* New York: Wiley, 1998.

Bernstein, R. S., D. C. Sokal, S. T. Seitz, B. Auvert, J. Stover, and W. Naamara. Simulating the Control of a Heterosexual HIV Epidemic in a Severely Affected East African City. *Interfaces,* 28:3, 1998, 101–126.

Bertsimas, D., and R. M. Freund. *Data, Models, and Decision: The Fundamentals of Management Science.* Cincinnati, OH: South-Western, 2000.

Camm, J. D., and J. R. Evans. *Management Science and Decision Technology.* Cincinnati, OH: South-Western, 2000.

Crystal Ball 2000 User Manual. Denver, CO: Decisioneering, Inc., 2000.

Evans, J., and D. Olson. *Introduction to Simulation and Risk Analysis.* Upper Saddle River, NJ: Prentice Hall, 1998.

Forrester, J. W. *World Dynamics.* Cambridge, MA: Write-Allen Press, 1971.

Hammond, J. *The Beer Game.* HBS Case #9-694-104, Boston: Harvard Business School Publishing, 1994.

Henshaw, R. C., and J. R. Jackson. *The Executive Game.* 5th ed. Homewood, IL: Irwin, 1990.

Law, A. M., and W. D. Kelton. *Simulation Modeling and Analysis.* 3rd ed. Burr Ridge, IL: Irwin/McGraw-Hill, 2000.

Lembersky, M. R., and U. H. Chi. Decision Simulators Speed Implementation and Improve Operations. *Interfaces,* July–August, 1984, 1–15.

Martin, E., IV. Centralized Bakery Reduces Distribution Costs Using Simulation. *Interfaces,* 28:4, 1998, 38–46.

Mollaghasemi, M., K. LeCroy, and M. Georgiopoulos. Application of Neural Networks and Simulation Modeling in Manufacturing System Design. *Interfaces,* 28:5, 1998, 100–114.

Pidd, M. *Computer Simulation in Management Science.* 4th ed. New York: Wiley, 1998.

Ragsdale, C. T. *Spreadsheet Modeling and Decision Analysis.* 3rd ed. Cincinnati, OH: South-Western, 2001.

Toland, R. J., J. M. Kloeber, Jr., and J. A. Jackson. A Comparative Analysis of Hazardous Waste Remediation Alternatives. *Interfaces,* 28:5, 1998, 70–85.

Winston, W. L. and S. C. Albright. *Practical Management Science.* Pacific Grove, CA: Duxbury, 2001.

Zeigler, B. P. et al. *Theory of Modeling and Simulation: Integrating Discrete Event and Continuous Complex Dynamic Systems.* New York: Academic Press, 2000.

8 Implementation and Project Management

INTRODUCTION

RecSys Inc. is a manufacturer of high-quality residential play equipment. A typical play system includes swings, ladders, monkey bars, slides, covered play decks, and numerous other accessories. Because of the large number of options and base models, thousands of potential configurations are possible. This greatly complicates the task of planning and controlling the manufacturing plant where the play equipment is produced. In fact, recently, RecSys has been having difficulty meeting its promised delivery time of 3 weeks to its dealers, with many of its orders taking 4 to 6 weeks to reach the dealers. The dealers have begun complaining that this is having a significant impact on customer satisfaction and believe that if the situation persists it will ultimately result in lost business due to unfavorable word of mouth, as well as potentially encouraging more competition.

To address these concerns, the plant manager hired Kathrayn Rand, a recent MBA graduate, to analyze the situation further and identify opportunities to improve the plant's on-time delivery performance. To help better understand and define the problem, Kathrayn began developing an influence diagram. Her progress to date is shown in Exhibit 8.1.

Kathrayn's most immediate issue is that the plant manager has asked her to estimate how long it will take her to analyze the situation and develop a set of recommendations to resolve the problem. As Kathrayn looked over her influence diagram, she started to become a bit overwhelmed by the challenge confronting her. Where should

EXHIBIT 8.1 Influence Diagram for RecSys, Inc.

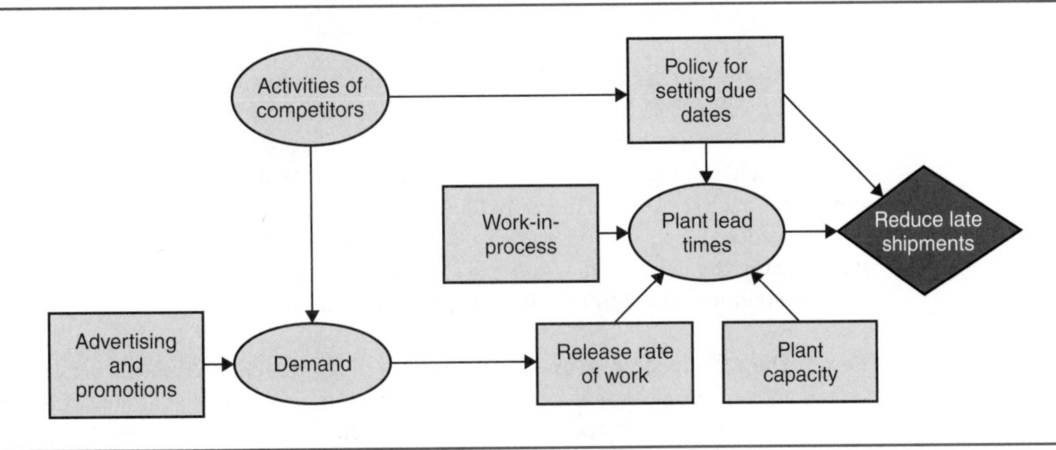

she start? Who should she talk to? What data are available? How much time should she tell the plant manager she needs to complete this study? What other resources will be required?

The purpose of this chapter is to illustrate how a number of project management tools and techniques can be used to help individuals like Kathrayn plan, schedule, monitor, and control what appears to be a complex, nonroutine, and unstructured set of tasks. Furthermore, you will discover that while a number of specialized models have been developed for project management, several of the other modeling approaches discussed earlier in this book (including optimization and simulation) also are quite useful in managing projects. We will return to Kathrayn's project in the detailed modeling example at the end of this chapter after better acquainting you with some of the models and techniques used to manage projects.

8.1 Implementation and Project Modeling

This chapter addresses the fifth step of the QBM process: implementation. In addition to a discussion of the general implementation process, we describe another QBM tool that has been found immensely helpful for achieving limited managerial objectives—project management. Although we use it here for the purpose of implementation, project management is just like the other quantitative models we have discussed in previous chapters in that it follows the same kind of modeling process, as we will describe. Of course, project management has a multitude of applications from new product development to moon landings to theatrical productions to political campaigns. However, its implementation in the context we describe here will result in the final realization of the QBM study.

The Project Modeling Process

The five steps described here apply to the modeling process for project management.

Step 1: Opportunity/Problem Recognition This step in our case is direct and is based on management's recognition that the analysis of a previously identified problem or opportunity is best organized and structured as a project. In managing projects, the primary objective is to achieve the project's goals, on time and on budget. For example, referring to the chapter opener, the objective of Kathrayn's project would be to identify ways to eliminate late shipments and for Kathrayn to accomplish this assignment within a specified time frame and with the resources allocated.

Step 2: Model Formulation The project model is constructed by laying out the tasks that need to be achieved to implement the QBM study, known as the work breakdown structure (WBS). In addition, the times required to achieve these tasks must be known or estimated, and the sequence of the tasks must be identified. In some cases, the use of resources for each task, and especially those resources required to expedite each task, may also be needed. Using this information, a variety of models are constructed, the most important being the network diagram and the Gantt chart. These two models are then used to plan, monitor, and control the implementation of the study.

Step 3: Data Collection The data items needed for the project models are largely described above. The WBS is derived by laying out the major tasks and then working with individuals familiar with these tasks to break them down into finer and finer subtasks until work elements at the individual project team member level are identified. It is commonly the case that exact estimates of the task times are unavailable, so either estimates or estimated distributions are used instead. The distributions are only approximate and are derived from the team member's estimate of an "optimistic" time and a "pessimistic" time, as well as a "most likely" time. The two extreme times are usually given as "one chance in a hundred" of occurring and, typically, the pessimistic time is much greater *above* the most likely time than the optimistic time is *below* the most likely time. That is, the distribution of task times is not symmetrical, but skewed.

Task sequences and resource needs are usually more easily specified. Even the resources required to expedite each task are fairly easily estimated. Considerable additional data may also be used in the project management process. Some examples are a linear responsibility chart showing who is responsible for what tasks and who else needs to be involved, task accounting information to construct an earned value chart to help monitor progress, and what other projects requiring the same resources might be ongoing and could potentially delay the project. (For more information on these aspects of project management, refer to Meredith and Mantel 2000.)

Step 4: Model Analysis Analyzing the network and Gantt models is relatively straightforward and will be described in detail in later sections. However, when massive, or even just large, projects are being analyzed, the manual approach becomes tedious and the use of software packages is recommended. Later in this chapter we will illustrate some examples of Microsoft Project 2000 printouts. For more information on this package and its application see Lowery and Stover (2001) and Mantel et al. (2001). However, other excellent packages also exist, some of which are specialized for certain situations such as construction projects, working with large amounts of limited resources, or handling multiple projects at the same time.

Step 5: Implementation The project model is implemented through the project management process. As will be illustrated in the following sections, project management is often divided into three phases: planning (model formulation and data collection), scheduling (model analysis), and monitoring and control (implementation). This last phase is the execution stage of the project where action is initiated. Our generic discussion of implementation below relates primarily to this phase of the project.

Structure of the Chapter

We begin with a general, but limited, discussion of implementation. We discuss two sets of issues: the rational aspects of implementation, and the "soft" issues. The rational aspects include such things as costs and benefits. The soft issues include many of the standard aspects of project management: organizational considerations, top management support, fear of change, peer pressure, communication, cognitive style, and so on. Following this, we move into project management, explaining its role in implementation, some definitions, and the different project models such as PERT and CPM. Then we go into detail describing how to plan the project, schedule the project, and monitor and control the project. Finally, we return to the chapter opener in the detailed modeling example, Section 8.6.

8.2 Implementing the Modeling Study

Implementing the results of a QBM study can be a long and complex process, even more complex than conducting the study itself. It involves rational elements such as costs and possible benefits, but it also involves more subtle, or "soft," issues such as communication, behavior, power, and politics. The best way to treat implementation is not as an afterthought following the QBM study, but rather, as a continuing step in the QBM process, possibly the *most important* step. Thus, it is wise for the modeler to consider implementation right at the beginning of the study, and to keep it in mind as the study progresses and changes.

Oftentimes the goal of a study will evolve as the work progresses, or delays and speedups will occur without expectation. Typically, the study grows in size and complexity. As things shift during the execution of the study, it is important to keep in mind how these changes will affect the eventual implementation of the results. For instance, it is common for a study to experience *scope creep,* as it is known, where the objectives of the study keep expanding due to new information or the needs of the client, but sometimes without a concomitant increase in the budget or resources to achieve the new objectives. If the modeler keeps eventual implementation in mind, it will become clear that completing the study is going to require additional cost, and may make the eventual implementation not worth the additional costs. Thus, these kinds of changes should always be evaluated in terms of their impact on the ultimate success of the implemented study.

In this section, we discuss some of these issues concerning implementation. We start with a discussion of some of these soft issues and then move to the more rational aspects—but only briefly, since entire books have been written on these topics.

Soft Aspects

Implementing the results of a study involves people taking action. But dealing with people can be the most difficult aspect of any study, involving cooperation, coordination, and continuous communication. It involves all the issues of project management (Meredith and Mantel 2000): selection of a credible and competent project manager, communication across all levels of the organization as well as with clients, dealing with conflict, negotiating with other managers and senior administrators, and so on. It also involves a variety of issues having to do with the organization as a whole, the environment, senior management, operating management, and day-to-day work groups.

The chances for success in implementing a QBM project depend to a great extent on the organization's *implementation climate.* This involves both the attitudes of the members of the organization and the setting, or "environment," that acts as a stage for the implementation of the study. Some organizations tend to study many issues but don't implement the results very often, or possibly with much success. Others never study something they don't plan to implement, and take great pride in the implementation success of their projects.

Another organizational aspect of project implementation is that of **organizational validity**—the concept that for a QBM project to be implemented, it must be compatible with, or "fit," the organization. This fit must occur at three levels: individual, small group, and organizational. If a project requires an extraordinary amount of change in individual attitudes, small-group dynamics, or organizational structure, then the probability of successful implementation will be reduced.

Top management support is also crucial for successful implementation of a QBM project. Support is demonstrated by the status and backing given by senior management to the project, by the resources allocated to it, and their willingness to wait for results. Without sufficient top management support, the cooperation of operating functional managers will be difficult to obtain and the fate of the project will be in jeopardy.

One might suspect that the quantitative nature of QBM would be a problem in communicating the results of a study to management because they would have to accept the study results largely on faith. To that extent, if the modeler uses simpler or more easily understood models, like simulation, the communication difficulty is lessened. However, experimentation over the years indicates that communication *per se* does not appear to be a major barrier to implementation. Of course, poor communication does not help implementation; but even when study recommendations were made in terms that the manager could clearly understand, the project was still often not implemented.

Frequently, poor communication takes the blame for a more basic reason for implementation failure: a difference in **cognitive styles** between the manager and the analyst. Cognitive style refers to the fact that people differ in the approach they take in addressing opportunities and problems. The differences in style may reflect underlying differences in personality, education, culture, or combinations of such traits. Two basic styles are relevant to our discussion here: Some people are *analytically oriented* and tend to address situations by seeking out cause–effect relationships in a step-by-step manner. Others are *intuitively oriented* and tend to utilize experience, common sense, and subconscious feelings in arriving at solutions.

While it is difficult to generalize, it is probably true that, on the whole, modelers tend to be analytically oriented while managers incline toward the intuitive end of the scale. This being the case, one would imagine that the manager and the modeler might have difficulty working together because the way in which they analyze situations is different. Thus, to the extent possible, study results and implementation proposals should be couched in the manager's cognitive style, amplifying the intuitive meanings of the study results and tying them to other experiences of the manager.

The implementation of a QBM study can be viewed as an introduction of change into an organization. The change can be social, technical, psychological, structural, or a combination of all these. Thus, if a manager resists the logical arguments presented in support of a QBM proposal, he or she might not be resisting the technical aspects of the proposed change as much as the perceived social or job-related ramifications. This fear of change may originate from various sources: that the job will be eliminated, that previous performance will be proven inefficient relative to the new approach, or that there will be a downgrading in the status or satisfaction of the job.

Large organizations especially tend to resist change. As the organization grows, relationships tend to stabilize, authority tends to become permanent, and a sense of security sets in. The possibility of change upsets this equilibrium and naturally arouses forces, including office politics, to oppose the change. The prevalence of politics in organizations is often underestimated. Some people ignore politics and may get hurt by their ignorance of what is happening around them. A few others try to use politics to further their own agenda. In most organizations it is not possible to take a neutral stance on all issues. Thus, the modeler is advised to take a defensive position by becoming involved, learning the rules, and determining what is going on and where the power centers are.

Few factors have a more pronounced influence on an employee's behavior than the pressures of formal and informal work groups. This is as true at the managerial level, where these groups tend to take on the characteristics of political coalitions, as it is in the lower echelons of the organization. Work groups can influence employee behavior in a variety of ways. Furthermore, group pressure is an important determinant of its members' output. In particular, there is a significant amount of pressure for conformity.

One of the best ways to encourage implementation is via participation of the people involved. The importance of participation in achieving significant change has been demonstrated through many research studies such as the classic Coch and French (1948) experiment. The researchers worked with four comparable groups of clothing factory operators. The groups differed mainly in the method by which they were exposed to a change in work procedures. Coch and French reported a marked difference between the results achieved by the methods of introducing the change. The most striking difference was between the "no participation" group and the "total participation" group. Immediately after the change was implemented, the output of the first group dropped to about two-thirds of its previous level and resistance developed almost immediately after the change occurred. Marked expressions of aggression against management occurred (e.g., conflict with the methods engineer, hostility toward the supervisor, deliberate restriction of production, and lack of cooperation with the supervisor). Seventeen percent of the group quit in the first 40 days.

In contrast, the output in the latter group showed a smaller initial drop followed by a very rapid recovery to a level that exceeded that prior to the change. The researchers found no sign of hostility toward the staff or toward the supervisors in these groups, and no one quit during the experimental period. Findings such as these suggest that participation is extremely important for implementing certain types of change.

Rational Issues and Reconsideration

The implementation of a QBM project requires an investment of resources to attain some expected benefit. Thus, the study results should show a clear advantage for the project over any other alternative, including that of doing nothing. Therefore, implementation is highly dependent on the ability of the modeler to show such an advantage. Of course, timing can change many aspects of a project. What might have been urgent when the study was initiated might be immaterial by the time the results are ready. Thus, every proposed course of action should include a **cost–benefit** or **cost–effectiveness analysis.**

The costs may appear to be straightforward to determine, but this is probably misleading (Meredith 1981). Many costs in an organization—insurance, top administration, overhead, sales—are typically "allocated" to products, services, or projects according to historical accounting procedures and might not fairly represent actual costs of the QBM project. Other cost issues are just as difficult: life-cycle costs, discount rates, taxation, loss/gain in goodwill, and so on.

Even more difficult than cost determination is the benefit side of the proposal. First, there is the time issue—when do the benefits accrue? Are only short-term benefits considered? In addition, how are benefits allocated to projects? And how are benefits such as goodwill, improved competitive ability, faster response, higher quality, or reduced confusion calculated? Must a monetary value be placed on these abstract benefits?

All required resources and personnel, and their timing, should be included in the proposal. Commitments should be secured so that all resources will be available when needed. Lack of appropriate resources has frequently been a major obstacle to successful implementation. The proposal should also include answers to questions concerning the project itself:

- Who will be responsible for executing each portion of the project?
- When must each part be completed?
- What resources will be required for each of the parts?
- What information will be needed, and where will it come from?

In brief, a complete project planning document (Meredith and Mantel 2000) for implementation should be prepared. Based on the answers to these questions, operating procedures can be designed and any necessary training and transitions can be planned beforehand so they do not become problems to implementation later.

Last, as mentioned in the first chapter long ago, an opportunity or problem rarely is permanently resolved. An opportunity may be partially capitalized on, but competitors may later take advantage of other segments of it. And problems are almost never completely solved. More often, they shift into another area, or pop up in another guise later on. They may not be as severe as the initial problem (although sometimes the "solution" has actually worsened the problem), but they usually remain fairly substantial. Thus, it behooves modelers to make sure their proposals are not just a way of shifting a problem to somewhere else in the organization.

The Role of Project Management

As noted earlier, project management has evolved specifically to achieve limited objectives of management. That is, project management is specifically meant for situations with an explicit goal or objectives, a fixed budget, and a set due date, such as might be appropriate for a construction or R&D project, a space launch, the analysis of congestion at an airport, or even a wedding. It is *not* meant to handle the ongoing tasks of an organization, such as regular production, sales, cost accounting, or hiring. It is more than just a model because it includes monitoring and control for implementation, as well as planning and scheduling. (Moreover, with large projects it may even include its own termination, which can also be handled as a project.) As an example of the type of situation that would be amenable to project management, consider the following situation.

EXAMPLE

Moose Lake

The morning mail held exciting news for Judi Kosen, special projects manager for Restoration, Inc. As she anxiously opened the letter from the Proposal Committee, Environmental Projects Branch, Department of the Interior, her eyes spotted the words ". . . invite you to bid . . . ," " . . . activity network required . . ." Then, to be sure there was no mistake, she reread the good news. After rereading the letter, Judi settled back in her chair to contemplate the task before her.

Four months ago, Restoration, Inc., had requested that the Department of the Interior include them in the competition to revitalize Moose Lake in the northern part of the state. The company had developed a new technique to combat water pollution by increasing the basic amount of oxygen in a lake, called *oxygenation.* Oxygenation of a lake is a very complex project involving several of the company's departments as well as outside suppliers. It requires specialized equipment and supplies and specially trained personnel. The contract would be for a period of more than half a year, a long enough time for significant changes to occur in the economy, in prices, and in the availability of resources. The Department of the Interior wanted the project to be completed on time. Therefore, the contract would contain a $2,000 penalty clause for each week beyond the 30-week completion time. Judi realized that there were many factors that could cause a delay in such a complex project and wondered if it would be possible to use an activity network to minimize any delays.

As she examined all of the various factors, operations, and activities involved in the project, she felt that planning, monitoring, and controlling this job could be staggering. Then she remembered a course she had taken several years ago. The professor talked about the planning and control of large, complex projects using a network tool that had a long name and had to be abbreviated. It finally came back to her; yes—it was PERT.

Characteristics of Project Management The Moose Lake situation is an example of **project management.** Project management is distinguished from other types of organizational processes primarily by the nonrepetitive nature of the work; a *project* is usually a one-time effort. Although similar work may have been done previously, or may be done in the future (Restoration, Inc., may receive a contract for the oxygenation of other lakes), it is not usually repeated in the identical manner the way that cars or TV sets are manufactured on a production line or the way mortgage applications are processed at banks. The management of projects is more complicated than the management of traditional operational and administrative processes due to the following characteristics, generally typical of all projects to a greater or lesser degree.

1. The duration of a project may be weeks, months, or even years. During such a long period, many changes may occur, most of which are difficult to predict. Such changes may have a significant impact on project costs, technology, and resources. The longer the duration of the project, the more uncertain are the execution times and costs.
2. A project is complex in nature, involving many interrelated activities and participants from both within the organization and outside (e.g., suppliers, subcontractors). (Our Moose Lake example is highly simplified for the purpose of easier demonstration.)
3. Delays in completion time may be very costly. Penalties for delays may also amount to thousands of dollars per day. Completing projects late may result in lost opportunities and ill will.
4. Most of the project activities are sequential. Some activities cannot start until others are completed.
5. Projects are typically a unique undertaking, something that has not been encountered previously.

As a consequence, the management of projects is rather complicated. Exhibit 8.2 summarizes the major characteristics of projects. Until the mid-1950s, there were no generally accepted formal techniques to aid in project management. Each manager had his or her own management scheme. However, the need for formal tools soon became apparent. Two of the best-known tools that fill this need are **Program Evaluation Review Technique (PERT)** and **Critical Path Method (CPM).** PERT was developed by the U.S. Navy with Booz, Allen & Hamilton, Inc. and Lockheed Corporation to accelerate development of the Polaris missile system in the late 1950s. CPM was developed by DuPont in the same time period as PERT.

Definitions Used in PERT and CPM In order to explain the purpose, structure, and operation of PERT and CPM, it is helpful to define the following terms:

Activity An activity is a task that requires resources and takes a certain amount of time for completion. Examples of activities are studying for an examination, designing a part, connecting bridge girders, or training an employee.

Event An event is a specific accomplishment at a recognizable point in time. Typically, events correspond to the start or completion of one or more activities. Events *do not* have a time duration, per se. To reach an event, all the activities that precede it must

Exhibit 8.2 Common Project Characteristics

Uniqueness	Wide visibility
Fixed duration	Extensive interactions
Complexity	Uncertainty
High interdependence	

be completed. An event can be viewed as a goal attained, whereas the activities leading to it can be viewed as the means of achieving it.

Milestones Special events that correspond to significant accomplishments and are used to gauge the progress of the project. Examples of milestones include passing a course at a university, submission of engineering drafts, completion of a span on a bridge, or the arrival of a new machine.

Project A project is a collection of activities and events with a definable beginning and a definable end (the goal). Examples include getting a college degree, patenting an invention, building a bridge, or installing new machinery.

Network A network is a logical and chronological set of activities and events, graphically illustrating relationships among the various activities and events of the project.

Critical Activity A critical activity is an activity that, if even slightly delayed, will hold up the scheduled completion time of the entire project.

Path A path is a sequence of adjacent activities that form a continuous path between two events.

Critical Path A critical path is a sequence of critical activities that forms a continuous path between the start of a project and its completion. The critical path is the path with the longest duration. It is critical in the sense that if it is delayed, the completion of the entire project will be delayed.

The Major Differences and Similarities Between PERT and CPM PERT and CPM are very similar in their approach; however, two distinctions are usually made between them. The first relates to the way in which activity durations are estimated. In PERT, three estimates are used to form a weighted average of the expected completion time of each activity, based on a probability distribution of completion times. Therefore, PERT is considered a probabilistic tool. In CPM, there is only one estimate of duration; that is, CPM is a deterministic tool. The second difference is that CPM allows an explicit estimate of costs in addition to time. Thus, while PERT is basically a tool for planning and control of time, CPM can be used to control both the time and the cost of the project. Extensions of both PERT and CPM allow the user to manage other resources in addition to time and money, to trade off resources, to analyze different types of schedules, and to balance the use of resources.

Due to their complex nature, it is very difficult to completely eliminate the delays and cost overruns typically associated with large projects. However, with the appropriate management systems for planning, organizing, and controlling, it is possible to reduce them to a reasonable level. The problem is that the cost of implementing and executing such systems can exceed their benefits because of the large amount of monitoring and reporting that is required.

For example, overhauling a Boeing 757 airplane may involve 8,000 different activities (work orders). To be completely in control, management must plan, organize, monitor, report, and act on each of these 8,000 work orders, either daily or perhaps even on a shift-by-shift basis. This will require an extreme amount of reporting. However, using the concept of *management by exception,* management may elect to exercise tight control over only the most critical activities. The number of critical activities may be only 5 percent of the total activities; less control is then exercised over the remaining activities.

The major purpose of PERT and CPM is to objectively identify these critical activities. Further, these techniques can tell us how close the remaining activities are to becoming critical. (This available delay is called *slack* or *float.*) Accordingly, any PERT or CPM program provides management with the following information at the minimum:

1. Which activities are critical
2. Which activities are noncritical

3. The amount of slack on each noncritical activity and therefore how long the start of the activity can be delayed without delaying the completion of the project

Computerized programs, as will be shown later, may also provide other valuable information.

The project management process is conducted in six steps, the first three of which constitute planning the project, the next two scheduling the project, and the final step monitoring and controlling the project. These three groups also represent the next three sections of the chapter. We start with planning, of course.

8.3 Planning the Project

There are many issues to address in planning a project: selecting the project manager, deciding how to organize the project and to whom it should report, finding a project team, determining the tasks that need to be executed and how long they will take and the resources they will need, and so on. For example, it is common to organize small projects as a part of a functional department within an organization. However, it is also common to form a special task force reporting to, say, the vice-president of marketing (see Exhibit 8.3*a*). For large projects, an entire division may be formed and permanently placed within the organization. And some organizations that work with projects on a regular basis form "matrixed" organizations where the many projects report to both a functional manager as well as a "program" manager responsible for a variety of similar projects (see Exhibit 8.3*b*).

Similarly, the identification of the tasks to be accomplished, and their subtasks, and their work package elements, and so on down to the most elemental level—the result of which is known as the **work breakdown structure (WBS)**—is another major element of project management. Here, we take a very simplified approach to this crucial aspect of project planning but it is a major piece of the project plan because all schedules and budgets are tied to the WBS. We will not have time to discuss all these topics here; entire

EXHIBIT 8.3 Functional Versus Matrix Organizations

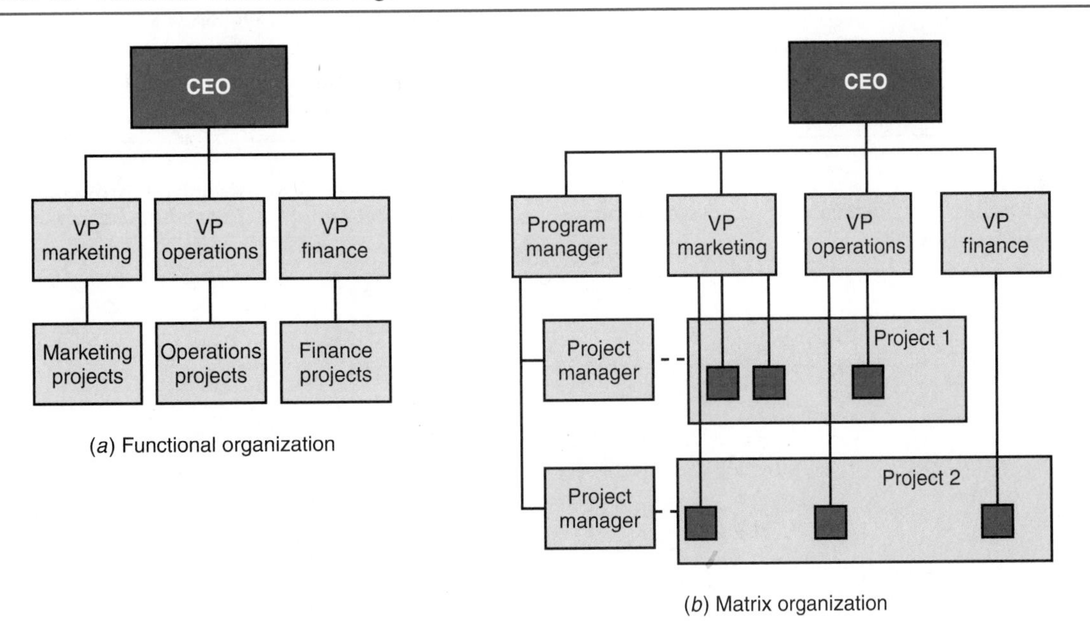

(*a*) Functional organization

(*b*) Matrix organization

books have been written on these topics, and even on some of the subtopics. The interested reader is referred to the bibliography.

As noted earlier, we will address only three major activities within this stage of project management: analysis, sequencing, and estimation.

Step 1: Analysis of the Project

After consultation with all department heads, a list of activities is agreed on (see first two columns of Exhibit 8.4 for Moose Lake). Each activity is clearly defined and responsibility is assigned to the proper department heads (not shown here). Note that we have not yet determined the activity durations or which activities feed into or precede other activities, to be done in steps 2 and 3.

Step 2: Sequence the Activities

Once the content of each activity is defined in the WBS, the sequence of execution is determined. For example, personnel cannot be hired before proper authorization is granted (administrative setup) and equipment cannot be assembled before all parts and materials are on site. Such information is then added in Exhibit 8.4 (in the "Required Immediately Preceding Activities" column).

Step 3: Estimate Activity Times and Costs

The next step is to determine or estimate the required duration (elapsed time) for each activity, designated as *expected time*, t_e. We distinguish two cases: in CPM, the activity duration is considered to be deterministic (certain), as in Exhibit 8.4. This assumption occurs when there is a wealth of experience regarding activity times (e.g., in many construction projects) or the activity actually has a fixed duration (e.g., lab tests and chemical processes). In the case of PERT, however, we assume that the duration is unknown, so we use three estimates with an averaging procedure (discussed after the deterministic illustration). In both cases, we also assume that there are sufficient resources to complete the activities.

In addition to time, we may also enter other information such as cost per activity (labor, parts), trade-offs between time and cost, the availability of resources, and milestone dates. The more information entered, the more output information can be provided to management.

EXHIBIT 8.4
Moose Lake Project Activities

Activity	Description	Duration (Weeks)	Required Immediately Preceding Activities
a	Set up administration	3	None
b	Hire personnel	4	*a*
c	Obtain materials	4	*a*
d	Transport materials to Moose Lake	2	*c*
e	Gather measuring team	4	*a*
f	Develop schedule	6	*c*
g	Assemble equipment	3	*d, b*
h	Plan evaluation	1	*e*
i	Oxygenate	12	*f, g*
j	Measure and evaluate	2	*i, h*

8.4 Scheduling the Project

To schedule the project we first construct it (step 4) and then analyze it (step 5). This section looks at these two steps, as well as some additional characteristics of the PERT/CPM network.

Step 4: Construct the Network

The PERT (or CPM) network is a graphical representation of information such as that in Exhibit 8.4. It shows the interrelationships among the activities, the events, and the entire project. To construct a PERT network, start by viewing an activity as an arrow (arc) between two events (circles or nodes). For example, the first activity of the Moose Lake Project, "Set up administration," is shown in Exhibit 8.5.

The arrow points in the direction of the time flow, but its length is *not* related to the duration of the activity (it is arbitrarily set at a suitable length for drawing the diagram). The number circled in front of the arrow, **1** in Exhibit 8.5, is the event that *precedes* the activity. The number circled after the arrow, **2** in Exhibit 8.5, is the *succeeding* event. The numbering of the events is somewhat arbitrary—several methods are used in real-life projects. The activity between events **1** and **2** is labeled *a.* It can also be labeled **1–2.**

The construction of the network starts with event **1** (the beginning of the project), which precedes the first activity, *a.* This will always be the activity (or activities) that *does not* require any preceding activity. It is placed at the left side of the diagram (similar to the start of a decision tree); the event before this activity is marked **1,** and the one after it is marked **2.** Next, the data show that activities *b, c,* and *e* must all be preceded by activity *a,* whose conclusion is event **2.** Therefore, all of these activities can start only after event **2** has been accomplished. This is shown in Exhibit 8.6.

At the end of each activity, a number is assigned to designate the forthcoming event. The assignment of the numbers **3, 4,** and **5** is made as the network progresses from left to right. The representation in Exhibit 8.6 shows that activities *b, c,* and *e* can be conducted simultaneously, but none can start until activity *a* has been completed. Note that activity *c* was placed above activity *b* in the diagram; this was done merely as a matter of convenience for drawing the remaining diagram. Note that the development of network diagrams is iterative and that creating a satisfactory one often requires revisiting earlier decisions regarding the placement of nodes and activities.

EXHIBIT 8.5 An Activity

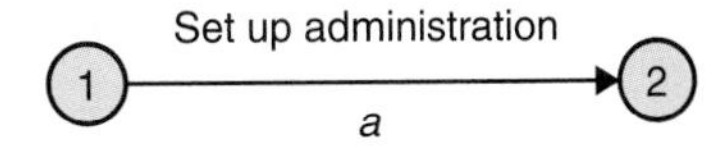

EXHIBIT 8.6 Precedence Requirements

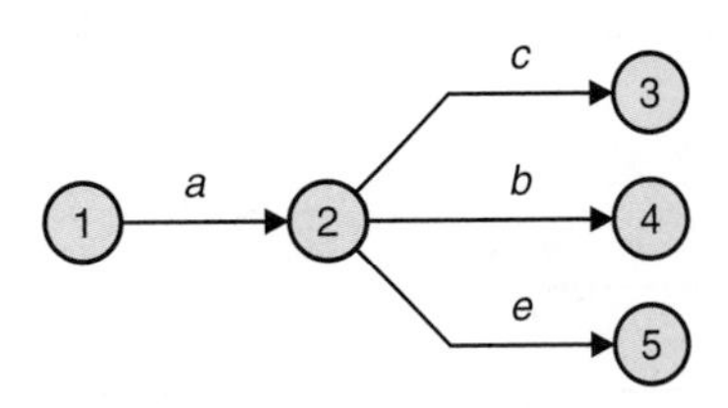

The construction of the entire network continues in the same manner. Out of event **3,** succeeding activities *d* and *f* are extended (Exhibit 8.7). Out of event **4,** the succeeding activity *g* is drawn, and out of event **5,** the succeeding activity *h* is drawn. The diagram grows to the right until all activities and events are depicted.

Some typical network relationships are illustrated in Exhibit 8.8. The last two include a dummy activity.

Dummy Activities In the construction of a network, care must be taken to assure that the activities and events are in proper sequence. One device that helps proper sequencing is **dummy activities.** Dummy activities are characterized by their use of zero time and zero resources; their only function is to designate a precedence relationship. Graphically, such activities are shown as broken lines. For example, consider the activities listed at the top of the next page.

EXHIBIT 8.7 PERT Network for Moose Lake Project

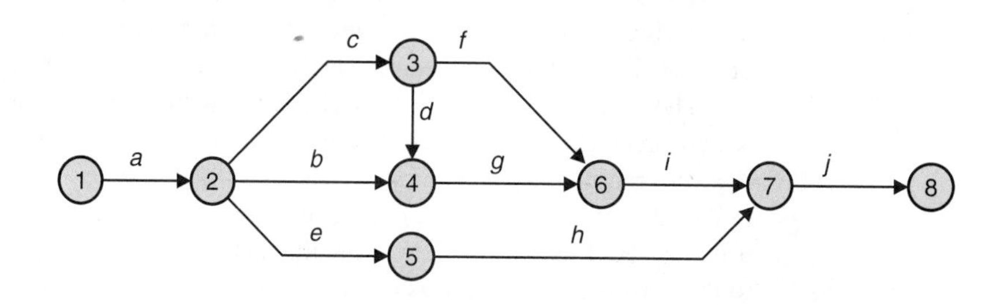

EXHIBIT 8.8 Network Relationships

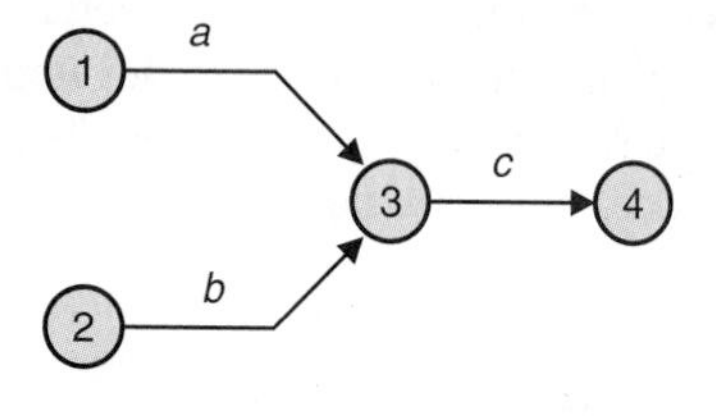

a and *b* can start simultaneously, but *c* can start only after both *a* and *b* are finished.

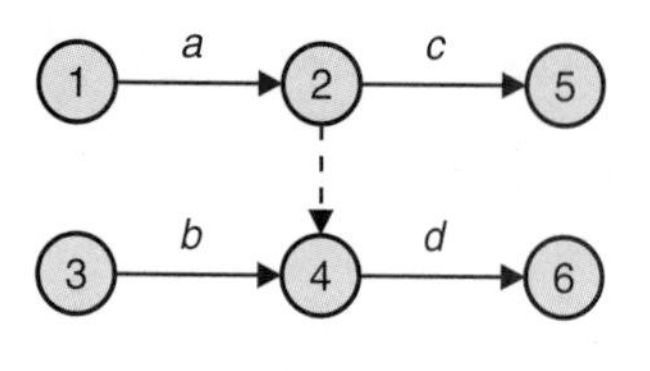

a and *b* can start simultaneously, *c* can start after *a* is completed, but *d* must wait for both *a* and *b* to finish.

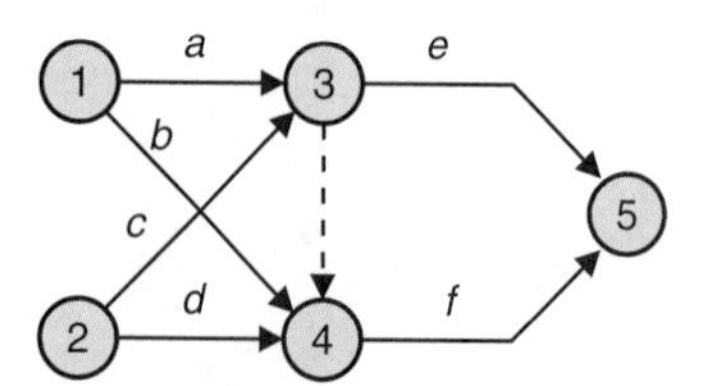

e can start after *a* and *c* are completed, but *f* must wait until *a, b, c,* and *d* are completed.

Activity	Required Preceding Activities
a	None
b	None
c	*b*
d	*a, c*
e	*a*
f	*d, e*

To diagram this network, it is necessary to use a dummy activity, as shown in Exhibit 8.9. The reason the dummy activity is needed is that without it, activities *d* and *e* would have exactly the same starting and ending nodes. The problem with two or more activities having the same starting and ending nodes is that they lose their unique identities, complicating the task of planning, monitoring, and controlling them. Thus, whenever two or more activities have the same starting and ending node, dummy activities are added to preserve the precedence relationships and provide each activity with a unique identity.

Gantt Charts A **Gantt chart** is a time-scaled diagram in which the activities are represented by bars whose length is proportional to the activity's duration. Exhibit 8.10 illustrates a network diagram and its equivalent bar chart. The Gantt chart in Exhibit 8.10*b* shows that all activities in this project can be completed by time 6. Activity **2–4** does not start until activity **1–2** has been completed, and activity **3–4** does not start until **1–3** has been completed. Activities **1–2** and **2–4,** however, can take place at the same time that **1–3** and **3–4** are taking place, as long as the precedence relationships are maintained. Software programs produce Gantt charts with dates inserted instead of elapsed time. There are several ways to depict the activities on a chart, depending on when the activities are scheduled to "start," if they are not critical. Many of these programs also show the precedence relationships on the Gantt chart, as is illustrated in Exhibit 8.11.

Step 5: Event Analysis

Event-oriented analysis is used primarily where activity times are assumed to be known with certainty. The following procedure is used.

1. Enter time estimates on the network.
2. Compute the earliest and latest dates for all events.
3. Find the slack on the events and identify critical events.
4. Find the slack on the activities and identify critical activities.
5. Find the critical path.

The details of this procedure are illustrated on the following pages.

EXHIBIT 8.9 Dummy Activity

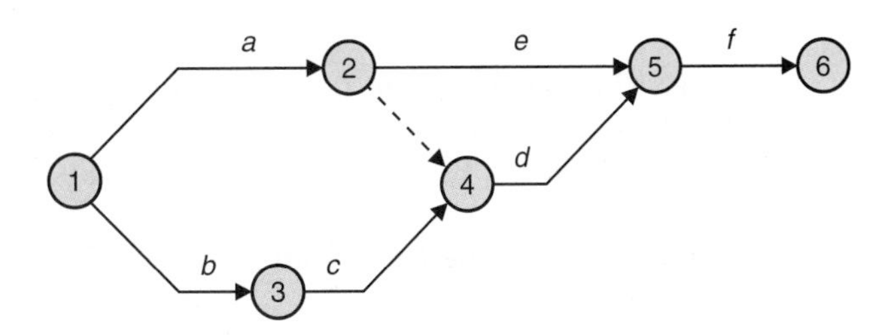

EXHIBIT 8.10 Comparison of PERT and a Gantt Chart

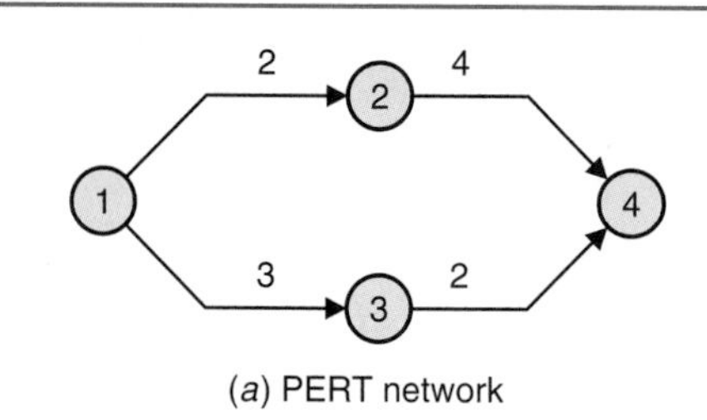

(*a*) PERT network

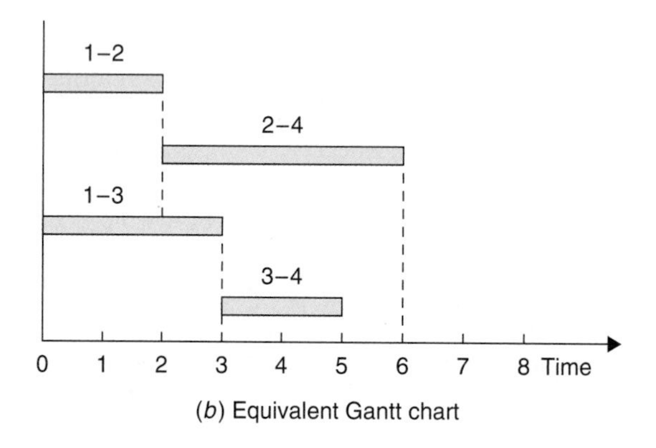

(*b*) Equivalent Gantt chart

EXHIBIT 8.11 Gantt Chart Created by Microsoft Project for Figure 8.10*a*

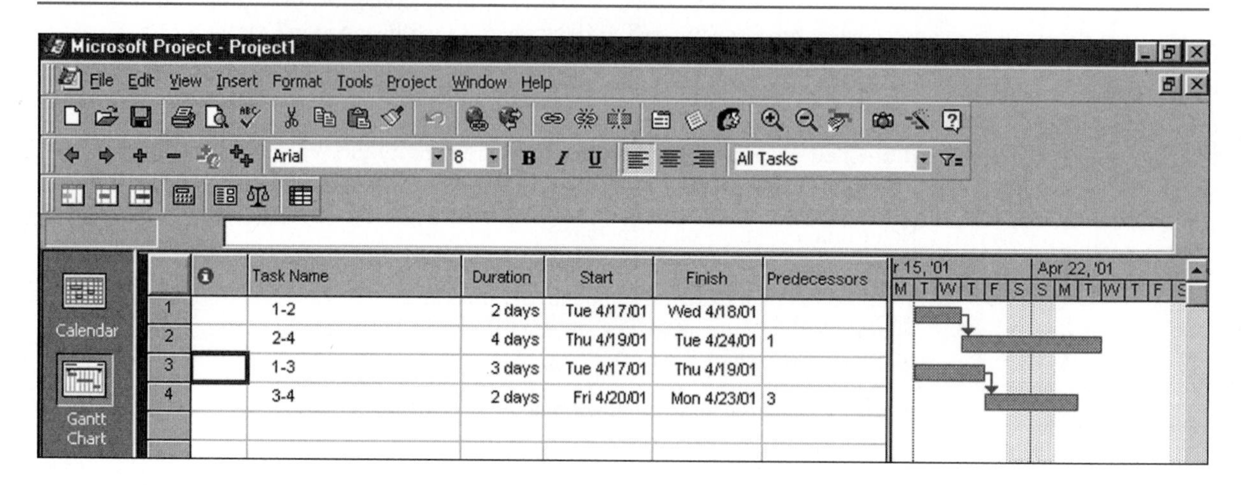

Enter Time Estimates on the Network Once the network is completed, the activity durations (t_e) are entered on the diagram (in parentheses, above the arcs or arrows) as shown in Exhibit 8.12. (We discuss time estimating in PERT in a later section.)

Compute the Earliest and Latest Dates This approach is based on two important concepts:

- **The Earliest Date (T_E)** By definition, the earliest date, T_E, for an event to occur is immediately after *all* the preceding activities have been completed. For example, a

EXHIBIT 8.12 Time Estimates for the Project

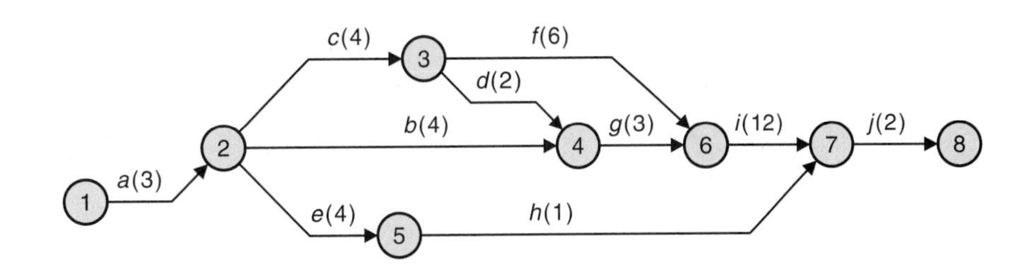

EXHIBIT 8.13 Calculating Earliest and Latest Dates

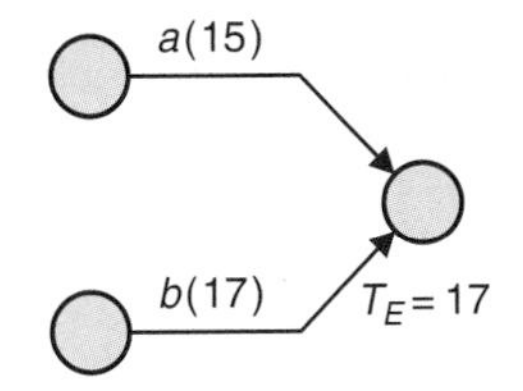

(*a*) Earliest date of latest activity

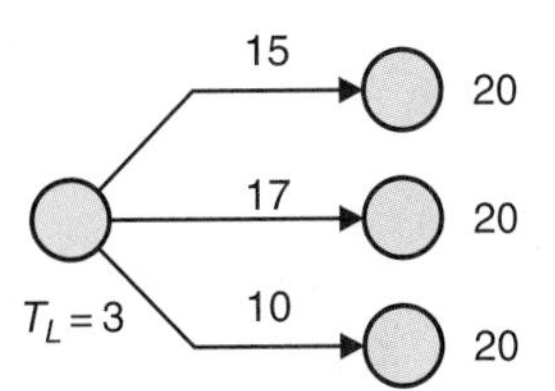

(*b*) Earliest date without delay

certain event is preceded by two activities, as shown in Exhibit 8.13*a*. If the earliest date that activity *a* can be completed is 15 weeks, and the earliest date that activity *b* can be completed is 17 weeks, then the earliest time that the event can occur is at the conclusion of 17 weeks. Because this rule is true for every event (including the last one), then *the earliest possible date for completing the entire project is the earliest date of the last completed event.*

- **The Latest Allowable Date (T_L)** The latest allowable date for each event, T_L, is the latest date that an event can occur *without causing a delay* in the already-determined project's completion date. In Exhibit 8.13*b* the T_L for each ending event is 20. While the top and bottom ending events can finish earlier than time 20, they can finish as late as 20 without causing a delay in the project's completion time. The completion date for the project can be either the earliest possible completion date or any other agreed-upon date. Unless otherwise stated, we will use the earliest possible completion date for our calculations. The computation is done as follows.

Step 1. Conduct a Forward Pass: Find T_E for Each Event. In order to compute T_E for an event, the duration of each path leading to the event is computed. If several paths lead to an event, then the path with the *largest elapsed time* is selected.

We will now find T_E for all events. To begin with, Exhibit 8.12 is reproduced as Exhibit 8.14.

Event 1 This is the event at the beginning of the project. The T_E for this event is set to zero. This information is written above the event (Exhibit 8.14).

Event 2 There is only one activity from event **1** to **2;** its duration is 3 weeks. The T_E for event **2** is thus 3.

Event 3 Similarly, T_E for event **3** is 7 weeks.

Event 4 Note that T_E is to be determined by the longest (timewise) path leading to an event. However, it is not necessary to return to the beginning of the network to compute all paths leading to an event to figure the longest one. With the following formula, use can be made of existing information.

Length of a path = duration of the last activity on the path + T_E of the preceding event

Exhibit 8.15 shows the portion of the network leading to event **4.** Event **4** represents the completion of both activity *d* and activity *b*. The length of the path **1–2–3–4** is determined as

T_E for event **3**	= 7
+ duration of activity *d*	= 2
Total	9

EXHIBIT 8.14 Computation of T_E and T_L

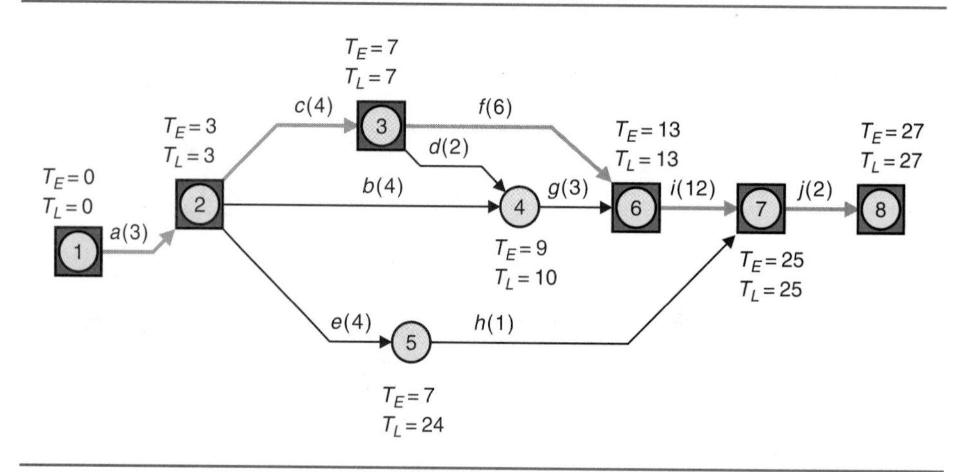

EXHIBIT 8.15 Event 4

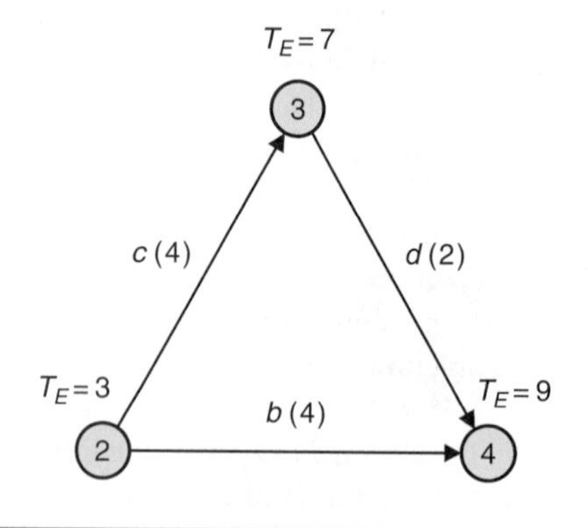

The length of path **1–2–4** is determined as:

$$\begin{array}{ll} T_E \text{ for event } \mathbf{2} & = 3 \\ + \text{ duration of activity } b & = \underline{4} \\ \text{Total} & \ 7 \end{array}$$

All paths leading to the event are now compared. Because 9 is the larger number, T_E for event **4** will be 9.

Event 5 T_E for this event is 7 weeks.
Event 6 Two paths are considered:

- Path **1–2–3–6** whose length is $T_E = 7$ (for event **3**) + 6 = 13
- Path **1–2–3–4–6** whose length is 9 + 3 = 12 (*Note:* Path **2–3–4** is longer than **2–4.**)

Thus, the larger is 13 weeks.

Event 7 Checking the two paths leading to **7,** we find T_E to be 13 + 12 = 25.
Event 8 This event designates the *end* of the project, because no activities emerge from it. Therefore, the earliest date for this event, 27 weeks, is the earliest date that the entire project can be completed. (This is good news for Judi, as project director, because the required completion time was 30 weeks.)

Step 2. Conduct a Backward Pass: Find T_L for Each Event. (Refer to Exhibit 8.14.) To compute each T_L, start from the last event (**8**) and work backward (i.e., right to left) all the way to event **1.**

For Event 8 T_L for the last event is set equal to the computed earliest completion time of the project (27 weeks).
For Event 7 Because the latest that event **8** can occur is 27 weeks, and because it takes 2 weeks to complete activity *j,* the latest allowable date that event **7** can occur is 27 – 2 = 25 weeks.
For Event 6 Because the latest that event **7** can occur is week 25 and because activity *i* lasts 12 weeks, the latest time for event **6** is 25 – 12 = 13.
For Event 5 In a similar manner, T_L is found to be 24 (T_L for **7** is 25 minus 1 week for activity $h = 24$).
For Event 4 In a similar manner, T_L is computed as 10.
For Event 3 Here, two activities, *d* and *f,* must be considered. Because activity *d* lasts 2 weeks, and because it must be completed no later than the tenth week (the latest allowable time for event **4**), then activity *d* must start not later than 10 – 2 = 8. Activity *f* takes 6 weeks; it must be completed, at the latest, by week 13 (which is T_L for event **6**). Therefore, activity *f* must be started *not later than* 13 – 6 = 7.

Now, to enable both activities to start on time so that there will be no delay in the entire project, event **3** must occur, *at the latest,* by week 7, which is the *smaller* of the two T_L's. Computation is continued in the same manner, event by event, until event **1** is reached. Of special interest is event **2.** Here, three T_L's and activities must be considered. For *c,* 7 – 4 = 3; for *b,* 10 – 4 = 6; and for *e,* 24 – 4 = 20. The *smallest one,* 3 weeks, is selected as T_L for event **2.** For event **1,** T_L is zero. *Note:* For the first event, T_L must be zero if $T_E = T_L$ for the last event.

Find the Slack on the Events and Identify Critical Events The difference between the T_L and the T_E, for each event, is defined as *slack* (*S*).

$$S = T_L - T_E$$

Two cases are distinguished:

1. **When $T_L = T_E$ for the Last Event (the End of the Project)** In this case, slacks in the network can either be zero, whereupon the events are called **critical events,** or larger than zero, whereupon the events are considered to have positive slack. In our example (Exhibit 8.14), we assumed $T_L = T_E$ for the final event and all critical events, which have zero slack. These are shown with a box around them to aid in quick recognition. Note that only events **4** and **5** are not critical here.
2. **When $T_L \neq T_E$ for the Last Event** In this case, the *critical* events are defined as those events with the *minimum slack,* which *can* be negative (when $T_L < T_E$).

What is the meaning of slack? Because T_E is the earliest that an event can be reached and T_L is the latest that the event can occur without delaying the entire project, then the difference, the slack, tells how long the event can "linger" without delaying the entire project. Any delay in a critical event will cause a delay in the entire project.

Let us examine event **5.** For this event, the slack is $T_L - T_E = 24 - 7 = 17$ weeks. The meaning of this is that although event **5** can be reached in 7 weeks, management has the flexibility to reach this event at any time during the ensuing 17 weeks (up to the 24th week) without causing a delay in the entire project.

Find the Slack on the Activities and Identify Critical Activities Similar to the slack on an event, there is also slack on activities. This slack tells us how long the activity can linger without delaying the entire project. The following equation can be used to find the amount of slack:

$$\text{Activity slack} = \begin{bmatrix} T_L \text{ for the event} \\ \text{at the end} \\ \text{of the activity} \end{bmatrix} - \begin{bmatrix} T_E \text{ for the event} \\ \text{at the beginning} \\ \text{of the activity} \end{bmatrix} - \begin{bmatrix} t_e \\ \text{duration} \\ \text{of the activity} \end{bmatrix}$$

This slack is also called **total float (TF).** It is distinguished from a type of slack called **free float (FF),** which will be discussed later. In our example, we get the following results shown in Exhibit 8.16.

Again, two cases are distinguished:

1. **When $T_L = T_E$ for the Last Event** In this case, an activity with zero slack is defined as a *critical activity.*

EXHIBIT 8.16
Calculating Activity Slacks

Activity	T_L	Minus	T_E	Minus	t_e	=	Slack (TF)
a	3	–	0	–	3	=	0
b	10		3		4		3
c	7		3		4		0
d	10		7		2		1
e	24		3		4		17
f	13		7		6		0
g	13		9		3		1
h	25		7		1		17
i	25		13		12		0
j	27		25		2		0

2. **When $T_L \neq T_E$ for the Last Event** In this case, the activities with the *minimum slack* are the critical ones.

Find the Critical Path The critical path is the path(s) in the network, leading from the beginning of the project to its end, *all* of whose activities and events are critical. This definition implies that if $T_L = T_E$ for the last event, then there is *zero slack* on the critical path. Otherwise, the critical path is that path with the minimum slack on it.

The critical path has certain additional characteristics:

1. There can be more than one critical path in the network.
2. The critical path is the longest (timewise) path in the network.

In our example, the (one) critical path is **1–2–3–6–7–8.**

The importance of identifying the critical path is that it points out those activities and events that are critical and, as such, must be carefully monitored and controlled. Before getting into these topics, however, let us note some additional characteristics of the PERT/CPM network.

PERT/CPM Network Characteristics

Regular Slack (Total Float) Such a case is shown in Exhibit 8.17. Activity **1–3** has a duration of 6 weeks, whereas path **1–2–3** has a duration of 13 weeks. Therefore, activity **1–3** can linger 13 – 6 = 7 weeks; that is, there is a 7-week slack on the activity. Regular slack denotes the *maximum* amount of slack available, some of which may be shared.

Shared Slack (Slack on a Noncritical Path) Whenever there are two or more noncritical activities or noncritical events connected in series, their slack is called *shared.* An example of shared slack is shown in Exhibit 8.18, where activities *e* and *h* are connected in series. The critical portion of the path between events **2** and **7** requires 22 weeks.

Activity *e* requires 4 weeks and activity *h* requires 1 week, a total of 5 weeks. Therefore, there is a slack of 22 – 5 = 17 weeks, which can be distributed between *e* and *h* in any combination. For example, a slack of 17 on *e* and zero on *h,* 16 on *e* and 1 on *h,* and so on.

As another example of shared slack, Exhibit 8.19 shows a situation with two noncritical events, **2** and **3,** and three noncritical activities *a, b,* and *c.* Activities *a, b,* and *c* together require 3 + 4 + 5 = 12 weeks. Therefore, there is a regular slack of 30 – 12 = 18 on both events **2** and **3** (e.g., 18 on **2,** zero on **3;** 17 on **2** and 1 on **3,** and so on). That is, there is a shared slack of 18 on events **2** and **3.**

There is also a regular slack of 30 – 12 = 18 on each of activities *a, b,* and *c,* which is shared among the three. Again, one may use the 18 weeks on activity *a* alone, or 6 weeks

EXHIBIT 8.17 Regular Slack

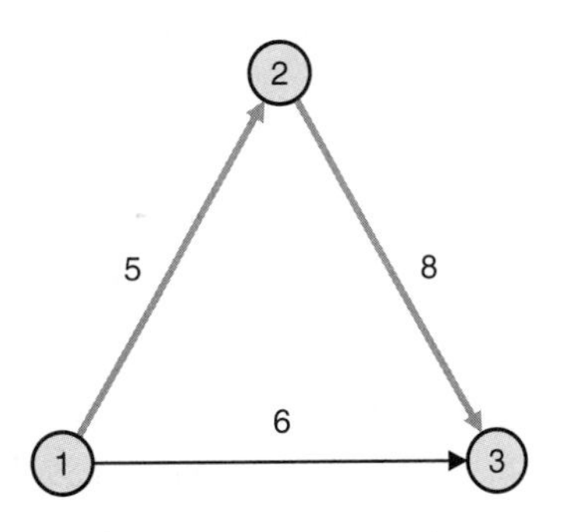

EXHIBIT 8.18 Shared Slack, Example 1

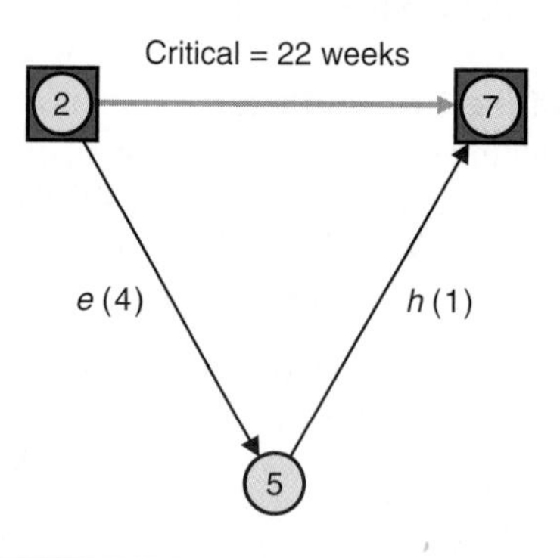

EXHIBIT 8.19 Shared Slack, Example 2

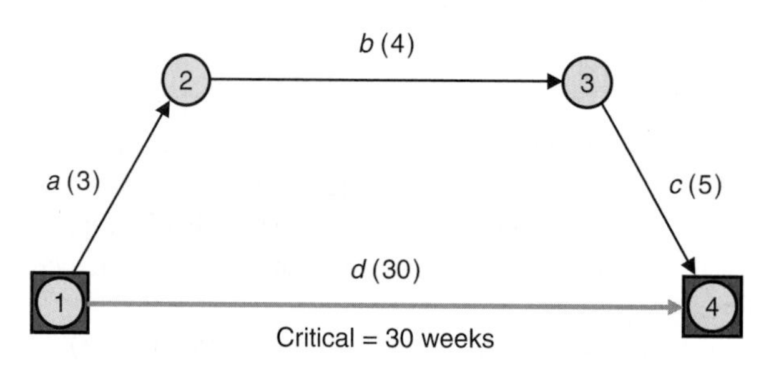

on each activity, and so on. (There are as many possibilities as there are ways of allocating 18 among three recipients.)

In a situation of shared slack, it can be viewed as being on a path, rather than on an individual activity. For example, in Exhibit 8.19, there is a slack of 18 on path **1–2–3–4.**

Free Float Free float (FF) is a slack that represents the time any activity can be delayed before it delays the earliest start time of any activity immediately following. For example, in Exhibit 8.17, the slack of 7 on **1–3** is free, as is the slack of 17 on **5–7** in Exhibit 8.18. Notice that the free float on activity *e* in Exhibit 8.18 is zero. In general, if there are several noncritical activities in a series, only the *last one* will have a free float.

Free float is important in the case of shared slack. It means that if a slack has not been shared, it can all be used in the last sharing activity. Computer printouts typically give both the TF and the FF for each activity, as will be demonstrated later.

The Case When $T_L \neq T_E$ for the Project In the previous computations, we assumed that $T_L = T_E$ for the last event. However, this may not always be the case. If T_L is larger than T_E for the last event, then the slack for the last event will be positive.

EXAMPLE

Moose Lake Revisited

In the Moose Lake project, Judi can consider T_L as 30 (the agreed-on completion time). Thus, the slack on event **8** will be 30 – 27 = 3, and so will the slacks on all critical events. Further, the slack on all noncritical events will be three weeks larger, too. For example, for event **5,** the new T_L = 27; because T_E = 7 (unchanged), then the slack for event **5** = 27 – 7 = 20 weeks.

If T_L is smaller than T_E for the last event, a *negative slack* will result, indicating that the desired date cannot be achieved and a delay of the magnitude of the negative value is expected.

A Critical Path Leading to an Event The critical path procedure outlined previously can also be used to find the critical path leading to any desired event. For example, the critical path to event **4** is **1–2–3–4.** (See Exhibit 8.14.)

Use of Complete Enumeration In small problems, like the one in the example, the critical path may be identified by listing all possible paths leading from the beginning of the project to its end. The path with the *largest* duration is the critical path (composed of all critical activities and events). The reason for this is that in order to complete the project, all activities *must* be accomplished. Because the longest path is longer than any other, its completion gives enough elapsed time to complete *all other paths.* This guarantees that every single activity in the network will be accomplished. In the Moose Lake example, the following four paths are identified:

		Total Duration (Weeks)	
Path 1	1–2–3–4–6–7–8	26	
Path 2	1–2–3–6–7–8	27	←*Maximum*
Path 3	1–2–4–6–7–8	24	
Path 4	1–2–5–7–8	10	

When *all* paths are compared (complete enumeration approach), the longest path is found to be Path 2, with a duration of 27 weeks.

In large, complex networks with hundreds of activities, especially when continuous updating is required, the complete enumeration approach may take a long time. In such cases, the use of specialized software packages is recommended.

Multiple Critical Paths In our example, there is a single critical path for the project. However, in other problems, multiple paths may occur. For example, if activity *d* were 3 weeks, there would have been *two* critical paths: the one already identified, **1–2–3–6–7–8,** and another one, **1–2–3–4–6–7–8.**

Several Starting Activities Many of the examples given so far exhibit a single starting activity. However, this is not necessary, and real projects often do have multiple starting activities. For example, a project to redesign an organization's business process may have two starting activities: (1) assessing the capabilities of new technologies, and (2) assessing customer requirements.

Estimating Activity Times in PERT

Because the Moose Lake project was an experimental project, it was suitable for a PERT analysis. Judi asked each department to submit three estimates of duration time for each activity the department was responsible for, using the following guidelines:

- **Optimistic Estimate (t_o)** An estimate of the *shortest possible time* (duration) in which the activity can be accomplished. The probability that the activity will take less than this time is 0.01.

- **Most Likely Estimate (t_m)** The duration that would occur most often if the activity were repeated under exactly the same conditions many times. Equivalently, it is the time that would be estimated most often by experts.
- **Pessimistic Estimate (t_p)** The longest time that the activity could take "when everything goes wrong." The probability that the activity will exceed this duration is 0.01.

All three estimates are entered in Exhibit 8.20. Notice that in some cases, $t_o = t_m = t_p$, that is, the exact time duration is known.

Computing the Weighted Average Once the three time estimates are obtained, their weighted average is computed. This average, which is called the mean time of an activity, t_e, is a *weighted average* of the three time estimates. It is computed using

$$t_e = \frac{t_o + 4t_m + t_p}{6}$$

where t_e is the expected duration of the activity.

The formula gives four times more weight to the most likely estimate than to the pessimistic or optimistic estimates. The division by 6, the sum of the weights, is to obtain a weighted average resulting in that shown in the "Duration" column of Exhibit 8.4. *(Note:* This equation is based on the assumption that the beta distribution is the probability distribution of duration times. Other weights are used in real-life problems based on experience.) For example, in Exhibit 8.20, for activity *a,* the weighted average is

$$t_e = [1(1) + 4(3) + 1(5)]/6 = 3 \text{ weeks}$$

Finding the Probabilities of Completion in PERT

PERT has more capabilities than just as a planning and control tool. It can also be used to give management an indication of risk in terms of project completion. This is a crucial analysis that considers the chance of completing the project on, before, or after scheduled dates.

The three estimates of activity duration in PERT, t_o, t_m, and t_p, are assumed to follow a probability distribution called the beta distribution, shown in Exhibit 8.21 for activity *b*

EXHIBIT 8.20
Moose Lake Project Activities (Weeks)

Activity	Description	Optimistic (t_o)	Most Likely (t_m)	Pessimistic (t_p)	Weighted Average (t_e)
a	Set up administration	1	3	5	3
b	Hire personnel	1	3	11	4
c	Obtain materials	3	4	5	4
d	Transport materials to Moose Lake	1	2	3	2
e	Gather measuring team	3	3	9	4
f	Develop schedule	2	5	14	6
g	Assemble equipment	2	3	4	3
h	Plan evaluation	1	1	1	1
i	Oxygenate	12	12	12	12
j	Measure and evaluate	1	2	3	2

as an example ($t_o = 1$, $t_m = 3$, $t_p = 11$, and their weighted average $t_e = 4$). Even though estimates of each activity duration follow the beta distribution, the estimate of the combined duration of several activities (such as those on the critical path) approaches the *normal* distribution. (*Note:* The justification for the use of the normal distribution is based on the central limit theorem: the sum of n independent variables tends to be normally distributed as n approaches infinity (becomes "large enough").

The project estimated completion time, T_S, is computed as the expected time, T_E, for the last event. (The T_E for the last event is the same as the sum of the t_e's along the critical path if the project starts at time 0.) Therefore, there is a 50 percent chance that the *entire* project will be completed by its earliest projected time (27 weeks in our earlier example). However, a "50 percent chance" is usually too low a confidence level for managerial planning. Management may want to know the chances of completing the project in other time periods, say 25 or 30 weeks. To answer such questions, an analysis involving the probability associated with the duration times is conducted.

To calculate the probabilities, it is necessary to find the *standard deviation* of the activity durations. The standard deviation of the activity durations is calculated as

$$\text{Standard deviation of an activity} = \sigma = \frac{t_p - t_o}{6}$$

The variance of the activity's distribution is given by

$$\text{Variance of activity} = \sigma^2 = \left(\frac{t_p - t_o}{6}\right)^2$$

EXHIBIT 8.21 Activity Time Distribution for Activity *b*

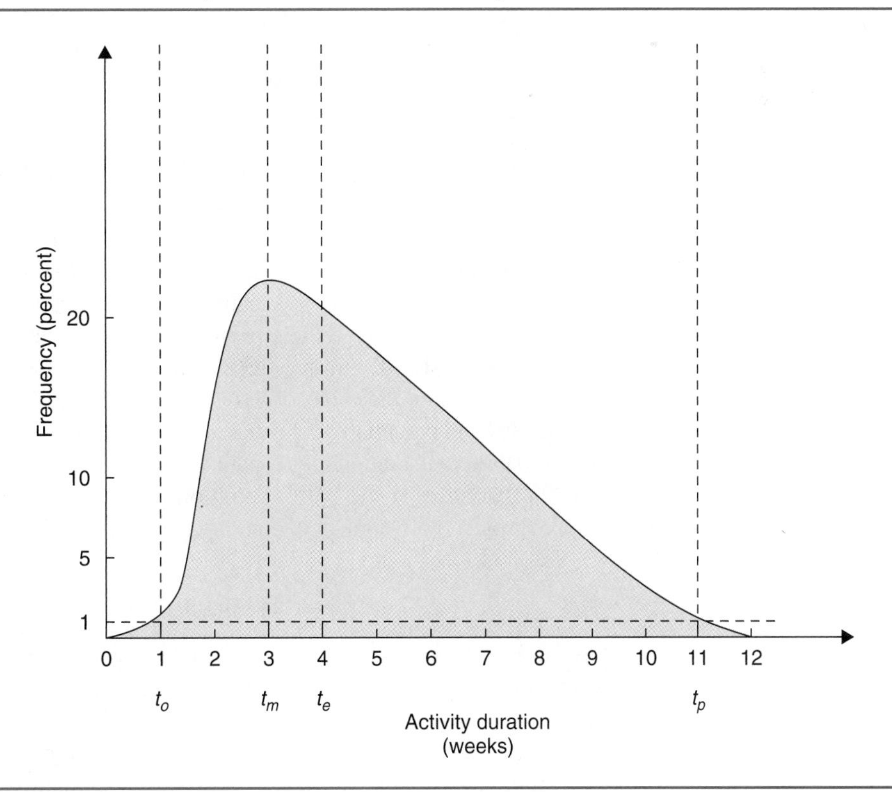

For example, for activity b, the standard deviation is

$$\sigma_b = (11 - 1)/6 = 1.67 \text{ weeks}$$

and the variance is

$$\sigma_b^2 = 1.67^2 = 2.78 \text{ weeks squared}$$

For activities h and i, the variance is zero, because $t_o = t_m = t_p$ for these activities. This means that no uncertainty is involved in their estimates. The larger the variance, the greater the degree of uncertainty involved.

Assuming that the durations of the activities are independent of each other, the variance of a *group* of activities (designated by V) can be computed by adding the variances of the activities in that group. The value of V is then expressed by:

$$V = \sigma_1^2 + \sigma_2^2 + \cdots + \sigma_n^2$$

where n is the number of activities in the group.

Earlier we defined the critical path as the path with the longest duration. When activities have uncertain durations it is typically not possible to identify the critical path prior to the completion of the project. For example, referring to the Moose Lake project, the expected time of path *a–c–f–i–j* is 27 weeks. Likewise the expected times of paths *a–b–g–i–j* and *a–e–h–j* are 24 and 10, respectively. (For ease of demonstration, we will ignore path *a–c–d–g–i–j* in the remainder of our analysis in this section.) Based on these calculations, one might conclude that path *a–c–f–i–j* is the critical path. However, if each activity's actual duration is the same as its most likely duration except for activities c and f, which take their optimistic amount of time to complete, and activities b and g, which finish in their pessimistic amount of time, then the time to complete paths *a–c–f–i–j* and *a–b–g–i–j* would be 22 weeks and 32 weeks, respectively. Thus, we see that it is possible for either of these paths to be the critical path.

Furthermore, in real-world projects, it can be dangerous to try to identify the critical path prior to the completion of the project when activity times are uncertain. In these cases the critical path tends to shift as the project progresses and the completion of some activities goes better than expected while others go much worse than expected. This can create confusion for the project team members. One week they are under the close scrutiny of the project manager because the activities they have been assigned to are on the "critical path," while the next week another path has been identified as the critical path. In fact human behavior likely plays a role in the shifting of the critical path. On the one hand, because of the close scrutiny project team members are under when assigned to activities identified as critical, these activities tend to be completed by the most likely time estimate or earlier. On the other hand, project team members operating under the assumption that their activities are not critical feel less pressure and therefore tend to complete their activities closer to the pessimistic time estimate. In these situations, the critical path is often found to shift each time the project's progress is updated. The managerial implication is that all paths that can potentially delay the project should be appropriately managed.

The probability of completing a path by a specified time can be calculated. First, however, the variance of the path's duration must be calculated. For example, in the Moose Lake project, the variance of path *a–c–f–i–j* is calculated as

$$\begin{aligned} V &= \sigma_a^2 + \sigma_c^2 + \sigma_f^2 + \sigma_i^2 + \sigma_j^2 \\ &= 0.44 + 0.11 + 4.00 + 0 + 0.11 = 4.66 \end{aligned}$$

The value of V can be computed, in a similar manner, for *any event* in the network by considering the group of activities along the path leading to the event. *Note:* The method described here is valid only if the following assumptions hold: (1) there is a large number of activities (at least 25) on the path and (2) the activities' completion times are independent of each other. If the above assumptions are not valid, simulation (described later) must be used.

Managerial Applications The managerial questions raised at the beginning of this section—the chance of completing the project in a certain desired time and the duration necessary for obtaining any desired probability of completion—can now be answered. Let

T_S = the project's estimated completion time (27 weeks in the example)
D = the desired completion time (30 weeks in the example)
z = the number of standard deviations of a normal distribution corresponding to the probability of completing the project by the desired completion time

$$z = \frac{X - \mu}{\sigma} = \frac{D - T_s}{\sqrt{V}} = \frac{D - T_s}{\sigma}$$

EXAMPLE

Finding the Probability of Completion within a Desired Time, D

Management wishes to know the probability of completing the Moose Lake project *on or before* the 30th week, as specified in the contract. Thus: $D = 30$ and $T_S = 27$ (as computed). The t_e, variance, and standard deviation for each Moose Lake activity can be calculated as shown in the spreadsheet in Exhibit 8.22. The T_S, variance, and standard deviation for each path are calculated in the spreadsheet shown in Exhibit 8.23 based on the values calculated in Exhibit 8.22. Since the longest that path *a–e–h–j* can take is 18 weeks (5 + 9 + 1 + 3), it has virtually a 100 percent chance of being completed by week 30 and therefore does not have to be considered further.

The probability of completing path *a–c–f–i–j* by week 30 can be calculated as follows:

$$z = (30 - 27) / \sqrt{4.66} = 3 / 2.16 = 1.39$$

The probability equivalent to z = 1.39 is 0.9177. Therefore, there is a 91.77 percent chance of completing this path within 30 weeks. (Remember that there is a 50 percent chance of completing this path by week 27.) Exhibit 8.24 depicts the situation.

EXHIBIT 8.22 Calculating Each Moose Lake Activity t_e, Variance, and Standard Deviation

	A	B	C	D	E	F	G	H	I
1			Most				Standard		
2	Activity	Optimistic	Likely	Pessimistic	t_e	Variance	Deviation		
3	a	1	3	5	3	0.44	0.67		
4	b	1	3	11	4	2.78	1.67		
5	c	3	4	5	4	0.11	0.33		
6	d	1	2	3	2	0.11	0.33		
7	e	3	3	9	4	1.00	1.00		
8	f	2	5	14	6	4.00	2.00		
9	g	2	4	4	3	0.11	0.33		
10	h	1	1	1	1	0.00	0.00	=SQRT(F12)	
11	i	12	12	12	12	0.00	0.00		
12	j	1	2	3	2	0.11	0.33		
13									
14			=(B12+(4*C12)+D12)/6						
15									
16					=((D12-B12)/6)^2				
17									

EXHIBIT 8.23 Calculating Moose Lake Path T_S's, Variances, and Standard Deviations

	A	B	C	D
1				Path
2		Path	Path	Standard
3	Path	T_S	Variance	Deviation
4	*a-c-f-i-j*	27	4.66	2.16
5	*a-b-g-i-j*	24	3.44	1.85
6	*a-e-h-j*	10	1.55	1.24
7				
8				
9				
10				
11				
12				

=0.44+1+0+0.11

=SQRT(C6)

EXHIBIT 8.24 Chance of Completing Path *a–c–f–i–j* in 25, 27, and 30 Weeks

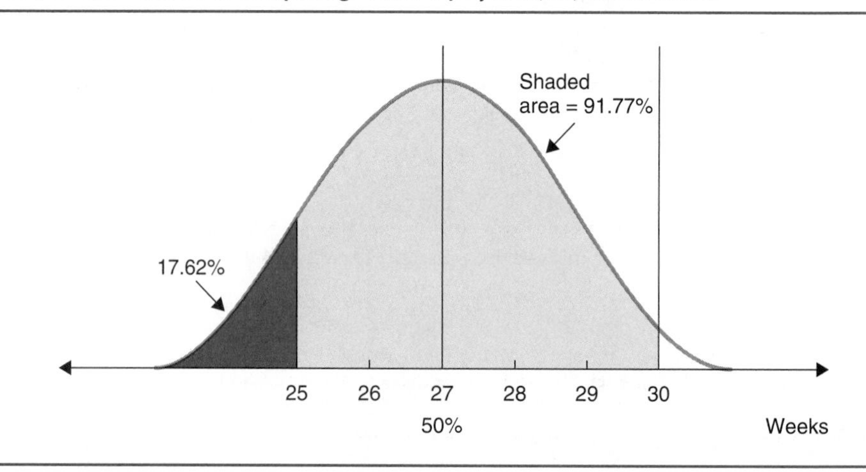

In a similar manner, the z for completing path *a–c–f–i–j* within 25 weeks is

$$z = (25 - 27) / \sqrt{4.66} = -0.93$$

from which we find the probability of 0.1762. Thus, there is only a 17.62 percent chance of completing the project in 25 weeks or less (see Exhibit 8.24). (*Note:* If z is negative, the corresponding probability is always less than 50 percent. If z is positive, the corresponding probability is always more than 50 percent; and if z is 0, the corresponding probability is exactly 50 percent.)

To calculate the probability that the project is completed by the specified time, similar calculations are made to determine the probability that the other paths are completed by the specified time. Finally, since all paths must be completed for the project to be completed, the probability that the project is completed by the specified time is calculated by multiplying together all the probabilities that each path finishes by the specified time. This is based on our earlier assumption *that the paths are independent* of one another. Referring to the spreadsheet shown in Exhibit 8.25, we see that the probability of the Moose Lake project being completed by week 30 is 91.71 percent. Also, note how Excel's NORMDIST function was used to simplify calculating the probability that each path finished by the specified time.

EXHIBIT 8.25 Calculating the Probability of Finishing the Moose Lake Project by Week 30

	A	B	C	D	E	F	G	H	I
1	***D***	30							
2									
3				**Path**	**Probability of**				
4		**Path**	**Path**	**Standard**	**Path Finishing**				
5	**Path**	T_S	**Variance**	**Deviation**	**by Time *D***				
6	*a-c-f-i-j*	27	4.66	2.16	91.77%				
7	*a-b-g-i-j*	24	3.44	1.85	99.94%				
8	*a-e-h-j*	10	1.55	1.24	100.00%				
9									
10				**Probabilty of project**					
11				**finishing by *D***	**91.71%**				
12									
13									
14									

=NORMDIST(B$1,B6,D6,TRUE)
copy to cells E7:E8

=E6*E7*E8

EXAMPLE

Finding the Duration Associated with a Desired Probability

In the previous example, a chance of 91.77 percent of completing path *a–c–f–i–j* in 30 weeks was computed. Suppose that management would like to know for what duration they can be 80 percent sure of completion. The value of z associated with 80 percent is 0.845. Solving the previous equation with *D* as the unknown, we obtain:

$$D = z\sigma + T_S$$

or

$$D = (0.845 \times 2.16) + 27 = 28.83 \text{ weeks}$$

That is, there is an 80 percent chance of completing path *a–c–f–i–j* in 28.83 weeks. Alternatively, we could have arrived at the same 28.8 weeks using Excel's NORMINV function as follows:

NORMINV(0.8,27,2.16)

where the first parameter corresponds to the probability of completing the path, the second parameter to the path's T_S, and the third parameter to the path's standard deviation of completion time. The computation of *D* enables management to make delivery commitments knowing the degree of risk assumed.

Determining the Distribution of Project Completion Times with Simulation

Here we take a different approach to assess the risk associated with completing a project by a specified time, building on the probabilistic approach just discussed. As we will demonstrate, computer simulation can be used to approximate the distribution of project completion times and then, from this distribution, the risk of completing the project within various time frames assessed. The primary advantage of using simulation is that the assumption of path independence is not required. This is significant because, in reality, paths often have common tasks (for example, *a* and *j* in Exhibit 8.25) and share resources, making this assumption rather unrealistic.

To simulate the completion of the Moose Lake project, consider the spreadsheet shown in Exhibit 8.26. Simulating the completion of the project is accomplished by generating random activity times for the ten activities (i.e., activities *a* to *j*) and then adding up the activity times that make up each path. The path completion times are next calculated based on the randomly generated activity times. Finally, the longest path on a given replication determines the project's overall duration. Given the size limitations of pages in this book, the spreadsheet shown in Exhibit 8.26 was developed to simulate the completion of the Moose Lake project 25 times. In reality, the simulation model should be replicated hundreds and perhaps even thousands of times. Fortunately, replicating the model additional times only requires copying the formulas down to additional rows, a rather trivial task.

As you can see in Exhibit 8.26, columns A through J are used to store the randomly generated activity times for activities *a* to *j*, respectively. In column K, the time to complete path *a–c–f–i–j* is calculated based on the activity times generated in columns A through J. For example, the formula =A3+C3+F3+I3+J3 was entered in cell K3. Columns L and M are used to calculate the completion time of the other two paths, respectively. Once entered, the formulas in cells K3:M3 can be easily copied down to replicate the completion of the project multiple times. Finally, the project's duration on a given replication is calculated in column N by finding which path on the particular replication took the longest to complete. For example, =MAX(K3:M3) was entered in cell N3 and then copied down to the remaining cells in the range.

For the purpose of this example, we make one simplifying assumption. Namely, we will assume that the activity times are normally distributed rather than having a beta distribution. We make this assumption because generating random numbers from a normal distribution is more straightforward in Excel than trying to generate random numbers from a beta distribution. We do note, however, that there are add-ins available, such as

EXHIBIT 8.26 Spreadsheet Model to Simulate Moose Lake Project Completion

	A	B	C	D	E	F	G	H	I	J	K	L	M	N
1	Activity	Activity	Activity	Activity	Activity	Activity	Activity	Activity	Activity	Activity	Path	Path	Path	Project
2	A	B	C	D	E	F	G	H	I	J	*a-c-f-i-j*	*a-b-g-i-j*	*a-e-h-j*	Fin. Time
3											0	0	0	0
4											0	0	0	0
5											0	0	0	0
6											0	0	0	0
7											0	0	0	0
8											0	0	0	0
9											0	0	0	0
10											0	0	0	0
11											0	0	0	0
12											0	0	0	0
13											0	0	0	0
14											0	0	0	0
15											0	0	0	0
16											0	0	0	0
17											0	0	0	0
18											0	0	0	0
19											0	0	0	0
20											0	0	0	0
21											0	0	0	0
22											0	0	0	0
23											0	0	0	0
24											0	0	0	0
25											0	0	0	0
26											0	0	0	0
27											0	0	0	0
28										Min	**0**	**0**	**0**	**0**
29										Max	**0**	**0**	**0**	**0**
30										Average	**0**	**0**	**0**	**0**

=A3+C3+F3+I3+J3
copy to K4:K27

=A3+B3+G3+I3+J3
copy to L4:L27

=A3+E3+H3+J3
copy to M4:M27

=MAX(K3:M3)
copy to N4:N27

Crystal Ball (see Chapter 7 Appendix), that can be used to facilitate the task of generating random numbers from many of the more complex probability distributions, including the beta distribution. Given our purpose here of illustrating the process of simulating the completion of the project, our assumption of normal activity times is acceptable.

To generate the random activity times for activity *a,* first select Tools from Excel's main menu. Then select Data Analysis and Random Number Generation. After selecting Random Number Generation, fill in the fields of the Random Number Generation dialog box as shown in Exhibit 8.27 and click on the OK button. Refer to Exhibit 8.22 for each activity's mean and standard deviation. (For more information on using Excel's Random Number Generation tool, see Chapter 7.) Repeat these steps for the other activities.

This procedure is repeated to generate the activity times for activities *b* through *g,* and *j.* Activities *h* and *i* have no variance and therefore their durations can be entered as constants. Exhibit 8.28 shows the final spreadsheet after all the random activity times have been generated.

Upon reviewing Exhibit 8.28, several insights emerge. First, across the 25 replications of the project, the fastest the project was completed was 23.5 weeks and the longest was 29.5 weeks. On average, the project required 26.8 weeks, which is relatively close to the 27 weeks we calculated earlier. Furthermore, in 92 percent (23/25) of the cases, path *a–c–f–i–j* was the critical path while path *a–b–g–i–j* was the critical path 8 percent (2/25) of the time.

We can take this analysis a step further and investigate the actual distribution of project completion times. To do this, the simulation model was extended to simulate the completion of the project 100 times. Based on the larger sample size of 100, the average project completion time was 27.1 weeks with a standard deviation of 1.8 weeks. Furthermore, across these 100 replications, the fastest project completion time was 23.1 weeks and the

EXHIBIT 8.27 Excel's Random Number Generation Box for Activity *a*

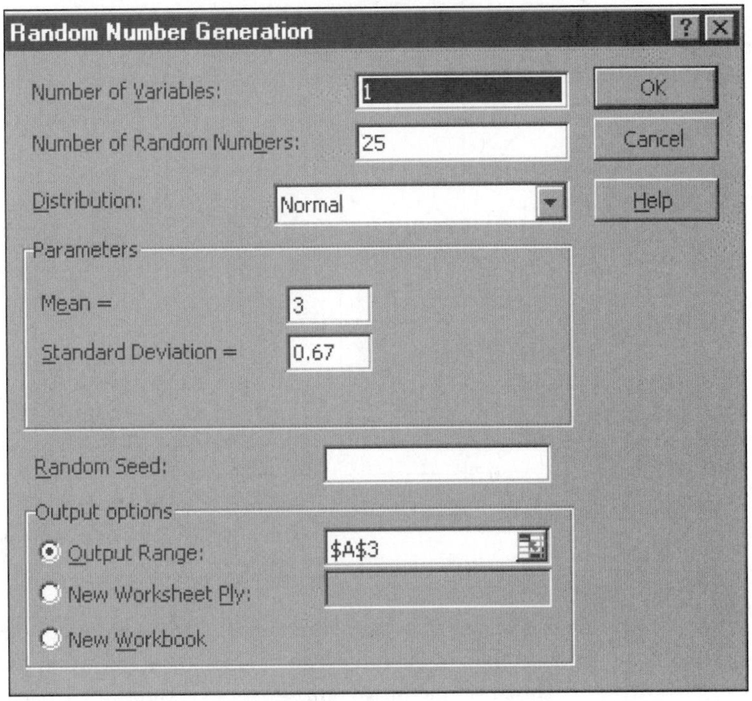

Random Number Generation

Number of Variables: 1

Number of Random Numbers: 25

Distribution: Normal

Parameters

Mean = 3

Standard Deviation = 0.67

Random Seed:

Output options

Output Range: $A$3

New Worksheet Ply:

New Workbook

OK

Cancel

Help

EXHIBIT 8.28 Final Spreadsheet Simulating Completion of Project 25 Times

	A	B	C	D	E	F	G	H	I	J	K	L	M	N
1	**Activity**	**Activity**	**Activity**	**Activity**	**Activity**	**Activity**	**Activity**	**Activity**	**Activity**	**Activity**	**Path**	**Path**	**Path**	**Project**
2	**A**	**B**	**C**	**D**	**E**	**F**	**G**	**H**	**I**	**J**	***a-c-f-i-j***	***a-b-g-i-j***	***a-e-h-j***	**Fin. Time**
3	2.8	2.9	4.0	2.0	4.1	4.8	2.6	1.0	12.0	1.9	25.5	22.2	9.8	25.5
4	2.1	2.7	3.9	2.5	3.0	7.7	2.9	1.0	12.0	1.9	27.6	21.7	8.1	27.6
5	3.2	4.8	3.6	1.9	4.4	6.3	3.0	1.0	12.0	1.8	26.9	24.7	10.4	26.9
6	3.9	5.7	4.3	2.1	4.0	5.1	2.4	1.0	12.0	2.1	27.4	26.0	11.0	27.4
7	3.8	4.7	4.8	2.2	2.5	3.0	3.2	1.0	12.0	2.0	25.6	25.8	9.3	25.8
8	4.2	6.3	3.8	2.2	3.4	7.2	3.0	1.0	12.0	2.3	29.5	27.7	10.9	29.5
9	1.5	3.1	3.6	2.0	4.6	6.4	3.0	1.0	12.0	2.0	25.6	21.6	9.2	25.6
10	2.8	3.4	4.0	2.1	3.5	7.2	3.3	1.0	12.0	1.7	27.7	23.3	9.1	27.7
11	3.7	3.8	3.5	2.2	4.1	4.7	3.4	1.0	12.0	2.5	26.5	25.5	11.3	26.5
12	2.3	5.7	4.4	1.7	3.1	8.8	2.8	1.0	12.0	1.9	29.4	24.7	8.3	29.4
13	2.5	2.9	3.7	2.2	2.8	9.3	3.1	1.0	12.0	1.8	29.3	22.4	8.1	29.3
14	1.9	2.7	3.7	2.1	4.4	8.6	2.8	1.0	12.0	2.1	28.3	21.5	9.4	28.3
15	1.8	4.6	3.9	2.4	5.4	4.9	2.7	1.0	12.0	1.7	24.3	22.7	9.9	24.3
16	2.3	3.9	4.4	1.6	3.9	6.9	3.0	1.0	12.0	2.0	27.7	23.2	9.3	27.7
17	2.5	3.1	3.8	1.5	2.6	5.4	3.2	1.0	12.0	1.9	25.5	22.6	8.0	25.5
18	1.6	3.9	4.4	2.0	5.1	6.4	2.9	1.0	12.0	2.4	26.8	22.7	10.1	26.8
19	2.6	4.3	4.6	2.2	7.1	3.9	2.9	1.0	12.0	2.2	25.3	24.1	12.9	25.3
20	2.7	6.5	4.3	1.9	5.5	4.1	3.0	1.0	12.0	1.8	24.8	26.0	11.0	26.0
21	3.1	3.8	4.2	1.8	3.5	6.5	3.0	1.0	12.0	2.0	27.9	24.0	9.7	27.9
22	2.8	6.2	3.5	2.5	3.8	4.9	3.0	1.0	12.0	1.8	25.0	25.8	9.4	25.8
23	2.8	5.3	4.4	2.2	3.4	5.1	2.7	1.0	12.0	1.5	25.8	24.4	8.7	25.8
24	2.8	1.8	4.0	1.8	4.1	2.2	3.1	1.0	12.0	2.5	23.5	22.1	10.3	23.5
25	3.9	5.4	3.7	2.3	5.2	7.1	2.5	1.0	12.0	2.1	28.8	25.8	12.1	28.8
26	2.9	0.7	3.9	2.0	4.2	4.6	3.3	1.0	12.0	2.1	25.6	21.1	10.3	25.6
27	2.9	1.7	3.7	2.2	4.0	5.9	3.1	1.0	12.0	1.9	26.4	21.6	9.8	26.4
28			**3.48127**	**1.541**	**2.5**	**2.2**	**2.4**	**1.0**	**12.0**	**Min**	**23.5**	**21.4**	**8.0**	**23.5**
29										**Max**	**29.5**	**27.7**	**12.9**	**29.5**
30										**Average**	**26.7**	**23.7**	**9.8**	**26.8**

EXHIBIT 8.29 Distribution of Project Completion Times

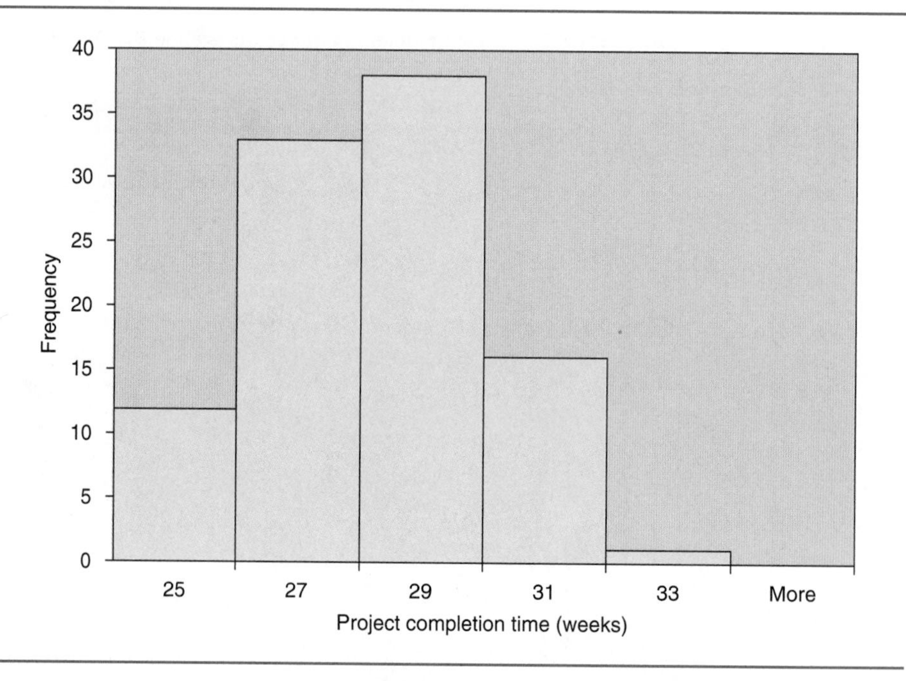

longest was 32.2 weeks. Next, Excel's Histogram tool (see Chapter 2) was used to create the histogram of the project completion times shown in Exhibit 8.29.

Based on the histogram shown in Exhibit 8.29, the project completion times appear to be normally distributed (in the detailed modeling example we verify this using the chi-square goodness of fit test discussed in Chapter 2). We can use this information similar to the way the information was used about individual paths to calculate the probability that the project is completed by some specified time. For example, the probability that the project is completed within 30 days can be calculated using Excel as

=NORMDIST(30,27.1,1.8,TRUE)

which equals 94.64 percent. Therefore, there is almost a 95 percent chance of completing this project in 30 weeks or less. We can also reverse the question and determine what project completion time has an 80 percent chance of being completed as follows:

=NORMINV(0.80,27.1,1.8)

which equals 28.6 weeks. Thus, there is an 80 percent chance that the Moose Lake project can be completed in 28.6 weeks or less.

8.5 Step 6: Monitoring and Controlling the Project

We discuss monitoring briefly, noting some common techniques used by project managers, and then turn our attention to means for controlling projects.

Monitoring the Project

The preceding sections were concerned with planning functions. Once the project has been initiated, management's attention turns to *monitoring* the progress of the project and taking action to bring reality into conformance with plan, known as *control.* One of the primary means for monitoring is the use of cost variance reports. The costs for the activities have been determined beforehand, and deviations from these costs give insight into how the project is progressing. In general, costs that are running too high may be an indication that the budget will be exceeded, resulting in a loss for the organization. If the costs are running too low, it may be an indication that the project is behind schedule.

Clearly, monitoring costs is helpful but not definitive. First, it only monitors the cost aspect of the project, leaving performance and schedule unmonitored. Second, even if costs are precisely what they should be, it may well be that activity costs are indeed excessive, but progress is slower than planned so the resulting cost figures do not show any problem. To alleviate this difficulty somewhat, it is common to see project managers using some form of *critical ratio* such as (actual progress divided by scheduled progress) × (budgeted cost divided by actual cost). Although an improvement over simply monitoring costs, this ratio, too, can be misleading.

A better approach developed for project management is the use of *earned value* to denote progress on a project. A baseline plan that combines planned schedule and cost values is derived and used to compare with both actual costs and actual earned value completed. This then gives management three variances to help monitor progress: a spending variance (when costs exceed value completed), a schedule variance (when value completed falls below the baseline plan), and a time variance (when value completed falls behind schedule). Other variances and ratios are also commonly derived from the earned value data to help management monitor the progress of the project. (For more information, refer to Meredith and Mantel 2000.)

Controlling the Project

The previous two sections showed how to construct a network and find the critical activities, the noncritical activities, and the slack. Based on this information, and using the principle of management by exception, it is possible to construct a management system with tighter control over the critical activities and less control over the noncritical activities. Alternatively, one can use very tight control over critical activities, less control on activities with small slack, and the least control over activities with large slack. PERT/CPM can also be used to generate additional information, as will be shown.

Suppose that the Moose Lake project started on schedule. However, the very first activity, the administrative setup, is delayed. Although the duration of this critical activity had been estimated as three weeks, it is now clear that it will take 4 weeks to handle all the administrative details. Thus, when the time comes for event **2,** its T_E will be 4, rather than 3. The slack in event **2** will thus be:

$$S = T_L - T_E = 3 - 4 = -1$$

That is, the slack has a negative value and is labeled as *negative slack.* A negative value for a slack means that the project is behind schedule. If T_E's are now computed for all the remaining critical events, including the ending event, there will be a negative slack of 1. This implies that the *entire project* will be delayed by one week. What can management do about this?

Look, for a moment, at event **5,** which is not critical. The previous slack for this event was computed as 24 – 7 = 17; now it will be 24 – 8 = 16, still a positive slack. Activities *b, d, e,* and *h* likewise possess positive slack. This means that these activities can still be delayed without delaying the entire project. Slowing down noncritical activities may release resources (such as labor, tools, and equipment) that can then be transferred to one (or more) of the critical activities. If such a transfer could reduce the completion time of any critical activity by one week, the delay could be eliminated and the project would still be completed on schedule.

A similar situation may develop if a noncritical activity such as *g* requires 6, rather than 3, weeks for completion. The T_E for event **6** would then be 15 weeks, and a negative slack of 2 would be formed at event **6.** Notice that the critical path will be changed to **1–2–3–4–6–7–8,** and activities *d* and *g* will become critical. In general, any deviation of the actual time from the computed duration should be reported to the project manager, who in turn will re-compute the critical path. Previously noncritical events and activities may become critical, and vice versa.

In addition to transferring resources to critical activities, management can correct delays by some other actions:

- Relaxing (making less strict) the technical specifications or the required quality
- Changing the scope of the project by reducing the desired goals and thereby the amount of work
- Changing the sequence of activities
- Obtaining additional resources for the project
- Expediting activities by various incentives
- Starting activities while preceding ones are still being worked on

The regular PERT/CPM analysis is limited to planning the elapsed time. Because the planning is done prior to actual project execution, it is not always possible to know the precise resource availability, so we assume that there are sufficient resources for executing the activities as planned.

Suppose, however, that a project includes two activities that have the same early start date. In addition, they both are noncritical, with 3 days' slack and a duration of 7 days. Assume that each activity requires a bulldozer throughout the duration, and it is then discovered that only one of the two bulldozers is operable. It is therefore apparent that both activities *cannot* proceed simultaneously. Because neither of them is critical, neither has a priority, and because there is not sufficient slack, then either one must be postponed, the second bulldozer must be quickly repaired, or the network must be rearranged to move one of these activities forward in time. In other words, the way that the company allocates one of its resources may affect the critical path and the completion date.

This example can be extended to cover other resources (e.g., labor, money). In general, whenever several activities need a limited resource, it is necessary to decide how to allocate this resource. Because the regular PERT/CPM analysis essentially contains only early and late start dates for the activities, the project manager is free to decide on an actual start date between these limits. In contrast, a **resource allocation schedule** contains a scheduled start date for every activity, taking into consideration the availability of resources.

EXAMPLE

Resource Allocation Schedule

The following example involves the management of labor. Assume that a project is given as shown in Exhibit 8.30 (times in weeks). The critical path is shown as a heavy line on the network; the earliest and latest dates for each activity are given in Exhibit 8.31. The last column of the table lists the number of employees required each week to work on each activity. For simplicity, we assume that all the employees possess the same skill.

Solution. The project shown in Exhibit 8.30 is transferred to a bar chart in Exhibit 8.32 with each activity at its earliest starting time. In Exhibit 8.33, the weekly labor requirements indicate an *unbalanced demand* for labor, ranging from one to seven employees per week, which may be difficult to arrange. A more leveled schedule can be derived by delaying the noncritical activities **1–3, 1–4,** or **4–5.** Such an arrangement is shown in Exhibit 8.34. This arrangement reduces the peak manpower requirements to five without affecting the completion date of the project. But if only four employees are available, it would be necessary to delay the completion date of the entire project or use overtime (at an increased cost). A trial-and-error approach is used in such cases.

EXHIBIT 8.30 Example Project

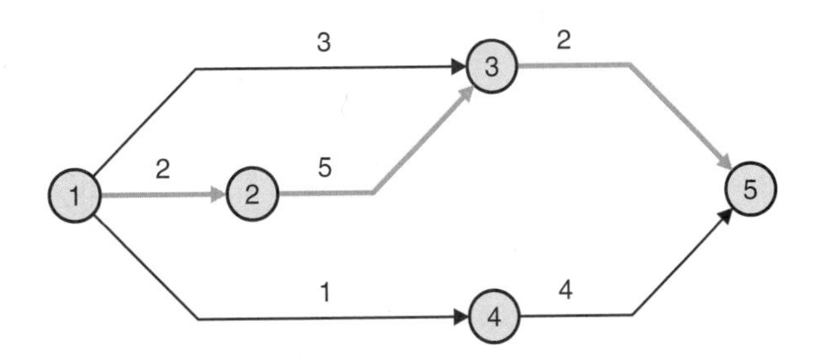

EXHIBIT 8.31 Earliest and Latest Dates

Activity	Earliest Start	Latest Start	Slack (T_F)	Employees
1–2	0	0	0	2
1–3	0	4	4	2
1–4	0	4	4	3
2–3	2	2	0	2
3–5	7	7	0	1
4–5	1	5	4	2

EXHIBIT 8.32 Earliest Start Schedule

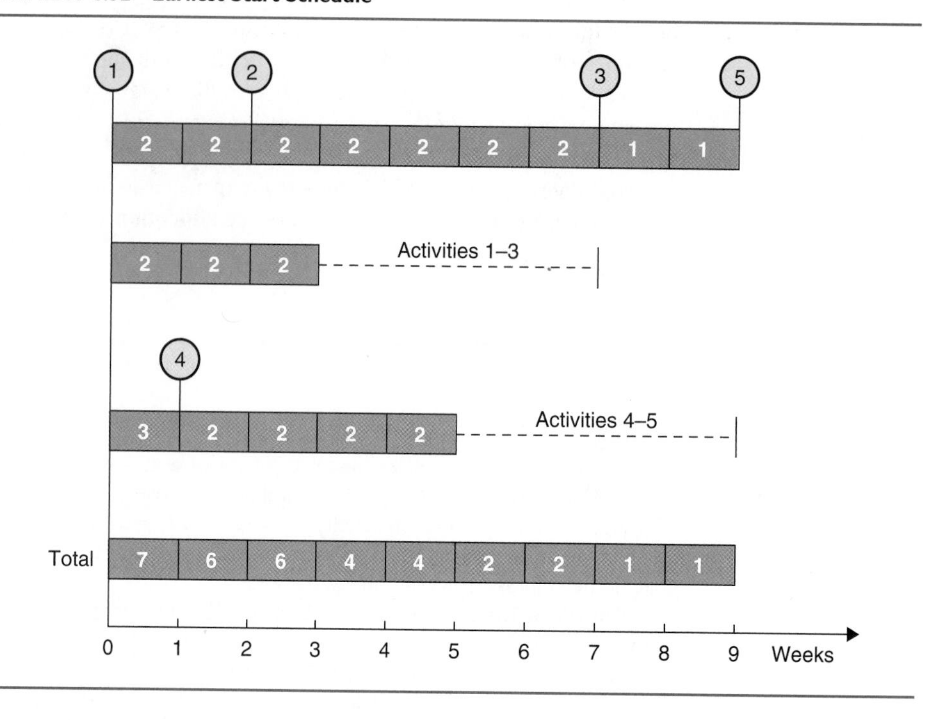

EXHIBIT 8.33 Manpower Requirement for ES Schedule

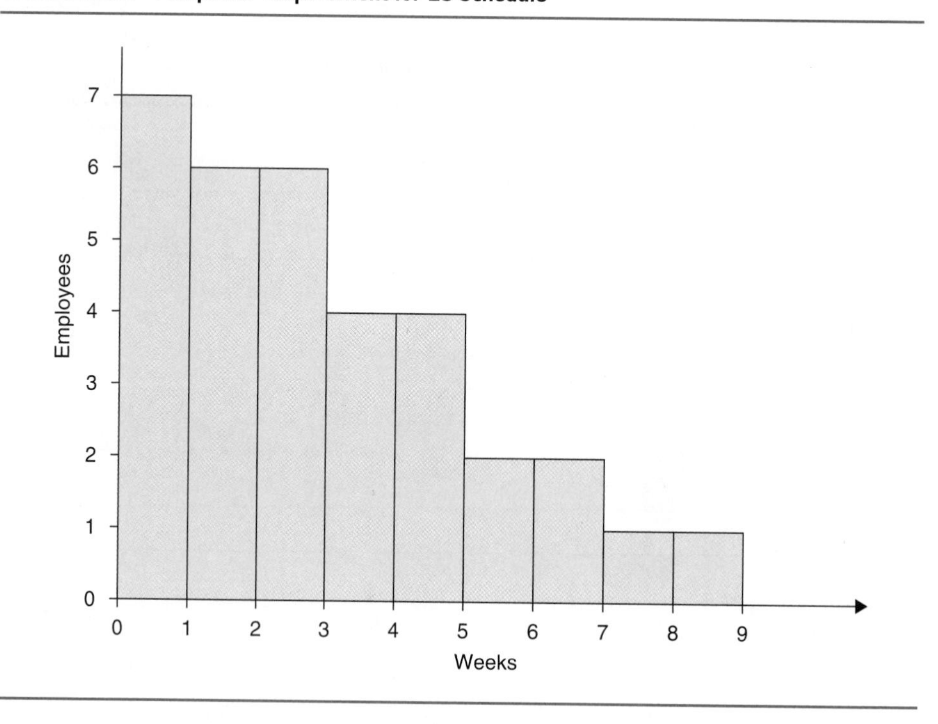

Critical Path Method (CPM): Cost–Time Trade-Offs

CPM analysis is used to evaluate various alternatives of executing projects in those cases where it is possible to *expedite* the execution of some or all of the project's activities. Expediting activities requires additional resources, which means increasing the cost of the

EXHIBIT 8.34 Leveled Schedule of Labor Demands

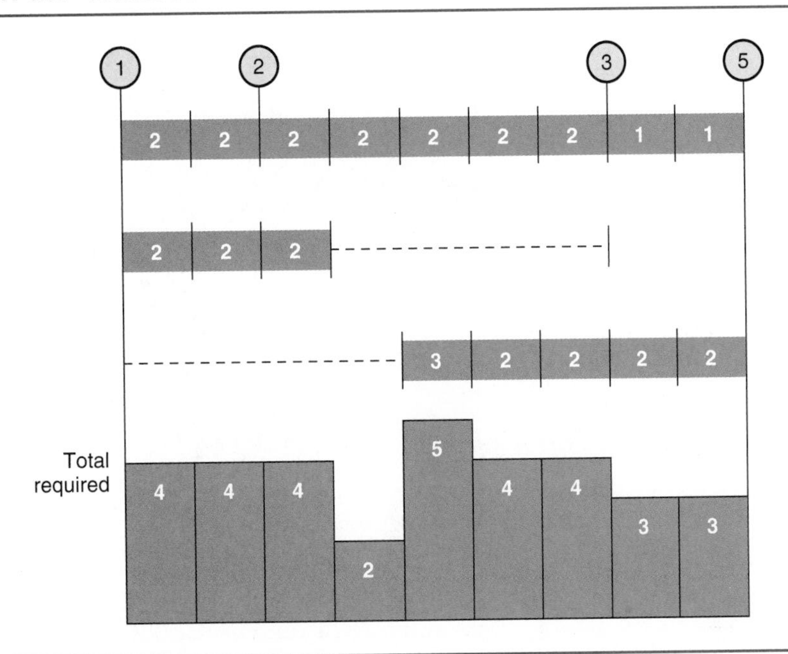

project. However, considerable savings may be realized in projects finished ahead of schedule. An example of this was observed in Phoenix, Arizona, where two builders constructed two large condominium projects. When an economic slump hit, the demand for condominiums dropped considerably. One of the builders decided to expedite construction, at a considerable cost, in order to finish first. He sold 240 units in a short time, exhausting the demand. When the second builder completed his project, he could not sell the units and was forced to file for bankruptcy. Thus, the decision of how much to expedite may be of great importance to management. The tool that enables such an analysis is CPM.

The CPM Concept Exhibit 8.35 presents the relationship in CPM between cost and time. An activity can be performed in a *normal manner* (normal point in Exhibit 8.35) requiring T_n units of time and C_n units of money (where n designates *normal*). In an extreme case, the activity can be performed on a *crash* basis (e.g., using overtime, special services, extra tools) at a time T_c and a cost C_c (where c designates *crash*). No activity can be executed in less time than T_c or more than T_n, but one can take any value between.

The crash point and the normal point can be connected by a *straight line*. Any intermediate point X on the straight line will involve T_x time and C_x cost. The slope of the straight line gives the relationship between time and cost, as shown in Exhibit 8.35.

$$\text{Slope} = \frac{C_c - C_n}{T_n - T_c}$$

The slope gives us the cost *increase* associated with a reduction of one unit of the activity duration. The assumption of a linear relationship between cost and time is not valid in all cases. In some cases, the relationships are described by a nonlinear function (the broken curve in Exhibit 8.35) and the solid line only approximates the broken line. In other cases, a stepwise curve is applicable. It is customary to write the normal and crash data for each activity directly on the diagram as shown in Exhibit 8.36 (time above the line, cost underneath). For example, for activity **1–3,** the normal time is 5 weeks, at a cost of $4,000; the crash time is 3 weeks, at a cost of $5,200.

EXHIBIT 8.35 CPM Cost–Time Trade-Offs for an Activity

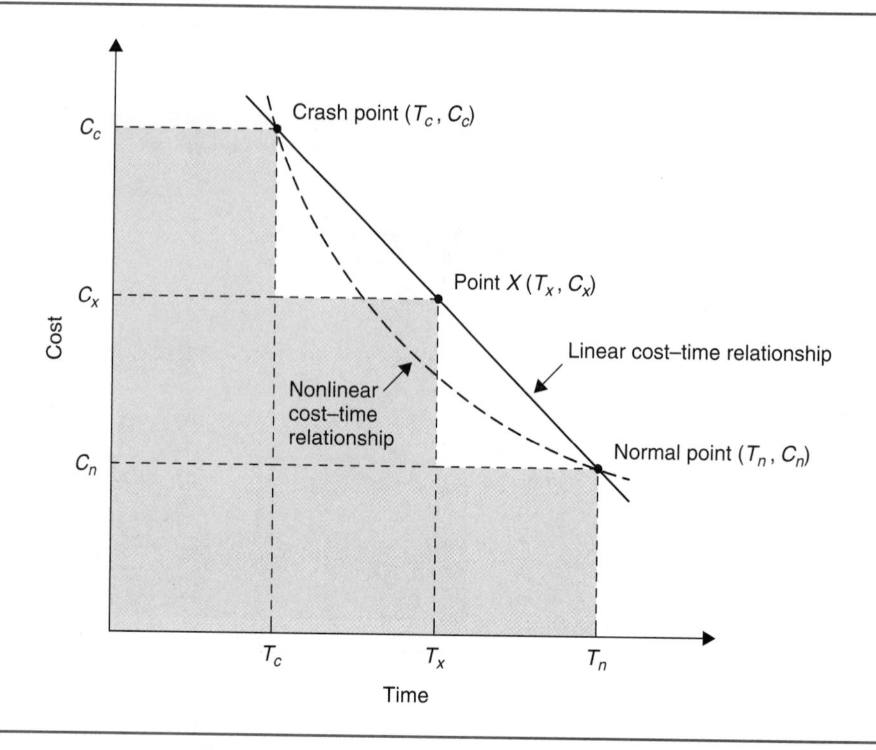

EXHIBIT 8.36 CPM Labeling

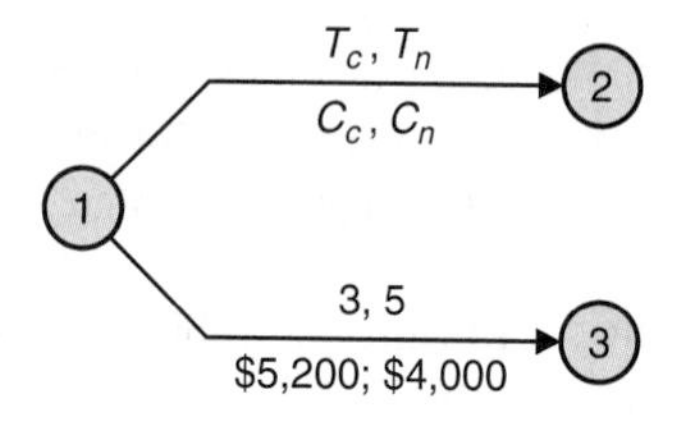

The slope of activity **1–3** is

$$\frac{\$5,200 - \$4,000}{5 - 3}$$

This is the cost required to expedite the activity by one week. The linear relationship means that it will cost $1,200 to expedite the activity by 2 weeks. The true slope will actually be a negative number because the direction of the line is from northwest to southeast; however, our equation yields a *positive* number that we use as a cost *increase.*

The CPM Analysis The CPM analysis examines the total cost involved in executing the project at various scheduled times, starting with either the lowest cost–longest duration alternative or with the higher cost–shortest duration alternative. The additional cost of expediting the project can then be compared with the possible savings from the expedited completion (e.g., a client may pay a bonus for completion ahead of schedule).

The CPM analysis starts by solving the problem twice. First, attention is paid only to *normal times.* Using the procedure outlined earlier and assuming that the normal times are the t_e's, a solution is derived, and its cost is also computed. Second, by considering only the crash times as t_e's, another solution is derived, and its cost is also computed. Once the two solutions are computed, the cost–time trade-offs are used to find the least-cost plan for any number of weeks (days) between the *all crash* and *all normal* plans. This cost can then be compared with the anticipated benefits.

EXAMPLE

Finding the Least-Cost Plan

A network of activities for a maintenance project is shown in Exhibit 8.37. The problem is to find the least-cost plan for various project durations. The normal time (in days) and cost, as well as the crash time and cost, are shown in Exhibit 8.38. The column "Cost slope" indicates the incremental *increase* in cost when the duration of the project is decreased by one day. For example, for activity *D:*

Slope = \$30 per day

Solution. The first observation that can be made from Exhibit 8.38 is that if all activities are performed in the normal duration, the total cost will be \$1,860. Second, if all activities are performed on a crash basis, the total cost will be \$2,860. The *times* required to complete the project on an *all-normal* basis and on an *all-crash* basis should be determined next.

Considering first *all-normal* times (disregard the normal costs, the crash time, and the crash costs), the critical path can be computed using the procedure shown earlier in this chapter. The results are shown in Exhibit 8.39. The critical path is **0–1–2–4–5** for a duration of 25 days and a cost of \$1,860.

EXHIBIT 8.37 A Maintenance Project

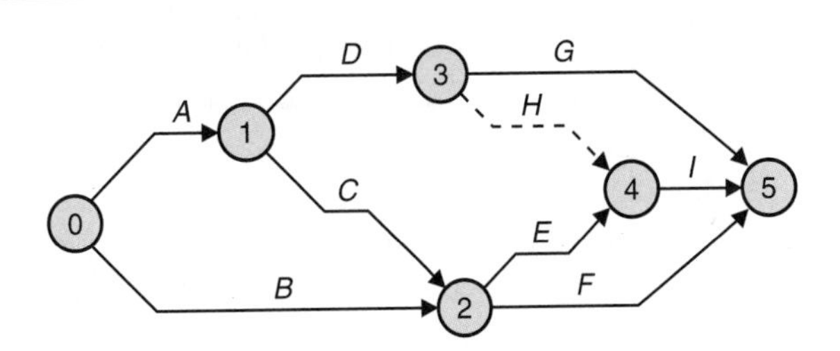

EXHIBIT 8.38 Time and Cost Information

Activity	Normal Time	Normal Cost	Crash Time	Crash Cost	Cost Slope
A	5	\$ 100	4	\$ 140	40
B	9	200	7	300	50
C	7	250	4	340	30
D	9	280	7	340	30
E	5	250	2	460	70
F	11	400	7	720	80
G	6	300	4	420	60
I	8	80	6	140	30
Total		\$1,860		\$2,860	

EXHIBIT 8.39 All-Normal Solution, 25 Days

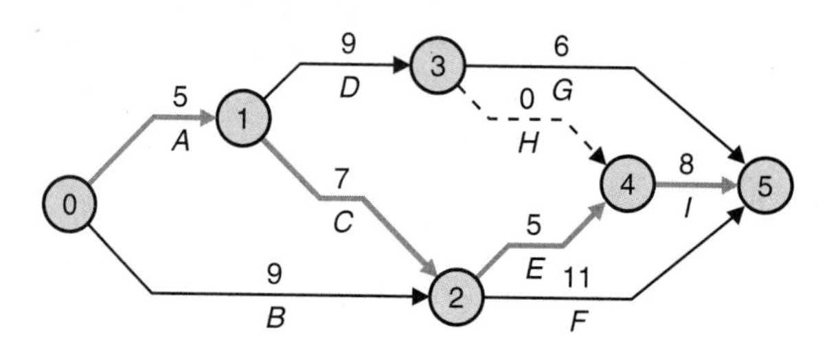

EXHIBIT 8.40 All-Crash Solution, 17 Days

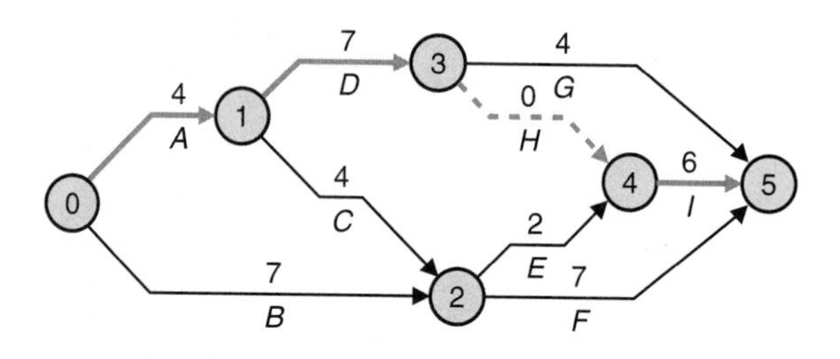

In a similar manner, the critical path of *all-crash* duration is computed (Exhibit 8.40). The critical path is **0–1–3–4–5** for a duration of 17 days and a cost of **$2,860**. At this stage, the following analysis is performed:

1. Determine the *minimum* cost for the crash time of 17 days.
2. Determine the least-cost plan for any desired number of days between all-normal to all-crash.

Find the Minimum Cost for the Crash Time So far, it was found that it is possible to perform the project in 17 days at a cost of $2,860. The question is whether it is possible to perform the project in 17 days but at a lower cost. To achieve a cost reduction, the noncritical activities could be performed at a slower pace. (This is called *expanding* the activities.) There is a simple procedure for this.

Step 1 All *noncritical* activities found in Exhibit 8.40 are listed with their appropriate cost slopes:

Noncritical Activities	Cost Slope
B. 0–2	50
C. 1–2	30
E. 2–4	70
G. 3–5	60
F. 2–5	80

Step 2 The activity with the *largest* slope is selected (activity *F*) first. The largest savings can be made if this activity is expanded first. It would be desirable to expand it *as much as possible* to achieve as large a cost reduction as possible. Because activity *F is* on two noncritical paths, **0–1–2–5** and **0–2–5,** it can be expanded until one of these becomes critical. Because path **0–2–5** now takes 14 days, it can be ex-

EXHIBIT 8.41 Least-Cost Crash Schedule of 17 Days ($2,460)

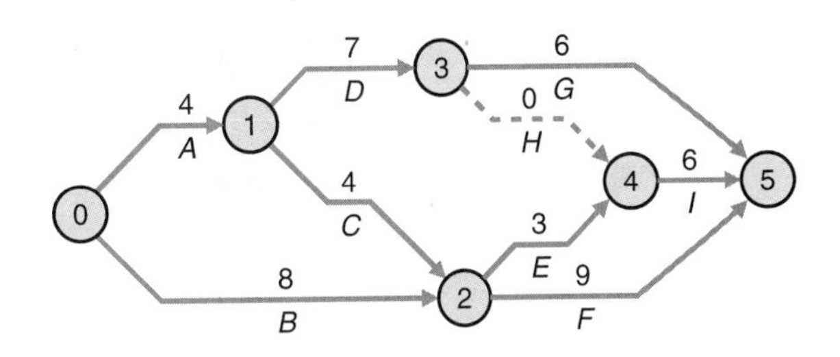

panded by 3 days to make it critical (up to 17 days). However, path **0–1–2–5** takes 15 days, and therefore only 2 days can be added to it to make it critical. Therefore, the maximum number of days that can be added to activity *F* is two (the smaller of the two).

There is another point that should be checked in expanding an activity. The crash time of activity *F* is 7 days. The normal time is given as 11 days. Therefore, expansion by 2 days is feasible. In other cases, it *may not be feasible* to expand up to the maximum length allowed by the length of the noncritical path because of the normal time limitations that are imposed on an individual activity.

The expansion of activity *F* now yields an additional critical path, **0–1–2–5,** which will take 17 days at a cost reduction of $160.

Step 3 Activity *E,* which has the *second largest* cost reduction potential, is expanded next. Here, an expansion of only one day is possible at a $70 saving. In a similar manner, the expansion of activity *G* by 2 days will yield an additional $120, and finally, activity *B* can be expanded by 1 day, resulting in a $50 saving. Notice that because activity *F* has been expanded to 9 days, the maximum that activity *B* can be expanded is to 8 days (17 – 9 = 8). It is impossible to expand activity *C,* because it became *critical* as a result of the expansion of activity *F.*

The total cost savings are 160 + 70 + 120 + 50 = 400. Thus, the revised plan calls for a 17-day project, at a total cost of $2,860 – $400 = $2,460. This new schedule is shown in Exhibit 8.41. Notice that all activities are now *critical;* that is, no further expansion is possible. The information is then entered into a cost–time diagram (Exhibit 8.42) as point *A.*

Determine the Least-Cost Plan for Any Desired Number of Days The normal schedule is the *longest* (slowest) schedule for carrying out the project and costs the *least.* On the other hand, the all-crash schedule is *the fastest,* but is also the most expensive. In certain cases, management needs to know the cost of carrying out the project at some point between the fastest and the slowest. Such a situation may develop, for example, when a customer offers to pay a certain amount as a bonus for finishing ahead of schedule.

EXAMPLE

Least-Cost Plan for 22 Days

Let us assume that management would like to find the least-cost plan for 22 days. Two approaches are available: either *compressing* the project from 25 days (all normal) to 22 days, or *expanding* the project from 17 days (all crash) up to 22. The former approach will be illustrated here.

EXHIBIT 8.42 Cost–Time Trade-Offs

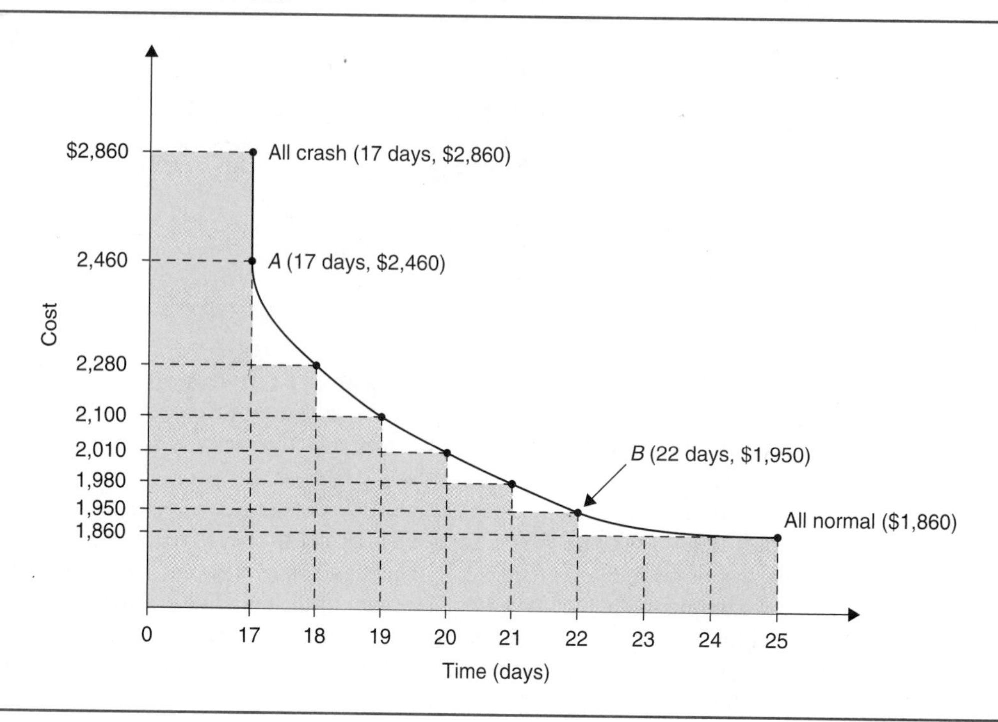

EXHIBIT 8.43 A 22-Day, Least-Cost Schedule

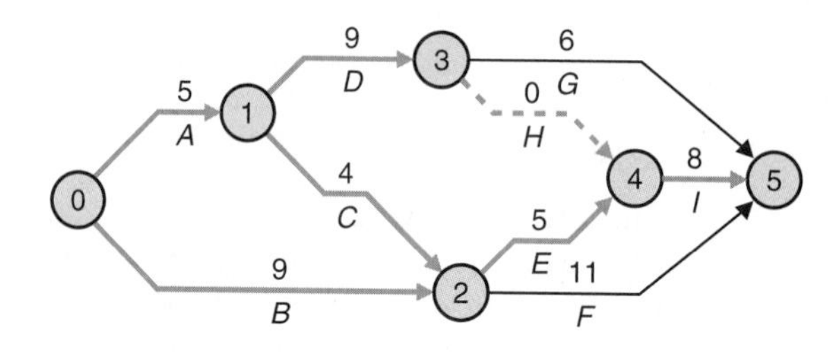

The first step is to list all *critical activities* of the all-normal schedule (Exhibit 8.39). The list of these activities and their slopes follows:

Critical Activities	Slope
A. 0–1	40
C. 1–2	30
E. 2–4	70
I. 4–5	30

The activity with the *least* slope will be compressed first, because decreasing the project time by one day will result in the smallest increase in cost. In the example, either activity *C* or activity *I* can be selected, because both have the smallest slope (lowest cost). Arbitrarily, activity *C* is selected. How much can activity *C* be compressed? The most an activity can be compressed is up to its *crash time* (4 days here). However, such a reduction may create one (or more) additional critical paths. The minute an additional critical path is created, the compression should be stopped and a cost reevaluation should be made. In the example, two additional critical paths are formed after activity *C* is reduced from 7 to 4 days. Thus, the maximum project compression is to 22 days (see Exhibit 8.43) at a cost of $1,860 + 90 = $1,950 (point *B* in Exhibit 8.42).

EXHIBIT 8.44 A 20-Day, Least-Cost Schedule

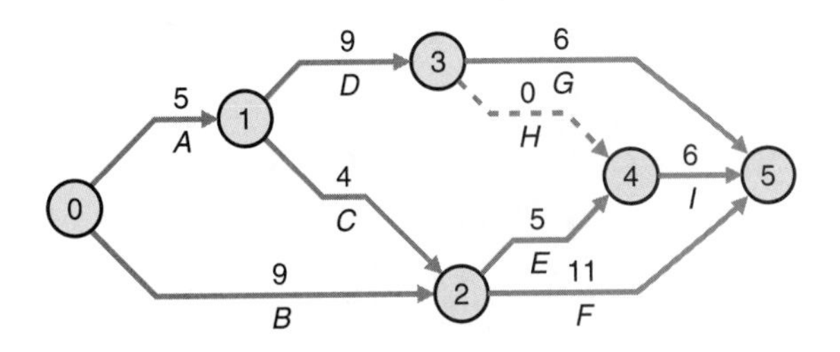

EXHIBIT 8.45 A 19-Day, Least-Cost Schedule

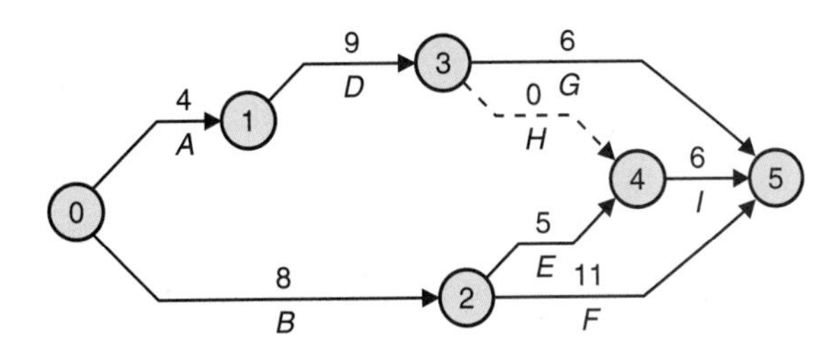

Notice that the compression to 22 days also tells us that the best plan for achieving 24 days is to cut 1 day from activity *C* and the best plan for 23 days is cutting two days from activity *C*.

Compression to 20 Days

In a similar manner, the best plan for 21 days can be found. Starting with 22 days, the *critical* activity *I* (Exhibit 8.43) is expedited, because its cost increase is now the smallest. Compressing by 1 day yields a 21-day schedule with a cost of $1,980. This activity could be compressed by 2 days to its crash time of 6 days. After compressing it by 2 days, we get a 20-day schedule at a cost of $2,010 (see Exhibit 8.44). This information is now entered in Exhibit 8.42. Note that the entire network is now critical.

Additional Compression

At this stage, a single activity can no longer be considered by itself, because there are several critical paths involved. For example, if activity *D* is reduced by one day, the critical path **0–1–3–5** will be reduced to 19 days, but other critical paths will also have to be reduced by 1 day. In this case, activities *E* and *F* have to be compressed by 1 day each and the cost effect will be felt in two places. Therefore, it is necessary to check all combinations of possible reductions to make sure that the smallest total cost is added. This is done by taking the smallest slope on a path (rather than an activity) first and adding the resultant cost impact to the other paths.

Next, calculate the least slope on the next path, taking into consideration the impact on the resultant cost, and so on. Finally, all alternatives are compared, and the one with the least-cost increase is selected. In the example, a compression of activities *A* and *B* by 1 day results in a 19-day schedule at a cost of $2,010 + 40 + 50 = $2,100 (see Exhibit 8.45).

Further compression is done in the same manner. An 18-day schedule can be obtained with a cost of $2,280, and a 17-day schedule has a cost of $2,460. All these results are entered in Exhibit 8.42 for the purpose of evaluation of the anticipated benefits.

Once Exhibit 8.42 is completed, it can be used to facilitate decisions regarding the expediting of projects. This is especially useful when one must decide whether to permit a delay or reduce it. Because the cost curves of delays and expediting run contrary to each other (see Exhibit 8.46), it is necessary to find the optimal strategy each time a delay develops. If, for example, the cost of delay is smaller than that of expediting, then the delay should be permitted.

EXHIBIT 8.46 Cost Analysis

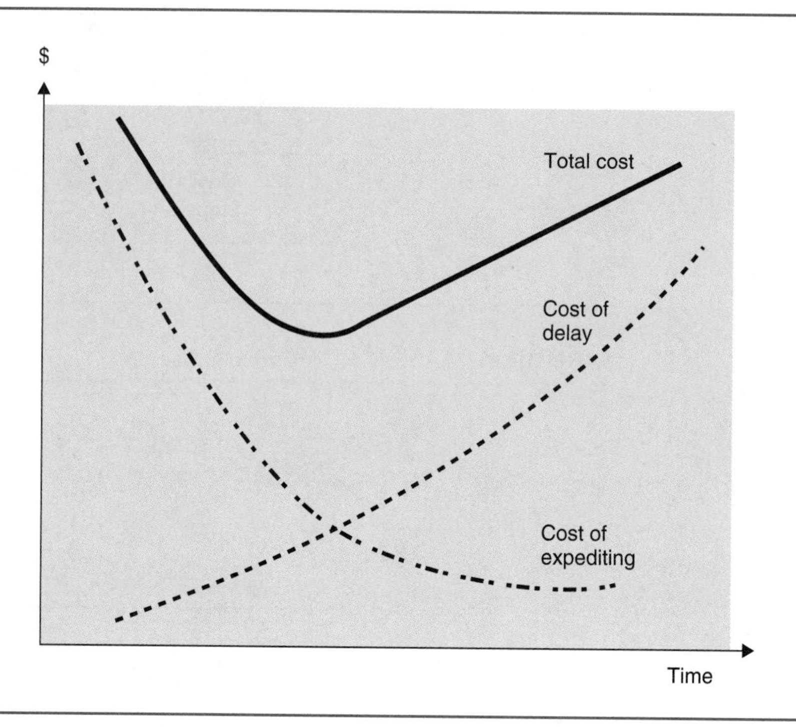

Analyzing Cost–Time Trade-Offs with Excel's Solver

As you worked through and read the preceding example, you might have been wondering whether spreadsheets or other software packages might have been able to facilitate the task of crashing a project. Clearly, for realistically sized projects, the manual approach just described could become quite tedious. Therefore, in this section, we demonstrate how Excel's Solver can be used to greatly facilitate the task of choosing the activities to expedite such that the project is completed by the specified deadline and the expediting costs are minimized.

To illustrate this, the example in Exhibit 8.37 will be used. First the data from Exhibit 8.38 were entered into the spreadsheet shown in Exhibit 8.47. The desired project deadline is entered in cell B1 and the total cost of meeting this deadline, including both normal and crash cost, is calculated in cell B2. Below this, columns A to F contain the information given in Exhibit 8.38. In column G, the maximum amount each activity can be crashed is calculated by subtracting the Crash Time (column C) from the Normal Time (column B). Column H corresponds to one of our decision variables, namely, how much to crash each activity. Based on the values entered in column H, the cost of crashing each activity is calculated in column I. Finally, the actual time to complete each activity is calculated in column J by subtracting the amount the activity is crashed (column H) from the activity's normal time (column B).

In cells A18:B22, another table was created to keep track of the time each event occurs. Node 0 was excluded because we assume that the project is started at time zero. As you will see, we need to keep track of the time each event occurs to ensure that the precedence relationships in the network diagram are not violated. For example, we need to make sure that node 3 does not occur until after node 1 occurs plus the time it takes to complete activity *D*.

To illustrate the use of Solver, assume that we desire to complete the project in 22 days. We begin by selecting Tools from Excel's main menu and then Solver from the next menu

EXHIBIT 8.47 Analyzing Cost–Time Trade-Offs with Excel

	A	B	C	D	E	F	G	H	I	J
1	Deadline		Goal							
2	Total Cost	$1,860.00	To be minimized							
3										
4		Normal	Crash	Normal	Crash	Cost	Maximum	Amount	Crashing	Actual
5	Activity	Time	Time	Cost	Cost	Slope	Crash Amt.	to Crash	Cost	Time
6	*A*	5	4	$100	$140	40	1		$0.00	5
7	*B*	9	7	$200	$300	50	2		$0.00	9
8	*C*	7	4	$250	$340	30	3		$0.00	7
9	*D*	9	7	$280	$340	30	2		$0.00	9
10	*E*	5	2	$250	$460	70	3		$0.00	5
11	*F*	11	7	$400	$720	80	4		$0.00	11
12	*G*	6	4	$300	$420	60	2		$0.00	6
13	*I*	8	6	$80	$140	30	2		$0.00	8
14	Total			$1,860	$2,860				$0.00	
15										
16		Time of								
17	Node	Event								
18	1									
19	2									
20	3									
21	4									
22	5									
23										

Callouts: "Decision variables" (cells H6:H13); "Decision variables" (cells B18:B22).

Key Formulas	
Cell B2	=D14+I14
Cell F6	=(E6-D6)/(B6-C6) {copy to cells F7:F13}
Cell G6	=B6-C6 {copy to cells G7:G13}
Cell I6	=F6*H6 {copy to cells I7:I13}
Cell J6	=B6-H6 {copy to J7:J13}

that appears (refer to Chapter 4 if you need a refresher on Excel's Solver). In our case, our goal is to minimize the total cost of the project. Thus, we specify cell B2 as the Target Cell and select the Min radio button in the Solver Parameters dialog box.

Next we specify the cells Excel can change in order to finish the project within 22 days at the least cost. In the spreadsheet shown in Exhibit 8.47, the decision variables or cells that can be changed are the amount of time each activity is crashed (cells H6:H13) and the time each event occurs (cells B18:B22). Thus, these ranges are specified for the By Changing Cells field.

Finally, we must enter the constraints associated with this situation. Perhaps the most obvious constraint is that we want to complete the project within 22 days (cell B1). Since node 5 (cell B22) corresponds to the event of the project being completed, we specify this constraint as follows:

$$B22 \leq B1$$

Another important set of constraints is needed to ensure that we do not crash an activity more than the maximum number of days that it can be crashed. Constraints to ensure this could be entered as follows:

$$H6 \leq G6 \text{ (activity } A)$$
$$H7 \leq G7 \text{ (activity } B)$$
$$H8 \leq G8 \text{ (activity } C)$$
$$H9 \leq G9 \text{ (activity } D)$$
$$H10 \leq G10 \text{ (activity } E)$$
$$H11 \leq G11 \text{ (activity } F)$$
$$H12 \leq G12 \text{ (activity } G)$$
$$H13 \leq G13 \text{ (activity } I)$$

Alternatively, by employing Excel's ability to work with ranges, these eight constraints can be entered as the following single constraint:

$$H6{:}H13 \leq G6{:}G13$$

Another set of constraints is needed to ensure that the precedence relationships associated with the project are not violated. We do this by keeping track of the event times of the nodes. For example, the time of the event corresponding to node 1 cannot occur until after activity *A* has been completed (assuming the project begins at time zero). The time to complete activity *A* is its normal time (cell B6) minus the amount it is crashed (cell H6). Since the time of the event corresponding to node 1 is recorded in cell B18, the constraint to ensure that node 1 does not occur until after activity *A* has been completed can be written as follows:

$$B18 \geq B6 - H6$$

In effect, this constraint says that node 1 (cell B18) cannot occur until activity *A* has been completed.

Constraints for the other nodes are developed in a similar fashion. Moving on to node 2, we observe that this node has 2 arrows leading into it. Thus, we will need the following two constraints for node 2:

$$B19 \geq B7 - H7$$
$$B19 \geq B18 + B8 - H8$$

The first constraint ensures that the time of event 2 (cell B19) does not occur until after activity *B* is completed. The second constraint ensures that event 2 does not occur until after node 1 is complete (cell B18) plus the time it takes to complete activity *C*. The remaining constraints are developed in a similar manner, one constraint for each arrow leading into the node:

$B20 \geq B18 + B9 - H9$ (node 3, activity *D*)
$B21 \geq B20$ (node 4, dummy activity)
$B21 \geq B19 + B10 - H10$ (node 4, activity *E*)
$B22 \geq B19 + B11 - H11$ (node 5, activity *F*)
$B22 \geq B20 + B12 - H12$ (node 5, activity *G*)
$B22 \geq B21 + B13 - H13$ (node 5, activity *I*)

Finally, since it does not make sense to crash an activity a negative amount of time, nor does it make sense for a node to occur at a time before time zero, we add the following nonnegativity constraints:

$$H6{:}H13 \geq 0$$
$$B18{:}B22 \geq 0$$

After specifying this information, the Solver Parameters' dialog box appears as shown in Exhibit 8.48. Prior to solving, don't forgot to select the Options . . . button and then select the Assume Linear Model check box. The solution Solver develops is shown in Exhibit 8.49. According to the solution, activity *C* should be crashed 1 day (cell H8) and activity *I* crashed 2 days (cell H13). This results in a total cost of $1,950 (cell B2). Incidentally, this is exactly the same cost we obtained earlier with the manual approach, but Solver found an alternate solution.

Having developed this spreadsheet, we can now quickly and easily evaluate the cost of incrementally increasing the amount the project is crashed. Doing this simply requires changing the deadline entered in cell B1 and then running Solver again. A summary of doing this is provided in Exhibit 8.50.

EXHIBIT 8.48 Solver Parameters Dialog Box for Example Crashing Problem

Solver Parameters

Set Target Cell: $B$2

Equal To: ○ Max ◉ Min ○ Value of: 0

By Changing Cells:

$H$6:$H$13,$B$18:$B$22

Subject to the Constraints:

$B$18 >= $B$6-$H$6
$B$18:$B$22 >= 0
$B$19 >= $B$18+$B$8-$H$8
$B$19 >= $B$7-$H$7
$B$20 >= $B$18+$B$9-$H$9
$B$21 >= $B$19+$B$10-$H$10

Solve | Close | Guess | Options | Add | Change | Delete | Reset All | Help

EXHIBIT 8.49 Solver's Optimal Solution for 22-Day Project Duration

	A	B	C	D	E	F	G	H	I	J
1	Deadline	22								
2	Total Cost	$1,950.00								
3										
4		Normal	Crash	Normal	Crash	Cost	Maximum	Amount	Crashing	Actual
5	Activity	Time	Time	Cost	Cost	Slope	Crash Amt.	to Crash	Cost	Time
6	*A*	5	4	$100	$140	40	1	0	$0.00	5
7	*B*	9	7	$200	$300	50	2	0	$0.00	9
8	*C*	7	4	$250	$340	30	3	1	$30.00	6
9	*D*	9	7	$280	$340	30	2	0	$0.00	9
10	*E*	5	2	$250	$460	70	3	0	$0.00	5
11	*F*	11	7	$400	$720	80	4	0	$0.00	11
12	*G*	6	4	$300	$420	60	2	0	$0.00	6
13	*I*	8	6	$80	$140	30	2	2	$60.00	6
14	Total			$1,860	$2,860				$90.00	
15										
16		Time of								
17	Node	Event								
18	15					*Key Formulas*				
19	21	1				Cell B2	=D14+I14			
20	31	6				Cell F6	=(E6-D6)/(B6-C6) {copy to cells F7:F13}			
21	41	6				Cell G6	=B6-C6 {copy to cells G7:G13}			
22	52	2				Cell I6	=F6*H6 {copy to cells I7:I13}			
23						Cell J6	=B6-H6 {copy to J7:J13}			

EXHIBIT 8.50 Summary of Time–Cost Trade-Off Analysis

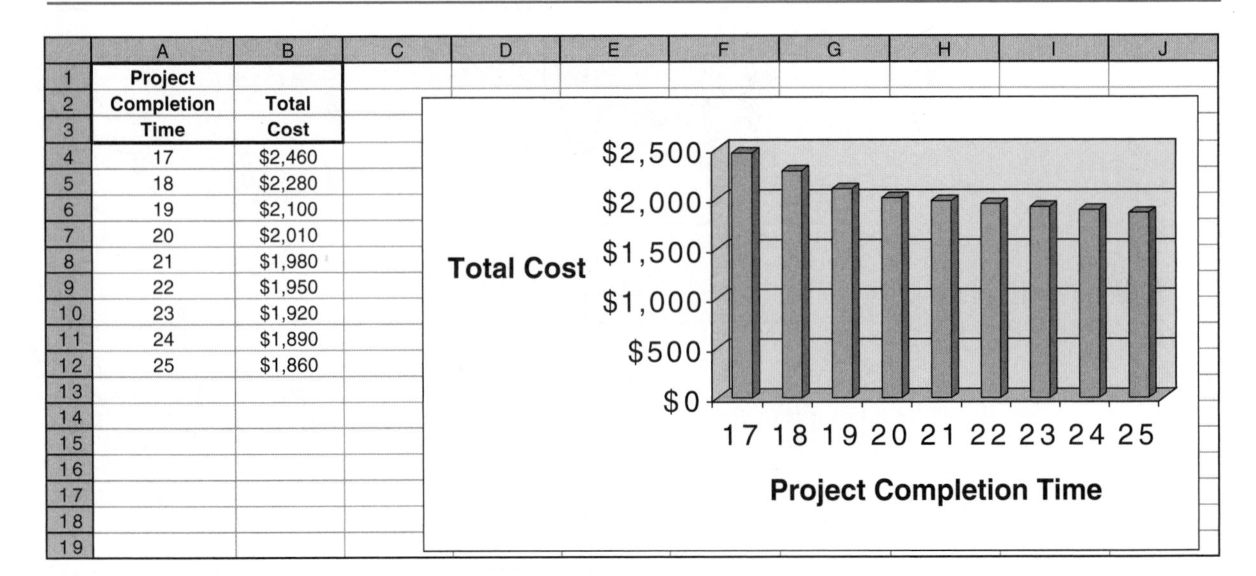

Project Completion Time	Total Cost
17	$2,460
18	$2,280
19	$2,100
20	$2,010
21	$1,980
22	$1,950
23	$1,920
24	$1,890
25	$1,860

8.6 Detailed Modeling Example

In this section we return to Kathrayn Rand's project to model and study RecSys Inc.'s problem of late shipments discussed at the beginning of this chapter.

Step 1: Opportunity/Problem Recognition

Recently, RecSys Inc.'s lead times for its residential recreational equipment have increased from three weeks to four to six weeks. This is creating problems for its distributors as well as reducing overall customer satisfaction. Kathrayn Rand has been assigned the project of modeling and studying the situation. The influence diagram she initially developed was shown in Exhibit 8.1. The most pressing issue now facing Kathrayn is developing an estimate of the project's duration for the plant manager's approval.

Step 2: Model Formulation

As we proceed through this example, we will utilize a variety of models. Our first task is to develop a work breakdown structure to identify the scope of the tasks required to complete this project, as shown in Exhibit 8.51. Once the tasks associated with the project have been identified, the next step is to develop time estimates for each task and to identify precedence relationships among the tasks. Kathrayn developed three time estimates for each activity and then, based on this, calculated the expected time, variance, and standard deviation for each activity as shown in Exhibit 8.52. Precedence relationships among the tasks are also included in Exhibit 8.52. Based on the information in Exhibit 8.52, the network diagram for this modeling project was developed as shown in Exhibit 8.53. Analysis of the network diagram reveals that there are 10 paths:

EXHIBIT 8.51 Work Breakdown Structure for RecSys Modeling Project

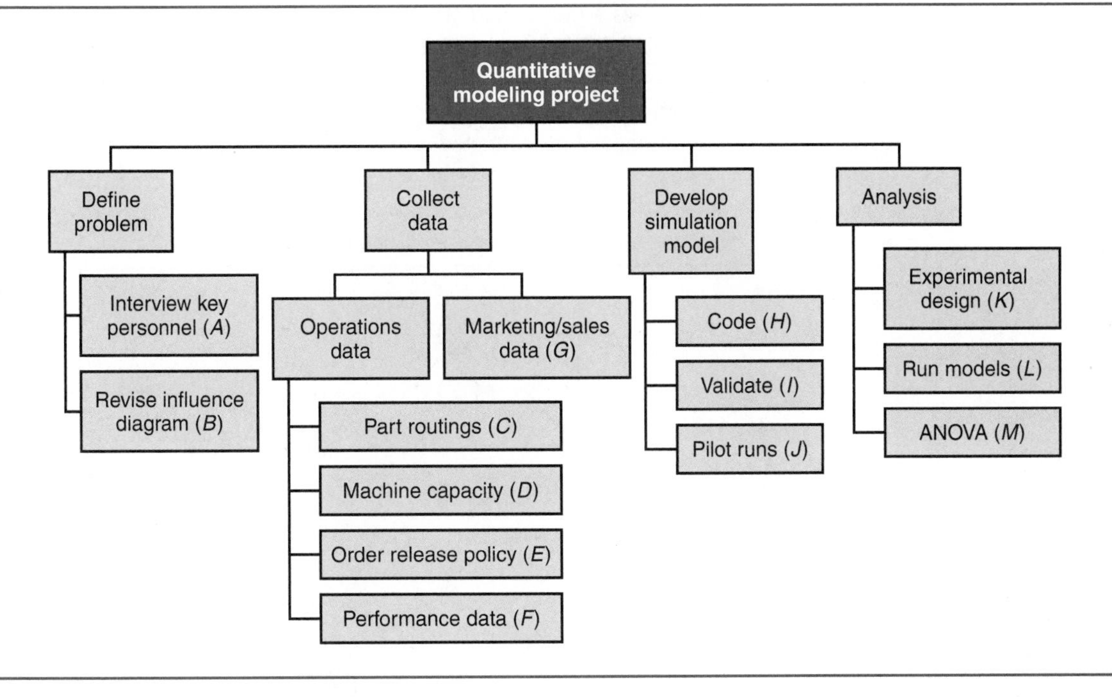

EXHIBIT 8.52 Activity Time Estimates and Precedence Information

	A	B	C	D	E	F	G	H	I
1					Most				
2		Activity	Preceded	Optimistic	Likely	Pessimistic	Expected		Standard
3	Activity	Code	by	Duration	Duration	Duration	Time	Variance	Deviation
4	Interview personnel	*A*	--	3	5	10	5.5	1.36	1.17
5	Influence diagram	*B*	*A*	1	2	4	2.2	0.25	0.50
6	Part routings	*C*	*B*	3	5	12	5.8	2.25	1.50
7	Machine capacity	*D*	*B*	2	3	7	3.5	0.69	0.83
8	Order release	*E*	*B*	1	1	2	1.2	0.03	0.17
9	Performance data	*F*	*B*	3	6	15	7.0	4.00	2.00
10	Marketing data	*G*	*B*	2	4	5	3.8	0.25	0.50
11	Code model	*H*	*C,D,E,F,G*	4	6	10	6.3	1.00	1.00
12	Validate model	*I*	*H*	1	3	5	3.0	0.44	0.67
13	Pilot runs	*J*	*I*	1	2	4	2.2	0.25	0.50
14	Experimental design	*K*	*I*	1	1	1	1.0	0.00	0.00
15	Run models	*L*	*J,K*	5	10	15	10.0	2.78	1.67
16	ANOVA	*M*	*L*	1	1	2	1.2	0.03	0.17

A–B–C–H–I–J–L–M *A–B–C–H–I–K–L–M*
A–B–D–H–I–J–L–M *A–B–D–H–I–K–L–M*
A–B–E–H–I–J–L–M *A–B–E–H–I–K–L–M*
A–B–F–H–I–J–L–M *A–B–F–H–I–K–L–M*
A–B–G–H–I–J–L–M *A–B–G–H–I–K–L–M*

Note that in both Exhibits 8.52 and 8.53 only those activities that could not be broken down further are listed. For example, the task "Define Problem" is not listed in these figures because this task is made up of the tasks "Interview Key Personnel" and "Revise Influence Diagram." Kathrayn also entered the information into Microsoft Project (MSP).

MSP provides a wide variety of ways to view project information, including as a Gantt chart (Exhibit 8.54) and as a PERT chart (Exhibit 8.55). In Exhibit 8.54 notice how MSP included precedence relationships in its Gantt chart view. In Exhibit 8.55 notice that MSP displays activities on the nodes as opposed to the convention used in this book of placing the activities on the arrows. Each node in the diagram lists the activity's name, its number, its estimated duration, and its estimated start and finish days.

EXHIBIT 8.53 Network Diagram for RecSys Modeling Project

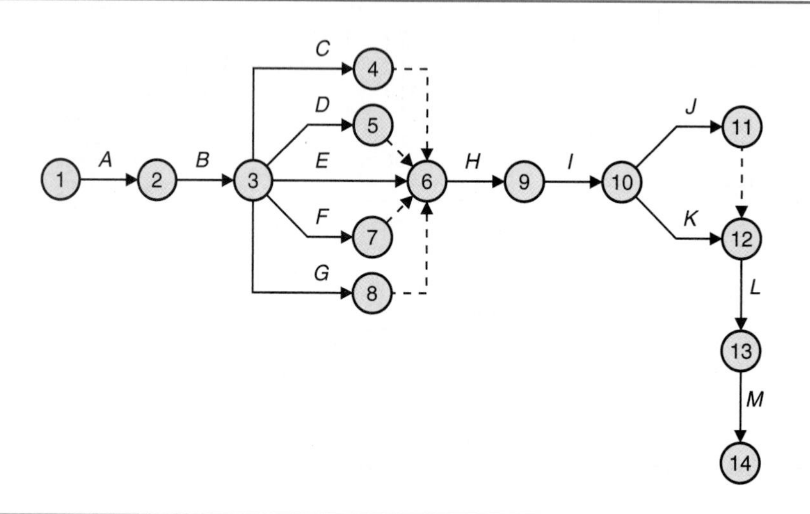

EXHIBIT 8.54 Microsoft Project's Gantt Chart View

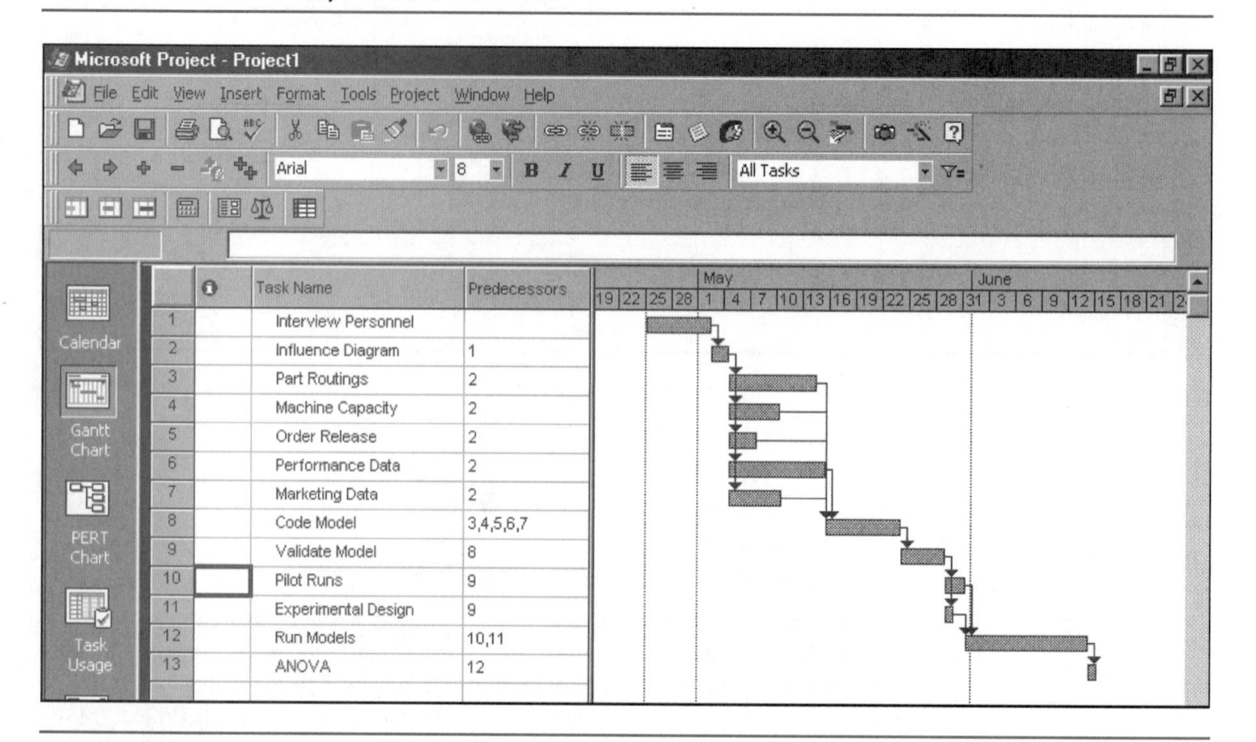

Step 3: Data Collection

Data are needed to both plan this modeling project and then to execute it. In terms of planning the project, the required tasks need to be enumerated, precedence relationships identified, and activity durations estimated, as previously discussed. In addition to this data, modeling this situation with a simulation model requires a variety of operational data including part routings, marketing and sales data, and data on performance measures such as work-in-process levels and shop lead times. Developing the influence diagram earlier helps identify the data requirements for this project. In this case, most of the data should be obtainable via corporate databases (e.g., part routings, machine capacity, and work-in-process levels) or through interviews of key personnel (e.g., advertising and promotional activities, order release policies, and the policy for setting due dates). Other data may need to be compiled by one or more project team members (e.g., actions of competitors and shop lead times).

The issue currently confronting Kathrayn relates to determining the amount of time it will take to complete this project. Once the project planning data have been collected, including the identification of the activities, their precedence relationships, and their estimated durations, a model similar to the one shown in Exhibit 8.56 can be developed to simulate the completion of the project in order to better understand the distribution of the

EXHIBIT 8.55 Microsoft Project's PERT Chart View

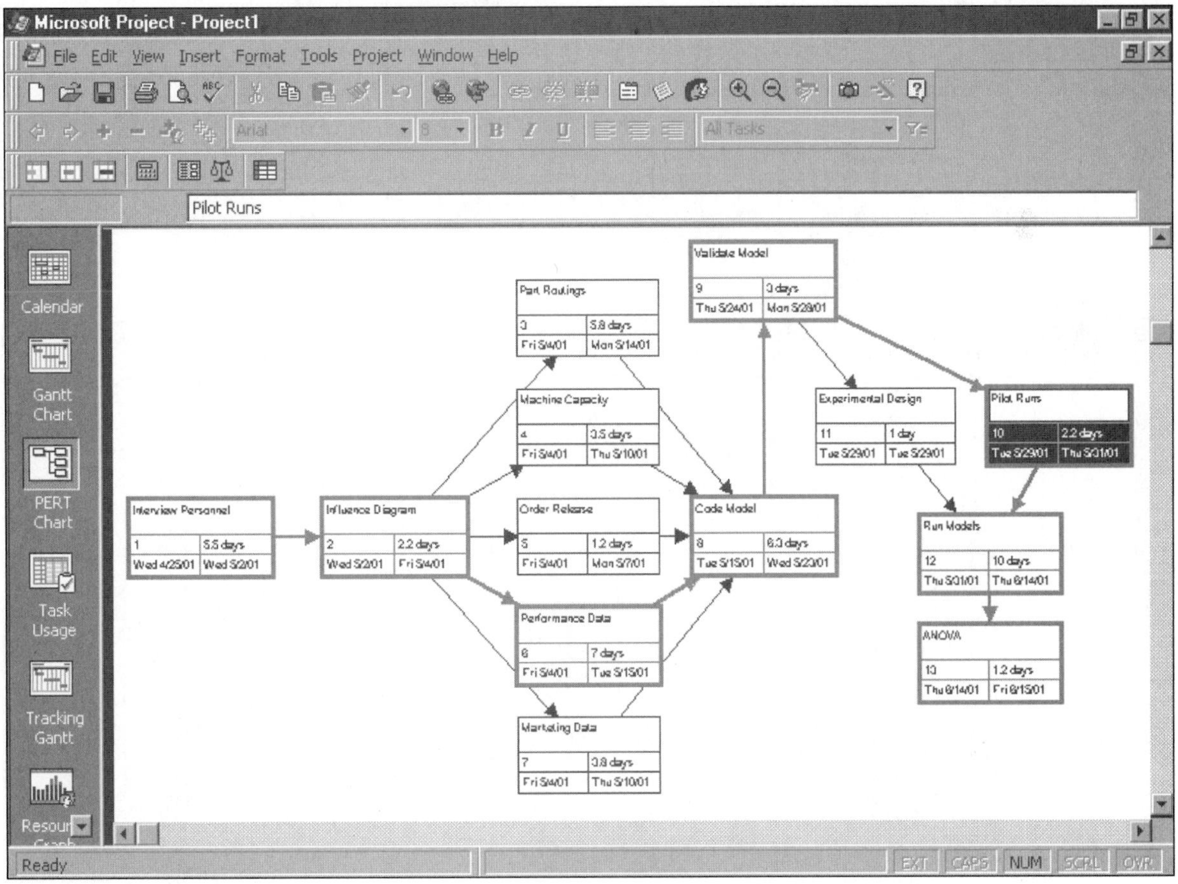

project's completion time. Kathrayn can use this information to negotiate with management regarding the amount of time she will be allocated to complete the modeling project.

Following the procedure described earlier in the chapter, the spreadsheet shown in Exhibit 8.56 was created. Columns A through M store the randomly generated activity times for the 13 activities in this project, respectively. As was done earlier, the activity times were assumed to be normally distributed. The time to complete each of the ten paths based on the randomly generated activity times is calculated in Columns N through W, respectively. Finally, the time to complete the project is calculated in column X.

EXHIBIT 8.56 Spreadsheet to Simulate Completion of Project and Help Determine Distribution of Project Completion Times

	A	B	C	D	E	F	G	H	I	J	K	L	M
1	Activity	Activity	Activity	Activity	Activity	Activity	Activity	Activity	Activity	Activity	Activity	Activity	Activity
2	A	B	C	D	E	F	G	H	I	J	K	L	M
3	5.1	2.8	6.4	3.3	1.4	4.2	2.9	6.2	3.4	1.5	1.0	9.6	1.0
4	4.0	2.6	4.0	3.7	1.1	7.4	3.6	6.5	2.6	1.7	1.0	9.8	1.4
5	5.8	1.8	15.4	4.2	1.3	10.6	3.2	5.6	2.6	1.5	1.0	9.4	0.8
6	7.0	1.6	6.2	3.5	1.2	7.1	3.5	6.3	3.1	2.3	1.0	5.1	1.2
7	6.9	3.1	5.8	4.2	1.1	5.2	3.4	6.3	2.8	2.2	1.0	8.5	0.9
8	7.5	3.1	6.2	3.9	1.0	5.1	3.3	6.8	2.3	1.7	1.0	9.3	1.3
9	2.9	2.2	4.3	2.8	0.9	9.0	2.6	6.1	2.6	3.1	1.0	7.4	1.2
10	5.2	2.4	6.8	2.8	1.2	7.3	4.2	7.7	4.1	2.7	1.0	11.9	1.0
11	6.8	2.8	4.1	2.8	0.9	7.8	3.3	6.8	3.1	2.4	1.0	10.1	1.2
12	4.2	1.7	6.0	3.1	1.4	9.7	4.1	6.3	4.0	1.9	1.0	12.0	1.1
13	4.7	2.7	7.2	3.8	1.1	5.4	3.6	7.0	3.0	2.2	1.0	11.4	1.0
14	3.5	2.1	4.0	2.6	1.0	7.2	3.9	6.2	3.5	2.0	1.0	11.9	1.4
15	3.3	1.3	5.2	3.0	1.2	4.3	4.3	7.7	3.4	2.6	1.0	12.0	1.0
16	4.4	2.7	8.0	3.4	1.5	8.7	3.2	5.7	4.7	1.8	1.0	10.6	1.0
17	4.6	1.3	3.6	3.4	1.2	5.6	3.9	7.8	4.0	1.9	1.0	8.5	1.0
18	3.0	2.7	2.7	3.7	1.2	8.5	4.3	6.2	2.8	2.9	1.0	11.3	1.0
19	4.8	2.1	5.5	4.2	0.7	6.6	3.7	4.2	3.2	1.8	1.0	10.8	0.9
20	5.0	2.1	5.5	4.0	1.4	8.4	3.8	6.3	3.5	2.9	1.0	10.4	0.9
21	5.7	2.4	6.2	4.8	1.1	6.4	3.0	5.8	2.8	2.9	1.0	10.3	1.3
22	5.1	2.8	5.8	3.1	1.0	7.2	4.9	5.4	3.8	1.2	1.0	9.6	1.5
23	5.1	2.1	7.1	4.4	1.1	6.5	3.0	6.7	4.3	2.8	1.0	9.1	1.2
24	5.1	2.6	3.6	2.7	1.1	6.2	3.3	5.2	2.4	2.8	1.0	11.5	1.3
25	7.1	2.4	5.6	5.8	0.9	3.9	2.8	6.0	2.7	1.6	1.0	9.2	1.2

	N	O	P	Q	R	S	T	U	V	W	X
1	Path	Path	Path	Path	Path	Path	Path	Path	Path	Path	Completion
2	ABCHIJLM	ABDHIJLM	ABEHIJLM	ABFHIJLM	ABGHIJLM	ABCHIKLM	ABDHIKLM	ABEHIKLM	ABFHIKLM	ABGHIKLM	Time
3	36.1	32.9	31.1	33.8	32.6	35.6	32.4	30.5	33.3	32.1	36.1
4	32.6	32.3	29.7	36.0	32.3	31.9	31.6	29.0	35.3	31.5	36.0
5	32.8	31.6	28.7	38.0	30.6	32.3	31.1	28.2	37.4	30.1	38.0
6	32.8	30.1	27.8	33.7	30.1	31.5	28.7	26.5	32.4	28.8	33.7
7	36.5	34.9	31.8	35.9	34.1	35.3	33.7	30.6	34.7	32.9	36.5
8	38.3	36.0	33.1	37.2	35.4	37.6	35.2	32.4	36.5	34.7	38.3
9	29.8	28.3	26.4	34.5	28.1	27.7	26.2	24.3	32.4	26.0	34.5
10	41.8	37.8	36.2	42.3	39.2	40.2	36.2	34.5	40.7	37.6	42.3
11	37.3	35.9	34.1	40.9	36.4	35.9	34.5	32.7	39.5	35.0	40.9
12	37.3	34.4	32.7	41.0	35.4	36.4	33.5	31.7	40.1	34.4	41.0
13	39.1	35.7	33.0	37.3	35.5	37.9	34.5	31.8	36.1	34.3	39.1
14	34.6	33.2	31.6	37.7	34.5	33.6	32.2	30.6	36.7	33.5	37.7
15	36.6	34.4	32.5	35.7	35.6	34.9	32.8	30.9	34.1	34.0	36.6
16	38.9	34.4	32.4	39.6	34.2	38.2	33.6	31.6	38.8	33.4	39.6
17	32.6	32.4	30.2	34.6	33.0	31.7	31.5	29.4	33.7	32.1	34.6
18	35.1	33.5	30.9	38.3	34.0	33.3	31.6	29.1	36.4	32.2	38.3
19	33.3	32.1	28.6	34.4	31.6	32.5	31.3	27.8	33.6	30.7	34.4
20	36.4	34.9	32.3	39.4	34.7	34.5	33.0	30.4	37.5	32.8	39.4
21	37.5	36.1	32.4	37.7	34.3	35.5	34.2	30.5	35.8	32.4	37.7
22	35.1	32.3	30.2	36.4	34.1	34.9	32.2	30.1	36.2	34.0	36.4
23	38.4	35.7	32.3	37.7	34.2	36.6	33.9	30.6	36.0	32.5	38.4
24	34.5	33.5	31.9	37.0	34.1	32.7	31.8	30.2	35.3	32.4	37.0
25	35.8	36.0	31.2	34.1	33.0	35.2	35.4	30.5	33.5	32.4	36.0

Step 4: Analysis of the Model

The simulation model was developed to simulate the completion of the project 200 times (because of page size limitations, only a portion of the spreadsheet is shown in Exhibit 8.56). Across the 200 replications, the project's average duration was 37.9 days, with a standard deviation of 2.9 days. The longest project duration was 44.4 days and the shortest time required to complete the project was 29.6 days. Further analysis of the results indicates that path *A–B–F–M–I–J–L–M* was the critical path 72.5 percent of the time and path *A–B–C–H–I–J–L–M* was critical 26 percent of the time. Paths *A–B–D–H–I–J–L–M, A–B–G–H–I–J–L–M,* and *A–B–F–H–I–K–L–M* were each critical 0.5 percent of the time.

A histogram of the project completion times for the 200 replications of the project was created and is shown in Exhibit 8.57. The shape of the histogram suggests that the distribution for the project's completion time may be normal. As is demonstrated in Exhibit 8.58, a chi-squared goodness of fit test (see Section 2.6) confirms that a normal distribution with mean of 37.9 and standard deviation of 2.9 provides a reasonably good approximation for the distribution of project completion times.

Based on these results, we can extend this analysis one step further and make statistical inferences about the project. For example, if management provides Kathrayn with 40 days to complete the project, we can calculate the probability that the project will be completed within this time frame using Excel's NORMDIST function as follows:

=NORMDIST(40,37.9,2.9,TRUE)

In this case, the probability that the project could be completed in 40 days or less is 76.6 percent. Alternatively, we could reverse the question, and determine what deadline would provide Kathrayn with a 90 percent chance of completing the project on time using Excel's NORMINV function as follows:

=NORMINV(0.90,37.9,2.9)

EXHIBIT 8.57 Distribution of Project Completion Times Appears Approximately Normally Distributed

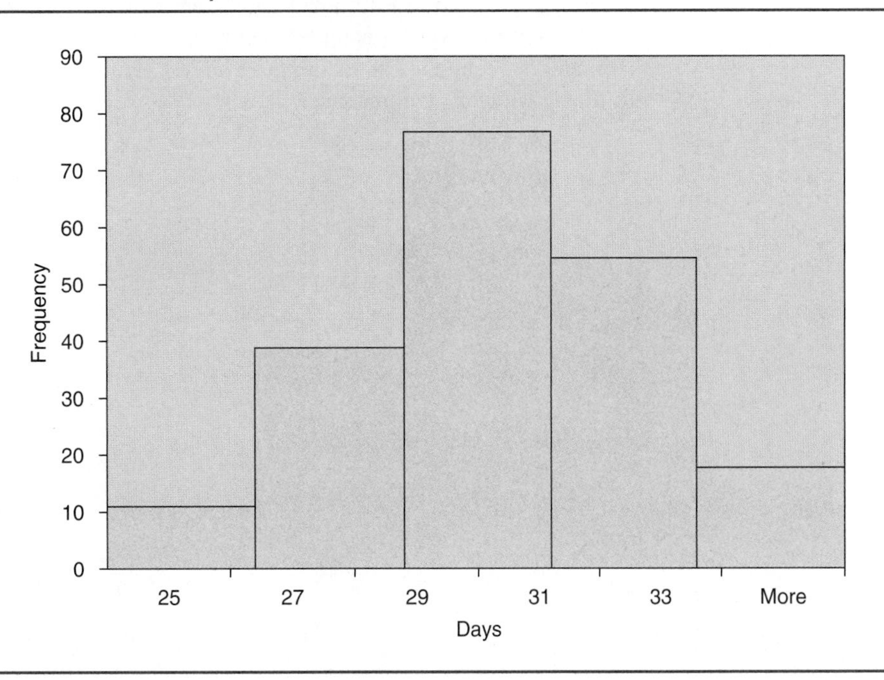

EXHIBIT 8.58 Chi-Squared Goodness of Fit Test to Test Hypothesis that Distribution of Project Completion Times Is Normally Distributed

	A	B	C	D	E	F	G	H	I
1	Upper	Observed		Expected					
2	Limit of	Frequency		Frequency					
3	Range	f_o	Probability	f_e	$(f_o-f_e)^2/f_e$				
4	33	11	4.55%	9.1	0.392				
5	36	39	21.06%	42.1	0.232				
6	39	77	39.16%	78.3	0.022				
7	42	55	27.35%	54.7	0.002				
8	>42	18	7.87%	15.7	0.324				
9									
10				Calc. Val.	0.972				
11				Crit. Val.	5.991				
12									
13			Key Formulas						
14	Cell C4	=NORMDIST(A4,37.9,2.9,TRUE)							
15	Cell C5	=NORMDIST(A5,37.9,2.9,TRUE)-C4							
16	Cell C6	=NORMDIST(A6,37.9,2.9,TRUE)-C4-C5							
17	Cell C7	=NORMDIST(A7,37.9,2.9,TRUE)-C4-C5-C6							
18	Cell C8	=1-NORMDIST(42,37.9,2.9,TRUE)							
19	Cell D4	=C4*200 {copy to cells D5:D8}							
20	Cell E4	=((B4-D4)^2)/D4 {copy to cells E5:E8}							
21	Cell E10	=SUM(E4:E8)							
22	Cell E11	=CHIINV(0.05,2)							

Cannot reject hypothesis that distribution of project completion times is normally distributed with mean of 37.9 and standard deviation of 2.9

As it turns out, there is a 90 percent chance that Kathrayn could complete this project in 41.6 days.

Step 5: Implementation

At this stage, it is up to Kathrayn to negotiate with management regarding the amount of time she will be given to complete the project as well as for other additional resources that might be required. The memo in Exhibit 8.59 illustrates how Kathrayn might initiate this dialog with the plant manager.

EXHIBIT 8.59 Memo from Kathrayn

MEMO

To: Ronald Wilson, plant manager
From: Kathrayn Rand, senior analyst
Date: 4/26/02
Re: Project to investigate late delivery problem

Introduction

I have completed my initial planning for the project to investigate the late delivery problem. To appropriately study this problem, I recommend that you allow me to work full time on this project for a period of 42 days. This will provide me with sufficient time to clearly define the problem, collect the required data, conduct computer simulation experiments, analyze the results, and identify opportunities to rectify the situation.

Analysis

Based on my preliminary analysis of the situation, I have developed a work breakdown structure for the project. Accordingly, I envision that this project will entail four major phases: problem definition, data collection, model development, and analysis. I have also developed time estimates for the duration of each major task in the project and identified constraints regarding the sequencing of these tasks. Most likely the project can be completed in 38 days, although there is a 50 percent chance it will take longer than this.

Recommendations

I recommend that 42 days be allocated to complete this project. This time frame is realistic and should be sufficiently timely to address the problem at hand.

Assumptions/Limitations

There were a number of assumptions made in my initial planning. First, assumptions were made regarding the availability of key data. The duration of the project will be extended if the project team has to collect/compile data assumed to be available. Likewise, it was assumed that the project team would have access to key employees. Violations of these assumptions could increase the project's duration and/or decrease the validity and quality of the results. Furthermore, it was assumed that the project team would consist of three full-time employees—one senior analyst and two analysts. Changing the levels of these resources will impact the project's duration. Finally, the scope of the project was defined to be the identification opportunities to reduce shop lead times. Increasing the scope of this project to investigate other issues that arise in the course of this study will likely increase the project's duration and cost.

QUESTIONS

1. Given the likelihood of project scope creep and its associated danger, what might be done to control this common phenomenon?
2. What kinds of steps could the modeler take to couch the study and its results in the cognitive style of the manager?
3. What position is best for the modeler to take relative to politics in the workplace?
4. Is there any alternative to monitarizing, or even quantifying, abstract benefits in a cost–benefit analysis?
5. Why is it important to identify the slack in a project?
6. The work breakdown structure (WBS) has often been considered to be the most important of all the project management documents. Why do you think this is so?
7. We spend a considerable amount of time in the chapter discussing the ways to determine the true probability of completing the project by a specified time. One way is to simply use the pre-project critical path. Another way is to multiply the probabilities of all the paths in the network being completed by some specified time. The third way shown was to run a simulation study. How do you think the first and second method would compare to the simulated results?
8. This chapter did not describe solutions to the reality of resource requirements for activities and the conflict between activities needing the same resources. How do you think this is handled in a real project environment?
9. What data are needed by the project manager to accurately monitor project progress through the use of a critical ratio? Through the use of earned value?
10. How might PERT be used for controlling a project? How might CPM be used for planning a project?

EXPERIENTIAL EXERCISE

Identify an activity in your life that could be managed as a project (e.g., planning a wedding, planning a family reunion, building a house). Complete the following for your project.

1. Develop a work breakdown structure for your project.
2. Estimate the optimistic, pessimistic, and most likely duration for each activity. Calculate the expected time, variance, and standard deviation for each activity.
3. Identify the precedence relationships among the activities and then develop a network diagram for your project.
4. Enumerate all the paths through your network diagram.
5. Calculate the expected time and standard deviation for each path. What is your best guess of how long it will take to complete the project? Which paths are likely to delay the completion of the project?
6. Simulate the completion of the project 100 times and develop a histogram showing the distribution of project completion times.
7. If you wanted to have a 95 percent chance of completing the project on time, how much time should you allow to complete the project?

MODELING EXERCISES

Unless stated otherwise, slack means total float, TF, in all problems.

1. The events of the project listed at right are designated as **1**, **2**, and so on.
   a. Draw the network.
   b. Find the critical path by complete enumeration.
   c. Find, for all events, the earliest and latest dates.
   d. Find the slacks on all the events and activities.
   e. Find the critical path, using the T_E's and T_L's.

Activity	Preceding Event	Succeeding Event	t_e (Weeks)	Preceding Activities
a	**1**	**2**	3	None
b	**1**	**3**	6	None
c	**1**	**4**	8	None
d	**2**	**5**	7	*a*
e	**3**	**5**	5	*b*
f	**4**	**5**	10	*c*
g	**4**	**6**	4	*c*
h	**5**	**7**	5	*d, e, f*
i	**6**	**7**	6	*g*

2. Given the following PERT network (times are in weeks):
   Determine:
   a. The T_E and T_L for each event.
   b. The slacks on all events and activities.
   c. The critical activities and path.
   d. The shared slacks.

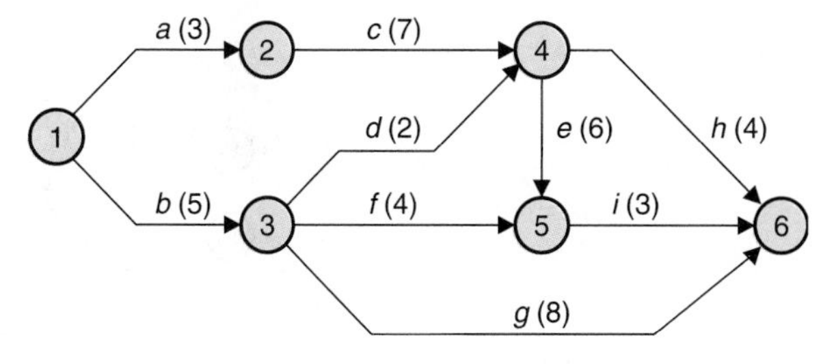

3. Suppose that management has a contract to finish the project in Problem 2 in 22 weeks.
   Determine:
   a. The slack on event **3**.
   b. The slack on activity *g*.
4. Given the following schedule for a liability work package done as part of an accounting audit in a corporation.
   Find:
   a. The critical path.
   b. The slack time on *f* (process confirmations).
   c. The slack time on *c* (test pension plan).
   d. The slack time on *h* (verify debt restriction compliance).

Activity	Duration (Days)	Preceding Activities
a. Obtain schedule of liabilities	3	None
b. Mail confirmation	15	*a*
c. Test pension plan	5	*a*
d. Vouch selected liabilities	60	*a*
e. Test accruals and amortization	6	*d*
f. Process confirmations	40	*b*
g. Reconcile interest expense to debt	10	*c*, *e*
h. Verify debt restriction compliance	7	*f*
i. Investigate debit balances	6	*g*
j. Review subsequent payments	12	*h*, *i*

5. In the project network shown in the figure, the number alongside each activity designates the activity duration (t_e) in weeks.
   Determine:
   a. The T_E and T_L for each event.
   b. The earliest time that the project can be completed.
   c. The slack on all events and activities.
   d. The critical events and activities.
   e. The critical path.
   f. The shared slacks.

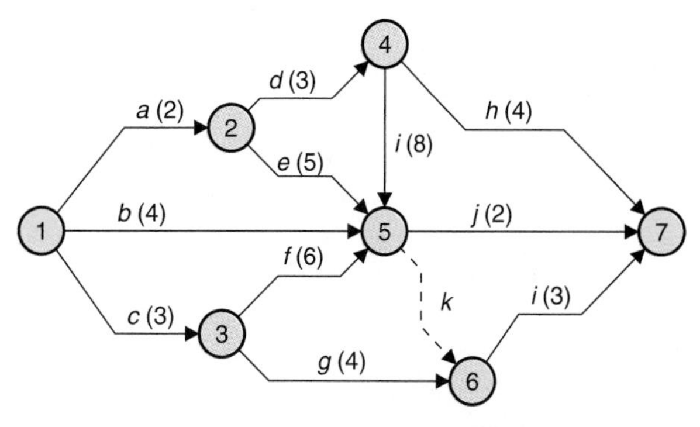

6. Given the following information regarding a project:
   a. Draw the network.
   b. What is the critical path?
   c. What will the scheduled (earliest completion) time be for the entire project?
   d. What is the critical path to event **4** (end of activities *c* and *e*)? What is the earliest time this event can be reached?
   e. What is the effect on the project if activity *e* takes an extra week? Two extra weeks? Three extra weeks?

Activity	t_e (Weeks)	Preceding Activities
a	3	None
b	1	None
c	3	*a*
d	4	*a*
e	4	*b*
f	5	*b*
g	2	*c*, *e*
h	3	*f*

7. Construct a network for the project below and find its critical path. (Use a complete enumeration approach.)

Activity	t_e (Weeks)	Preceding Activities
a	3	None
b	5	*a*
c	3	*a*
d	1	*c*
e	3	*b*
f	4	*b*, *d*
g	2	*c*
h	3	*g*, *f*
i	1	*e*, *h*

8. Construct a network for the project:
   a. Draw the network.
   b. Find the critical path by complete enumeration.
   c. Assume activity *a* took 5 weeks. Replan the project.
   d. From where would you suggest transferring resources, and to what activities, such that the original target date may be maintained?

Activity	t_e (Weeks)	Preceding Activities
a	3	None
b	5	None
c	14	*a*
d	5	*a*
e	4	*b*
f	7	*b*
g	8	*d, e*
h	5	*g, f*

9. Given a PERT network:

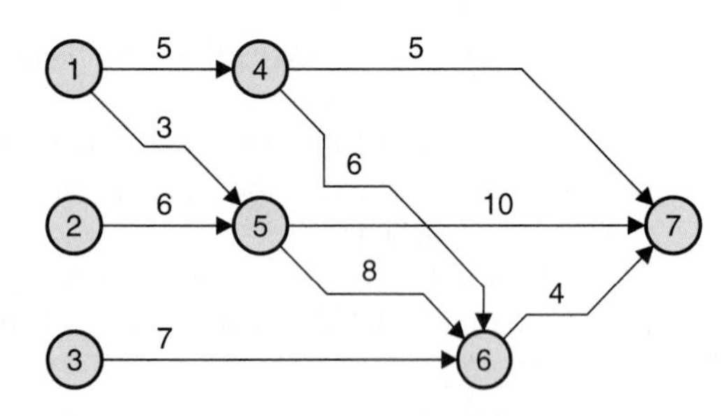

   Note that three activities can start immediately. Find:
   a. The critical path.
   b. The earliest time to complete the project.
   c. The slack on activities **4–6, 5–6,** and **4–7.**

10. Assume that in Problem 2 you need two employees for each of the activities **1–2, 1–3,** and **5–6;** three employees for each of the activities **4–5** and **3–6;** and one employee for each of the remaining activities.
   a. Prepare the labor demands if all activities start at their earliest times.
   b. Prepare a plan that will level the labor demand, over time, as much as possible. Do not "split" jobs. Once started, they must be completed.

11. Solve the maintenance work project network below. Find the critical path and the regular slack.

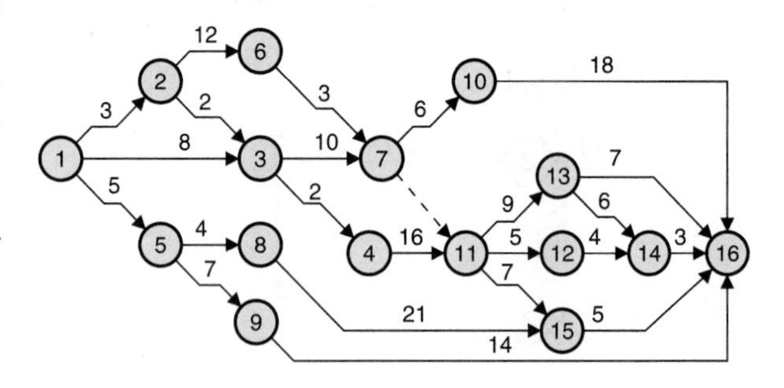

12. Build a PERT network given the following:

Activity	Time	Immediately Preceding Activity
Drill	2	None
Cut	1	None
Punch	3	Drill
Bend	2	Drill
Inspect	6	Cut, punch
Assemble	3	Cut, punch
Test	1	Assemble
Paint	2	Bend, inspect, test

13. Given a project with the following information from a computer printout:

Activity	Standard Deviation	Critical	Duration
a	2	Yes	2
b	1	No	3
c	0	Yes	4
d	3	No	2
e	1	Yes	1
f	2	No	6
g	2	Yes	4
h	0	Yes	2

   Find:
   a. The probability of completing this project in 12 weeks (or less).
   b. The probability of completing this project in 16 weeks (or less).
   c. The probability of completing this project in 13 weeks (or less).

d. The number of weeks required to assure a 92.5 percent chance of completion.

14. Given the following project:

Activity	Times (Weeks) Optimistic	Most Likely	Pessimistic
1–2	5	11	11
1–3	10	10	10
1–4	2	5	8
2–6	1	7	13
3–6	4	4	10
3–7	4	7	10
3–5	2	2	2
4–5	0	6	6
5–7	2	8	14
6–7	1	4	7

a. Find all "earliest dates," including project completion (T_E's for all events).
b. Find all "latest dates" (T_L's for all events).
c. Determine the critical path and the values of the event slacks.
d. What is the critical path leading to event **5**?
e. What will happen if activity **4–5**'s actual time slips to 9?
f. What will be the slack on activity **3–5** if activity **4–5** slips to 9 weeks and activity **5–7** takes 6 weeks?

15.
a. Find the probability of finishing the project in Problem 14 in 19 weeks. In 17 weeks. In 24 weeks.
b. What is the probability of completing event **5** in Problem 14 by 9 weeks?
c. If management wants to be 80 percent sure that the project will be completed by a "guaranteed" date, what date should be quoted?

16. A contracting firm has estimated the following event completion times:

Activity	Times Optimistic	Most Likely	Pessimistic
1–2	3	6	9
1–3	1	4	7
3–2	0	3	6
3–4	3	3	3
3–5	2	2	8
2–4	0	0	6
2–5	2	5	8
4–6	4	4	10
4–5	1	1	1
5–6	1	4	7

If the firm can complete the project within 14 days, it will be given a $20,000 bonus. If not, it must pay a one-time penalty of $3,500. Should the firm accept the contract? What other factors are probably relevant? Are there any noncritical paths whose variance might become important?

17. Given a PERT network:

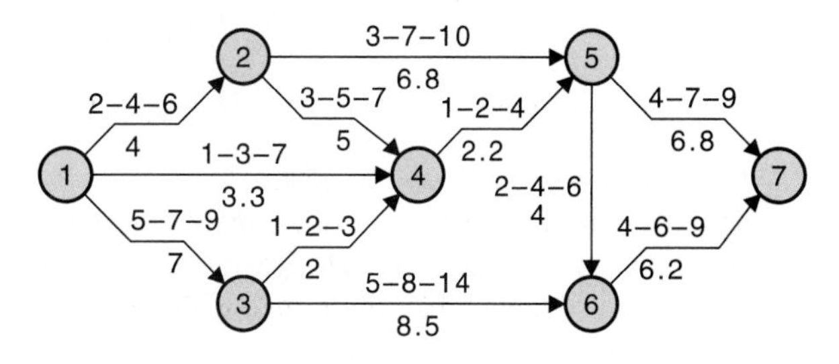

Find:
a. The estimated project completion time.
b. The critical path.
c. The slack on events **2** and **3.**
d. The slack on activities **1–4** and **2–5.**
e. The probability the project will be completed in 20 weeks or less.
f. The probability the project will be completed in 30 weeks or less.
g. The number of weeks required to complete the project with 95 percent certainty.

18. The following data were obtained from a study of the time required to overhaul a small power plant:

Activity	Crash Schedule Time	Cost	Normal Schedule Time	Cost
1–2	3	6	5	4
1–3	1	5	5	3
2–4	5	7	10	4
3–4	2	6	7	4
2–6	2	5	6	3
4–6	5	9	11	6
4–5	4	6	6	3
6–7	1	4	5	2
5–7	1	5	4	2

Note: Costs are given in thousands of dollars; time in weeks.

a. Find the all-normal schedule and cost.
b. Find the all-crash schedule and cost.
c. Find the total cost required to expedite all activities from all-normal (case *a*) to all-crash (case *b*).
d. Find the *least-cost* plan for the all-crash time schedule. Start from the all-crash problem (*b*).
e. Find the least cost for an intermediate time schedule of 17 weeks.

19. Reconsider Problem 1 under the constraint that the project *must* be completed in 16 weeks. This time, however, activities *c, f, h,* and *i* might be crashed as follows:

Activity	Crash Time (Weeks)	Additional Cost per Week
c	7	40
f	6	20
h	2	10
i	3	30

Find the best schedule and its cost.

20. The CPM network below has a normal time and a fixed cost of $90 per day. The various activities can be reduced up to their crash time with the additional costs shown:

Activity	Crash Time	Cost Increase, per Day Reduction
1–2	4	30 first day, 50 second, 70 third
2–3	6	40 first day, 45 second, 65 third
1–3	10	60 each
2–4	9	35 first, 60 second
3–4	3	—

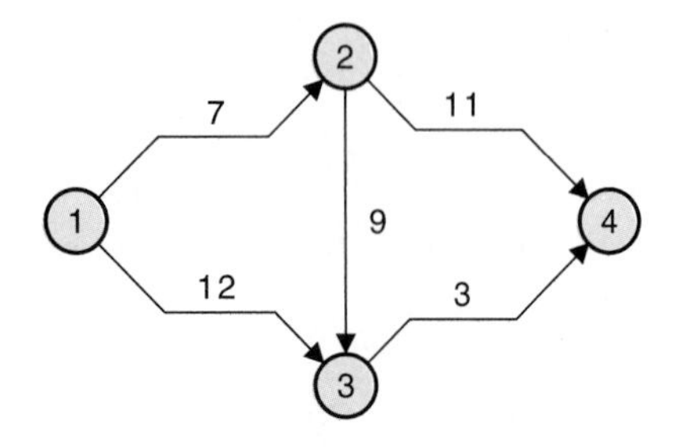

Find the least-cost schedule.

Hint: Start with normal time of 19 days (**1–2–3–4**). The total cost then is $19 \times 90 = \$1{,}710$. Then start cutting to 18. You save $90 fixed cost but have a cost increase of $30 when you cut activity **1–2** by 1 day. Continue until no further reductions are possible or the cost climbs.

21. Given the following network with normal times and crash times (in parentheses), find the optimal time–cost plan. Assume indirect costs are $100 per day. The crash data are:

Activity	Time Reduction, Direct Cost per Day
1–2	$30 first, $50 second.
2–3	$80 each
3–4	$25 first, $60 second
2–4	$30 first, $70 second, $90 third

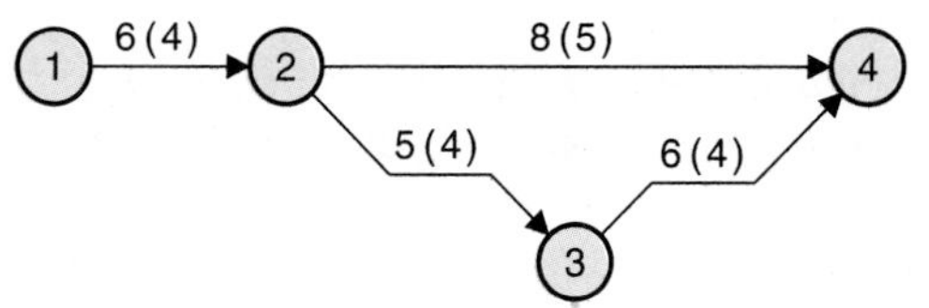

CASES

NutriTech

NutriTech produces a line of vitamins and nutritional supplements. It recently introduced its NT Sports Energy Bar, which is based on new scientific findings about the proper balance of macronutrients. The energy bar has become extremely popular among elite athletes and other people who follow the diet. One distinguishing feature of the NT Sports Energy Bar is that each bar contains 50 milligrams of eicosapentaenoic acid (EPA), a substance strongly linked to reducing the risk of cancer but found in only a few foods, such as salmon. NutriTech was able to include EPA in its sports bars because it had previously developed and patented a process to refine EPA for its line of fish-oil capsules.

Because of the success of the NT Energy Bar in the United States, NutriTech is considering offering it in Europe. With its domestic facility currently operating at capacity, the president of NutriTech has decided to investigate the option of building a new 15,000-square-foot production facility in Europe at a cost of $10 million.

The project to build the new European facility involves four major phases: (1) concept development, (2) definition of the plan, (3) design and construction, and (4) start-up and turnover. During the concept development phase, a program manager is chosen who will oversee all four phases of the project, and the manager is given a budget to develop a plan. The outcome of the concept development phase is a rough plan, feasibility estimates for the project, and a rough schedule. Also, a justification for the project and a budget for the next phase are developed.

In the plan definition phase, the program manager selects a project manager to oversee the activities associated with this phase. Plan definition consists of four major activities that are completed more or less concurrently: defining the project scope, developing a broad schedule of activities, developing detailed cost estimates, and developing a plan for staffing. The output of this phase is a detailed plan and proposal for management specifying how much the project will cost, how long it will take, and what the deliverables are.

If the project gets management's approval and management provides the appropriations, the project progresses to the third phase, design and construction. This phase consists of four major activities: detailed engineering, mobilization of the construction employees, procurement of production equipment, and construction of the facility. Typically, the detailed engineering and the mobilization of the construction employees are done concurrently. Once these activities are completed, construction of the facility and procurement of the production equipment are done concurrently. The outcome of this phase is the physical construction of the facility.

The final phase, start-up and turnover, consists of four major activities: pre–start-up inspection of the facility, recruiting and training the work force, solving start-up problems, and determining optimal operating parameters (called centerlining). Once the pre–start-up inspection is completed, the work force is recruited and trained at the same time that start-up problems are solved. Centerlining is initiated on the completion of these activities. The desired outcome of this phase is a facility operating at design requirements.

The following table provides optimistic, most likely, and pessimistic time estimates for the major activities.

Activity	Optimistic Time (Weeks)	Most Likely Time (Weeks)	Pessimistic Time (Weeks)
Concept development	12	36	48
Plan definition			
Define project scope	4	8	12
Develop broad schedule	1	2	4
Detailed cost estimates	1.5	2.5	3.5
Develop staffing plan	1.5	2.5	3
Design and construction			
Detailed engineering	8	12	24
Facility construction	32	36	96
Mobilization of employees	2	8	16
Procurement of equipment	4	12	48
Start-up and turnover			
Pre–start-up inspection	1	2	4
Recruiting and training	1	2	4
Solving start-up problems	0	4	8
Centerlining	0	4	16

Questions

1. Draw a network diagram for this project. Identify all the paths through the network.
2. Simulate the completion of this project 100 times, assuming that the activity times follow a normal distribution. Estimate the mean and standard deviation of the project completion time.
3. Develop a histogram to summarize the results of your simulation.
4. Calculate the probability that the project can be completed within 120 weeks. What is the probability that the project will take longer than 160 weeks? What is the probability that the project will take between 120 and 130 weeks?
5. If NutriTech has a policy of setting project deadlines based on providing the project with an 80 percent chance of successfully meeting this deadline, what deadline should be set for this project?

Dart Investments

Dart Investments provides investment advisory services to high-net-worth individuals. Scott Michaels, the firm's vice-president of information technology, has recently been charged by the president to develop a Web site to promote its investment services, to provide access to customer account information, and to allow individuals to establish new accounts online.

Scott decided to assign this project to Nicolette Connors, one of two directors in the information technology group. Since Dart Investments did not currently have a presence on the Web, Scott and Nicolette agreed that an appropriate starting point for the project would be for the project team to benchmark existing Web sites in order to gain a better understanding of the state of the art in this area. At the conclusion of their first meeting, Scott asked Nicolette to prepare a rough estimate of how long and how much this project would cost if it were pursued at a normal pace. Noting that the president appeared particularly anxious to launch the Web site, Scott also requested that she prepare a time and budget estimate related to launching the Web site as quickly as possible.

During the first project team meeting, the team identified seven major tasks associated with the project. The first task would be to benchmark existing Web sites. The team estimated that completing this task at a normal pace would likely require 10 days at a cost of $15,000. However, the team estimated that this task could be completed in as few as 7 days at a cost of $18,750 if the maximum allowable amount of overtime was used.

Once the benchmark study is completed, a project plan and project definition document can be prepared for top management approval. The team estimated that this task could be completed in 5 days at a cost of $3,750 working at a normal pace, or in 3 days at a cost of $4,500.

When the project receives the approval of top management, Web site design can begin. The team estimated that Web site design would require 15 days at a cost of $45,000 using no overtime or 10 days at a cost of $58,500 using all allowable overtime.

After the Web site design is complete, three tasks can be carried out simultaneously: (1) developing the Web site's database, (2) developing and coding the actual Web pages, and (3) developing and coding the Web sites forms. The team estimated that database development would require 10 days and cost $9,000 using no overtime, but could be completed in 7 days at a cost of $11,250 using overtime. Likewise, the team estimated that developing and coding the Web pages would require 10 days and cost $15,000 using no overtime, or could be reduced by 2 days at a total cost of $19,500. Developing the forms was to be subcontracted out and would take 7 days at a cost of $8,400. The organization that was to be used to create the forms does not provide an option for paying more for rush jobs.

Finally, once the database is developed, the Web pages coded, and the forms created, the entire Web site would need to be tested and debugged. The team estimated that this would require 3 days at a cost of $4,500. Using overtime, the team estimated that the testing and debugging task could be reduced by a day at a total cost of $6,750.

Questions

1. What is the cost of completing this project if no overtime is used? How long will it take to complete the project?
2. What is the shortest amount of time in which the project can be completed? What is the cost of completing the project in the shortest amount of time?
3. Suppose that the benchmarking study actually required 13 days as opposed to the 10 days originally estimated. What actions would you take to keep the project on a normal schedule?
4. Suppose the president wanted the Web site launched in 35 days. What actions would you take to meet this deadline? How much extra would it cost to complete the project in 35 days?

BIBLIOGRAPHY

Angus, R. B., N. A. Gundersen, and T. P. Cullinane. *Planning, Performing, and Controlling Projects: Principles and Applications.* 2nd ed. Upper Saddle River, NJ: Prentice Hall, 2000.

Chen, M. T. Simplified Project Economic Evaluation. *Cost Engineering,* 40 (January 1998): 31–35.

Coch, L., and J. R. P. French, Jr. Overcoming Resistance to Change. *Human Relations,* 1, 1948, 512–532.

Duncan, W. R. *A Guide to the Project Management Body of Knowledge.* Upper Darby, PA: Project Management Institute Publications, 1996.

Ghattas, R.G., and S. L. McKee. *Practical Project Management,* Upper Saddle River, NJ: Prentice Hall, 2001.

Graham, R. J., and R. L. Englund. *Creating an Environment for Successful Projects.* San Francisco: Jossey-Bass, 1997.

Gray, C. F., and E. W. Larson. *Project Management: The Managerial Process.* Upper Saddle River, NJ: Prentice Hall, 2000.

Gido, J., and J. P. Clements. *Successful Project Management.* Cincinnati, OH: South-Western Publishing Co., 1999.

Khang, D. B., and M. Yin. Time, Cost and Quality Tradeoff in Project Management. *International Journal of Project Management,* 17, 1999, 249–256.

Kolish, R. Resource Allocation Capabilities of Commercial Project Management Software Packages. *Interfaces,* 29, July–August 1999, 19–31.

Lowery, G., and T. S. Stover. *Managing Projects with Microsoft Project 2000: For Windows.* New York: Wiley, 2001.

Mantel, S. J., Jr., J. R. Meredith, S. M. Shafer, and M. M. Sutton. *Project Management in Practice.* New York: Wiley, 2001.

Meredith, J. R. The Implementation of Computer-Based Systems. *Journal of Operations Management,* 2, October 1981, 11–21.

Meredith, J. R., and S. J. Mantel, Jr. *Project Management: A Managerial Approach.* 4th ed. New York: Wiley, 2000.

Nicholas, J. M. *Project Management for Business and Technology.* Upper Saddle River, NJ: Prentice Hall, 2001.

Verma, V. K., and H. J. Thamhain. *Human Resource Skills for the Project Manager.* Upper Darby, PA: Project Management Institute Publications, 1997.

Appendix A

MATHEMATICS

The purpose of this appendix is to review the mathematical concepts that are used in this text.

A1 Definitions

Some notation is used throughout the text, independent of subject, and the student should be intimately familiar with these symbols:

$!$ Factorial: $n! = n(n-1)(n-2)\ .\ .\ .\ (1)$

Example

$$5! = 5 \cdot 4 \cdot 3 \cdot 2 \cdot 1 = 120$$

$\sum$ Summation

$\sum_i$ Sum over all values of the index i:

$$\sum_i x_i = x_1 + x_2 + \cdots + x_n, \quad \text{where } i = 1, 2, \ldots, n.$$

Alternatively, the symbol $\sum_{i=1}^{n} x_i$ can be used.

$\sum\sum$ Double summation

Example

$$\sum_{i=1}^{3}\sum_{j=1}^{2} x_{ij} = x_{11} + x_{12} + x_{21} + x_{22} + x_{31} + x_{32}$$

Constant: A constant is a quantity that always maintains a fixed value.
Parameter: A parameter is usually constant throughout a problem but may change from problem to problem.
Variable: A variable is a quantity whose value may change throughout a problem.
Continuous variable: The variable may assume *any* value (e.g., 14.7638 . . .) within its acceptable range.
Discrete variable: The variable may only take on certain (countable) values (e.g., $\frac{1}{4}$, $\frac{2}{4}$, $\frac{3}{4}$, and so on), frequently *integers* (1, 2, 3, . . .).
Independent variable: In an equation, this variable is known. It is usually shown on the x axis of graphs.
Dependent variable: In an equation, this variable, whose value is desired, is unknown. It is usually shown on the y axis of graphs.

Example

In the equation for a circle's circumference, $C = \pi D$:

C = continuous dependent variable
π = constant (3.14159)
D = continuous independent variable (the diameter)

A2 Functions

A function is a mathematical expression that states a relationship between at least two variables. The expression $y = f(x)$ is read as follows: y is a function of x. This means that given a value for x, y can be determined, although $y = f(x)$ does not tell us how. It states that some relationship exists. This relationship may take the form of a table or an equation. For example, $C = \pi D$ means that C is a function of D. That is, given D, it is possible to determine C. This example demonstrates a *single-valued* function, because for each value of D there exists only one value of C. Similarly, the equation $y = 4 + x^2$ is a single-valued function. However, the equation $y^2 = 4 + x$ is an example of a *multiple-valued function* because y, *the dependent variable,* may take on more than one value for each value of the *independent variable* x (if $x = 0$, then $y = 2$ or -2). In the above examples, y was a function of the single variable x. However, in expressions like $y = f(x, z) = x^2 + 3z$, y is a function of *several variables.* Again, x and z would be the independent variables and y the dependent variable.

As shown in Exhibit A1, the *slope* of a function measures how much the dependent variable changes for a small amount of increase in each of the independent variables. If the function is a straight line, then the slope of the function is the same everywhere. However, for functions that are not straight lines, it is necessary to specify *where* the slope is to be measured and in what direction, because it may be different at different values of the independent variables. If the dependent variable *increases* (*decreases*) with small increases in the independent variables, then we say that the function is *positively* (*negatively*) *sloped.*

EXHIBIT A1 Slope of Linear Functions

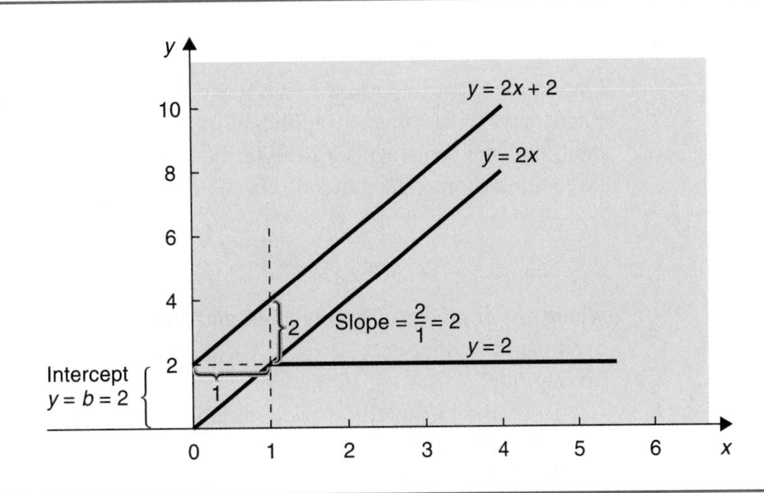

Continuous and Discrete Functions

In a manner similar to continuous and discrete variables, there exist continuous and discrete functions also. Examples of continuous functions are

$$(1)\ y = x^2 + 2x \text{ and } (2)\ y = 5$$

Some examples of discrete functions are:

$$(1)\quad y = 5x \qquad \textit{where } x = 0, 1, 2, \ldots$$

$$(2)\quad y = \begin{cases} 10 + 3x & \text{for} \quad 0 \le x \le 2 \\ 14 + x & \qquad\quad x \ge 2 \end{cases} \text{ and } x \text{ integer}$$

Equalities

Functional relationships where the value of the dependent variable *equals* certain values of the independent variable are termed equations. For example, $y = 4x$.

Inequalities

If functional relationships cannot be written as equations but it is known that one exceeds the other, then they must be unequal and can be expressed as inequalities. Such relationships can be designated by the symbol $\neq$. For example $y \neq 5x$ means that y is *not* equal to $5x$.

When relationships are not equal, they can take *one* of four possible forms:

Form	Symbol	Example
Less than	$<$	$y < 6x + 2$
Less than or equal to	$\le$	$y \le 4x - 1$
Greater than	$>$	$y > 2x + 9$
Greater than or equal to	$\ge$	$y \ge 5x$

A3 Linear Equations

One of the most important functional relationships is the *linear equation,* due to its simplicity and wide range of applicability. A linear equation of two variables (one dependent, one independent) is a straight line. A linear equation of three variables is a plane, in three dimensions. The general form of a linear equation is

$$y = a_1x_1 + a_2x_2 + \cdots + a_nx_n + b = \sum_i a_ix_i + b$$

where the a_i and b are constants and the x_i are different variables. Let V_i designate each variable; that is, $x_1, x_2, \ldots, x_n$. Then a linear equation always satisfies the following rule:

If kV_i is substituted for each variable V_i in the original equation, $y = f(V_i)$, where k is a constant, then the result will be ky.

Mathematically

$$\text{If } y = f(V_i) \text{ is linear, then } f(kV_i) = ky$$

Example

Is the function $y = 5x + 3(w - 4z)$ linear?

Solution Substitute k times each variable in the equation and find

$$5(kx) + 3[kw - 4(kz)] = 5kx + 3k(w - 4z) = ky$$

Thus, according to A2, the equation is linear.

The Slope of a Linear Equation

The slope of a linear equation is constant at all points of the function. In general, the equation of a straight line is given as

$$y = ax + b$$

where a and b are constants. The slope of such a line is always a. The constant b also has a special name: the *intercept.* This is because when x is set to 0 (which is where the line "intercepts" the y-axis), the value of y equals b. The slope and intercept for the equation $y = 2x + 2$ are shown in Figure A1. Also shown in the figure are parts of the linear functions $y = 2$ and $y = 2x$, both of which differ from $y = 2x + 2$. The slope of $y = 2$ is 0, and its intercept is 2. The slope of $y = 2x$ is 2 and its intercept is 0.

Nonlinear Functions

Any function that does not meet the linearity requirement is considered nonlinear.

A4 Rules of Manipulation

Inequalities

The rules of manipulation of inequalities are the same as for equations with one exception: when multiplying or dividing the inequality by a negative number, the inequality sign is reversed. For example, given

$$3 \geq -5$$

Multiplying by –1:

$$-3 \leq 5$$

As another example:

$$-4y + 10x \geq 2x - 8$$

Dividing by –2:

$$2y - 5x \leq -x + 4$$

Exponents

The definition of x^n (called x to the nth power) is x multiplied together n times, where n is called the *exponent.* The rules for manipulating exponents are

1. *Rule of multiplication:*

$$x^a x^b = x^{a+b}$$

   For example, $x^2x^3 = x^5$.

2. *Rule of division:*

$$x^a/x^b = x^{a-b}$$

The second rule holds for cases where $b > a$ as well. This states, for example, that $x^3/x^4 = x^{3-4} = x^{-1}$. Invoking rule 1, it is possible to work backwards through cross-multiplication to obtain

$$x^4x^{-1} = x^{4-1} = x^3$$

Implications

1.

$$x^n/x^n = x^0 = 1$$

   for any nonzero value of x.

2. It can be seen that $1/x^2 = x^0/x^2 = x^{0-2} = x^{-2}$; that is, in general:

$$x^{-a} = 1/x^a$$

3.

$$(x^a)^b = x^{a(b)}$$

Example

$$(x^2)^3 = (xx)(xx)(xx) = x^{2(3)} = x^6$$

4.

$$\sqrt[b]{x^a} = x^{a/b}$$

Example

$$\sqrt[3]{x^6} = x^{6/3} = x^2$$

Appendix B

TABLES

1 Binomial Probabilities
2 Poisson Probabilities
3 Area Under the Normal Curve
4 Area in Upper Tail *t* distribution
5 χ^2 Distribution
6 *F* Distribution

TABLE I Binomial Probabilities Probability of exactly x successes in n trials when probability of success on any trial is p. Excel function: BINOMDIST(x,n,p,false).

$n = 2$

	p, probability of success									
x	0.05	0.10	0.15	0.20	0.25	0.30	0.35	0.40	0.45	0.50
0	0.9025	0.8100	0.7225	0.6400	0.5625	0.4900	0.4225	0.3600	0.3025	0.2500
1	0.0950	0.1800	0.2550	0.3200	0.3750	0.4200	0.4550	0.4800	0.4950	0.5000
2	0.0025	0.0100	0.0225	0.0400	0.0625	0.0900	0.1225	0.1600	0.2025	0.2500

$n = 3$

	p, probability of success									
x	0.05	0.10	0.15	0.20	0.25	0.30	0.35	0.40	0.45	0.50
0	0.8574	0.7290	0.6141	0.5120	0.4219	0.3430	0.2746	0.2160	0.1664	0.1250
1	0.1354	0.2430	0.3251	0.3840	0.4219	0.4410	0.4436	0.4320	0.4084	0.3750
2	0.0071	0.0270	0.0574	0.0960	0.1406	0.1890	0.2389	0.2880	0.3341	0.3750
3	0.0001	0.0010	0.0034	0.0080	0.0156	0.0270	0.0429	0.0640	0.0911	0.1250

$n = 4$

	p, probability of success									
x	0.05	0.10	0.15	0.20	0.25	0.30	0.35	0.40	0.45	0.50
0	0.8145	0.6561	0.5220	0.4096	0.3164	0.2401	0.1785	0.1296	0.0915	0.0625
1	0.1715	0.2916	0.3685	0.4096	0.4219	0.4116	0.3845	0.3456	0.2995	0.2500
2	0.0135	0.0486	0.0975	0.1536	0.2109	0.2646	0.3105	0.3456	0.3675	0.3750
3	0.0005	0.0036	0.0115	0.0256	0.0469	0.0756	0.1115	0.1536	0.2005	0.2500
4	0.0000	0.0001	0.0005	0.0016	0.0039	0.0081	0.0150	0.0256	0.0410	0.0625

$n = 5$

	p, probability of success									
x	0.05	0.10	0.15	0.20	0.25	0.30	0.35	0.40	0.45	0.50
0	0.7738	0.5905	0.4437	0.3277	0.2373	0.1681	0.1160	0.0778	0.0503	0.0313
1	0.2036	0.3281	0.3915	0.4096	0.3955	0.3602	0.3124	0.2592	0.2059	0.1563
2	0.0214	0.0729	0.1382	0.2048	0.2637	0.3087	0.3364	0.3456	0.3369	0.3125
3	0.0011	0.0081	0.0244	0.0512	0.0879	0.1323	0.1811	0.2304	0.2757	0.3125
4	0.0000	0.0005	0.0022	0.0064	0.0146	0.0284	0.0488	0.0768	0.1128	0.1563
5	0.0000	0.0000	0.0001	0.0003	0.0010	0.0024	0.0053	0.0102	0.0185	0.0313

$n = 6$

	p, probability of success									
x	0.05	0.10	0.15	0.20	0.25	0.30	0.35	0.40	0.45	0.50
0	0.7351	0.5314	0.3771	0.2621	0.1780	0.1176	0.0754	0.0467	0.0277	0.0156
1	0.2321	0.3543	0.3993	0.3932	0.3560	0.3025	0.2437	0.1866	0.1359	0.0938
2	0.0305	0.0984	0.1762	0.2458	0.2966	0.3241	0.3280	0.3110	0.2780	0.2344
3	0.0021	0.0146	0.0415	0.0819	0.1318	0.1852	0.2355	0.2765	0.3032	0.3125
4	0.0001	0.0012	0.0055	0.0154	0.0330	0.0595	0.0951	0.1382	0.1861	0.2344
5	0.0000	0.0001	0.0004	0.0015	0.0044	0.0102	0.0205	0.0369	0.0609	0.0938
6	0.0000	0.0000	0.0000	0.0001	0.0002	0.0007	0.0018	0.0041	0.0083	0.0156

$n = 7$

	p, probability of success									
x	0.05	0.10	0.15	0.20	0.25	0.30	0.35	0.40	0.45	0.50
0	0.6983	0.4783	0.3206	0.2097	0.1335	0.0824	0.0490	0.0280	0.0152	0.0078
1	0.2573	0.3720	0.3960	0.3670	0.3115	0.2471	0.1848	0.1306	0.0872	0.0547
2	0.0406	0.1240	0.2097	0.2753	0.3115	0.3177	0.2985	0.2613	0.2140	0.1641
3	0.0036	0.0230	0.0617	0.1147	0.1730	0.2269	0.2679	0.2903	0.2918	0.2734
4	0.0002	0.0026	0.0109	0.0287	0.0577	0.0972	0.1442	0.1935	0.2388	0.2734
5	0.0000	0.0002	0.0012	0.0043	0.0115	0.0250	0.0466	0.0774	0.1172	0.1641
6	0.0000	0.0000	0.0001	0.0004	0.0013	0.0036	0.0084	0.0172	0.0320	0.0547
7	0.0000	0.0000	0.0000	0.0000	0.0001	0.0002	0.0006	0.0016	0.0037	0.0078

$n = 8$

	p, probability of success									
x	0.05	0.10	0.15	0.20	0.25	0.30	0.35	0.40	0.45	0.50
0	0.6634	0.4305	0.2725	0.1678	0.1001	0.0576	0.0319	0.0168	0.0084	0.0039
1	0.2793	0.3826	0.3847	0.3355	0.2670	0.1977	0.1373	0.0896	0.0548	0.0313
2	0.0515	0.1488	0.2376	0.2936	0.3115	0.2965	0.2587	0.2090	0.1569	0.1094
3	0.0054	0.0331	0.0839	0.1468	0.2076	0.2541	0.2786	0.2787	0.2568	0.2188
4	0.0004	0.0046	0.0185	0.0459	0.0865	0.1361	0.1875	0.2322	0.2627	0.2734
5	0.0000	0.0004	0.0026	0.0092	0.0231	0.0467	0.0808	0.1239	0.1719	0.2188
6	0.0000	0.0000	0.0002	0.0011	0.0038	0.0100	0.0217	0.0413	0.0703	0.1094
7	0.0000	0.0000	0.0000	0.0001	0.0004	0.0012	0.0033	0.0079	0.0164	0.0313
8	0.0000	0.0000	0.0000	0.0000	0.0000	0.0001	0.0002	0.0007	0.0017	0.0039

TABLE I ***Continued***

$n = 9$	p, probability of success									
x	0.05	0.10	0.15	0.20	0.25	0.30	0.35	0.40	0.45	0.50
0	0.6302	0.3874	0.2316	0.1342	0.0751	0.0404	0.0207	0.0101	0.0046	0.0020
1	0.2985	0.3874	0.3679	0.3020	0.2253	0.1556	0.1004	0.0605	0.0339	0.0176
2	0.0629	0.1722	0.2597	0.3020	0.3003	0.2668	0.2162	0.1612	0.1110	0.0703
3	0.0077	0.0446	0.1069	0.1762	0.2336	0.2668	0.2716	0.2508	0.2119	0.1641
4	0.0006	0.0074	0.0283	0.0661	0.1168	0.1715	0.2194	0.2508	0.2600	0.2461
5	0.0000	0.0008	0.0050	0.0165	0.0389	0.0735	0.1181	0.1672	0.2128	0.2461
6	0.0000	0.0001	0.0006	0.0028	0.0087	0.0210	0.0424	0.0743	0.1160	0.1641
7	0.0000	0.0000	0.0000	0.0003	0.0012	0.0039	0.0098	0.0212	0.0407	0.0703
8	0.0000	0.0000	0.0000	0.0000	0.0001	0.0004	0.0013	0.0035	0.0083	0.0176
9	0.0000	0.0000	0.0000	0.0000	0.0000	0.0000	0.0001	0.0003	0.0008	0.0020

$n = 10$	p, probability of success									
x	0.05	0.10	0.15	0.20	0.25	0.30	0.35	0.40	0.45	0.50
0	0.5987	0.3487	0.1969	0.1074	0.0563	0.0282	0.0135	0.0060	0.0025	0.0010
1	0.3151	0.3874	0.3474	0.2684	0.1877	0.1211	0.0725	0.0403	0.0207	0.0098
2	0.0746	0.1937	0.2759	0.3020	0.2816	0.2335	0.1757	0.1209	0.0763	0.0439
3	0.0105	0.0574	0.1298	0.2013	0.2503	0.2668	0.2522	0.2150	0.1665	0.1172
4	0.0010	0.0112	0.0401	0.0881	0.1460	0.2001	0.2377	0.2508	0.2384	0.2051
5	0.0001	0.0015	0.0085	0.0264	0.0584	0.1029	0.1536	0.2007	0.2340	0.2461
6	0.0000	0.0001	0.0012	0.0055	0.0162	0.0368	0.0689	0.1115	0.1596	0.2051
7	0.0000	0.0000	0.0001	0.0008	0.0031	0.0090	0.0212	0.0425	0.0746	0.1172
8	0.0000	0.0000	0.0000	0.0001	0.0004	0.0014	0.0043	0.0106	0.0229	0.0439
9	0.0000	0.0000	0.0000	0.0000	0.0000	0.0001	0.0005	0.0016	0.0042	0.0098
10	0.0000	0.0000	0.0000	0.0000	0.0000	0.0000	0.0000	0.0001	0.0003	0.0010

$n = 15$	p, probability of success									
x	0.05	0.10	0.15	0.20	0.25	0.30	0.35	0.40	0.45	0.50
0	0.4633	0.2059	0.0874	0.0352	0.0134	0.0047	0.0016	0.0005	0.0001	0.0000
1	0.3658	0.3432	0.2312	0.1319	0.0668	0.0305	0.0126	0.0047	0.0016	0.0005
2	0.1348	0.2669	0.2856	0.2309	0.1559	0.0916	0.0476	0.0219	0.0090	0.0032
3	0.0307	0.1285	0.2184	0.2501	0.2252	0.1700	0.1110	0.0634	0.0318	0.0139
4	0.0049	0.0428	0.1156	0.1876	0.2252	0.2186	0.1792	0.1268	0.0780	0.0417
5	0.0006	0.0105	0.0449	0.1032	0.1651	0.2061	0.2123	0.1859	0.1404	0.0916
6	0.0000	0.0019	0.0132	0.0430	0.0917	0.1472	0.1906	0.2066	0.1914	0.1527
7	0.0000	0.0003	0.0030	0.0138	0.0393	0.0811	0.1319	0.1771	0.2013	0.1964
8	0.0000	0.0000	0.0005	0.0035	0.0131	0.0348	0.0710	0.1181	0.1647	0.1964
9	0.0000	0.0000	0.0001	0.0007	0.0034	0.0116	0.0298	0.0612	0.1048	0.1527
10	0.0000	0.0000	0.0000	0.0001	0.0007	0.0030	0.0096	0.0245	0.0515	0.0916
11	0.0000	0.0000	0.0000	0.0000	0.0001	0.0006	0.0024	0.0074	0.0191	0.0417
12	0.0000	0.0000	0.0000	0.0000	0.0000	0.0001	0.0004	0.0016	0.0052	0.0139
13	0.0000	0.0000	0.0000	0.0000	0.0000	0.0000	0.0001	0.0003	0.0010	0.0032
14	0.0000	0.0000	0.0000	0.0000	0.0000	0.0000	0.0000	0.0000	0.0001	0.0005
15	0.0000	0.0000	0.0000	0.0000	0.0000	0.0000	0.0000	0.0000	0.0000	0.0000

$n = 20$	p, probability of success									
x	0.05	0.10	0.15	0.20	0.25	0.30	0.35	0.40	0.45	0.50
0	0.3585	0.1216	0.0388	0.0115	0.0032	0.0008	0.0002	0.0000	0.0000	0.0000
1	0.3774	0.2702	0.1368	0.0576	0.0211	0.0068	0.0020	0.0005	0.0001	0.0000
2	0.1887	0.2852	0.2293	0.1369	0.0669	0.0278	0.0100	0.0031	0.0008	0.0002
3	0.0596	0.1901	0.2428	0.2054	0.1339	0.0716	0.0323	0.0123	0.0040	0.0011
4	0.0133	0.0898	0.1821	0.2182	0.1897	0.1304	0.0738	0.0350	0.0139	0.0046
5	0.0022	0.0319	0.1028	0.1746	0.2023	0.1789	0.1272	0.0746	0.0365	0.0148
6	0.0003	0.0089	0.0454	0.1091	0.1686	0.1916	0.1712	0.1244	0.0746	0.0370
7	0.0000	0.0020	0.0160	0.0545	0.1124	0.1643	0.1844	0.1659	0.1221	0.0739
8	0.0000	0.0004	0.0046	0.0222	0.0609	0.1144	0.1614	0.1797	0.1623	0.1201
9	0.0000	0.0001	0.0011	0.0074	0.0271	0.0654	0.1158	0.1597	0.1771	0.1602
10	0.0000	0.0000	0.0002	0.0020	0.0099	0.0308	0.0686	0.1171	0.1593	0.1762
11	0.0000	0.0000	0.0000	0.0005	0.0030	0.0120	0.0336	0.0710	0.1185	0.1602
12	0.0000	0.0000	0.0000	0.0001	0.0008	0.0039	0.0136	0.0355	0.0727	0.1201
13	0.0000	0.0000	0.0000	0.0000	0.0002	0.0010	0.0045	0.0146	0.0366	0.0739
14	0.0000	0.0000	0.0000	0.0000	0.0000	0.0002	0.0012	0.0049	0.0150	0.0370
15	0.0000	0.0000	0.0000	0.0000	0.0000	0.0000	0.0003	0.0013	0.0049	0.0148
16	0.0000	0.0000	0.0000	0.0000	0.0000	0.0000	0.0000	0.0003	0.0013	0.0046
17	0.0000	0.0000	0.0000	0.0000	0.0000	0.0000	0.0000	0.0000	0.0002	0.0011
18	0.0000	0.0000	0.0000	0.0000	0.0000	0.0000	0.0000	0.0000	0.0000	0.0002
19	0.0000	0.0000	0.0000	0.0000	0.0000	0.0000	0.0000	0.0000	0.0000	0.0000
20	0.0000	0.0000	0.0000	0.0000	0.0000	0.0000	0.0000	0.0000	0.0000	0.0000

Source: Computed using Excel.

TABLE 2 Poisson Probabilities Probability of exactly x occurrences of an event when the average or expected number of events is λ. Excel function: POISSON(x,λ,false).

	Average or expected number of occurrences									
x	0.10	0.20	0.30	0.40	0.50	0.60	0.70	0.80	0.90	1.00
0	0.9048	0.8187	0.7408	0.6703	0.6065	0.5488	0.4966	0.4493	0.4066	0.3679
1	0.0905	0.1637	0.2222	0.2681	0.3033	0.3293	0.3476	0.3595	0.3659	0.3679
2	0.0045	0.0164	0.0333	0.0536	0.0758	0.0988	0.1217	0.1438	0.1647	0.1839
3	0.0002	0.0011	0.0033	0.0072	0.0126	0.0198	0.0284	0.0383	0.0494	0.0613
4	0.0000	0.0001	0.0003	0.0007	0.0016	0.0030	0.0050	0.0077	0.0111	0.0153
5	0.0000	0.0000	0.0000	0.0001	0.0002	0.0004	0.0007	0.0012	0.0020	0.0031
6	0.0000	0.0000	0.0000	0.0000	0.0000	0.0000	0.0001	0.0002	0.0003	0.0005
7	0.0000	0.0000	0.0000	0.0000	0.0000	0.0000	0.0000	0.0000	0.0000	0.0001
8	0.0000	0.0000	0.0000	0.0000	0.0000	0.0000	0.0000	0.0000	0.0000	0.0000

	Average or expected number of occurrences									
x	1.10	1.20	1.30	1.40	1.50	1.60	1.70	1.80	1.90	2.00
0	0.3329	0.3012	0.2725	0.2466	0.2231	0.2019	0.1827	0.1653	0.1496	0.1353
1	0.3662	0.3614	0.3543	0.3452	0.3347	0.3230	0.3106	0.2975	0.2842	0.2707
2	0.2014	0.2169	0.2303	0.2417	0.2510	0.2584	0.2640	0.2678	0.2700	0.2707
3	0.0738	0.0867	0.0998	0.1128	0.1255	0.1378	0.1496	0.1607	0.1710	0.1804
4	0.0203	0.0260	0.0324	0.0395	0.0471	0.0551	0.0636	0.0723	0.0812	0.0902
5	0.0045	0.0062	0.0084	0.0111	0.0141	0.0176	0.0216	0.0260	0.0309	0.0361
6	0.0008	0.0012	0.0018	0.0026	0.0035	0.0047	0.0061	0.0078	0.0098	0.0120
7	0.0001	0.0002	0.0003	0.0005	0.0008	0.0011	0.0015	0.0020	0.0027	0.0034
8	0.0000	0.0000	0.0001	0.0001	0.0001	0.0002	0.0003	0.0005	0.0006	0.0009
9	0.0000	0.0000	0.0000	0.0000	0.0000	0.0000	0.0001	0.0001	0.0001	0.0002
10	0.0000	0.0000	0.0000	0.0000	0.0000	0.0000	0.0000	0.0000	0.0000	0.0000

	Average or expected number of occurrences									
x	2.10	2.20	2.30	2.40	2.50	2.60	2.70	2.80	2.90	3.00
0	0.1225	0.1108	0.1003	0.0907	0.0821	0.0743	0.0672	0.0608	0.0550	0.0498
1	0.2572	0.2438	0.2306	0.2177	0.2052	0.1931	0.1815	0.1703	0.1596	0.1494
2	0.2700	0.2681	0.2652	0.2613	0.2565	0.2510	0.2450	0.2384	0.2314	0.2240
3	0.1890	0.1966	0.2033	0.2090	0.2138	0.2176	0.2205	0.2225	0.2237	0.2240
4	0.0992	0.1082	0.1169	0.1254	0.1336	0.1414	0.1488	0.1557	0.1622	0.1680
5	0.0417	0.0476	0.0538	0.0602	0.0668	0.0735	0.0804	0.0872	0.0940	0.1008
6	0.0146	0.0174	0.0206	0.0241	0.0278	0.0319	0.0362	0.0407	0.0455	0.0504
7	0.0044	0.0055	0.0068	0.0083	0.0099	0.0118	0.0139	0.0163	0.0188	0.0216
8	0.0011	0.0015	0.0019	0.0025	0.0031	0.0038	0.0047	0.0057	0.0068	0.0081
9	0.0003	0.0004	0.0005	0.0007	0.0009	0.0011	0.0014	0.0018	0.0022	0.0027
10	0.0001	0.0001	0.0001	0.0002	0.0002	0.0003	0.0004	0.0005	0.0006	0.0008
11	0.0000	0.0000	0.0000	0.0000	0.0000	0.0001	0.0001	0.0001	0.0002	0.0002
12	0.0000	0.0000	0.0000	0.0000	0.0000	0.0000	0.0000	0.0000	0.0000	0.0001
13	0.0000	0.0000	0.0000	0.0000	0.0000	0.0000	0.0000	0.0000	0.0000	0.0000

	Average or expected number of occurrences									
x	3.10	3.20	3.30	3.40	3.50	3.60	3.70	3.80	3.90	4.00
0	0.0450	0.0408	0.0369	0.0334	0.0302	0.0273	0.0247	0.0224	0.0202	0.0183
1	0.1397	0.1304	0.1217	0.1135	0.1057	0.0984	0.0915	0.0850	0.0789	0.0733
2	0.2165	0.2087	0.2008	0.1929	0.1850	0.1771	0.1692	0.1615	0.1539	0.1465
3	0.2237	0.2226	0.2209	0.2186	0.2158	0.2125	0.2087	0.2046	0.2001	0.1954
4	0.1733	0.1781	0.1823	0.1858	0.1888	0.1912	0.1931	0.1944	0.1951	0.1954
5	0.1075	0.1140	0.1203	0.1264	0.1322	0.1377	0.1429	0.1477	0.1522	0.1563
6	0.0555	0.0608	0.0662	0.0716	0.0771	0.0826	0.0881	0.0936	0.0989	0.1042
7	0.0246	0.0278	0.0312	0.0348	0.0385	0.0425	0.0466	0.0508	0.0551	0.0595
8	0.0095	0.0111	0.0129	0.0148	0.0169	0.0191	0.0215	0.0241	0.0269	0.0298
9	0.0033	0.0040	0.0047	0.0056	0.0066	0.0076	0.0089	0.0102	0.0116	0.0132
10	0.0010	0.0013	0.0016	0.0019	0.0023	0.0028	0.0033	0.0039	0.0045	0.0053
11	0.0003	0.0004	0.0005	0.0006	0.0007	0.0009	0.0011	0.0013	0.0016	0.0019
12	0.0001	0.0001	0.0001	0.0002	0.0002	0.0003	0.0003	0.0004	0.0005	0.0006
13	0.0000	0.0000	0.0000	0.0000	0.0001	0.0001	0.0001	0.0001	0.0002	0.0002
14	0.0000	0.0000	0.0000	0.0000	0.0000	0.0000	0.0000	0.0000	0.0000	0.0001
15	0.0000	0.0000	0.0000	0.0000	0.0000	0.0000	0.0000	0.0000	0.0000	0.0000

TABLE 2 Continued

	Average or expected number of occurrences									
x	4.10	4.20	4.30	4.40	4.50	4.60	4.70	4.80	4.90	5.00
0	0.0166	0.0150	0.0136	0.0123	0.0111	0.0101	0.0091	0.0082	0.0074	0.0067
1	0.0679	0.0630	0.0583	0.0540	0.0500	0.0462	0.0427	0.0395	0.0365	0.0337
2	0.1393	0.1323	0.1254	0.1188	0.1125	0.1063	0.1005	0.0948	0.0894	0.0842
3	0.1904	0.1852	0.1798	0.1743	0.1687	0.1631	0.1574	0.1517	0.1460	0.1404
4	0.1951	0.1944	0.1933	0.1917	0.1898	0.1875	0.1849	0.1820	0.1789	0.1755
5	0.1600	0.1633	0.1662	0.1687	0.1708	0.1725	0.1738	0.1747	0.1753	0.1755
6	0.1093	0.1143	0.1191	0.1237	0.1281	0.1323	0.1362	0.1398	0.1432	0.1462
7	0.0640	0.0686	0.0732	0.0778	0.0824	0.0869	0.0914	0.0959	0.1002	0.1044
8	0.0328	0.0360	0.0393	0.0428	0.0463	0.0500	0.0537	0.0575	0.0614	0.0653
9	0.0150	0.0168	0.0188	0.0209	0.0232	0.0255	0.0281	0.0307	0.0334	0.0363
10	0.0061	0.0071	0.0081	0.0092	0.0104	0.0118	0.0132	0.0147	0.0164	0.0181
11	0.0023	0.0027	0.0032	0.0037	0.0043	0.0049	0.0056	0.0064	0.0073	0.0082
12	0.0008	0.0009	0.0011	0.0013	0.0016	0.0019	0.0022	0.0026	0.0030	0.0034
13	0.0002	0.0003	0.0004	0.0005	0.0006	0.0007	0.0008	0.0009	0.0011	0.0013
14	0.0001	0.0001	0.0001	0.0001	0.0002	0.0002	0.0003	0.0003	0.0004	0.0005
15	0.0000	0.0000	0.0000	0.0000	0.0001	0.0001	0.0001	0.0001	0.0001	0.0002
16	0.0000	0.0000	0.0000	0.0000	0.0000	0.0000	0.0000	0.0000	0.0000	0.0000

	Average or expected number of occurrences									
x	6.00	8.00	10.00	12.00	14.00	16.00	18.00	20.00	22.00	24.00
0	0.0025	0.0003	0.0000	0.0000	0.0000	0.0000	0.0000	0.0000	0.0000	0.0000
1	0.0149	0.0027	0.0005	0.0001	0.0000	0.0000	0.0000	0.0000	0.0000	0.0000
2	0.0446	0.0107	0.0023	0.0004	0.0001	0.0000	0.0000	0.0000	0.0000	0.0000
3	0.0892	0.0286	0.0076	0.0018	0.0004	0.0001	0.0000	0.0000	0.0000	0.0000
4	0.1339	0.0573	0.0189	0.0053	0.0013	0.0003	0.0001	0.0000	0.0000	0.0000
5	0.1606	0.0916	0.0378	0.0127	0.0037	0.0010	0.0002	0.0001	0.0000	0.0000
6	0.1606	0.1221	0.0631	0.0255	0.0087	0.0026	0.0007	0.0002	0.0000	0.0000
7	0.1377	0.1396	0.0901	0.0437	0.0174	0.0060	0.0019	0.0005	0.0001	0.0000
8	0.1033	0.1396	0.1126	0.0655	0.0304	0.0120	0.0042	0.0013	0.0004	0.0001
9	0.0688	0.1241	0.1251	0.0874	0.0473	0.0213	0.0083	0.0029	0.0009	0.0003
10	0.0413	0.0993	0.1251	0.1048	0.0663	0.0341	0.0150	0.0058	0.0020	0.0007
11	0.0225	0.0722	0.1137	0.1144	0.0844	0.0496	0.0245	0.0106	0.0041	0.0014
12	0.0113	0.0481	0.0948	0.1144	0.0984	0.0661	0.0368	0.0176	0.0075	0.0029
13	0.0052	0.0296	0.0729	0.1056	0.1060	0.0814	0.0509	0.0271	0.0127	0.0053
14	0.0022	0.0169	0.0521	0.0905	0.1060	0.0930	0.0655	0.0387	0.0199	0.0091
15	0.0009	0.0090	0.0347	0.0724	0.0989	0.0992	0.0786	0.0516	0.0292	0.0146
16	0.0003	0.0045	0.0217	0.0543	0.0866	0.0992	0.0884	0.0646	0.0401	0.0219
17	0.0001	0.0021	0.0128	0.0383	0.0713	0.0934	0.0936	0.0760	0.0520	0.0309
18	0.0000	0.0009	0.0071	0.0255	0.0554	0.0830	0.0936	0.0844	0.0635	0.0412
19	0.0000	0.0004	0.0037	0.0161	0.0409	0.0699	0.0887	0.0888	0.0735	0.0520
20	0.0000	0.0002	0.0019	0.0097	0.0286	0.0559	0.0798	0.0888	0.0809	0.0624
21	0.0000	0.0001	0.0009	0.0055	0.0191	0.0426	0.0684	0.0846	0.0847	0.0713
22	0.0000	0.0000	0.0004	0.0030	0.0121	0.0310	0.0560	0.0769	0.0847	0.0778
23	0.0000	0.0000	0.0002	0.0016	0.0074	0.0216	0.0438	0.0669	0.0810	0.0812
24	0.0000	0.0000	0.0001	0.0008	0.0043	0.0144	0.0328	0.0557	0.0743	0.0812
25	0.0000	0.0000	0.0000	0.0004	0.0024	0.0092	0.0237	0.0446	0.0654	0.0779
26	0.0000	0.0000	0.0000	0.0002	0.0013	0.0057	0.0164	0.0343	0.0553	0.0719
27	0.0000	0.0000	0.0000	0.0001	0.0007	0.0034	0.0109	0.0254	0.0451	0.0639
28	0.0000	0.0000	0.0000	0.0000	0.0003	0.0019	0.0070	0.0181	0.0354	0.0548
29	0.0000	0.0000	0.0000	0.0000	0.0002	0.0011	0.0044	0.0125	0.0269	0.0453
30	0.0000	0.0000	0.0000	0.0000	0.0001	0.0006	0.0026	0.0083	0.0197	0.0363
31	0.0000	0.0000	0.0000	0.0000	0.0000	0.0003	0.0015	0.0054	0.0140	0.0281
32	0.0000	0.0000	0.0000	0.0000	0.0000	0.0001	0.0009	0.0034	0.0096	0.0211
33	0.0000	0.0000	0.0000	0.0000	0.0000	0.0001	0.0005	0.0020	0.0064	0.0153
34	0.0000	0.0000	0.0000	0.0000	0.0000	0.0000	0.0002	0.0012	0.0041	0.0108
35	0.0000	0.0000	0.0000	0.0000	0.0000	0.0000	0.0001	0.0007	0.0026	0.0074
36	0.0000	0.0000	0.0000	0.0000	0.0000	0.0000	0.0001	0.0004	0.0016	0.0049
37	0.0000	0.0000	0.0000	0.0000	0.0000	0.0000	0.0000	0.0002	0.0009	0.0032
38	0.0000	0.0000	0.0000	0.0000	0.0000	0.0000	0.0000	0.0001	0.0005	0.0020
39	0.0000	0.0000	0.0000	0.0000	0.0000	0.0000	0.0000	0.0001	0.0003	0.0012
40	0.0000	0.0000	0.0000	0.0000	0.0000	0.0000	0.0000	0.0000	0.0002	0.0007
41	0.0000	0.0000	0.0000	0.0000	0.0000	0.0000	0.0000	0.0000	0.0001	0.0004
42	0.0000	0.0000	0.0000	0.0000	0.0000	0.0000	0.0000	0.0000	0.0000	0.0003
43	0.0000	0.0000	0.0000	0.0000	0.0000	0.0000	0.0000	0.0000	0.0000	0.0001
44	0.0000	0.0000	0.0000	0.0000	0.0000	0.0000	0.0000	0.0000	0.0000	0.0001
45	0.0000	0.0000	0.0000	0.0000	0.0000	0.0000	0.0000	0.0000	0.0000	0.0000

Source: Computed using Excel.

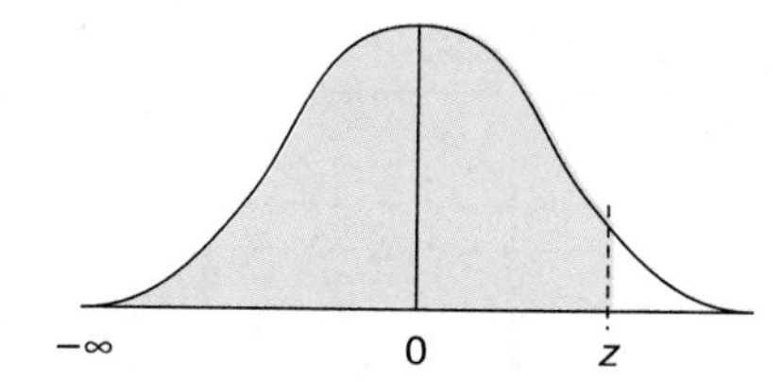

TABLE 3 Area Under the Normal Curve Area from negative infinity to z. Excel function: NORMSDIST(z).

z	**0.00**	**0.01**	**0.02**	**0.03**	**0.04**	**0.05**	**0.06**	**0.07**	**0.08**	**0.09**
0.00	0.5000	0.5040	0.5080	0.5120	0.5160	0.5199	0.5239	0.5279	0.5319	0.5359
0.10	0.5398	0.5438	0.5478	0.5517	0.5557	0.5596	0.5636	0.5675	0.5714	0.5753
0.20	0.5793	0.5832	0.5871	0.5910	0.5948	0.5987	0.6026	0.6064	0.6103	0.6141
0.30	0.6179	0.6217	0.6255	0.6293	0.6331	0.6368	0.6406	0.6443	0.6480	0.6517
0.40	0.6554	0.6591	0.6628	0.6664	0.6700	0.6736	0.6772	0.6808	0.6844	0.6879
0.50	0.6915	0.6950	0.6985	0.7019	0.7054	0.7088	0.7123	0.7157	0.7190	0.7224
0.60	0.7257	0.7291	0.7324	0.7357	0.7389	0.7422	0.7454	0.7486	0.7517	0.7549
0.70	0.7580	0.7611	0.7642	0.7673	0.7704	0.7734	0.7764	0.7794	0.7823	0.7852
0.80	0.7881	0.7910	0.7939	0.7967	0.7995	0.8023	0.8051	0.8078	0.8106	0.8133
0.90	0.8159	0.8186	0.8212	0.8238	0.8264	0.8289	0.8315	0.8340	0.8365	0.8389
1.00	0.8413	0.8438	0.8461	0.8485	0.8508	0.8531	0.8554	0.8577	0.8599	0.8621
1.10	0.8643	0.8665	0.8686	0.8708	0.8729	0.8749	0.8770	0.8790	0.8810	0.8830
1.20	0.8849	0.8869	0.8888	0.8907	0.8925	0.8944	0.8962	0.8980	0.8997	0.9015
1.30	0.9032	0.9049	0.9066	0.9082	0.9099	0.9115	0.9131	0.9147	0.9162	0.9177
1.40	0.9192	0.9207	0.9222	0.9236	0.9251	0.9265	0.9279	0.9292	0.9306	0.9319
1.50	0.9332	0.9345	0.9357	0.9370	0.9382	0.9394	0.9406	0.9418	0.9429	0.9441
1.60	0.9452	0.9463	0.9474	0.9484	0.9495	0.9505	0.9515	0.9525	0.9535	0.9545
1.70	0.9554	0.9564	0.9573	0.9582	0.9591	0.9599	0.9608	0.9616	0.9625	0.9633
1.80	0.9641	0.9649	0.9656	0.9664	0.9671	0.9678	0.9686	0.9693	0.9699	0.9706
1.90	0.9713	0.9719	0.9726	0.9732	0.9738	0.9744	0.9750	0.9756	0.9761	0.9767
2.00	0.9772	0.9778	0.9783	0.9788	0.9793	0.9798	0.9803	0.9808	0.9812	0.9817
2.10	0.9821	0.9826	0.9830	0.9834	0.9838	0.9842	0.9846	0.9850	0.9854	0.9857
2.20	0.9861	0.9864	0.9868	0.9871	0.9875	0.9878	0.9881	0.9884	0.9887	0.9890
2.30	0.9893	0.9896	0.9898	0.9901	0.9904	0.9906	0.9909	0.9911	0.9913	0.9916
2.40	0.9918	0.9920	0.9922	0.9925	0.9927	0.9929	0.9931	0.9932	0.9934	0.9936
2.50	0.9938	0.9940	0.9941	0.9943	0.9945	0.9946	0.9948	0.9949	0.9951	0.9952
2.60	0.9953	0.9955	0.9956	0.9957	0.9959	0.9960	0.9961	0.9962	0.9963	0.9964
2.70	0.9965	0.9966	0.9967	0.9968	0.9969	0.9970	0.9971	0.9972	0.9973	0.9974
2.80	0.9974	0.9975	0.9976	0.9977	0.9977	0.9978	0.9979	0.9979	0.9980	0.9981
2.90	0.9981	0.9982	0.9982	0.9983	0.9984	0.9984	0.9985	0.9985	0.9986	0.9986
3.00	0.9987	0.9987	0.9987	0.9988	0.9988	0.9989	0.9989	0.9989	0.9990	0.9990

Source: Computed using Excel.

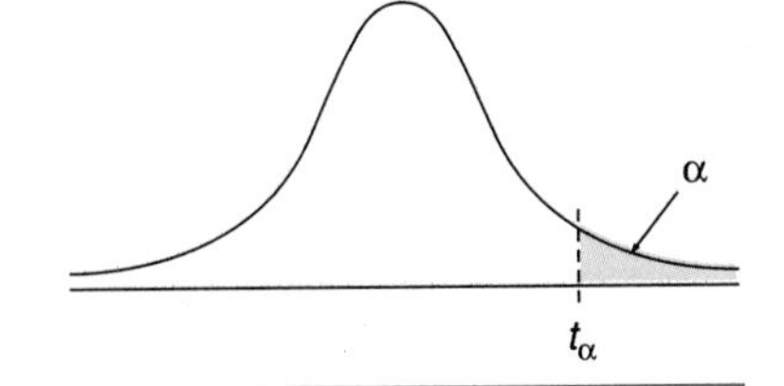

TABLE 4 Area in Upper Tail of t Distribution Area in upper tail (α) based on degrees of freedom (df). Excel function: TINV(α*2,df).

	Area in upper tail, α					
df	**0.1**	**0.05**	**0.025**	**0.01**	**0.005**	**0.001**
1	3.078	6.314	12.706	31.821	63.656	318.289
2	1.886	2.920	4.303	6.965	9.925	22.328
3	1.638	2.353	3.182	4.541	5.841	10.214
4	1.533	2.132	2.776	3.747	4.604	7.173
5	1.476	2.015	2.571	3.365	4.032	5.894
6	1.440	1.943	2.447	3.143	3.707	5.208
7	1.415	1.895	2.365	2.998	3.499	4.785
8	1.397	1.860	2.306	2.896	3.355	4.501
9	1.383	1.833	2.262	2.821	3.250	4.297
10	1.372	1.812	2.228	2.764	3.169	4.144
11	1.363	1.796	2.201	2.718	3.106	4.025
12	1.356	1.782	2.179	2.681	3.055	3.930
13	1.350	1.771	2.160	2.650	3.012	3.852
14	1.345	1.761	2.145	2.624	2.977	3.787
15	1.341	1.753	2.131	2.602	2.947	3.733
16	1.337	1.746	2.120	2.583	2.921	3.686
17	1.333	1.740	2.110	2.567	2.898	3.646
18	1.330	1.734	2.101	2.552	2.878	3.610
19	1.328	1.729	2.093	2.539	2.861	3.579
20	1.325	1.725	2.086	2.528	2.845	3.552
21	1.323	1.721	2.080	2.518	2.831	3.527
22	1.321	1.717	2.074	2.508	2.819	3.505
23	1.319	1.714	2.069	2.500	2.807	3.485
24	1.318	1.711	2.064	2.492	2.797	3.467
25	1.316	1.708	2.060	2.485	2.787	3.450
26	1.315	1.706	2.056	2.479	2.779	3.435
27	1.314	1.703	2.052	2.473	2.771	3.421
28	1.313	1.701	2.048	2.467	2.763	3.408
29	1.311	1.699	2.045	2.462	2.756	3.396
30	1.310	1.697	2.042	2.457	2.750	3.385
35	1.306	1.690	2.030	2.438	2.724	3.340
40	1.303	1.684	2.021	2.423	2.704	3.307
45	1.301	1.679	2.014	2.412	2.690	3.281
50	1.299	1.676	2.009	2.403	2.678	3.261
55	1.297	1.673	2.004	2.396	2.668	3.245
60	1.296	1.671	2.000	2.390	2.660	3.232
65	1.295	1.669	1.997	2.385	2.654	3.220
70	1.294	1.667	1.994	2.381	2.648	3.211
75	1.293	1.665	1.992	2.377	2.643	3.202
80	1.292	1.664	1.990	2.374	2.639	3.195
85	1.292	1.663	1.988	2.371	2.635	3.189
90	1.291	1.662	1.987	2.368	2.632	3.183
95	1.291	1.661	1.985	2.366	2.629	3.178
100	1.290	1.660	1.984	2.364	2.626	3.174
125	1.288	1.657	1.979	2.357	2.616	3.157
150	1.287	1.655	1.976	2.351	2.609	3.145
175	1.286	1.654	1.974	2.348	2.604	3.137
200	1.286	1.653	1.972	2.345	2.601	3.131
225	1.285	1.652	1.971	2.343	2.598	3.127
250	1.285	1.651	1.969	2.341	2.596	3.123
275	1.285	1.650	1.969	2.340	2.594	3.120
300	1.284	1.650	1.968	2.339	2.592	3.118

Source: Computed using Excel.

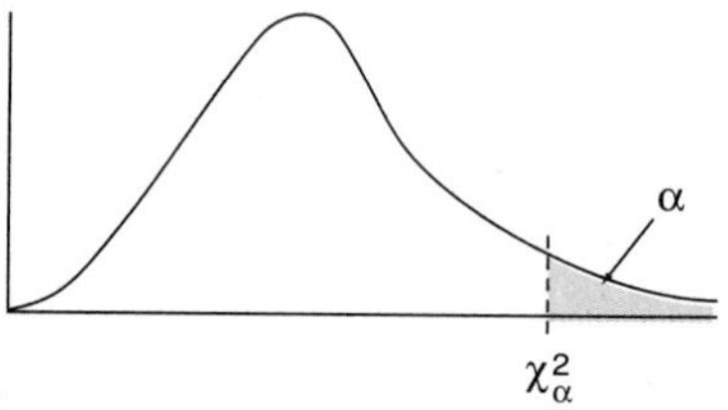

TABLE 5 χ^2 **Distribution** Area in upper tail (α) based on degrees of freedom (df). Excel function: CHIINV(α,df).

	Area in upper tail, α					
df	**0.1**	**0.05**	**0.025**	**0.01**	**0.005**	**0.001**
1	2.706	3.841	5.024	6.635	7.879	10.827
2	4.605	5.991	7.378	9.210	10.597	13.815
3	6.251	7.815	9.348	11.345	12.838	16.266
4	7.779	9.488	11.143	13.277	14.860	18.466
5	9.236	11.070	12.832	15.086	16.750	20.515
6	10.645	12.592	14.449	16.812	18.548	22.457
7	12.017	14.067	16.013	18.475	20.278	24.321
8	13.362	15.507	17.535	20.090	21.955	26.124
9	14.684	16.919	19.023	21.666	23.589	27.877
10	15.987	18.307	20.483	23.209	25.188	29.588
11	17.275	19.675	21.920	24.725	26.757	31.264
12	18.549	21.026	23.337	26.217	28.300	32.909
13	19.812	22.362	24.736	27.688	29.819	34.527
14	21.064	23.685	26.119	29.141	31.319	36.124
15	22.307	24.996	27.488	30.578	32.801	37.698
16	23.542	26.296	28.845	32.000	34.267	39.252
17	24.769	27.587	30.191	33.409	35.718	40.791
18	25.989	28.869	31.526	34.805	37.156	42.312
19	27.204	30.144	32.852	36.191	38.582	43.819
20	28.412	31.410	34.170	37.566	39.997	45.314
21	29.615	32.671	35.479	38.932	41.401	46.796
22	30.813	33.924	36.781	40.289	42.796	48.268
23	32.007	35.172	38.076	41.638	44.181	49.728
24	33.196	36.415	39.364	42.980	45.558	51.179
25	34.382	37.652	40.646	44.314	46.928	52.619
26	35.563	38.885	41.923	45.642	48.290	54.051
27	36.741	40.113	43.195	46.963	49.645	55.475
28	37.916	41.337	44.461	48.278	50.994	56.892
29	39.087	42.557	45.722	49.588	52.335	58.301
30	40.256	43.773	46.979	50.892	53.672	59.702
35	46.059	49.802	53.203	57.342	60.275	66.619
40	51.805	55.758	59.342	63.691	66.766	73.403
45	57.505	61.656	65.410	69.957	73.166	80.078
50	63.167	67.505	71.420	76.154	79.490	86.660
55	68.796	73.311	77.380	82.292	85.749	93.167
60	74.397	79.082	83.298	88.379	91.952	99.608
65	79.973	84.821	89.177	94.422	98.105	105.988
70	85.527	90.531	95.023	100.425	104.215	112.317
75	91.061	96.217	100.839	106.393	110.285	118.599
80	96.578	101.879	106.629	112.329	116.321	124.839
85	102.079	107.522	112.393	118.236	122.324	131.043
90	107.565	113.145	118.136	124.116	128.299	137.208
95	113.038	118.752	123.858	129.973	134.247	143.343
100	118.498	124.342	129.561	135.807	140.170	149.449
125	145.643	152.094	157.838	164.694	169.471	179.605
150	172.581	179.581	185.800	193.207	198.360	209.265
175	199.363	206.867	213.524	221.438	226.936	238.552
200	226.021	233.994	241.058	249.445	255.264	267.539
225	252.578	260.992	268.438	277.269	283.390	296.290
250	279.050	287.882	295.689	304.939	311.346	324.831
275	305.451	314.678	322.829	332.480	339.158	353.202
300	331.788	341.395	349.874	359.906	366.844	381.424

Source: Computed using Excel.

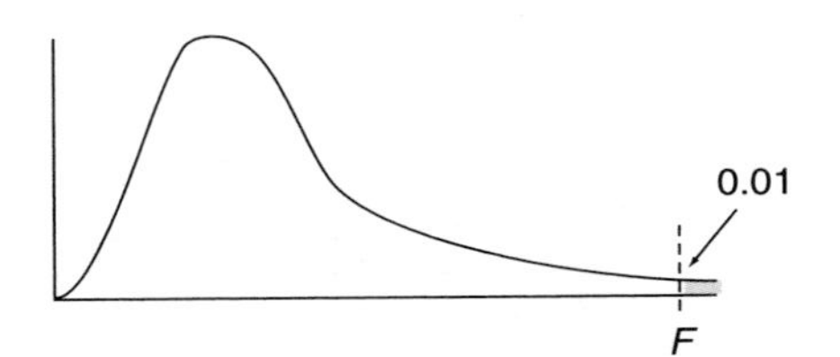

TABLE 6a ***F* Distribution** Area equal to 0.01 in upper tail. Excel function: FINV (0.01,df_1,df_2).

df_2 \ df_1	1	2	3	4	5	6	7	8	9	10	15	20	25	30	35	40	45	50	100	150	200	250	300
1	4052.2	4999.3	5403.5	5624.3	5764.0	5859.0	5928.3	5981.0	6022.4	6055.9	6157.0	6208.7	6239.9	6260.4	6275.3	6286.4	6295.7	6302.3	6333.9	6344.6	6349.8	6353.5	6355.3
2	98.5	99.0	99.2	99.3	99.3	99.3	99.4	99.4	99.4	99.4	99.4	99.4	99.5	99.5	99.5	99.5	99.5	99.5	99.5	99.5	99.5	99.5	99.5
3	34.1	30.8	29.5	28.7	28.2	27.9	27.7	27.5	27.3	27.2	26.9	26.7	26.6	26.5	26.5	26.4	26.4	26.4	26.2	26.2	26.2	26.2	26.2
4	21.2	18.0	16.7	16.0	15.5	15.2	15.0	14.8	14.7	14.5	14.2	14.0	13.9	13.8	13.8	13.7	13.7	13.7	13.6	13.5	13.5	13.5	13.5
5	16.3	13.3	12.1	11.4	11.0	10.7	10.5	10.3	10.2	10.1	9.7	9.6	9.4	9.4	9.3	9.3	9.3	9.2	9.1	9.1	9.1	9.1	9.1
6	13.7	10.9	9.8	9.1	8.7	8.5	8.3	8.1	8.0	7.9	7.6	7.4	7.3	7.2	7.2	7.1	7.1	7.1	7.0	7.0	6.9	6.9	6.9
7	12.2	9.5	8.5	7.8	7.5	7.2	7.0	6.8	6.7	6.6	6.3	6.2	6.1	6.0	5.9	5.9	5.9	5.9	5.8	5.7	5.7	5.7	5.7
8	11.3	8.6	7.6	7.0	6.6	6.4	6.2	6.0	5.9	5.8	5.5	5.4	5.3	5.2	5.2	5.1	5.1	5.1	5.0	4.9	4.9	4.9	4.9
9	10.6	8.0	7.0	6.4	6.1	5.8	5.6	5.5	5.4	5.3	5.0	4.8	4.7	4.6	4.6	4.6	4.5	4.5	4.4	4.4	4.4	4.4	4.3
10	10.0	7.6	6.6	6.0	5.6	5.4	5.2	5.1	4.9	4.8	4.6	4.4	4.3	4.2	4.2	4.2	4.1	4.1	4.0	4.0	4.0	4.0	3.9
15	8.7	6.4	5.4	4.9	4.6	4.3	4.1	4.0	3.9	3.8	3.5	3.4	3.3	3.2	3.2	3.1	3.1	3.1	3.0	2.9	2.9	2.9	2.9
20	8.1	5.8	4.9	4.4	4.1	3.9	3.7	3.6	3.5	3.4	3.1	2.9	2.8	2.8	2.7	2.7	2.7	2.6	2.5	2.5	2.5	2.5	2.5
25	7.8	5.6	4.7	4.2	3.9	3.6	3.5	3.3	3.2	3.1	2.9	2.7	2.6	2.5	2.5	2.5	2.4	2.4	2.3	2.3	2.2	2.2	2.2
30	7.6	5.4	4.5	4.0	3.7	3.5	3.3	3.2	3.1	3.0	2.7	2.5	2.5	2.4	2.3	2.3	2.3	2.2	2.1	2.1	2.1	2.1	2.0
35	7.4	5.3	4.4	3.9	3.6	3.4	3.2	3.1	3.0	2.9	2.6	2.4	2.3	2.3	2.2	2.2	2.2	2.1	2.0	2.0	2.0	1.9	1.9
40	7.3	5.2	4.3	3.8	3.5	3.3	3.1	3.0	2.9	2.8	2.5	2.4	2.3	2.2	2.2	2.1	2.1	2.1	1.9	1.9	1.9	1.9	1.9
45	7.2	5.1	4.2	3.8	3.5	3.2	3.1	2.9	2.8	2.7	2.5	2.3	2.2	2.1	2.1	2.1	2.0	2.0	1.9	1.8	1.8	1.8	1.8
50	7.2	5.1	4.2	3.7	3.4	3.2	3.0	2.9	2.8	2.7	2.4	2.3	2.2	2.1	2.0	2.0	2.0	1.9	1.8	1.8	1.8	1.7	1.7
100	6.9	4.8	4.0	3.5	3.2	3.0	2.8	2.7	2.6	2.5	2.2	2.1	2.0	1.9	1.8	1.8	1.8	1.7	1.6	1.5	1.5	1.5	1.5
150	6.8	4.7	3.9	3.4	3.1	2.9	2.8	2.6	2.5	2.4	2.2	2.0	1.9	1.8	1.8	1.7	1.7	1.7	1.5	1.5	1.4	1.4	1.4
200	6.8	4.7	3.9	3.4	3.1	2.9	2.7	2.6	2.5	2.4	2.1	2.0	1.9	1.8	1.7	1.7	1.7	1.6	1.5	1.4	1.4	1.4	1.4
250	6.7	4.7	3.9	3.4	3.1	2.9	2.7	2.6	2.5	2.4	2.1	2.0	1.8	1.8	1.7	1.7	1.6	1.6	1.5	1.4	1.4	1.3	1.3
300	6.7	4.7	3.8	3.4	3.1	2.9	2.7	2.6	2.5	2.4	2.1	1.9	1.8	1.8	1.7	1.7	1.6	1.6	1.4	1.4	1.3	1.3	1.3

TABLE 6b ***F* Distribution** Area equal to 0.05 in upper tail. Excel function: FINV(0.05,df_1,df_2).

	df_1																						
df_2	1	2	3	4	5	6	7	8	9	10	15	20	25	30	35	40	45	50	100	150	200	250	300
1	161.4	199.5	215.7	224.6	230.2	234.0	236.8	238.9	240.5	241.9	245.9	248.0	249.3	250.1	250.7	251.1	251.5	251.8	253.0	253.5	253.7	253.8	253.9
2	18.5	19.0	19.2	19.2	19.3	19.3	19.4	19.4	19.4	19.4	19.4	19.4	19.5	19.5	19.5	19.5	19.5	19.5	19.5	19.5	19.5	19.5	19.5
3	10.1	9.6	9.3	9.1	9.0	8.9	8.9	8.8	8.8	8.8	8.7	8.7	8.6	8.6	8.6	8.6	8.6	8.6	8.6	8.5	8.5	8.5	8.5
4	7.7	6.9	6.6	6.4	6.3	6.2	6.1	6.0	6.0	6.0	5.9	5.8	5.8	5.7	5.7	5.7	5.7	5.7	5.7	5.7	5.6	5.6	5.6
5	6.6	5.8	5.4	5.2	5.1	5.0	4.9	4.8	4.8	4.7	4.6	4.6	4.5	4.5	4.5	4.5	4.5	4.4	4.4	4.4	4.4	4.4	4.4
6	6.0	5.1	4.8	4.5	4.4	4.3	4.2	4.1	4.1	4.1	3.9	3.9	3.8	3.8	3.8	3.8	3.8	3.8	3.7	3.7	3.7	3.7	3.7
7	5.6	4.7	4.3	4.1	4.0	3.9	3.8	3.7	3.7	3.6	3.5	3.4	3.4	3.4	3.4	3.3	3.3	3.3	3.3	3.3	3.3	3.2	3.2
8	5.3	4.5	4.1	3.8	3.7	3.6	3.5	3.4	3.4	3.3	3.2	3.2	3.1	3.1	3.1	3.0	3.0	3.0	3.0	3.0	3.0	2.9	2.9
9	5.1	4.3	3.9	3.6	3.5	3.4	3.3	3.2	3.2	3.1	3.0	2.9	2.9	2.9	2.8	2.8	2.8	2.8	2.8	2.7	2.7	2.7	2.7
10	5.0	4.1	3.7	3.5	3.3	3.2	3.1	3.1	3.0	3.0	2.8	2.8	2.7	2.7	2.7	2.7	2.6	2.6	2.6	2.6	2.6	2.6	2.6
15	4.5	3.7	3.3	3.1	2.9	2.8	2.7	2.6	2.6	2.5	2.4	2.3	2.3	2.2	2.2	2.2	2.2	2.2	2.1	2.1	2.1	2.1	2.1
20	4.4	3.5	3.1	2.9	2.7	2.6	2.5	2.4	2.4	2.3	2.2	2.1	2.1	2.0	2.0	2.0	2.0	2.0	1.9	1.9	1.9	1.9	1.9
25	4.2	3.4	3.0	2.8	2.6	2.5	2.4	2.3	2.3	2.2	2.1	2.0	2.0	1.9	1.9	1.9	1.9	1.8	1.8	1.8	1.7	1.7	1.7
30	4.2	3.3	2.9	2.7	2.5	2.4	2.3	2.3	2.2	2.2	2.0	1.9	1.9	1.8	1.8	1.8	1.8	1.8	1.7	1.7	1.7	1.7	1.6
35	4.1	3.3	2.9	2.6	2.5	2.4	2.3	2.2	2.2	2.1	2.0	1.9	1.8	1.8	1.8	1.7	1.7	1.7	1.6	1.6	1.6	1.6	1.6
40	4.1	3.2	2.8	2.6	2.4	2.3	2.2	2.2	2.1	2.1	1.9	1.8	1.8	1.7	1.7	1.7	1.7	1.7	1.6	1.6	1.6	1.5	1.5
45	4.1	3.2	2.8	2.6	2.4	2.3	2.2	2.2	2.1	2.0	1.9	1.8	1.8	1.7	1.7	1.7	1.6	1.6	1.6	1.5	1.5	1.5	1.5
50	4.0	3.2	2.8	2.6	2.4	2.3	2.2	2.1	2.1	2.0	1.9	1.8	1.7	1.7	1.7	1.6	1.6	1.6	1.5	1.5	1.5	1.5	1.5
100	3.9	3.1	2.7	2.5	2.3	2.2	2.1	2.0	2.0	1.9	1.8	1.7	1.6	1.6	1.5	1.5	1.5	1.5	1.4	1.4	1.3	1.3	1.3
150	3.9	3.1	2.7	2.4	2.3	2.2	2.1	2.0	1.9	1.9	1.7	1.6	1.6	1.5	1.5	1.5	1.5	1.4	1.3	1.3	1.3	1.3	1.3
200	3.9	3.0	2.6	2.4	2.3	2.1	2.1	2.0	1.9	1.9	1.7	1.6	1.6	1.5	1.5	1.5	1.4	1.4	1.3	1.3	1.3	1.2	1.2
250	3.9	3.0	2.6	2.4	2.3	2.1	2.0	2.0	1.9	1.9	1.7	1.6	1.6	1.5	1.5	1.4	1.4	1.4	1.3	1.3	1.2	1.2	1.2
300	3.9	3.0	2.6	2.4	2.2	2.1	2.0	2.0	1.9	1.9	1.7	1.6	1.5	1.5	1.5	1.4	1.4	1.4	1.3	1.3	1.2	1.2	1.2

Source: Computed using Excel.

Index

A

Adjusted R^2, 118
Activity, 379
  critical, 380, 390
  distribution of completion time, 395
  dummy, 384
  estimating times, 393–394
  expected time, 382
  most likely estimate, 394
  optimistic estimate, 393
  pessimistic estimate, 394
  slack, 390
  standard deviation of completion time, 395
  variance of completion time, 395
Algorithms, 12
Allocation situations, 19
Alternative hypothesis, 71
Analysis of models, 23–25
  alternate or multiple solutions, 24
  robust solutions, 25
  unique solutions, 24
Analysis of variance, 77–81
  interaction effects, 81
ANOVA. *See* Analysis of variance
Assignment problem, 165
Average. *See* Mean
Autocorrelation, 119

B

Bar charts, 44
Batch means approach, 326
Bayes' theorem, 253
Binomial distribution, 53, 70
Binomial probabilities, 442–443
Blending problem, 157, 160–161
Branch and bound, 197
Business games, 321

C

Categorical data, 41
Central limit theorem, 60
Certainty, 22, 228–230
Chi-square distribution, 448
Chi-square goodness of fit test, 60, 85
Coefficient of determination, 106–107
Cognitive styles, 376
Combinatorial problems, 230
Common random numbers, 325, 347
Complete enumeration, 197, 229, 393
Conditional probability, 49, 253
Confidence interval, 65
  determining sample size, 68–69
Constant, 436
Constraints, 153, 158
Continuous data, 42
Continuous distribution, 53
Continuous function, 437
Continuous variable, 436
Correlation coefficient, 108–109
Cost-benefit analysis, 377
Cost-effectiveness analysis, 377
Cost-time trade-offs, 406–418
  using Excel's Solver, 414–418
Cost variance reports, 403
CPM. *See* Critical Path Method
Crashing. *See* Cost-time trade-offs
Critical activity, 380, 390
Critical event, 390
Critical path, 380
  finding, 391
Critical Path Method, 379, 383
  contrasted with PERT, 380
Critical probability, 243
Critical ratio, 403
Cross-sectional data, 41
Crystal Ball, 350–358
  assumption cells, 351
  distribution gallery, 351–352
  forecast cells, 351
Cumulative distribution function (cdf), 52

D

Data collection, 21–23, 41
Data table, 233–237
Decision analysis, 8, 20, 223
  alternative courses of action, 225
  classification of decision situations, 228
  critical probability, 243
  decision horizon, 227
  decision tables, 20, 225
  most probable state of nature criterion, 240
  nonrepetitive decisions, 242
  payoff variability, 242
  states of nature, 226
  uncertainty criteria, 241
Decision tree, 20, 243–250
  chance point, 243–244
  constructing, 244
  decision node, 243
  decision points, 243
  evaluating, 245–246
  multiperiod, sequential decisions, 246–250
  relationship to decision table, 244
  structure, 243–244
Decision variables, 158, 162
Dependent variables, 12, 18, 100, 436
Deterministic, 20, 228
Deterministic simulation, 321
Discrete distribution, 53
Discrete function, 437
Discrete number, 42
Discrete simulation, 332–339
Discrete variable, 436
Dominance, 242
Dummy activity, 384
Dummy variable, 115

E

Earliest date, 386
Earned value, 403
Earned value chart, 374
Enumeration, 12
Equalities, 437
Erlang, A. K., 285
Events, 47, 48, 379
  earliest date, 386
  addition rule, 51
  collectively exhaustive, 49, 51
  complements law, 49
  conditional probability, 49, 50
  critical, 389–390
  definition, project management, 379
  dependent, 48
  independent, 48, 49
  intersection, 49
  joint probability, 49, 50
  latest allowable date, 387
  multiplication rule, 50
  mutually exclusive, 49, 51
  slack, 389–390
  union, 49
Excel
  absolute cell addresses, 5
  array functions, 104
  AVEDEV, 114
  BINOMDIST, 53
  Chart Wizard, 44, 83
  CHIINV, 62
  CORREL, 109, 114
  Data analysis toolpak add-in, 82
  Data table, 233–237
  descriptive statistics tool, 82–83
  EXP, 118
  EXPONDIST, 55, 56
  fitting a trendline, 104
  Goal Seek, 298
  Histogram tool, 83, 85
  IF, 75, 335
  LINEST, 104, 105, 113, 115, 127
  LN, 118
  MIN, 336
  NORMDIST, 86, 398, 423
  NORMINV, 86, 365, 399, 423
  NORMSINV, 74, 76
  OR, 75
  POISSON, 54
  Random number generation, 331–332
  Regression tool, 119
  relative cell addresses, 5
  ROUNDUP, 69
  SOLVER, 177–189, 414–418
  STANDARDIZE, 58
  SUMPRODUCT, 5, 177, 183
  TINV, 67, 76
  TREND, 105, 114, 129
Expected opportunity loss, 238, 240
Expected payoff criterion, 239–240
Expected value, 52, 238
Expected value of imperfect information, 256
Expected value of perfect information, 252
Experimental design, 325–326
Experimental probability, 48
Exponential distribution, 55
Exponential smoothing, 125–127
Exponents, 439–440
External validity, 324
Extrapolation, 110

F

F distribution, 449–450
Factorial, 435
Float. *See* Slack
Forecasting. *See also* Time series analysis
  bias, 126
  errors, 126
Free float, 392
Frequency distribution, 83
Frequency table, 44
Function, 436
  continuous, 437
  discrete, 437
  nonlinear, 438

G

Gantt chart, 385
Gomory cutting plane, 197
Graphical method, 165–177
Grouped items, 41

H

Heteroscedasticity, 109
Histogram, 47
Hurwicz criterion, 233
Hypothesis testing, 71–81
  alternative hypothesis, 71, 73
  for means, 73–74
  multiple means, 77–81
  null hypothesis, 71, 73
  p value, 72
  population proportion, 76–77
  power, 72
  regression, 119–121
  rejection region, 71–73

I

Ignorance, 230
Imperfect information, 253–261
Implementation, 25
Independent variables, 12, 18, 99–100, 436
Inequalities, 437
  rules of manipulation, 439
Infeasible solution, 175
Influence diagram
  defined, 2
  elements of, 2–3
Integer programming, 196–203
Interaction effects, 81
Internal validity, 324
Interval estimation, 64–70
  of mean, 65–68
  proportions, 69–70

J

Joint probability, 49

K

Kendall, D. G., 288

L

Laplace, 231
Latest allowable date, 388
Least squares regression, 103
Linear equation, 438
Linear programming, 8, 12, 156–165
  advantages of, 162
  assumptions, 163
  constraint coefficients, 158, 162
  constraints, 158
  cost coefficients, 158
  decision variables, 157–158, 162
  general model, 161–162
  graphical method, 163, 165–177
  isoprofit (isocost) lines, 170
  model components, 157–158
  multiple optimal solutions, 176
  objective function coefficients, 162
  profit coefficients, 158
  right-hand-side constants, 158, 162
  sensitivity analysis, 189–196
  solving, 163–164
  solving large problems, 181–185
  solving with Excel, 177–189
  surplus variables, 175
  unbounded solutions, 175
  value of objective function, 162
Linear responsibility chart, 374
Logical probability, 48

M

Mathematical model, 11
  components of, 18–19
Maximax, 233
Maximin, 231–232
Mean, 42
  of population, 42
  of sample, 43
  sampling distribution, 59
  standard error, 59
Median, 44
Milestones, 380
Minimax, 231–232
Minimin, 232
Mixed integer model, 197
Mode, 44
Model
  analog, 11
  defined, 9
  descriptive, 12
  internal validity, 23
  intuitive or subjective, 12
  deterministic, 12
  mathematical, 11
  normative, 12
  physical, 11
  prescriptive, 12, 19
  probabilistic, 12
  solving, 12, 13
  types of, 11
  usability, 9
  validity, 9, 23
Modeling
  benefits, 10
  drawbacks, 10–11
  implementing modeling studies, 375–381
Modeling process, 14–25
  characteristics of effective modeling process, 14

example, 6–7
five steps, 16
model formulation, 17–21
opportunity/problem recognition, 17
Monitoring and controlling projects, 403–418
cost variance reports, 403
Monte Carlo simulation, 241, 323, 327–332
Most likely time estimate, 394
Moving averages, 123–124
Multicollinearity, 114, 115, 133
Multifacility queues, 299–304
Multiple coefficient of determination, 114, 119
Multiple regression model, 100, 112–115
adjusted R^2, 118
error, 112
multiple coefficient of determination, 114, 119
parsimonious models, 133

N

Network, 380
Nominal data, 41
Nonlinear functions, 438
Nonnegativity constraints, 162
Normal distribution, 56
area under, 446
Normalized data, 45
Null hypothesis, 71

O

Objective function, 158, 162
Ontology, 9
Optimal solution, 12, 154
Optimistic time estimate, 393
Optimization, 20, 153
assumptions, 154
Ordinal data, 41
Outliers, 105
Output mix problem, 157, 159–160

P

p value, 72
Parameter, 436
Pareto chart, 44
Parsimonious models, 133
Payoff tables. *See* Decision analysis, decision tables
Payoffs, 227
Perfect information, 250–251
value of, 251–253
PERT. *See* Program Evaluation Review Technique
Pessimistic time estimate, 394
Point estimators, 64
characteristics of estimators, 65
consistency, 65
efficiency, 65
robustness, 65
Poisson distribution, 54
Poisson probabilities, 444–445
Population, 286
Postivists, 9
Postoptimality analysis. *See* Sensitivity analysis
Power of hypothesis test, 72
Prior probabilities, 253
Probability, 47, 238
experimental, 48
logical, 48
objective, 48, 238
subjective, 48, 238
Probability distributions, 51–52
Probability function, 51–52
Program Evaluation Review Technique, 379, 383
contrasted with CPM, 380
Project crashing. *See* Cost-time trade-offs
Project expediting. *See* Cost-time trade-offs
Project goals, 373
Project management
activity, 379
characteristics, 379
critical activity, 380
event, 379
monitoring and control, 403–418
milestones, 380
planning, 381–382
project defined, 379, 380
role of, 378
scheduling, 383–403
Project network, 380
Project planning, 381–382
Project scheduling, 383–403

Q

Queue, 286
Queuing theory, 8, 20–21, 284
arrival process, 286, 290–292
complex situations, 298–305
costs, 287–288
deterministic systems, 289–290
managerial issues, 286–287
(*M*/*M*/1 *FCFS*/∞/∞), 295–298
(*M*/*M*/*K FCFS*/∞/∞), 299–304
notation, 289
operating characteristics, 293–294, 296–297, 300
performance measures, 293–294, 296–297, 300
queue behavior, 295
queue discipline, 294
serial queues, 304–305
service facilities, 286, 292–293
service process, 286, 292–293
source, 286
structure of queuing system, 285–286
transient state, 295
utilization, 295

R

Random number, 328
generating exponential random numbers, 360
Random variable, 47
Random variate, 328
Range, 44
Regression. *See also* Simple linear regression and Multiple regression model
adjusted R^2, 118
analysis of residuals, 118–119
backward elimination, 118
data transformations, 117–118
dummy variable, 115
forward selection, 118
generalization, 110–111
identifying candidate independent variables, 115–117
lagged values, 115
model assumptions, 109
selecting variables, 118
stepwise procedure, 118
Regret criterion, 237
Regular slack, 391
Relativists, 10
Residual, 112
Risk, 20, 228, 237–243
Risk analysis, 241, 342–343
Risk profiles, 241

S

Sample, 39
Sample proportion, 70
Sampling distribution of mean, 59
Sampling error, 65
Scatter diagram, 98, 100
Scope creep, 375
Sensitivity analysis, 4, 24–25, 162, 189–196, 242–243, 326
coefficient limits, 191–192
objective function, 190–191
right-hand-sides, 192
with Excel, 192–196
Shared slack, 391
Significance level, 71, 72, 73
Simple linear regression, 100–112
calculating model parameters, 103
error, 102
mathematical form, 102
outliers, 105
residual, 102
Simplex method, 163
Simulation, 8, 13, 81, 241, 319
advantages, 322
batch means approach, 326
common random numbers, 325, 347
corporate and financial simulations, 321
Crystal Ball, 350–358
disadvantages, 323
experimental design, 325–326
general overview, 319–323
generating exponential random numbers, 360
independent replications, 326
probabilistic simulation, 321
random number, 328
random number generation, 329
random variate, 328
sensitivity analysis, 326
simulation run, 329
start-up, 326, 338
stopping rule, 325

terminating simulation, 325
time compression, 321, 322
time dependent simulation, 320, 339–342
time independent simulation, 320, 332–339
types of, 320–321
uses of, 322
variance reduction, 325
Slack, 380
activities, 390
events, 389–390
negative, 404
shared, 391
total, 391
Slack variables, 174–175
Slope, 436, 438
Standard error of mean, 59
Significance level, 71, 72
Standard deviation, 42, 43
Standard normal distribution, 57, 397
States of nature, 226
Steady state, 283, 338
Stepwise procedure, 118
Subjective probability, 48
Summarizing data, 42–47
Surplus variables, 175
Survey reliability, 254
System dynamics, 321

T

t distribution, 58, 67
area in upper tail, 447
Time series analysis, 121–130
difficulties of, 125
linear trend, multiplicative model, 127–130
time series data, 41, 121
Time series data, 41, 121
components of, 121–122
cycle, 123
random variations, 123
seasonal fluctuations, 123, 128–129
trend, 121–122
Total float, 390, 391
Transshipment problem, 164–165
Transportation problem, 164
Trend, 121–122
Type I error, 72–74
Type II error, 72

U

Uncertainty, 20, 228, 230–237
Uncontrollable parameters, 12, 18
Uniform distribution, 328

V

Validity, 324
Value of perfect information, 251–253
Variable, 436
continuous, 436
discrete, 436
Variable relationships, 100–101
Variance, 42
of population, 43
of random variable, 52
of sample, 43
Variance reduction, 325
Venn diagram, 49
Visual interactive simulation, 321–322

W

Waiting lines. *See* Queuing theory
WBS. *See* Work breakdown structure
Weighted moving average, 125
Weighted scoring model, 7
Willemain, T. R., 9, 14
What-if analysis. *See* Sensitivity analysis
Work breakdown structure, 374, 381, 418

Z

Zero-one integer model, 197, 200–203